Aircraft Engineering

Aircraft Engineering

Edited by Natalie Spagner

Clanrye International
www.clanryeinternational.com

Clanrye International,
750 Third Avenue, 9th Floor,
New York, NY 10017, USA

ISBN: 978-1-63240-819-8

Cataloging-in-Publication Data

Aircraft engineering / edited by Natalie Spagner.
p. cm.
Includes bibliographical references and index.
ISBN 978-1-63240-819-8
1. Aerospace engineering. 2. Aeronautics. I. Spagner, Natalie.
TL545 .A37 2019
629.1--dc23

For information on all Clanrye International publications visit our website at www.clanryeinternational.com

CLANRYE INTERNATIONAL

Printed in China.

Contents

Preface

This book has been a concerted effort by a group of academicians, researchers and scientists, who have contributed their research works for the realization of the book. This book has materialized in the wake of emerging advancements and innovations in this field. Therefore, the need of the hour was to compile all the required researches and disseminate the knowledge to a broad spectrum of people comprising of students, researchers and specialists of the field.

Aircraft is a machine that counters the gravitational pull of the earth by using either static lift, dynamic lift or downward thrust. Common aircraft include airplanes, helicopters, gliders, etc. Different aircraft have different methods of lift. Aerostats use buoyancy for floating in the air, while airplanes produce dynamic upthrust through aerodynamic lift or powered lift in the form of engine thrust. Aircraft can be classified as unpowered and powered aircraft. Unpowered aircraft include gliders, balloons and kites. Powered aircraft include propeller aircraft, jet aircraft and rotorcraft. The study of aircraft and their engineering is studied under the domain of aerospace engineering. It integrates the technological and engineering principles of aerodynamics, propulsion, avionics, structural analysis and manufacturing. This book traces the progress of aircraft engineering and highlights some of its key concepts and applications. It includes some of the vital pieces of work being conducted across the world, on various topics related to this discipline. This book aims to equip students and experts with the advanced topics and upcoming trends in this area of study.

At the end of the preface, I would like to thank the authors for their brilliant chapters and the publisher for guiding us all-through the making of the book till its final stage. Also, I would like to thank my family for providing the support and encouragement throughout my academic career and research projects.

Editor

1

A Model Predictive Guidance Strategy for a Multirotor Aerial Vehicle

Igor Afonso Acampora Prado[1], Davi Antônio dos Santos[1]

ABSTRACT: The present study faces the problem of safely controlling the position trajectory of a multirotor aerial vehicle subjected to a conic constraint on the total thrust vector and a linear convex constraint on the position vector. The problem is solved using a linear state-space model predictive control strategy, whose optimization is made handy by replacing the original conic constraint set on the thrust vector by an inscribed pyramidal space, which renders a linear set of inequalities. The proposed method is evaluated on the basis of Monte Carlo simulations taking into account a random disturbance force. The simulation results show the effectiveness of the method in tracking the commanded trajectory while respecting the constraints. They also predict the effect of both the speed command and the maximum allowed inclination angle on the system performance.

KEYWORDS: Multirotor aerial vehicle, Model predictive control, Position control, Guidance.

INTRODUCTION

Unmanned Aerial Vehicles (UAVs) have motivated and stimulated many researches in different fields of knowledge such as sensors fusion (Cheviron *et al.* 2007; Nemra and Aouf 2010; Gonçalves *et al.* 2013), computer vision (Saripalli *et al.* 2003; Xu *et al.* 2009; Xiao-Hong *et al.* 2012) and control strategies (Mian and Daobo 2008; Santos *et al.* 2013). A few years ago, building a low-cost miniature UAV was a challenge due to limitation imposed by equipment such as sensors, efficient motors, batteries and on-board computers. However, thanks to technological advances in actuators, small scale sensors, data processing and energy storage, the conditions improved significantly.

Among the different types of UAVs such as blimps (Elfes *et al.* 1998), fixed-wings (Beard *et al.* 2005) and rotary wings aircrafts (Bouabdallah *et al.* 2004), the present study focuses on the multirotor aerial vehicles (MAVs) (Gupte *et al.* 2012; Er *et al.* 2013). The interest for MAVs has increased in the last decade due to their low cost, high maneuverability, simplified mechanics, capability to perform vertical take-off and landing as well as hovering flight. These characteristics make them suitable for a wide range of applications, such as surveillance of indoor and urban environments, object delivery, building inspection, and agriculture monitoring. In all the aforementioned applications, a precise position tracking controller (or a guidance law) is crucial for autonomous operation.

Although there is a massive amount of concluded and ongoing research studies on MAVs, the design of control laws for such vehicles still has challenges to overcome. Most of those challenges are related to safety of the MAV itself and for people close to its operation. Mistler *et al.* (2013) and

1.Departamento de Ciência e Tecnologia Aeroespacial – Instituto Tecnológico de Aeronáutica – Divisão de Engenharia Mecânica – São José dos Campos/SP – Brazil.

Author for correspondence: Igor Afonso Acampora Prado | Departamento de Ciência e Tecnologia Aeroespacial – Instituto Tecnológico de Aeronáutica – Divisão de Engenharia Mecânica | Praça Marechal Eduardo Gomes, 50 – Vila das Acácias | CEP: 12.228-904 – São José dos Campos/SP – Brazil | Email: igoracampora@gmail.com

Mahony *et al.* (2012) propose linear control laws combined with the feedback linearization technique for guiding a quadrotor through a reference trajectory. Bouabdallah and Siegwart (2005) designed 2 control laws using, respectively, the sliding mode and the backstepping methods; the authors showed by simulations that the backstepping method outperforms the sliding mode controller. Madani and Benallegue (2007) propose position control scheme combining a backstepping controller with a sliding mode observer. Castillo *et al.* (2014) and Hua *et al.* (2009) present robust control law design methods considering that the system is subjected to external disturbance and model uncertainty. It is worth mentioning that none of the aforementioned methods have considered constraints on the control force.

In fact, few MAV control methods available in the literature have dealt with constraints issues (Castillo *et al.* 2005; Cunha *et al.* 2009). In Castillo *et al.* (2005), the authors divided the system into smaller subsystems with 2 degrees of freedom (DOF). For each subsystem, they applied a nested saturated controller (Teel 1992) to achieve global stability while respecting a maximum constraint on the total thrust magnitude. However, the subsystems were assumed to have uncoupled dynamics, which is not true in general. In Cunha *et al.* (2009), to avoid an unbounded growth of the actuation commands, the authors presented an asymptotically stable controller that limits the maximum thrust magnitude by saturating the position control error. These 2 studies only considered constraints on the maximum value of the magnitude of the total thrust vector.

More recently, Santos *et al.* (2013) and Yan *et al.* (2014) tackled the problem of controlling the position of an MAV under constraints both on the inclination of the rotor plane and on the magnitude of the total thrust. Santos *et al.* (2013) presented a simple but effective control method derived using feedback linearization and a proportional-derivative control law. Yan *et al.* (2014) solved the same problem by using the retrospective cost adaptive control (RCAC) strategy.

The model predictive control (MPC) strategy appears as the most interesting choice whenever constraints are a concern. In this method, a future control sequence is obtained by minimizing a cost function on predicted values of the controlled variable along a finite horizon, typically subjected to constraints (Camacho and Bordons 1998). Applications of MPC to position control of MAVs can be found in Raffo *et al.* (2010), Lopes *et al.* (2011), Alexis *et al.* (2012), and Chen *et al.* (2013). Raffo *et al.* (2010) proposed a control scheme consisting of an unconstrained MPC for position tracking and a non-linear H_∞ controller for attitude stabilization under aerodynamic disturbances and parametric as well as structural uncertainties. Lopes *et al.* (2011) designed a single MPC controller for both position control and attitude stabilization, considering constraints on both the pitch and the roll angles. Alexis *et al.* (2012) proposed a cascade MPC scheme, formulated over a set of piecewise affine models originated from both attitude and translation dynamics. In order to guide an MAV through a desired position trajectory, Chen *et al.* (2013) designed 2 separate MPCs, one for position control and the other for attitude control, the latter considering maximum constraints on the attitude angles.

In order to ensure a safe flight, it is essential to design a control law which avoids excessive accelerations and unexpected flips. The present study proposes an MPC for position control (guidance) of an MAV, constraining the total thrust vector within a conic set as well as the position vector within a parallelepiped set.

PROBLEM STATEMENT

PRELIMINARY DEFINITIONS

The motion of an MAV has 6 DOFs: 3 in translation and other 3 in rotation. However, this vehicle has only 4 independent control inputs: 3 torque components and the magnitude of the total thrust. Therefore, an MAV has an underactuated dynamics. At a first glance, it could be seem a challenge to deal with this characteristic. However, in practice, one needs to independently control only 4 DOF: the 3-dimensional position and the heading angle.

The block diagram of Fig. 1 describes a control system for controlling the 3-dimensional position $\mathbf{r} \in \mathbb{R}^3$ of an MAV to follow a time-varying position command $\mathbf{r}_c \in \mathbb{R}^3$. This

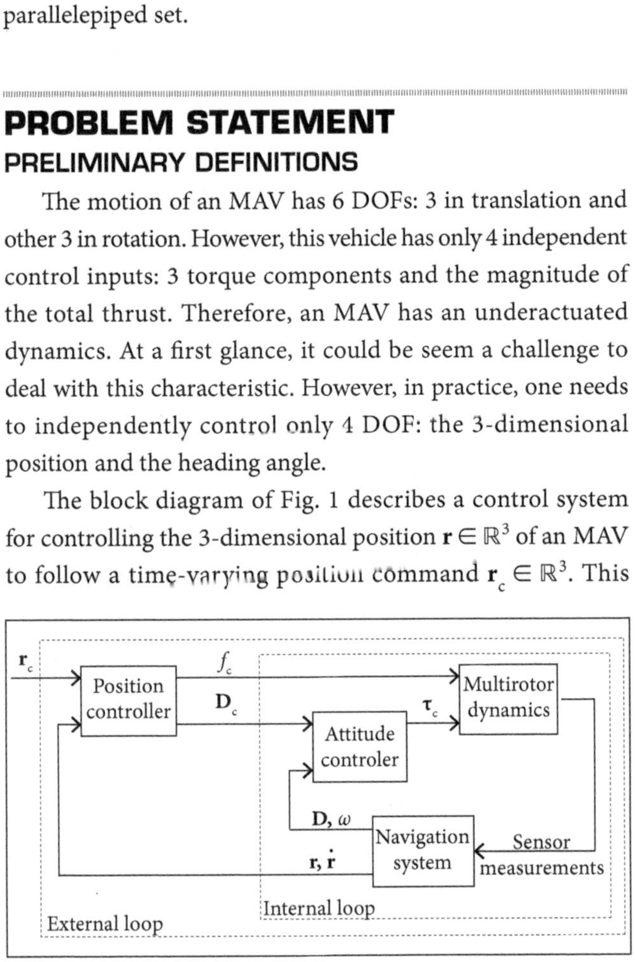

Figure 1. A typical structure of an MAV flight control system.

system is structured into 2 control loops: an internal loop for attitude control and an external one for position control. The Navigation System is generally responsible for estimating the vehicle's attitude $\mathbf{D} \in SO(3)$, angular velocity $\omega \in \mathbb{R}^3$, position $\mathbf{r}$, and linear velocity $\dot{\mathbf{r}} \in \mathbb{R}^3$. The Attitude Controller receives an attitude command $\mathbf{D}_c \in SO(3)$ and produces the control torque $\boldsymbol{\tau}_c \in \mathbb{R}^3$ necessary to rotate the rotor plane with respect to the local horizontal plane. The Position Controller has the function of computing a command $f_c \in \mathbb{R}^3$ for the total thrust magnitude and a command $\mathbf{D}_c \in SO(3)$ for the attitude. These commands are computed in such a way to produce suitable lateral acceleration on the vehicle. The present study is concerned with the design of a position controller.

POSITION CONTROL PROBLEM

Consider the MAV and the 3 Cartesian coordinate systems (CCS) illustrated in Fig. 2. Assume that the vehicle has a rigid structure. The body CCS $S_B \triangleq \{X_B, Y_B, Z_B\}$ is fixed to the structure and its origin coincides with the vehicle's center of mass (CM). The reference CCS $S_R \triangleq \{X_R, Y_R, Z_R\}$ is Earth-fixed and its origin is at a known point O. Finally, a second reference CCS $S_{R'} \triangleq \{X_{R'}, Y_{R'}, Z_{R'}\}$ is defined to be parallel to S_R, but with origin at CM. Assume that S_R is an inertial frame.

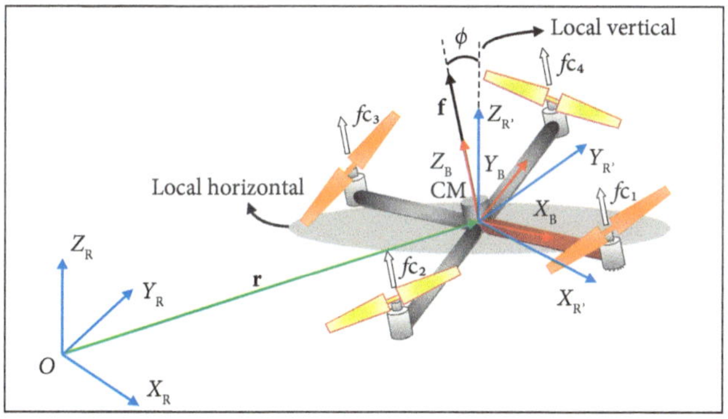

Figure 2. A multirotor aerial vehicle and 3 Cartesian coordinate systems.

Invoking the second Newton's law, the translational dynamics of the MAV illustrated in Fig. 2 can be immediately described in S_R by the following second-order differential equation:

$$\ddot{\mathbf{r}} = \frac{1}{m}(\mathbf{f} + \mathbf{f}_d) + \begin{bmatrix} 0 \\ 0 \\ -g \end{bmatrix}, \quad (1)$$

where: $\mathbf{r} \triangleq [r_x, r_y, r_z]^T \in \mathbb{R}^3$ is the CM position vector with respect to S_R; $\mathbf{f} \triangleq [f_x, f_y, f_z]^T \in \mathbb{R}^3$ is the total thrust vector represented in S_R; $\mathbf{f}_d \in \mathbb{R}^3$ is the disturbance force vector represented in S_R, m is the vehicle's mass; g is the gravitational acceleration.

If, instead of a quadrotor, we had considered a hexa-rotor or an octo-rotor, the model in Eq. 1 would not change in any aspects, except for the origin of $\mathbf{f}$ that would receive contributions of either 6 or 8, instead of 4 propellers. As illustrated in Fig. 2, $\mathbf{f}$ is perpendicular to the rotor plane.

Define the inclination angle $\phi \in \mathbb{R}$ of the rotor plane as the angle between Z_B and $Z_{R'}$. It can be expressed by

$$\phi \triangleq \cos^{-1} \frac{f_z}{f_c}, \quad (2)$$

where: $\mathbf{f}_c \triangleq ||\mathbf{f}||$.

Define the position control error $\tilde{\mathbf{r}} \in \mathbb{R}^3$ as

$$\tilde{\mathbf{r}} \triangleq \mathbf{r} - \mathbf{r}_c, \quad (3)$$

where: $\mathbf{r}_c \triangleq [r_{c,x}\ r_{c,y}\ r_{c,z}]^T$ is a position command.

Problem 1: $\phi_{max} \in \mathbb{R}$ denote the maximum allowable value of ϕ, $f_{min} \in \mathbb{R}$ and $f_{max} \in \mathbb{R}$ denote, respectively, the minimum and maximum allowable values of f_c, and $\mathbf{r}_{min} \in \mathbb{R}^3$ and $\mathbf{r}_{max} \in \mathbb{R}^3$ denote, respectively, the minimum and maximum allowable values of $\mathbf{r}$. The MAV guidance problem is to find a control law for $\mathbf{f}$ that minimizes $\tilde{\mathbf{r}}$ subjected to the inclination constraint $\phi \leq \phi_{max}$, to the force magnitude constraint $f_{min} \leq f_c \leq f_{max}$, and to the position constraint $\mathbf{r}_{min} \leq \mathbf{r} \leq \mathbf{r}_{max}$.

Remark 1: the control force $\mathbf{f}$ of Problem 1 is not the effective control force undergone by the vehicle. In fact, its magnitude f_c represents a command for the power electronics to drive the motors, while the inclination angle ϕ is used to compute the attitude command $\mathbf{D}_c$ for the attitude control loop to orient the rotor plane (Fig. 1). Nevertheless, f_c and $\mathbf{D}_c$ are assumed here to be identical to the respective actual variables. The assumption about f_c is reasonable if a precise model for the thrust force is available. On the other hand, the assumption about $\mathbf{D}_c$ can also be approximated in practice if the controllers are tuned to allow the internal loop to have a much faster dynamics than the external one.

Remark 2: during the design of the controller, disturbance force and model uncertainty will not be considered. However, in "Computational Simulations" section, the proposed control method will be evaluated under such non-ideal conditions.

Remark 3: the position $\mathbf{r}$ and the velocity $\dot{\mathbf{r}}$ of CM are assumed to be available for feedback. In practice, these variables are provided by a navigation system (Fig. 1), which is not the focus of the present study.

In Problem 1, the parallelepiped constraint imposed on the vehicle's position $\mathbf{r}$ is considered so as to avoid collisions with the bounds of a box-shaped indoor environment.

On the other hand, one can visualize the corresponding constraint space on $\mathbf{f}$ as a conic space with an inferior and a superior spherical lid, as illustrated in Fig. 3. Note that the so-generated constraint space is non-linear and non-convex. The constraints on both magnitude and inclination of $\mathbf{f}$ are directly connected to the vertical and lateral accelerations of the vehicle. As one can see in Fig. 4, the component f_z is responsible for controlling the altitude of the vehicle, while $\mathbf{f}_{xy}$ produces the lateral acceleration that guides it along the X_R and Y_R directions, where $\mathbf{f}_{xy} \overset{\Delta}{=} [f_x\ f_y] \in \mathbb{R}^2$ denotes the horizontal projection of $\mathbf{f}$. As ϕ increases, the lateral acceleration of the vehicle also increases. If the constraint f_{max} is not sufficiently high, the vehicle could suffer a loss of lift. Furthermore, it is interesting to choose a ϕ_{max} that avoids unexpected flips as well as large lateral accelerations.

The choice of suitable values for the constraints on $\mathbf{f}$ and $\mathbf{r}$ can enhance the flight safety, since it avoids abrupt motion and delimits the flight region, respectively. Such characteristics are quite desirable, for example, in MAVs used to support people in indoor and urban environments.

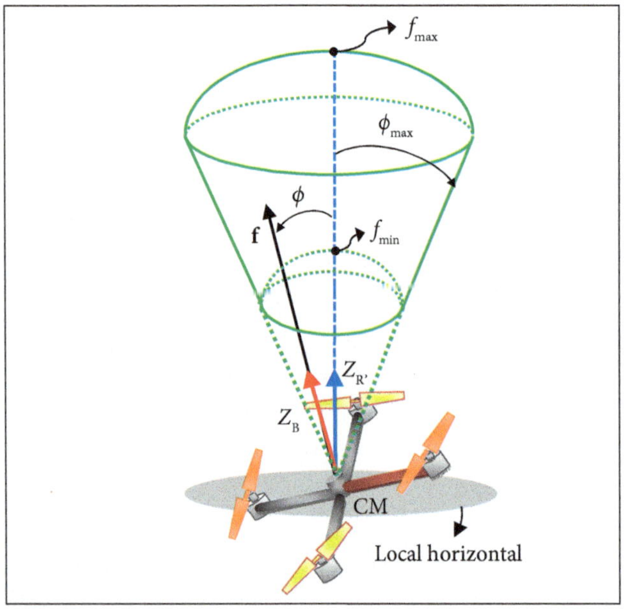

Figure 3. The original conic control space.

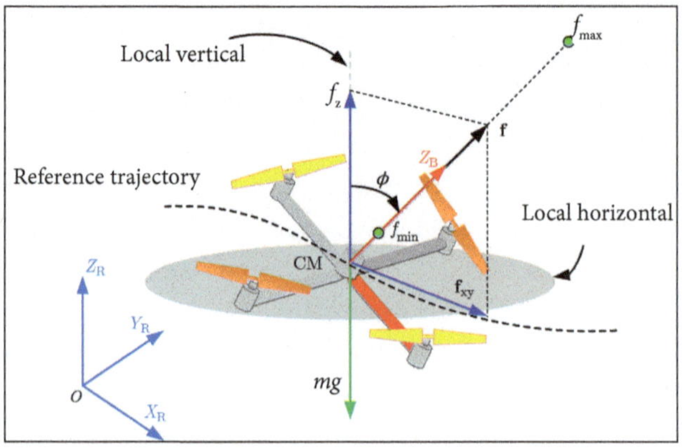

Figure 4. Analysis of **f** with respect to its constraints on inclination and magnitude.

PROBLEM SOLUTION

STATE-SPACE MODEL FOR TRANSLATION

Define the state vector $\mathbf{x} \overset{\Delta}{=} [r_x\ \dot{r}_x\ r_y\ \dot{r}_y\ r_z\ \dot{r}_z]^T \in \mathbb{R}^6$ and the control input vector $\mathbf{u} \overset{\Delta}{=} [u_x\ u_y\ u_z]^T \in \mathbb{R}^3$,

$$\mathbf{u} \triangleq \frac{1}{m}\mathbf{f} - \begin{bmatrix} 0 \\ 0 \\ g \end{bmatrix}. \tag{4}$$

Using Eq. 4, Eq. 1 can be immediately rewritten as a continuous-time linear state equation of the form $\dot{\mathbf{x}} = \mathbf{A}\mathbf{x} + \mathbf{B}\mathbf{u}$. Similarly, by defining the controlled output vector to be the MAV position $\mathbf{y} \overset{\Delta}{=} [r_x\ r_y\ r_z]^T \in \mathbb{R}^3$, one can obtain a continuous-time output equation of the form $\mathbf{y} = \mathbf{C}\mathbf{x}$. Let $\mathbf{x}(k) \in \mathbb{R}^6$, $\mathbf{u}(k) \in \mathbb{R}^3$ and $\mathbf{y}(k) \in \mathbb{R}^3$ denote, respectively, the state vector, the control input vector and the controlled output vector, all in the discrete-time domain. Using the ZOH method with a sampling time of $T_s = 20$ ms, a discrete-time version of the above state-space model is obtained as

$$\mathbf{x}(k+1) = \mathbf{A}_d\mathbf{x}(k) + \mathbf{B}_d\mathbf{u}(k), \tag{5}$$

$$\mathbf{y}(k) = \mathbf{C}_d\mathbf{x}(k), \tag{6}$$

where

$$\mathbf{A}_d = \begin{bmatrix} 1 & 0.02 & 0 & 0 & 0 & 0 \\ 0 & 1 & 0 & 0 & 0 & 0 \\ 0 & 0 & 1 & 0.02 & 0 & 0 \\ 0 & 0 & 0 & 1 & 0 & 0 \\ 0 & 0 & 0 & 0 & 1 & 0.02 \\ 0 & 0 & 0 & 0 & 0 & 1 \end{bmatrix} \in \mathbb{R}^{6\times 6}, \tag{7}$$

$$\mathbf{B}_\mathrm{d} = \begin{bmatrix} 2\times10^{-4} & 0 & 0 \\ 0.02 & 0 & 0 \\ 0 & 2\times10^{-4} & 0 \\ 0 & 0.02 & 0 \\ 0 & 0 & 2\times10^{-4} \\ 0 & 0 & 0.02 \end{bmatrix} \in \mathbb{R}^{6\times3}, \tag{8}$$

$$\mathbf{C}_\mathrm{d} = \begin{bmatrix} 1 & 0 & 0 & 0 & 0 & 0 \\ 0 & 0 & 1 & 0 & 0 & 0 \\ 0 & 0 & 0 & 0 & 1 & 0 \end{bmatrix} \in \mathbb{R}^{3\times6}. \tag{9}$$

Consider the discrete-time state-space model of Eqs. 5 and 6. It can be rewritten in the incremental-input form as (Maciejowski 2002)

$$\begin{aligned} \boldsymbol{\xi}(k+1) &= \tilde{\mathbf{A}}\boldsymbol{\xi}(k) + \tilde{\mathbf{B}}\Delta\mathbf{u}(k) \\ \mathbf{y}(k) &= \tilde{\mathbf{C}}\boldsymbol{\xi}(k) \end{aligned} , \tag{10}$$

with

$$\boldsymbol{\xi}(k) = \begin{bmatrix} \Delta\mathbf{x}(k) \\ \mathbf{y}(k) \end{bmatrix} \in \mathbb{R}^9, \tag{11}$$

$$\tilde{\mathbf{A}} = \begin{bmatrix} \mathbf{A}_\mathrm{d} & \mathbf{0}_{6\times3} \\ \mathbf{C}_\mathrm{d}\mathbf{A}_\mathrm{d} & \mathbf{I}_3 \end{bmatrix} \in \mathbb{R}^{9\times9}, \tag{12}$$

$$\tilde{\mathbf{B}} = \begin{bmatrix} \mathbf{B}_\mathrm{d} \\ \mathbf{C}_\mathrm{d}\mathbf{B}_\mathrm{d} \end{bmatrix} \in \mathbb{R}^{9\times3}, \tag{13}$$

$$\tilde{\mathbf{C}} = \begin{bmatrix} \mathbf{0}_{3\times6} & \mathbf{I}_3 \end{bmatrix} \in \mathbb{R}^{3\times9}, \tag{14}$$

where: $\Delta\mathbf{x}(k) \overset{\Delta}{=} \mathbf{x}(k) - \mathbf{x}(k-1) \in \mathbb{R}^6$ denotes the incremental state vector; $\Delta\mathbf{u}(k) \overset{\Delta}{=} \mathbf{u}(k) - \mathbf{u}(k-1) \in \mathbb{R}^3$ is the incremental control input vector; $\mathbf{I}_3$ is the identity matrix with dimensions 3×3; $\mathbf{0}_{3\times6}$ is a zero matrix with dimensions 3×6.

PREDICTION MODEL

Using Eq. 10, the prediction model can be obtained as (see Maciejowski 2002, p. 50)

$$\mathcal{Y}_N = \mathcal{G}\Delta\mathcal{U}_M + \mathcal{F}, \tag{15}$$

where: $\mathcal{Y}_N \in \mathbb{R}^{3N\times1}$ stacks the controlled outputs along a prediction horizon of length N; $\Delta\mathcal{U}_M \in \mathbb{R}^{3M\times1}$ stacks the incremental control inputs along a control horizon of length M,

$$\mathcal{G} \triangleq \begin{bmatrix} \tilde{\mathbf{C}}\tilde{\mathbf{B}} & \mathbf{0}_{3\times3} & \cdots & \mathbf{0}_{3\times3} \\ \tilde{\mathbf{C}}\tilde{\mathbf{A}}\tilde{\mathbf{B}} & \tilde{\mathbf{C}}\tilde{\mathbf{B}} & \cdots & \mathbf{0}_{3\times3} \\ \vdots & \vdots & \ddots & \vdots \\ \tilde{\mathbf{C}}\tilde{\mathbf{A}}^{M-1}\tilde{\mathbf{B}} & \tilde{\mathbf{C}}\tilde{\mathbf{A}}^{M-2}\tilde{\mathbf{B}} & \cdots & \tilde{\mathbf{C}}\tilde{\mathbf{B}} \\ \vdots & \vdots & & \vdots \\ \tilde{\mathbf{C}}\tilde{\mathbf{A}}^{N-1}\tilde{\mathbf{B}} & \tilde{\mathbf{C}}\tilde{\mathbf{A}}^{N-2}\tilde{\mathbf{B}} & \cdots & \tilde{\mathbf{C}}\tilde{\mathbf{A}}^{N-M}\tilde{\mathbf{B}} \end{bmatrix} \in \mathbb{R}^{3N\times3M} \tag{16}$$

and

$$\mathcal{F} \triangleq \begin{bmatrix} \tilde{\mathbf{C}}\tilde{\mathbf{A}} \\ \tilde{\mathbf{C}}\tilde{\mathbf{A}}^2 \\ \vdots \\ \tilde{\mathbf{C}}\tilde{\mathbf{A}}^N \end{bmatrix} \boldsymbol{\xi}(k) \in \mathbb{R}^{3N\times1}. \tag{17}$$

THRUST VECTOR CONSTRAINTS

Using Eq. 4, the thrust magnitude constraint inequation $f_{min} \le f_c \le f_{max}$ can be rewritten in terms of $\mathbf{u}$ as

$$f_\mathrm{min} \le m\sqrt{u_\mathrm{x}^2 + u_\mathrm{y}^2 + (u_\mathrm{z}+g)^2} \le f_\mathrm{max}. \tag{18}$$

Assuming that $0 \le \phi_{max} < \pi/2$ rad, the inclination constraint $\phi \le \phi_{max}$ established in Problem 1 can be replaced by $\cos\phi \ge \cos\phi_{max}$. Using Eqs. 2 and 4, the last inequation can be rewritten in terms of the components of $\mathbf{u}$ as

$$\frac{u_\mathrm{z}+g}{\sqrt{u_\mathrm{x}^2 + u_\mathrm{y}^2 + (u_\mathrm{z}+g)^2}} \ge \cos\phi_\mathrm{max}. \tag{19}$$

In order to obtain linear approximations for Eqs. 18 and 19, consider the rectangular pyramid inscribed in the original conic control space depicted in Fig. 3. The new control space is illustrated in Fig. 5a. By inspection of this figure, the new constraint on f_z can be expressed as

$$f_\mathrm{min} \le f_\mathrm{z} \le f_\mathrm{max}\cos\phi_\mathrm{max}. \tag{20}$$

Consider an arbitrary section of the pyramid depicted in Fig. 5a. Let f_z denote its $Z_{R'}$ coordinate. The projection of this section on the $X_{R'} - Y_{R'}$ plane consists of a square of dimensions α × α, as illustrated in Fig. 5b. From the geometry of this figure, one can write

$$\alpha = \sqrt{2} f_z \tan \phi_{\max}. \tag{21}$$

By inspection of Fig. 5b, it can be seen that the implication of **f** to be bounded inside the pyramidal space is that $f_x \in [-\alpha/2, \alpha/2]$ and $f_y \in [-\alpha/2, \alpha/2]$. By substituting Eq. 21 into these intervals, one can obtain

$$-\frac{\sqrt{2}}{2} f_z \tan \phi_{\max} \leq f_x \leq \frac{\sqrt{2}}{2} f_z \tan \phi_{\max} \tag{22}$$

$$-\frac{\sqrt{2}}{2} f_z \tan \phi_{\max} \leq f_y \leq \frac{\sqrt{2}}{2} f_z \tan \phi_{\max}. \tag{23}$$

Rewriting Eqs. 20, 22 and 23 in matrix form, one yields

$$\boldsymbol{\Lambda}\, \mathbf{f} \leq \boldsymbol{\lambda}, \tag{24}$$

where

$$\boldsymbol{\Lambda} \triangleq \begin{bmatrix} -1 & 0 & -\frac{\sqrt{2}}{2} \tan \phi_{\max} \\ 1 & 0 & -\frac{\sqrt{2}}{2} \tan \phi_{\max} \\ 0 & -1 & -\frac{\sqrt{2}}{2} \tan \phi_{\max} \\ 0 & 1 & -\frac{\sqrt{2}}{2} \tan \phi_{\max} \\ 0 & 0 & -1 \\ 0 & 0 & 1 \end{bmatrix} \tag{25}$$

and

$$\boldsymbol{\lambda} \triangleq \begin{bmatrix} 0 \\ 0 \\ 0 \\ 0 \\ -f_{\min} \\ f_{\max} \cos \phi_{\max} \end{bmatrix}. \tag{26}$$

Now using Eq. 4, Eq. 24 can be rewritten in terms of **u** as

$$\bar{\boldsymbol{\Lambda}} \mathbf{u} \leq \bar{\boldsymbol{\lambda}}, \tag{27}$$

where $\bar{\Lambda} \triangleq m\Lambda$ and

$$\bar{\boldsymbol{\lambda}} \triangleq \boldsymbol{\lambda} - m\boldsymbol{\Lambda} \begin{bmatrix} 0 \\ 0 \\ g \end{bmatrix}. \tag{28}$$

Finally, replacing $\mathbf{u}(k) = \Delta\mathbf{u}(k) + \mathbf{u}(k-1)$ into Eq. 27 and taking it at M future instants starting from k, the thrust vector constraint is obtained as

$$\mathcal{T} \Delta \mathcal{U}_M \leq \boldsymbol{\varrho}, \tag{29}$$

where $\mathcal{T} = [\bar{\lambda}]_M - \mathrm{diag}(\bar{\Lambda})[\mathbf{u}(k-1)]_M$ and

$$\mathcal{T} = \begin{bmatrix} \bar{\boldsymbol{\Lambda}} & \mathbf{0}_{6\times3} & \cdots & \mathbf{0}_{6\times3} \\ \bar{\boldsymbol{\Lambda}} & \bar{\boldsymbol{\Lambda}} & \cdots & \mathbf{0}_{6\times3} \\ \vdots & \vdots & \ddots & \vdots \\ \bar{\boldsymbol{\Lambda}} & \bar{\boldsymbol{\Lambda}} & \cdots & \bar{\boldsymbol{\Lambda}} \end{bmatrix} \in \mathbb{R}^{6M\times3M}. \tag{30}$$

The matrix $\mathcal{T}$ is lower block-triangular, $[\bullet]_M$ is an operator that stacks M copies of a column vector in an augmented vector, and $\mathbf{u}(k-1) \in \mathbb{R}^3$ is the control input at instant $k-1$.

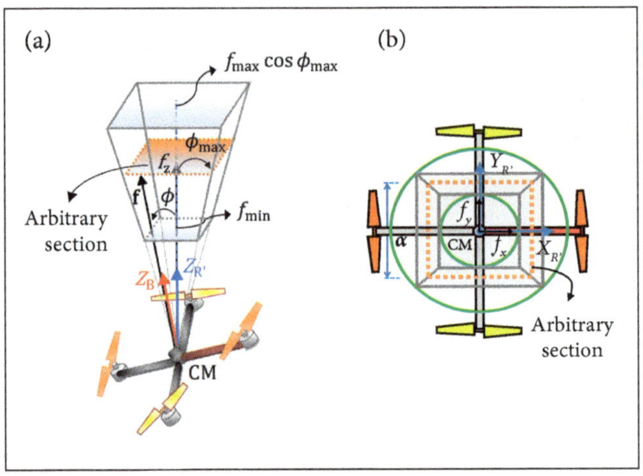

Figure 5. Pyramidal control space.

POSITION VECTOR CONSTRAINTS

Now consider the constraints on the MAV position $\mathbf{r}_{\min} \leq \mathbf{r} \leq \mathbf{r}_{\max}$. Taking them along the prediction horizon N, an augmented set of constraints is obtained as

$$[\mathbf{r}_{\min}]_N \leq \mathcal{Y}_N \leq [\mathbf{r}_{\max}]_N, \tag{31}$$

which can be rewritten in terms of Δu_M by means of Eq. 15, resulting in

$$\mathcal{A}\Delta\mathcal{U}_M \leq \gamma, \tag{32}$$

where

$$\mathcal{A} \triangleq \begin{bmatrix} \mathcal{G} \\ -\mathcal{G} \end{bmatrix} \in \mathbb{R}^{6N \times 3M} \tag{33}$$

and

$$\gamma \triangleq \begin{bmatrix} [\mathbf{r}_{\max}]_N - \mathcal{F} \\ \mathcal{F} - [\mathbf{r}_{\min}]_N \end{bmatrix} \in \mathbb{R}^{6N}. \tag{34}$$

MODEL PREDICTIVE CONTROLLER

The optimal control vector $\mathbf{u}^*(k) \in \mathbb{R}^3$ computed at the discrete-time instant k is given by $\mathbf{u}^*(k) = \Delta\mathbf{u}^*(k) + \mathbf{u}^*(k-1)$, where $\Delta\mathbf{u}^*(k) \in \mathbb{R}^3$ is the first control vector in Δu^*_M, which, in turn, is obtained by minimizing the following quadratic cost function:

$$J(\Delta\mathcal{U}_M) = \frac{1}{2}\Delta\mathcal{U}_M^{\mathrm{T}}\mathcal{H}\Delta\mathcal{U}_M + \mathcal{M}^{\mathrm{T}}\Delta\mathcal{U}_M + c, \tag{35}$$

subjected to

$$\mathcal{S}\Delta\mathcal{U}_M \leq \boldsymbol{\vartheta}, \tag{36}$$

where: $1/2H = \mathcal{G}^{\mathrm{T}}Q\mathcal{G} + R$, $M^{\mathrm{T}} = 2(\mathcal{F} - [r_c]_N)^{\mathrm{T}}Q\mathcal{G}$, $c = (\mathcal{F} - [r_c]_N)^{\mathrm{T}}Q(\mathcal{F} - [r_c]_N)$,

$$\mathcal{S} \triangleq \begin{bmatrix} \mathcal{T} \\ \mathcal{A} \end{bmatrix} \in \mathbb{R}^{(6M+6N)\times 3M}, \tag{37}$$

and

$$\boldsymbol{\vartheta} \triangleq \begin{bmatrix} \varrho \\ \gamma \end{bmatrix} \in \mathbb{R}^{(6M+6N)}. \tag{38}$$

In this study, the controlled output weighting matrix is set to $Q = \eta \times \mathbf{I}_{3N}$ and the control input weighting matrix is set to $R = \rho \times \mathbf{I}_{3M}$. The above optimization problem is in the conventional quadratic programming form, for which there are efficient numerical solution methods (Rossiter 2003). The Δ-input MPC formulation presented here has an intrinsic integral action, which allows to track a constant set point with 0 steady-state error and reject a constant disturbance input (Maciejowski 2002).

THRUST MAGNITUDE AND ATTITUDE COMMANDS

Now we need to compute the total thrust magnitude command f_c and the attitude command $\mathbf{D}_c$ from the control input $\mathbf{u}(k)$ provided by the MPC. The first command is obtained by simply taking the Euclidian norm of $\mathbf{f}$. Using Eq. 4, f_c is obtained as

$$f_c = m\sqrt{u_x^2 + u_y^2 + (u_z + g)^2}. \tag{39}$$

The attitude command $\mathbf{D}_c \in SO(3)$ for the internal (attitude) control loop (Fig. 1) is also computed from $\mathbf{f}$, which contains information about the orientation of the rotor plane with respect to the local horizontal. In order to provide a unique 3-dimensional attitude command, it is necessary to specify a heading angle. For example, one can choose a 0 heading angle, which is just equivalent to the attitude represented by the principal Euler angle/axis (ϕ; $\mathbf{e}$), where ϕ is computed from Eq. 2 and $\mathbf{e}$ is a unit vector given by

$$\mathbf{e} = \frac{\mathbf{f} \times \mathbf{f}_{xy}}{\|\mathbf{f} \times \mathbf{f}_{xy}\|}, \tag{40}$$

where: $\mathbf{f}_{xy}$ is the projection of $\mathbf{f}$ on the $X_{R'} - Y_{R'}$ plane.

The corresponding attitude command is given by (Shuster 1993)

$$\mathbf{D}(\phi; \mathbf{e}) = \cos\phi\mathbf{I}_3 + (1 - \cos\phi)\mathbf{e}\mathbf{e}^{\mathrm{T}} - \sin\phi[\mathbf{e}\times]. \tag{41}$$

Now considering an arbitrary heading angle ψ, the attitude command $\mathbf{D}_c$ results from 2 successive rotations. The first one is that represented by Eq. 41, while the second one is an elementary rotation $\mathbf{D}(\psi; Z_B)$of an angle ψ about the Z_B axis, *i.e.*

$$\mathbf{D}_c = \mathbf{D}(\psi; Z_B)\mathbf{D}(\phi; \mathbf{e}), \tag{42}$$

where

$$\mathbf{D}(\psi, Z_B) = \begin{bmatrix} \cos\psi & \sin\psi & 0 \\ -\sin\psi & \cos\psi & 0 \\ 0 & 0 & 1 \end{bmatrix}. \tag{43}$$

COMPUTATIONAL SIMULATIONS

SIMULATION PARAMETERS

The simulations are implemented in MATLAB/Simulink. The non-linear 6 DOF dynamics of an MAV is integrated using the Runge-Kutta-4 method with an integration step of 0.001 s. The vehicle's attitude is modeled using Euler angles in the rotation sequence 1-2-3. The vehicle's mass is $m = 1$ kg and the gravitational acceleration is assumed to be $g = 9.81$ m/s^2. The vehicle's inertia matrix is

$$\mathbf{J} = \begin{bmatrix} 0.0171 & 0 & 0 \\ 0 & 0.0171 & 0 \\ 0 & 0 & 0.0286 \end{bmatrix} \text{kgm}^2. \quad (44)$$

The interior-point method is adopted to solve the MPC optimization. The control input weighting matrix Q and controlled output weighting matrix R are adjusted with $\rho = 0.01$ and $\eta = 1$, respectively. The prediction and control horizons are set to $N = 80$ and $M = 5$, respectively. The maximum and minimum position bounds are $\mathbf{r}_{max} = [6\ 6\ 3]^T$ m and $\mathbf{r}_{min} = [0\ 0\ 0]^T$, respectively. The maximum and minimum constraints on the force magnitude are set in $f_{max} = 20$ N and $f_{min} = 2$ N, respectively. A non-zero minimum bound on the force magnitude f_{min} was chosen in order to avoid loss of attitude control. On the other hand, the maximum bound f_{max} is set sufficiently large to allow lateral accelerations without loss of lift. In order to verify the system robustness against an unknown input, a 0-mean Gaussian disturbance force with covariance $\mathbf{Q}_d = 0.3\times \mathbf{I}_3$ N^2 is taken into account.

The position commands consist of a sequence of line segments from the initial position $\mathbf{r}_i = [1\ 1\ 0]^T$ m to the final one $\mathbf{r}_f = [5\ 5\ 1]^T$ m passing by the waypoints $\mathbf{w}_1 = [1\ 1\ 1]^T$ m; $\mathbf{w}_2 = [2\ 2\ 2]^T$ m; $\mathbf{w}_3 = [4\ 2\ 2]^T$ m; $\mathbf{w}_4 = [5\ 3\ 2]^T$ m and $\mathbf{w}_5 = [5\ 5\ 2]^T$ m. A total of 15 MC simulations with 300 realizations are carried out considering all the combinations of 3 different speed command v values (0.5; 1.0 and 2.0 m/s) and 5 different maximum inclination ϕ_{max} values (10°; 15°; 20°; 25° and 30°).

The following figure of merit is used to evaluate the position control error:

$$e_q \triangleq \frac{1}{N_r\sqrt{k_f}} \sum_{i=1}^{N_r} \sqrt{\sum_{k=1}^{k_f} \left(r_{d,q}(k) - r_q^{(i)}(k)\right)^2}, \quad (45)$$

for q equal to x, y or z; $r_q^{(i)}(k)$ denotes the ith realization of $r_q(k)$.

For evaluating the frequency of constraint violation, the following figure of merit is adopted:

$$I_l \triangleq \frac{1}{N_r k_f} \sum_{i=1}^{N_r} M_l^{(i)}, \quad (46)$$

where: $M_l^{(i)}$ is the number of discrete-time instants (of the ith realization) in which constraint l is violated, for l equal to x, y, z, ϕ, f_{min} or f_{max}.

The attitude controllers chosen for the present simulation are uncoupled proportional-derivative control laws tuned so as to make the attitude dynamics have a bandwidth significantly larger than the bandwidth of the position control dynamics.

SIMULATION RESULTS

The Monte Carlo simulation results are summarized in Table 1 in terms of e_q and I_l. First, one can observe that the control error increases as the speed command v is increased or as the maximum inclination ϕ_{max} is decreased. For example, for $v = 0.5$ m/s, the position errors stay below 5 cm, whereas they approach 25 cm when the speed is set to $v = 2.0$ m/s. Regarding the violation of position constraints, no occurrence is observed with any of the 3 speed commands. Concerning the maximum inclination constraint ϕ_{max}, for all speed commands v, the number of violations reduces as ϕ_{max} is increased. Finally, the frequency of violations of f_{min} and f_{max} increases as the speed command is increased, but decreases as ϕ_{max} is increased.

Figure 6 shows the MC realizations of the MAV position together with the corresponding mean and standard deviation curves for $v = 1.0$ m/s and different values of ϕ_{max} (10°; 20°; 30°). One can see that the standard deviation decreases as ϕ_{max} increases. This behavior is due to the fact that a smaller value of ϕ_{max} results in a smaller horizontal projection $\mathbf{f}_{xy}$ of the total thrust on the horizontal plane, which, in turn, reduces the vehicle's maneuverability and capability to reject horizontal disturbance forces. On the contrary, as ϕ_{max} increases, $\mathbf{f}_{xy}$ becomes larger, improving the maneuverability, which, in turn, provides a better disturbance rejection.

On the other hand, Fig. 7 shows the MC realizations of the MAV position together with the corresponding mean and standard deviation curves for $\phi_{max} = 15°$ and different values of v (0.5; 1.0 and 2.0 m/s).

One can see that, as the speed command increases, the standard deviation also becomes larger. The main reason is the fact that larger speed commands require better maneuverability and larger horizontal acceleration to ensure that the vehicle follows the reference trajectory with acceptable performance.

The worst performance observed in Table 1 occurs with $\phi_{max} = 10°$ and $v = 2.0$ m/s. This scenario combines low maneuverability (due to a small ϕ_{max}) with a large speed command, which results in a large amount of control input constraint violations (Fig. 8). With a small ϕ_{max}, the vehicle does not reach sufficient lateral acceleration to follow a large speed command and simultaneously reject the disturbance force. It causes a large position error (Fig. 9).

Table 1. Monte Carlo simulation results for different values of v and ϕ_{max}.

v (m/s)	ϕ_{max} (deg)	e_x (m)	e_y (m)	e_z (m)	I_x (%)	I_y (%)	I_z (%)	I_ϕ (%)	I_{fmax} (%)	I_{fmin} (%)
0.5	10	0.028	0.037	0.021	0.00	0.00	0.00	0.58	0.00	0.00
	15	0.010	0.014	0.014	0.00	0.00	0.00	0.00	0.00	0.00
	20	0.009	0.013	0.014	0.00	0.00	0.00	0.00	0.00	0.00
	25	0.009	0.012	0.013	0.00	0.00	0.00	0.00	0.00	0.00
	30	0.009	0.012	0.013	0.00	0.00	0.00	0.00	0.00	0.00
1.0	10	0.057	0.071	0.060	0.00	0.00	0.00	1.54	0.00	0.00
	15	0.021	0.032	0.039	0.00	0.00	0.00	0.00	0.00	0.00
	20	0.021	0.031	0.039	0.00	0.00	0.00	0.00	0.00	0.00
	25	0.020	0.030	0.039	0.00	0.00	0.00	0.00	0.00	0.00
	30	0.020	0.030	0.039	0.00	0.00	0.00	0.00	0.00	0.00
2.0	10	0.211	0.258	0.216	0.00	0.00	0.00	15.6	0.37	0.36
	15	0.052	0.109	0.097	0.00	0.00	0.00	3.38	0.00	0.00
	20	0.042	0.097	0.095	0.00	0.00	0.00	0.00	0.00	0.00
	25	0.042	0.093	0.093	0.00	0.00	0.00	0.00	0.00	0.00
	30	0.042	0.088	0.092	0.00	0.00	0.00	0.00	0.00	0.00

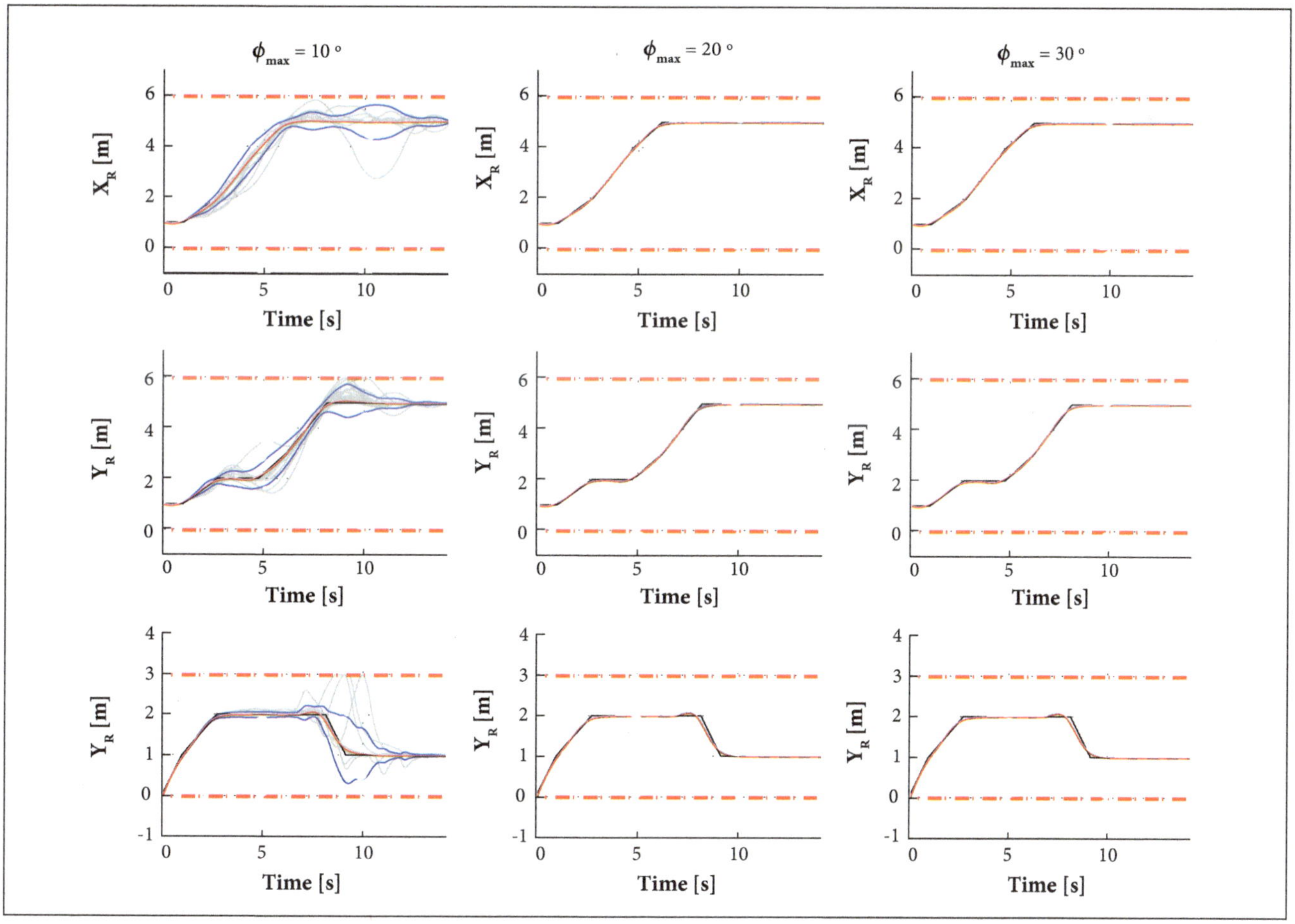

Figure 6. Performance of position control varying the maximum inclination constraint. The solid gray lines are MC realizations; the dashed red lines are the position constraints; the solid black lines are the position command; the solid red lines are the sample means; the solid blue lines are the sample means plus/minus the standard deviation.

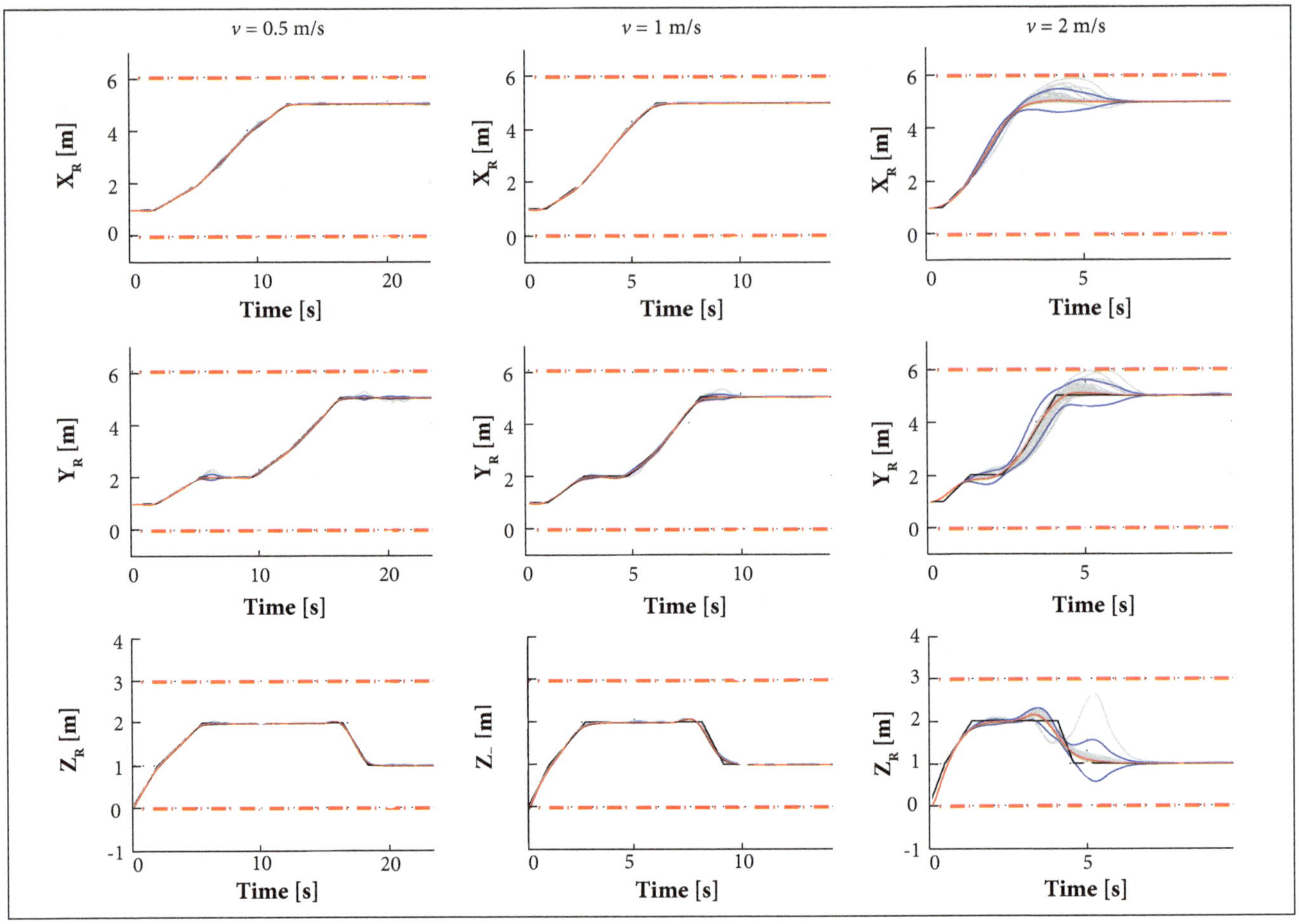

Figure 7. Performance of position control varying the speed command. The solid gray lines are MC realizations; the dashed red lines are the position constraints; the solid black lines are the position command; the solid red lines are the sample means; the solid blue lines are the sample means plus/minus the standard deviation.

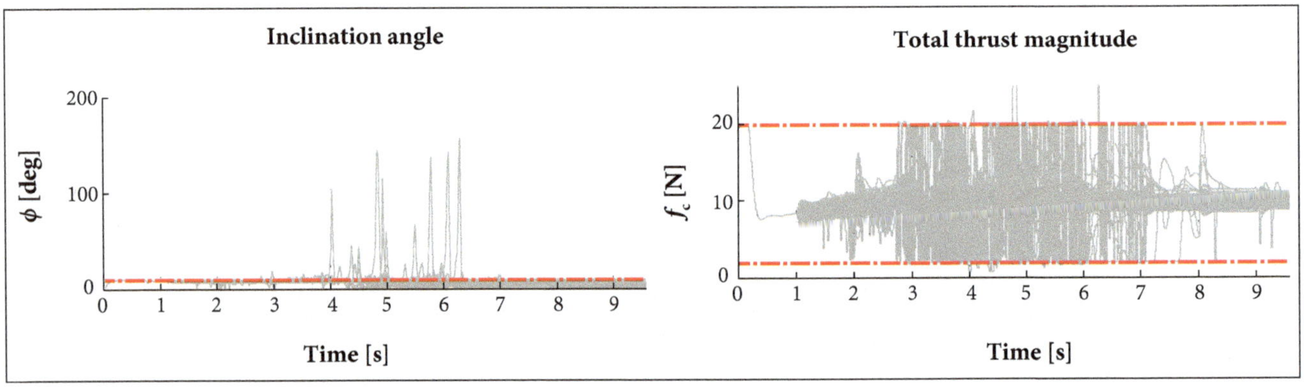

Figure 8. Inclination angle ϕ and thrust magnitude f_c in the worst case (ϕ_{max} = 10° and v = 2.0 m/s). The gray lines are the MC realizations and the dashed red lines are the constraints.

On the other hand, the best performance observed in Table 1 occurs with ϕ_{max} = 30° and v = 0.5 m/s. This scenario combines high maneuverability (due to a large ϕ_{max}) with a small speed command. In this case, the vehicle does not suffer a significant influence of the disturbance forces and respects the constraint on both the inclination angle ϕ and thrust magnitude f_c. For details, Fig. 10 shows ϕ and f_c, while Fig. 11 shows the corresponding position control performance.

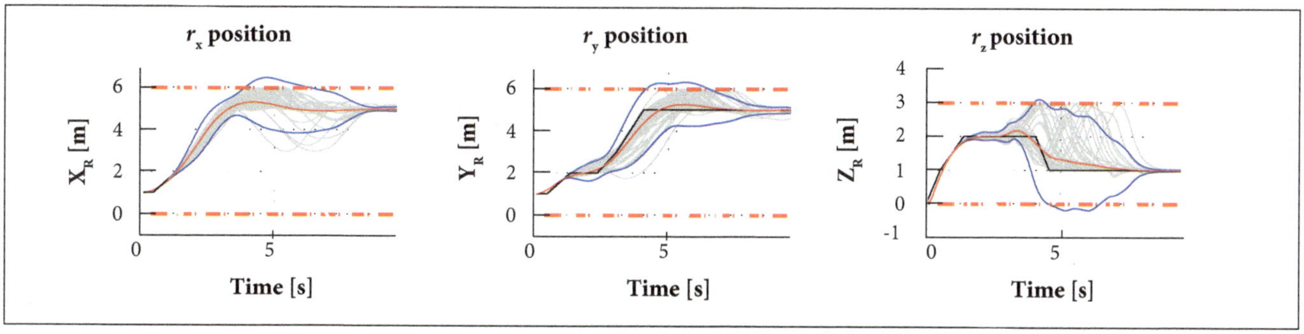

Figure 9. Performance of position control in the worst case (ϕ_{max} = 10° and v = 2.0 m/s) . The solid gray lines are MC realizations; the dashed red lines are the position constraints; the solid black line are the position commands; the solid red lines are the sample means; the solid blue lines are the sample means plus/minus the standard deviations.

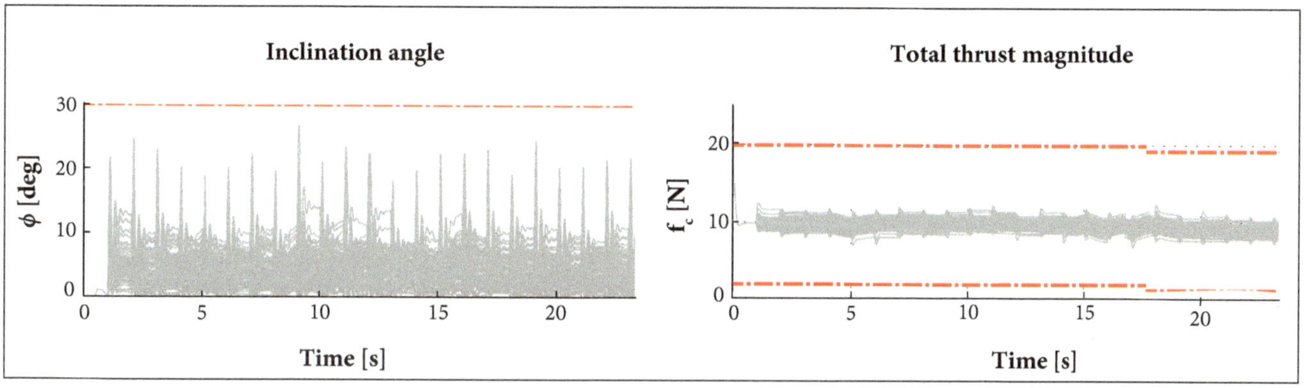

Figure 10. Inclination angle ϕ and thrust magnitude f_c in the best case (ϕ_{max} = 30° and v = 0.5 m/s). The gray lines are the MC realizations; the dashed red lines are the constraints.

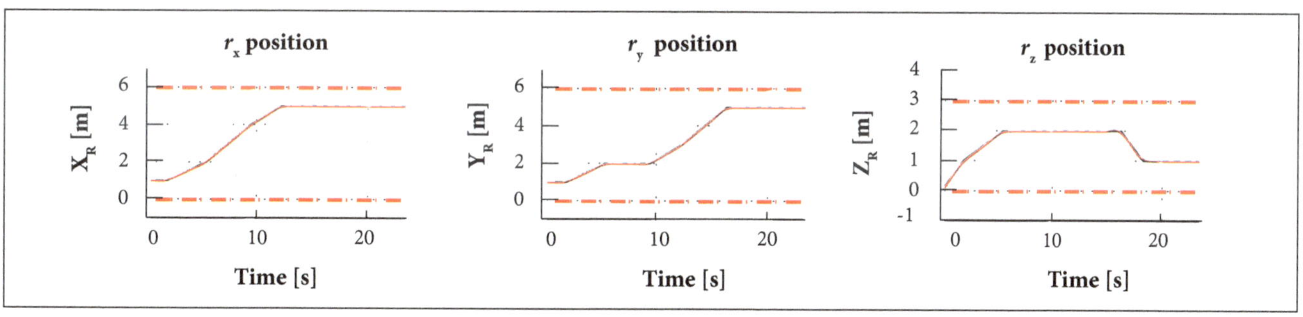

Figure 11. Performance of position control in the best case (ϕ_{max} = 30° and v = 0.5 m/s). The solid gray lines are MC realizations; the dashed red lines are the position constraints; the solid black lines are the position commands; the solid red lines are the sample means; the solid blue lines are the sample means plus/minus the standard deviations.

CONCLUSIONS

This study tackled the problem of controlling the position of a MAV subjected to constraints on the inclination of the rotor plane, on the total thrust magnitude, and on the position. The problem was solved using a conventional linear-quadratic state-space MPC formulation, which became possible thanks to the replacement of the original conic constraint space on the total thrust vector by an inscribed pyramid.

The method was evaluated by computational simulations considering that the vehicle was subjected to a Gaussian disturbance force. The proposed method showed able to control the vehicle's position, even under disturbance forces, while respecting the position and control constraints. However, if a large speed command is considered, it is necessary to relax the maximum inclination constraint in order to have sufficient lateral control force to overcome the disturbance forces.

The Δ-input MPC formulation used in this study appears as a good option for controlling the position of an MAV due to its ability of handling input and state constraints and, if suitably adjusted, it presents smooth responses to position commands. As the MPC has to solve an optimization problem at each update time, a drawback of this strategy is its high computational burden compared with traditional controllers such as the classic PID.

ACKNOWLEDGMENTS

The first author would like to thank Fundação de Amparo à Pesquisa do Estado do Amazonas for providing his master scholarship during the project. Furthermore, we are grateful to Conselho Nacional de Desenvolvimento Científico e Tecnológico for supporting the project under the grant 475251/2013.

AUTHOR'S CONTRIBUTION

Prado IAA and Santos DA conceived the idea and co-wrote the text. Prado IAA worked on the simulations results.

REFERENCES

Alexis K, Nikolakopoulos G, Tzes A (2012) Model predictive quadrotor control: attitude, altitude and position experimental studies. IET Control Theory Appl 6(12):1812-1827. doi: 10.1049/iet-cta.2011.0348

Beard R, Kingston D, Quigley M, Snyder D, Christiansen R, Johnson W, McLain T, Goodrich MA (2005) Autonomous vehicle technologies for small fixed-wing UAVS. J Aero Comput Inform Comm 2(1):92-108. doi: 10.2514/1.8371

Bouabdallah S, Murrieri P, Siegwart R (2004) Design and control of an indoor micro quadrotor. Proceedings of the IEEE International Conference on Robotics and Automation; New Orleans, USA.

Bouabdallah S, Siegwart R (2005) Backstepping and sliding-mode techniques applied to an indoor micro quadrotor. Proceedings of the IEEE International Conference on Robotics and Automation; Barcelona, USA.

Camacho EF, Bordons C (1998) Model predictive control. London: Springer.

Castillo P, Lozano R, Dzul A (2005) Stabilization of a mini rotorcraft with four rotors. IEEE Contr Syst Mag 25(6):45-55. doi: 10.1109/MCS.2005.1550152

Castillo P, Munoz LE, Santos O (2014) Robust control algorithm for a rotorcraft disturbed by crosswind. IEEE Trans Aero Electron Syst 50(1):756-763. doi: 10.1109/TAES.2013.110136

Chen X, Wang L, Xung J (2013) Cascaded model predictive control of a quadrotor UAV. Proceedings of the 3th Australian Control Conference; Fremantle, Australia.

Cheviron T, Hamel T, Mahony R, Baldwin G (2007) Robust nonlinear fusion of inertial and visual data for position, velocity and attitude estimation of UAV. Proceedings of the IEEE International Conference on Robotics and Automation; Rome, Italy.

Cunha R, Cabecinhas D, Silvestre C (2009) Nonlinear trajectory tracking control of a quadrotor vehicle. Proceedings of the 7th European Control Conference; Budapest, Hungary.

Elfes A, Bueno SS, Bergerman M, Ramos JG (1998) A semi-autonomous robotic airship for environmental monitoring missions. Proceedings of the IEEE International Conference on Robotics and Automation; Leuven, Belgium.

Er MJ, Yuan S, Wang N (2013) Development control and navigation of octocopter. Proceedings of the 10th International Conference on Control and Automation; Hangzhou, China.

Gonçalves PFSM, Brusnicki R, Santos DA (2013) Attitude determination of multirotors using camera vector measurements. Proceedings of the 22nd International Congress of Mechanical Engineering; Ribeirão Preto, Brazil.

Gupte S, Mohandas PIT, Conrad JM (2012) A survey of quadrotor unmanned aerial vehicles. Proceedings of the IEEE SoutheastCon; Orlando, USA.

Hua MD, Hamel T, Morin P, Samson C (2009) A control approach for thrust-propelled underactuated vehicles and its application to VTOL drones. IEEE Trans Automat Contr 54(8):1837-1853. doi: 10.1109/TAC.2009.2024569

Lopes RV, Santana PHRQA, Borges GA, Ishihara JY (2011) Model predictive control applied to tracking and attitude stabilization of a VTOL quadrotor aircraft. Proceedings of the 21st International Congress of Mechanical Engineering; Natal, Brazil.

Maciejowski JM (2002) Predictive control with constraints. Harlow; New York: Prentice-Hall.

Madani T, Benallegue A (2007) Sliding mode observer and backstepping control for a quadrotor unmanned aerial vehicles. Proceedings of the American Control Conference; New York, USA.

Mahony R, Kumar V, Corke P (2012) Multirotor aerial vehicles: modeling, estimation, and control of quadrotor. IEEE Robot Autom Mag 19(3):20-32. doi: 10.1109/MRA.2012.2206474

Mian AA, Daobo W (2008) Nonlinear flight control strategy for an underactuated quadrotor aerial robot. Proceedings of the International Conference on Networking, Sensing and Control; Sanya, China.

Mistler V, Benallegue A, M'Sirdi N (2013) Exact linearization and noninteracting control of a 4 rotors helicopter via dynamic feedback.

Proceedings of the 10th IEEE International Workshop on Robot and Human Interactive Communication; Bordeaux, France.

Nemra A, Aouf N (2010) Robust INS/GPS sensor fusion for UAV localization using SDRE nonlinear filtering. IEEE Sensor J 10(4):789-798. doi: 10.1109/JSEN.2009.2034730

Raffo GV, Ortega MG, Rubio FR (2010) An integral predictive/nonlinear H_∞ control structure for a quadrotor helicopter. Automatica 46(1):29-39. doi: 10.1016/j.automatica.2009.10.018

Rossiter JA (2003) Model-based predictive control: a practical approach. Boca Raton: CRC Press.

Santos DA, Saotome O, Cela A (2013) Trajectory control of multirotor helicopters with thrust vector constraints. Proceedings of the 21st Mediterranean Conference on Control and Automation; Chania, Greece.

Saripalli S, Montgomery JF, Sukhatme GS (2003) Visually-guided landing of an unmanned aerial vehicle. IEEE Trans Robot Autom 19(3):371-380. doi: 10.1109/TRA.2003.810239

Shuster MD (1993) A survey of attitude representations. J Astronaut Sci 41(4):439-517.

Teel AR (1992) Global stabilization and restricted tracking for multiple integrators with bounded controls. Syst Contr Lett 18(3):165-171. doi: 10.1016/0167-6911(92)90001-9

Xiao-Hong W, Gui-Li X, Yu-Peng T, Biao W, Jing-Dong W (2012) UAV's automatic landing in all weather based on the cooperative object and computer vision. Proceedings of the 2nd International Conference on Instrumentation, Measurement, Computer, Communication and Control; Harbin, China.

Xu G, Zhang Y, Ji S, Cheng Y, Tian Y (2009) Research on computer vision-based for UAV autonomous landing on a ship. Pattern Recogn Lett 30(6):600-605. doi: 10.1016/j.patrec.2008.12.011

Yan J, Santos DA, Bernstein DS (2014) Adaptive control with convex and saturation constraints. IET Control Theory Appl 8(12):1096-1104. doi: 10.1115/DSCC2013-3877

Numerical Study of Reduced Frequency Effect on Longitudinal Stability Derivatives of Airfoil under Pitching and Plunging Oscillations

Vahid Nasr Esfahani[1], Mahmoud Mani[1]

ABSTRACT: In this study, incompressible, unsteady and turbulent flow over an airfoil with pitching and plunging oscillations is numerically studied in order to investigate the effect of reduced frequency on stability derivatives of oscillating airfoil. Linear $k-\varepsilon$ model called Launder-Sharma and Rhie and Chow model are used for turbulence modeling and overcoming pressure checkerboard problem. This means that a co-located approach is proposed in this paper to study a moving grid problem, and the results demonstrate the high accuracy of the method. Control volume and Crank-Nicholson discretization method are also used for the numerical solution. It is shown that the longitudinal stability derivatives of plunging and pitching motions trend change intensively beyond the stall angle of attack while pitching rate has a completely opposite behavior. The results also show that increasing reduced frequency leads to stability reduction in plunging oscillation but it does not have significant effect on pitching oscillation case in pre-stall, stall and post-stall conditions. Grid convergence is examined to assess the accuracy of the numerical method that shows the high accuracy of it and this is a prominent achievement of the present study. The results of the proposed method in forces and moment show a good agreement with the experimental data.

KEYWORDS: Oscillating airfoil, Pitching oscillation, Plunging oscillation, Stability derivatives, Co-located method.

INTRODUCTION

One of the most important problems in the aerodynamics theory is analyzing the stability of aircraft in order to be assured of its controllability and safety. The accurate prediction of moment and forces applied on aircraft plays a prominent role in determining its stability. Unsteady aerodynamics of oscillating airfoils has received attention in order to analyze the aerodynamic forces on aircrafts.

Glauert (1935), in the 1930s, commenced studying oscillating airfoils. Following him, in 1940, Theodorsen and Garrick (1948) started much more accurate investigations on flutter problem of aircraft wings and paved the way for future studies on inviscid and incompressible flows over low-amplitude oscillating airfoils. Although plenty of numerical and experimental studies have been carried out to determine aerodynamic coefficients of static airfoils, more investigations on unsteady aerodynamics of oscillating airfoils are demanded. Among experimental studies, there are the papers by Panda and Zaman (1994), who investigated the flow field and estimation of lift from wake survey on a pitching airfoil. The Reynolds numbers were 44,000 and 22,000 based on chord length and the airfoil was NACA 0012 with sinusoidal oscillation. Tolouei *et al.* (2004) studied the flow around a pitching airfoil in a frequency range of 0.022 – 0.066. They particularly investigated the pressure distribution of an oscillating airfoil. Sadeghi *et al.* (2010) also experimentally studied unsteady wake of a pitching airfoil and investigated pitching amplitude effect on the wake thickness.

Tuncer (1986), in the 1980s, started a numerical investigation on rapidly pitched airfoils. He studied unsteady fields over

1.Amirkabir University of Technology – Aerospace Engineering Department – Tehran/Iran.

Author for correspondence: Vahid Nasr Esfahani | Amirkabir University of Technology – Aerospace Engineering Department | 1591634311, 424 Hafez Ave | Tehran/Iran | Email: vahid.esfahani@aut.ac.ir pitching airfoil and dynamic stall in his paper. In the 1990s, Tuncer (1997) used overset grid method to numerically study a plunging airfoil and compared the results with single grid solution. In that paper, he used Baldwin-Brath as a turbulence modeling and the Mach 0.3 flow was considered. Guilmineau and Queutey (1999) numerically studied dynamic stall on different airfoil sections. They used Baldwin-Brath and $k - \omega$ SST turbulence models. These studies are devoted to find a numerical approach to deal with oscillating airfoils. However, there is still a compelling desire to achieve highly accurate numerical method on this subject.

Additionally, plenty of experimental and numerical investigations have been carried out on stability derivatives and stability of aircrafts. Bak *et al.* (2000) experimentally investigated force coefficients and stability derivatives of NACA 63-415 airfoil in wind tunnel tests. Park *et al.* (2003) predicted stability coefficients using Euler equations in pitching motion. They used multi-grid method to calculate pitch and roll moments. Altun and İyigün (2004) measured the dynamic stability derivatives of a generic combat aircraft model in wind tunnel applying force oscillation on their model. Gopinath and Jameson (2005) used time-spectral method to study force and moment coefficients on 2-D and 3-D bodies. They used multi-grid method to solve Navier-Stokes equations for pitching airfoils and pitching wings. Schmidt and Newman (2010) also numerically investigated stability derivatives of a generic combat aircraft model. In recent studies Bhagwandin and Sahu (2014) numerically calculated stability derivatives for finned projectiles in subsonic and transonic regimes. Considering these papers, more investigations are required to study parameters that have significant effects on stability derivatives and oscillating motions such as oscillation amplitude or reduced frequency.

In the present research, a co-located numerical method is proposed to a moving grid problem. The purpose is to accurately estimate the stability derivatives of an oscillating airfoil and investigate the reduced frequency effect on the aerodynamics of oscillating airfoils. The proposed method is highly accurate; therefore, it can be a solution for the most common issue of numerical methods in unsteady aerodynamics, which is low accuracy. Moreover, the proposed method is applied to a 2-D problem and is able to be applied on more complex 3-D problems either. Numerical finite volume method with linear $k - \varepsilon$, Launder-Sharma turbulence model is used to study unsteady, incompressible and turbulent flow over a pitching and a plunging NACA 63-415 airfoil. After computing force and moment coefficient diagrams against time, longitudinal stability derivatives are obtained using curve fitting in both oscillations. This curve fitting is based on standard approaches that help the method to be able to model non-linear responses on flying vehicles. In another remarkable point of this numerical study, Rhie and Chow (1983) method is devoted to overcome pressure-velocity fluctuations (pressure checkerboard problem). After obtaining stability derivatives, the reduced frequency effects in pre-stall, stall and post-stall conditions on these derivatives are studied. Grid convergence index (GCI) and order of accuracy are also obtained to verify the numerical method and approve its accuracy.

GOVERNING EQUATIONS

Continuity and momentum equations in Reynolds-Averaged Navier-Stokes (RANS) equations for unsteady incompressible flow are as follows:

$$\frac{\partial U_i}{\partial x_i} = 0 \tag{1}$$

$$\frac{\partial}{\partial x_i}(U_i) + U_i \frac{\partial}{\partial x_i}(U_j) = \frac{-1}{\rho}\frac{\partial P}{\partial x_i}(\nu \frac{\partial U_i}{\partial x_j} - \overline{u_i u_j}) + S_i \tag{2}$$

where: t represents time; x is the space direction; U is the mean velocity; i and j are indexes, with $i, j = 1, 2, 3$; P is the pressure; U_i is the mean velocity of the flow; ρ is density; ν is viscosity; $u_i u_j$ is called Reynolds stress terms that are unknowns of the problem and have to be modeled.

Due to the oscillation motion, the source term S_i is zero. Boussinesq (1987) proposed a method in which Reynolds stress could be modeled using an equivalent turbulent viscosity and is expressed as:

$$\overline{u_i u_j} = -2\upsilon_t S_{ij} + \frac{2}{3}k\delta_{ij} \tag{3}$$

where: δ_{ij} is the Kronecker delta; ν_t is the turbulent kinematic viscosity and is calculated as:

$$\upsilon_t = C_\mu f_\mu \frac{k^2}{\varepsilon} \tag{4}$$

where: C_μ and f_μ values are given in Table 1.

Table 1. Turbulence modeling coefficients.

Coefficients	Launder-Sharma model
$C_{\varepsilon 1}$	1.44
$C_{\varepsilon 2}$	1.92
σ_k	1
σ_ε	1.3
C_μ	0.09
E	$2\nu\nu_t(\delta^2 u/\delta y^2)^2$
f_1	1
f_2	$1-0.3\exp(Re_t^2)$
f_μ	$\exp(-3.4/(1+Re_t/50)^2)$
Re_t	$k^2/\nu\varepsilon$

In order to model Reynolds stress terms, linear $k-\varepsilon$ turbulence model called Launder-Sharma is used. Turbulence kinetic energy (k) and dissipation rate (ε) equations are:

$$\frac{\partial k}{\partial t}+U_j\frac{\partial k}{\partial x_j}=\frac{\partial}{\partial x_j}((\nu+\frac{\nu_t}{\sigma_k})\frac{\partial k}{\partial x_j})+P^*-\varepsilon \quad (5)$$

$$\frac{\partial \varepsilon}{\partial t}+U_j\frac{\partial \varepsilon}{\partial x_j}=\frac{\partial}{\partial x_j}((\nu+\frac{\nu_t}{\sigma_k})\frac{\partial \varepsilon}{\partial x_j})+(C_1 f_1 P^*-C_{\varepsilon 2}f_2)\frac{\varepsilon}{k}+E \quad (6)$$

where: σ_k, σ_ε, $C_{\varepsilon 1}$, $C_{\varepsilon 2}$, f_1, f_2 and E are given in Table 1.

In these relations, P^* is the production term in turbulence flow and is calculated as:

$$P^*=-\overline{u_i u_j}\frac{\partial U_j}{\partial x_j}=2\nu_t S_{ij}S_{ij} \quad (7)$$

where: S_{ij} is the mean rate of stress tensor and is calculated this way:

$$S_{ij}=\frac{1}{2}(\frac{\partial U_i}{\partial x_j}+\frac{\partial U_j}{\partial x_i}) \quad (8)$$

NUMERICAL METHOD

In order to solve flow equations in this paper, control volume method is used. For stationary control volumes, integrated continuity and momentum equations for incompressible flow are:

$$\frac{d}{dt}\int_V dV+\int_S u_j dn_j=0 \quad (9)$$

$$\frac{d}{dt}\int_V u_i dV+\int_S u_j u_i dn_j=\frac{-1}{\rho}\int_S P dn_j+ +\int_S \nu(\frac{\partial u_i}{\partial x_j}+\frac{\partial u_j}{\partial x_i})dn_j+\int_V S_{u_i}dV \quad (10)$$

where: V and s are integral volume and surface, respectively; dn_j is the differential component of surface normal vector.

Note that, for a moving mesh problem, these equations are changed in some ways to model new control volume shapes. These corrections are made using Leibnitz rule. Let W_j be the boundary velocity of a control volume. Using Leibnitz rule, integrated continuity and momentum equations are considered just like the previous section; the only difference is that it is written for a moving mesh and a control volume.

$$\frac{d}{dt}\int_{V(t)} dV+\int_S (u_j-w_j)dn_j=0 \quad (11)$$

$$\frac{d}{dt}\int_{V(t)} u_i dV+\int_S (u_j-w_j)u_i dn_j=\frac{-1}{\rho}\int_S P dn_j+ +\int_S \nu(\frac{\partial u_i}{\partial x_j}+\frac{\partial u_j}{\partial x_i})dn_j+\int_V S_{u_i}dV \quad (12)$$

In the proposed numerical method, Crank-Nicholson method is used for discretization. To prevent pressure and velocity field fluctuation, Rhie and Chow method for incompressible flow is utilized. In this co-located method, velocity components on moving control volume boundaries are given as:

$$u_e=U_P-\frac{1}{2\rho\overline{\hat{u}_e}}(P_E-P_P)-\frac{\Delta x}{2\rho A_e}S_e \quad (13)$$

where: u shows the velocity on volume boundaries while U indicates the velocity at the center of the control volume; e indexes are related to the east of the control volume and E indexes are related to the neighboring control volume centers; P indexes indicate the particular control volume that is being solved.

Figure 1 shows the scheme of a cell.

Transporting velocity is also modeled to prevent pressure checkerboard problem:

$$\overline{\hat{u}_e}=\frac{U_P+U_E}{2}-\frac{k\prime\Delta x}{2\rho\overline{\hat{u}_e}A_e}\{\overline{\hat{u}_e}\left(S_e-\rho\overline{\hat{u}_e}\frac{A_E-A_P}{\Delta x}\right)+ +A_e\left(\frac{P_E-P_P}{\Delta x}\right)+\overline{\hat{u}_e}S_e\} \quad (14)$$

Comparing different solutions, the most appropriate value for k' is obtained; $k' = 1$ is selected for Eq. 14 for an accurate and smooth solution. For boundary conditions in this numerical method, at inlet, velocity, kinetic turbulent energy, dissipation rate, and molecular viscosity are set, and, at outlet, pressure is applied.

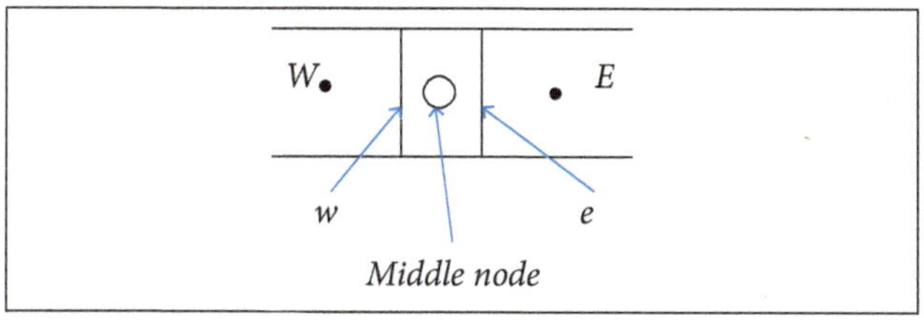

Figure 1. Control volume and its direction indexes. *W* and *E* indicate the center of control volumes on the western and eastern sides of the cell, respectively; *w* and *e* indicate the boundaries on the western and eastern sides of the cell, respectively.

SIMULATION OF PITCHING AND PLUNGING OSCILLATIONS

In order to calculate longitudinal stability derivatives, unsteady aerodynamics are considered. Moreover, force and moment coefficients against time in different angles of attack have to be computed. Two types of oscillating motions are used in this paper: (1) pitching and (2) plunging. Thus, a harmonic oscillating motion with constant amplitude is applied on airfoil. Pitching motion with sinusoidal variation in angle of attack is described as:

$$\alpha(t) = \alpha_m + \alpha_A \sin(\omega t) \tag{15}$$

where: α_m is the mean angle of attack; α_A is the oscillation amplitude; ω is the circular frequency and is determined using the reduced frequency relation $k_c = \omega l/2V$, where l is the chord length and V is the freestream velocity.

In addition, for plunging oscillation, the equation of motion is:

$$h(t) = \bar{a}\sin(\omega t) \tag{16}$$

where: $\bar{a}$ is the mean oscillation amplitude.

Using Taylor series for pitching moment and neglecting higher-than-2nd-order derivative terms, the pitching moment can be described based on reduced frequency for pitching airfoil (Ronch *et al.* 2011):

$$C_m(t) = C_{m\alpha}\alpha(t) + C_{m\dot{\alpha}}\frac{\dot{\alpha} k_c}{\omega} + C_{mq}\frac{q k_c}{\omega} \tag{17}$$

where: $\dot{\alpha}$ is the variation rate of angle of attack and q is the pitch rate; $C_{m\dot{\alpha}}$ is the pitch damping that can be neglected because of its small value.

Substituting Eq. 15 for $\alpha(t)$ in Eq. 17, Eq. 18 is obtained:

$$C_m(t) = C_{m\alpha} A\sin(\omega t) + (C_{m\dot{\alpha}} + C_{mq})k_c\alpha_A A\cos(\omega t) \tag{18}$$

After applying pitching and plunging oscillations to the airfoil, force and drag coefficients are computed in time; then, using curve fitting, the following equation is assumed for the pitching moment:

$$C_m(t) = C_{mo} + C_{mA}\sin(\omega t - \delta) \tag{19}$$

where: C_{mA} is the amplitude; C_{m0} is the initial pitching moment; δ is the phase difference that is assumed by moment coefficient variations and IEEE-std-1050 standard for curve fitting.

Comparing Eq. 19 with Eq. 18, the following equation is obtained:

$$C_{m\dot{\alpha}} + C_{mq} = \frac{-C_{mA}}{k_c\alpha_A}\sin(\delta) \tag{20}$$

Additionally, in order to use the previous equation for plunging airfoil, an equivalent angle of attack is defined:

$$\alpha(t) = \alpha_m + \frac{\bar{\alpha}\omega}{V}\cos(\omega t) \tag{21}$$

This way, damping stability derivatives $C_{m\dot{\alpha}} + C_{mq}$ are obtainable using curve fitting. Noting that the pitch rate is negligible for plunging airfoil, $C_{m\dot{\alpha}}$ can be calculated for plunging airfoil and net value of C_{mq} is achieved.

GRID AND INITIAL CONDITIONS

Airflow Reynolds number is . The NACA 63-415 airfoil that is used in this study has chord length of 0.6 m and the inlet air velocity is 40 m·s^{-1}. Density is considered constant and is set at 1.225 kg·m^{-3}; the pressure at the outlet is set at 1 bar. Since the reduced frequency effects on the stability longitudinal derivatives are investigated, constant mean angle of attack and oscillation amplitude are determined, and the values are 1.5° and 1.3°, respectively. The reduced frequency interval for this study is 0.09 to 0.16, C-type grid is used for the airfoil and the domain is shown in Fig. 2. Three different grids with 11,200, 21,600, and 25,200 cells are used for grid refinement. The 21,600-cells grid is used for numerical solution. Note that the y plus value for the grid is about 1.1 and the wall distance is about 10^{-5} m. The grid convergence study has also been done and it is proposed in the next section. The time step for transient study is 0.001 s.

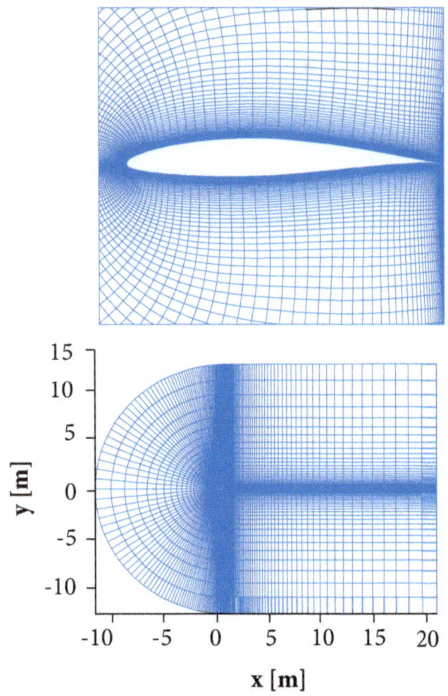

Figure 2. C-type mesh for numerical study.

RESULTS AND DISCUSSION

GEOMETRIC CONSERVATION LAW

Since a moving grid is considered in the present paper, in this section, a test case is introduced to demonstrate the correct performance of the algorithm on moving grids. The case is defined to prove that grid motions would not affect the solution. For this purpose, the fluid is initially stationary with the following parameters: $P = 1$ bar and $\rho = 1.225$ kg·m^{-3}. With these initial conditions, fluid flow equations are solved in the domain while the elements oscillate with the motion of displacement. After 3 s, with the time step of 0.001 s, the contours of dV ($V - 0$), variation of velocity magnitude after 3 s, that is physically expected to remain zero, are plotted in Fig. 3. As can be seen, they are almost zero. This demonstrates the correct implementation of Geometric Conservation Law in the solution.

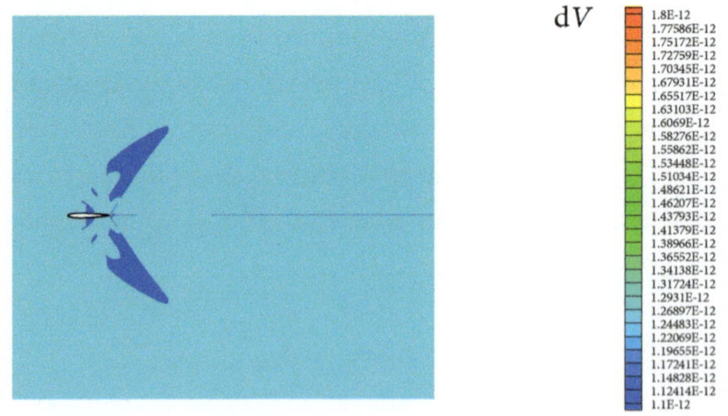

Figure 3. Contours of velocity magnitude variation after 3 s (dV = V – 0).

GRID CONVERGENCE STUDY

For grid study convergence, Richardson extrapolation is used to calculate a higher-order estimation of the flow fields from a series of lower-order discrete values ($f_1, f_2, \ldots, f_n$). For the case of grid refinement study, the value estimated from the Richardson extrapolation is the one that would result when the cell grid size is tended to zero.

$$f_{exact} = f_1 + (\frac{f_2 - f_1}{r^{p'} - 1}) \tag{22}$$

Considering h_1, h_2 2 grid spacing, with h_1 being the finer grid, r is the grid refinement ratio, $r = h_2/h_1$; p' is the order of accuracy of the numerical method and 3 different solutions are needed for this term, which is calculated as (Stern *et al.* 2001):

$$p' = \ln(\frac{f_3 - f_2}{f_2 - f_1}) / \ln(r) \tag{23}$$

Roache (1994) also suggests a GCI to provide a consistent manner in reporting the results of grid convergence. To compute the GCI, 3 levels are recommended in order to accurately estimate the order of convergence and to check that the solutions are within the asymptotic range of convergence:

$$GCI = (\frac{1.25(f_2 - f_1)}{r^{p'} - 1} \tag{24}$$

Table 2 shows grid convergence study results for the proposed numerical method. Since moment coefficient is a significant parameter in this paper, it is used to study grid convergence and obtain the order of accuracy. The convergence is studied for moment coefficient in 2.1° and 0.5° angles of attack. Subscripts 3, 2 and 1 represent, respectively, 11,200-, 21,600- and 25,200-cells grids and $\varepsilon_{ij} = f_i - f_j$.

As Table 2 shows, the order of accuracy for the numerical method is about 1.92 while the theoretical order of accuracy is 2. The difference is most likely due to grid stretching or turbulence modeling. There is a reduction in *GCI* value for the successive grid refinements ($GCI_{12} < GCI_{23}$). The *GCI* for finer grid (GCI_{12}) is relatively low if compared to the coarser grid (GCI_{23}), indicating that the dependency of the numerical method on the cell size has been reduced. Additionally, as the *GCI*reduction from the coarser grid to the finer grid is relatively high, the grid independent solution can be said to have been nearly achieved. Further refinement of the grid will not give much change in the results. So, it can be said that the solution on the proper grid resolution is nearly grid-independent.

Figure 4 indicates the independency of solution from grids for lift and moment coefficients of a plunging airfoil. In Fig. 5, lift and drag coefficient diagrams in different angles of attack for 3 different time steps are shown to demonstrate the independency of results from time step. As Fig. 5 shows, 0.001 s is assumed for the time step of the solution. Note that the results are shown for a pitching airfoil in this figure.

Table 2. Grid convergence index and order of accuracy for C_m.

Angle of attack	ε_{12}	ε_{23}	f_{exact}	GCI_{12} (%)	GCI_{23} (%)	p'
0.5°	0.0005	0.002	−0.076	0.312	1.203	1.926
2.1°	0.0008	0.003	−0.078	0.455	1.757	1.931

NUMERICAL RESULTS

As Figs. 4 and 5 show, lift, drag and moment coefficient diagrams against angle of attack have hysteresis loop in a 0.25 to 2.75 interval. Due to sinusoidal oscillations in airfoil motion, force and moment coefficients have hysteresis loops in particular intervals of angles of attack for both pitching and plunging airfoils. Figure 6 shows forces and moment coefficients for mean angle of attack of 1.5°, reduced frequency of 0.09 and pitching amplitude of 1.3° for a pitching NACA 63-415 airfoil. Hysteresis loops are clearly observed in these diagrams. This figure also compares the results of this research with wind tunnel test results found in Bak *et al.* (2000) to validate the proposed numerical method. It is shown that the results of the present study have good accuracy in comparison with the existing experimental results. Numerical method errors are 3.2, 1.4 and 3.9 percent for lift, moment and drag coefficient,

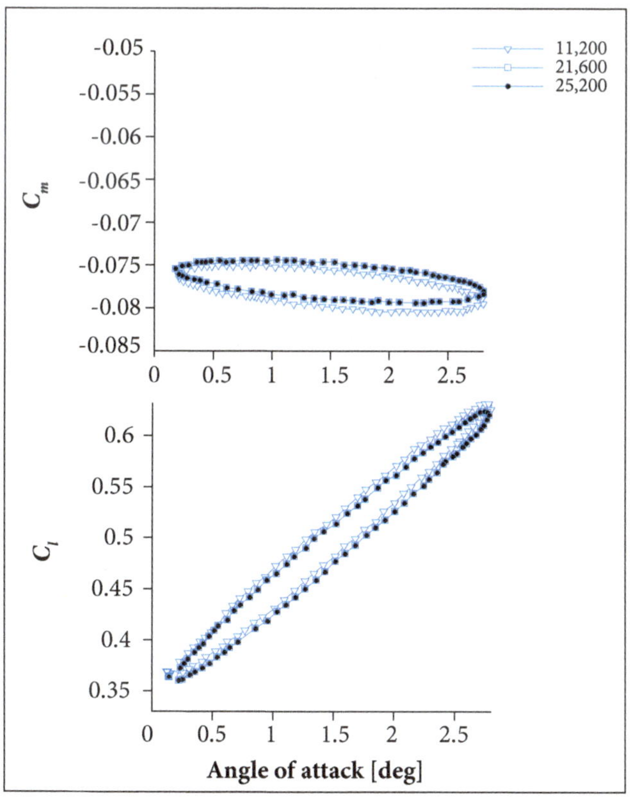

Figure 4. C_m and C_l against angle of attack for different grids.

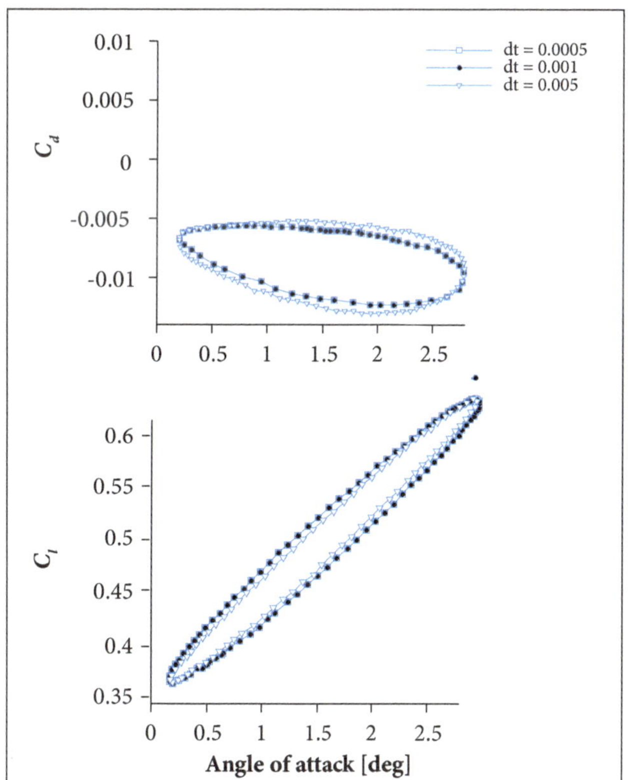

Figure 5. C_d and C_l against angle of attack for different time steps.

respectively. These errors are calculated by comparing areas of hysteresis loops.

In order to predict the time variation of longitudinal stability derivatives, force and moment diagrams against time have to be determined. Curve fitting these diagrams using Eq. 19, longitudinal stability derivatives can be calculated by Eq. 20. After this procedure, longitudinal stability derivative variations with different angles of attack can be obtained. Figure 6 shows the variation for longitudinal stability derivatives in different angles of attack. For this case, the mean angle of attack is 1.5° and the reduced frequency is 0.09 for both pitching and plunging airfoils. Table 3 also shows curve fitting data for stability derivatives in an angle of attack equal to 1.5°.

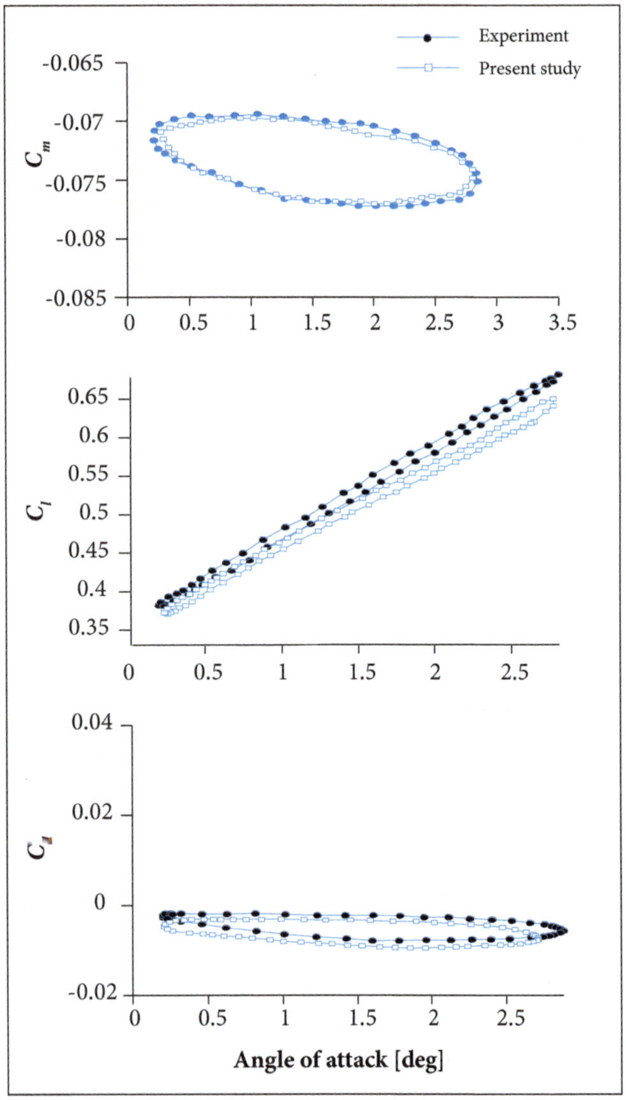

Figure 6. Comparison of present numerical and experimental results (Bak 2000).

Longitudinal stability derivative variations for pitching ($C_{m\dot{\alpha}}+C_{mq}$) and plunging airfoil ($C_{m\dot{\alpha}}$) with angle of attack are shown in Fig. 7. Numerical results show that the longitudinal stability derivatives of pitching and plunging airfoils have the same trend in different angles of attack, whilst the longitudinal stability of pitch rate (C_{mq}) has a completely different attitude in pre- and post-stall conditions. In other words, longitudinal stability derivatives for pitching and plunging airfoils values are increased in pre-stall condition and decreased in post-stall condition. Longitudinal stability derivatives of pitch rate have an opposite trend.

Figure 7 also shows that longitudinal stability derivatives of pitching and plunging airfoils have negative signs in all angles of attack, which describes the longitudinal stability in pitching and plunging motions. Longitudinal stability derivatives of pitch rate have negative sign after stall angle of attack and become more stable. In addition to that, in high angles of attack, longitudinal stability derivative values of plunging motion are positive, which means that this motion becomes unstable in post-stall condition. Longitudinal derivative values of pitching motion are still negative in this area, which indicates its stability.

Since reduced frequency is one of the main elements in oscillating motions, the effect of this parameter on stability derivatives of a plunging and pitching airfoil is investigated in this paper. Figures 8 to 10 show the reduced frequency effect on longitudinal stability derivatives. The reduced frequency is

Table 3. Curve fitting data for angle of attack of 1.5°.

Oscillation	Phase difference (δ)	C_{mA}	$C_{m\alpha} + C_{mq}$
Pitching	−0.949	0.002	−0.000
Plunging	−0.812	0.003	−0.172

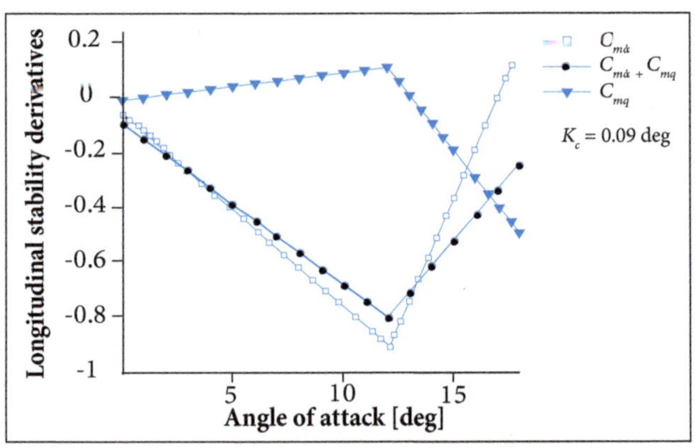

Figure 7. Variation of longitudinal stability derivatives with angle of attack.

increased from 0.09 to 0.16 and longitudinal stability derivative variations and trends are studied in 3 different conditions: pre-stall, stall and post-stall.

For these 3 conditions, 3 particular mean angles of attack are determined: 1.5°, 11° and 18° for pre-stall, stall and post-stall conditions, respectively. As shown in Fig. 8, by increasing

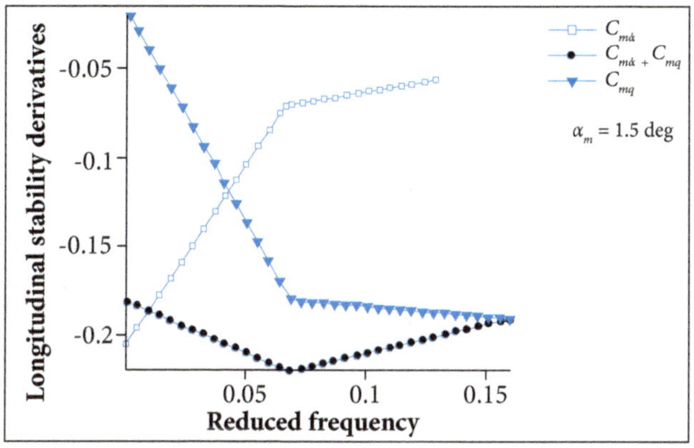

Figure 8. Reduced frequency effect on longitudinal stability derivatives in pre-stall condition.

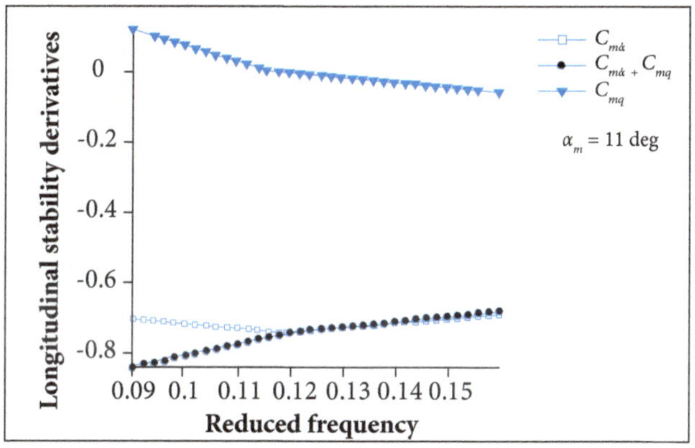

Figure 9. Reduced frequency effect on longitudinal stability derivatives in stall condition.

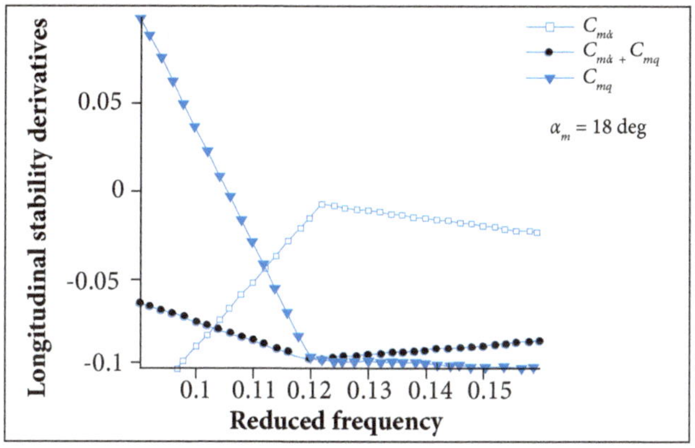

Figure 10. Reduced frequency effect on longitudinal stability derivatives in post-stall condition.

the reduced frequency from 0.09 to 0.16, longitudinal stability derivative values for plunging airfoil move toward positive values, which leads to unstable oscillations in higher reduced frequencies. This can be justified by fluid displacement effect on airfoil. On the other hand, the increasing in the reduced frequency does not have significant influence on longitudinal stability derivatives of pitching airfoil. Longitudinal stability of pitch rate is also stable in the studied reduced frequency interval. In plunging motion, by increasing the reduced frequency to 0.06, negative damping effects of wake on moment coefficient become evident and, by increasing further, these effects are more severe. In pitching airfoil, longitudinal stability derivatives in low reduced frequency have a different trend in comparison to plunging airfoil due to the pitch rate.

In Fig. 9, which is stall condition, with mean angle of attack equal to 11°, the situation is as the same as the pre-stall condition. The only notable difference is that the variation of longitudinal stability derivatives of plunging and pitching airfoils is not significant as in the pre-stall condition.

The mean angle of attack equal to 18° shows the post-stall condition in Fig. 10. In this condition, as in the previous ones, longitudinal stability derivatives of plunging and pitching airfoils have negative signs, which means they are stable. However, by increasing the reduced frequency, longitudinal stability derivatives of plunging airfoil move toward the positive values and become more unstable. It is important to note that the variation of longitudinal stability derivatives of plunging airfoil is more than that of the pitching airfoil. In other words, plunging airfoil is more sensitive to reduced frequency than pitching airfoil.

Studying pitch rate in all of the conditions shows the reduced frequency effect on the stability of the oscillation. Figure 8 shows that the pitch rate effect is elevated by increasing the reduced frequency and the negative value of C_{mq} is also increased, which means it is always in stable condition.

CONCLUSION

A numerical method was proposed in this paper to predict the stability derivatives of a NACA 63-415 airfoil under pitching and plunging oscillations and to study the effect of one of the most important oscillation parameters, which is the reduced frequency. This method was a co-located approach and was utilized to solve a moving grid problem. Grid convergence was also examined for numerical method and results indicated that the grid convergence index was below 5%, which demonstrates the high accuracy of the numerical approach. The Reynolds number was 1.6×10^6, linear $k - \varepsilon$ model was used as turbulence model and Rhie and Chow method was also devoted to overcome pressure checkerboard problem.

The paper investigated the behavior of stability derivatives in different conditions. Furthermore, it is shown that the stability of a plunging airfoil was decreased by increasing the reduced frequency, primarily in pre- and post-stall conditions. Finally, the research shows that the pitching airfoil stability is relatively independent of reduced frequency especially in low angles of attack due to the existence of the pitch rate in this motion.

REFERENCES

Altun M, İyigün İ (2004) Dynamic stability derivatives of a maneuvering combat aircraft model. Journal of Aeronautics and Space Technologies 1(3):19-27.

Bak C, Fuglsang P, Johansen J, Antoniou I (2000) Wind tunnel tests of the NACA 63-415 and a modified NACA 63-415 airfoil. Risø National Laboratory, Roskilde, Denmark; [accessed 2016 Jun 27]. http://orbit.dtu.dk/fedora/objects/orbit:88085/datastreams/file_7729395/content

Boussinesq, J. (1987) Essai sur la théorie des eaux courantes, Mémoires présentés par divers savants à l'Académie des Sciences 23 (1): 1-680.

Bhagwandin VA, Sahu J (2014) Numerical prediction of pitch damping stability derivatives for finned projectiles. J Spacecraft Rockets 51(5):1603-1618. doi: 10.2514/1.A32734

Glauert H (1935) Airplane propellers. In: Durand WF, editor. Division L of aerodynamic theory. Vol. 6. Berlin: Springer. p. 169-360.

Gopinath AK, Jameson A (2005) Time spectral method for periodic unsteady computations over two- and three-dimensional bodies. Proceedings of the 43rd AIAA Aerospace Sciences Meeting and Exhibit; Reno, USA.

Guilmineau E, Queutey P (1999) Numerical study of dynamic stall on several airfoil sections. AIAA J 37(1):128-130. doi: 10.2514/2.677

Panda J, Zaman KBMQ (1994) Experimental investigation of the flow field of an oscillating airfoil and estimation of lift from wake surveys. J Fluid Mech 265:65-95. doi: 10.1017/S0022112094000765

Park SH, Kim Y, Kwon JH (2003) Prediction of damping coefficients using the unsteady Euler equations. J Spacecraft Rockets 40(3):356-362. doi: 10.2514/2.3970

Rhie CM, Chow WL (1983) Numerical study of the turbulent flow past an airfoil with trailing edge separation. AIAA J 21(11):1525-1532. doi: 10.2514/3.8284

Roache PJ (1994) Perspective: a method for uniform reporting of grid refinement studies. J Fluids Eng 116(3):405-413. doi: 10.1115/1.2910291

Ronch AD, McCracken A, Badcock KJ (2011) Modeling of unsteady aerodynamic loads. AIAA Atmospheric Flight Mechanics Conference; Portland, USA.

Sadeghi H, Mani M, Alighanbari H (2010) An experimental study of the unsteady wake of an Eppler-361 airfoil. Proceedings of the 40th Fluid Dynamics Conference and Exhibit; Chicago, USA.

Schmidt S, Newman DM (2010) Estimation of dynamic stability derivatives of a generic aircraft. Proceedings of the 17th Australasian Fluid Mechanics Conference; Auckland, New Zealand.

Stern F, Wilson RV, Coleman HW, Paterson EG (2001) Comprehensive approach to verification and validation of CFD simulations — Part 1: Methodology and procedures. J Fluids Eng 123(4):793-802. doi: 10.1115/1.1412235

Theodorsen T, Garrick IE (1948) Mechanism of flutter: a theoretical and experimental investigation of the flutter problem. Technical Report N.A.C.A. Report No. 685.

Tolouei E, Mani M, Soltani MR, Boroomand M (2004) Flow analysis around a pitching airfoil. Proceedings of the 22nd Applied Aerodynamics Conference and Exhibit; Providence, USA.

Tuncer IH (1986) Unsteady aerodynamics of rapidly pitched airfoil. Proceedings of the 4th Joint Fluid Mechanics, Plasma Dynamics and Lasers Conference; Atlanta, USA.

Tuncer IH (1997) Two-dimensional unsteady Navier-Stokes solution method with moving overset grids. AIAA J 35(3):471-476. doi: 10.2514/2.153

Performance and Wake Analysis of Rotors in Axial Flight Using Computational Fluid Dynamics

Nik Ahmad Ridhwan Nik Mohd[1], George Barakos[2]

ABSTRACT: Flow field around rotors in axial flight is known to be complex especially in steep descent where the rotor is operating inside its own wake. It is often reported that, in this flight condition, the rotor is susceptible to severe wake interactions causing unsteady blade load, severe vibration, loss of performance, as well as poor control and handling. So far, there is little data from experimental and numerical analysis available for rotors in axial flight. In this paper, the steady Reynolds-Averaged Navier-Stokes Computational Fluid Dynamics solver Helicopter Multi-Block was used to predict the performance of rotors in axial flight. The main objective of this study was to improve the basic knowledge about the subject and to validate the flow solver used. The results obtained are presented in the form of surface pressure, rotor performance parameters, and vortex wake trajectories. The detailed velocity field of the tip vortex for a rotor in hover was also investigated, and a strong self-similarity of the swirl velocity profile was found. The predicted results obtained when compared with available experimental data showed a reasonably agreement for hover and descent rate, suggesting unsteady solution for rotors in vortex-ring state.

KEYWORDS: Axial flight, CFD, Performance, Rotor, Wake structure.

INTRODUCTION

In axial flight, particularly in steep descent where helicopter enters vortex-ring state (VRS), a rotor descends into its own wake. VRS is one of the extreme axial flight conditions and a potentially hazardous flight regime of operation where rotor blades are known to experience severe thrust fluctuations, highly unsteady blade airloads, aperiodic blade flapping, severe vibration, and sudden loss of aircraft altitude and control (Drees and Hendal 1951; Brinson and Ellenrieder 1998; Conlisk 2001; Leishman 2006; Brand *et al.* 2011). This problem is even more dangerous when helicopters operate close to the ground, which may cause unwanted accidents.

Several experiments and computational studies for low speed descending flight have been reported in literature, but relatively a few show detailed quantitative measurements of rotor performance and blade aerodynamic characteristics. Drees and Hendal (1951) are among the first to capture the flow field around rotors in axial flight. The airflow near a rotor installed in the wind tunnel was made visible by using a number of smoke generators. From the experiment, a large flow circulation was captured at the tip of the blade in VRS, and a periodic tumbling motion of the rotor was observed. Another study on measuring rotor performance in axial flight was carried out by Caradonna (1999). The climb performance of a scale model of UH-1H rotor was measured in the wind tunnel at a constant blade tip speed combined with various climb rates and blade collective pitch angles. The rotor cyclic was not considered throughout the test, and the results were

1.Universiti Teknologi Malaysia – Faculty of Mechanical Engineering – Aeronautical Laboratory – Johor Bahru/Johor – Malaysia. **2.**University of Glasgow – School of Engineering – CFD Laboratory – Glasgow – Scotland.

Author for correspondence: Nik Ahmad Ridhwan Nik Mohd | Universiti Teknologi Malaysia – Faculty of Mechanical Engineering – Aeronautical Laboratory | 81310 Skudai | Johor Bahru/Johor – Malaysia | Email: ridhwan@utm.my also compared to the HELIX-I code. The experimental results provided good quantitative rotor performance and a wake geometry database for Computational Fluid Dynamics (CFD) validation. The difficulties to get accurate rotor performance in hover from experiments, however, required an extrapolation technique to the limit of 0 climb speed. Instead of using a wind tunnel, Washizu *et al.* (1966) used a towing basin to measure a single rotor performance in axial flight. The use of large towing basin in their research was intended to eliminate blockage effects encountered in wind tunnels with closed test sections. Rotor performance for a range of descent speeds was presented in the form of a polar curve and compared with momentum theory. There are several other experiments on scaled rotor in axial flight (Green *et al.* 2005; McAlister *et al.* 2001). Brinson and Ellenrieder (1998) concluded that the entry to the fully developed vortex ring state was found to be sensitive to tunnel velocity and large changes in cyclic required to trim. Due to the extremely complex flow field near the rotor and the lack of data, rotor blade aerodynamics in axial flight is still not well understood.

A number of computer researches to model helicopter flow fields have also been reported. However, most of these focused on hover and forward flight, and only a few on axial flight. The PHIDIAS project provided a platform for the numerical simulation of the 6075 Dauphin VRS and windmill-brake states (WBS) (Bailly 2010). The HOST method was compared with the time-marching, unsteady wake model developed by MINT and flight test data conducted by ONERA at the CEV French Flight Centre. Most of computational fluid dynamics (CFD) solvers developed for rotor aerodynamic simulation are based on solving the Reynolds-Averaged Navier-Stokes equation and are susceptible to excessive numerical dissipation of vorticity on coarse meshes. If not controlled, this dissipation leads to the loss of coherence of vortical structures within the flow. To control this numerical dissipation and dispersion of vorticity in the flow field, a vorticity transport model was developed and tested on rotor in descent (Ahlin and Brown 2005; Line and Brown 2004).

In this study, the Helicopter Multi-block (HMB) CFD solver has been used to compute the performance of rotors at various rates of descent. The main objectives of this paper are to improve basic knowledge of rotor aerodynamics and to validate the flow solver. The results obtained are presented as surface pressure, rotor performance parameters, and vortex wake trajectories. The detailed velocity field of the tip vortex for rotor in hover is also investigated, and a strong self-similarity of the swirl velocity profile is found. The predicted results obtained were compared with available experimental data and showed a reasonably agreement for hovering and moderate rate of descent.

CFD FLOW SOLVER

The description of the HMB is discussed in detail by Barakos *et al.* (2005) and Steijl *et al.* (2006), being briefly presented in this paper. HMB is a computational fluid dynamic Navier-Stokes solver developed at the CFD Laboratory of the University of Liverpool and runs on parallel distributed memory computers. HMB solves the 3D Unsteady Reynolds-Averaged Navier-Stokes (URANS) equations on multi-block structured meshes using a cell-centred finite-volume method for spatial discretisation. In this study, the convective terms are discretised using the Osher's scheme. The Monotone Upstream-centred Schemes for Conservation Laws (MUSCL) interpolation is used to provide formally 3rd-order accuracy in the calculation of fluxes. The van Albada limiter is used to avoid spurious oscillation in the flow properties across shocks by locally reducing the accuracy of the numerical scheme to first order. The resulting linear system of equations is solved using a pre-conditioned Generalised Conjugate Gradient method in conjunction with a Lower Upper factorisation. For unsteady simulations, implicit dual-time stepping is used based on Jameson's pseudo-time integration approach. The viscous computations use the k-ω baseline (BSL) and the Shear Stress Transport (SST) turbulence model of Menter (1994). The SST blends the k-ω model for the inner boundary layer with the k-ε model for the outer boundary layer. Furthermore, the redefined SST eddy viscosity accounts for the effect of transport of turbulent shear stress. The SST model has been used for many external flows with adverse pressure gradients. The solver also has been used and validated for several fundamental flows apart from rotor cases (Badcock *et al.* 2000; Gagliardi and Barakos 2008).

Hovering rotor flow simulations are performed by assuming that the wake shed from the rotor is steady. The hover formulation of HMB allows the computation of hovering rotor flows to be treated as steady-state cases. The formulation uses a mesh that does not rotate and employs a transformation of the frame of reference to account for the rotation. Applying periodic boundaries, an n-bladed rotor can be approximated using a $1/n$ domain segment. The details of the hover formulation can be found in Steijl *et al.* (2006). For hovering and climbing rotor cases, the momentum based far field boundary model, assuming a three-dimension source-sink singularity at the rotor axis of rotation and at a rotor plane was used (Steijl *et al.* 2006; Srinivasan 1993). For ascending and descending rotor, the vertical velocity is added to the hover velocity. The Froude/Source-sink condition typically used (Steijl *et al.* 2006) for applying the boundary conditions in hover and descending flight at moderate speeds was not adequate for the extreme cases considered here. Therefore, in this paper, for extreme case in descending flight, a free-stream condition was assumed at the far field. For more efficient computations, the vortex-tube model that valid for most of axial flight condition can be used (Wang 1990; Perry *et al.* 2007; Nik Mohd and Barakos 2010).

CFD MESH GENERATION AND TEST CASES

A structured C-H-H CFD mesh was employed (Fig. 1). An H-type structure was used away from the blades with a C-type structure attached to them. This type of mesh topology allows for accurate viscous computations and can provide a mechanism for pitching the blades, with the near-blade mesh remaining in an undeformed state (Brocklehurst *et al.* 2008). For all meshes, viscous spacing in the direction normal to the blade surface with the first cell located at $10^{-5}c$ above the blade and exponential distribution was employed. For viscous cases, this mesh spacing constructed results in $y^+ < 1$. The rotor centre of rotation is located at the z-axis, the rotor blade is laid on the x-axis, and the quarter chord point is taken at $y = 0$ with the blade leading edge pointing at the positive y-axis. This geometry was used for the construction of the computational meshes for the rotors in axial flight. The rotor hub was modelled as a simple straight cylinder. Summary of rotor geometry and flow condition considered in this study are shown in Table 1.

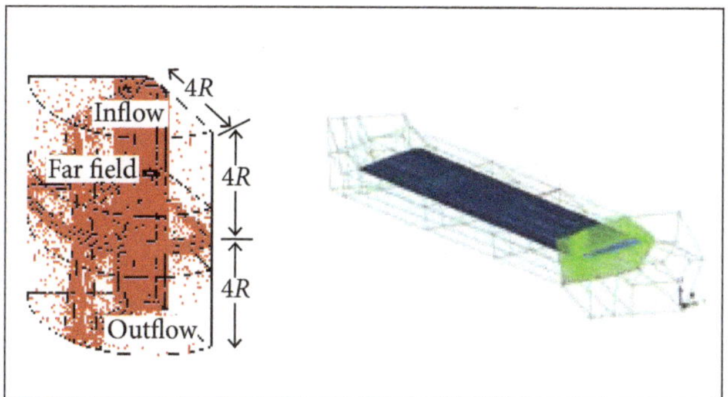

Figure 1. Multi-block mesh employed in HMB.

Table 1. Summary of rotor geometries for axial flight test cases.

Characteristic	UH-1H	Washizu
Rotor geometry		
Number of blades	2	3
Root cut-out	1 c	1 c
Rotor radius	13.67 c	16.67 c (1.1 m)
Chord	1 c	1 c (0.33 m)
Aspect ratio	13.67	16.67
Twist	Untwisted	−8.33°
Solidity	0.047	0.0573
Blade profile	NACA 0012	NACA 0012
Flow condition		
Tip Mach	0.5771	0.1693
Reynolds number (× 10^6)	1.028	0.1306
Grid size (× 10^6)	9	6.3
Number of blocks	170	272
Test case	Hover/climb	Hover/descent

RESULTS AND DISCUSSION

ASCENDING CARADONNA UH-1H ROTOR

For ascending case, the aerodynamic performance and wake geometry of the Caradonna UH-1H rotor were computed for a range of ascend rates, V_v/V_{tip}, from hover to ascending rate of 0.04. The results are presented for a blade at trimmed and untrimmed conditions. For untrimmed condition, the rotor blade collective pitch angle, θ_o, was set at 11° as in Caradonna (1999) experiment. For trimmed conditions, the thrust required for ascend was maintained constant closed to the hover state (Fig. 2a). This is done by considering the incremental of the blade collective pitch settings by 3/4 V_v/Vt_{ip} (Johnson 1980). The coning angle was set at 0 for all calculations. For the ascending test cases considered, the linear increments in the blade collective pitch resulting from the trimmed conditions are plotted in Fig. 2b. The blade tip Mach number was $M_{tip} = 0.5771$ and the Reynolds number, $Re = 1.028 \times 10^6$.

Rotor trimming requires an increment in blade collective pitch angle to maintain the rotor thrust as the ascend speed ratio is increased. This also results in an increment of the rotor torque required (Fig. 2c). For the test cases considered, the rotor losses its efficiency in ascending flight. Comparison with the experimental measurements (Fig. 2d) shows that HMB underpredicts rotor figure of merit (FM) but is in good agreement with HELIX-I data for ascending cases.

Predictions of hover FM were successfully presented in Nik Mohd *et al.* (2011) for the case of Caradonna and Tung (1981). There are, however, some discrepancies for the UH-1H rotor for both HMB and HELIX-I data. In experiments, the drop in the value of FM is expected due to excessive flow unsteadiness (Caradonna 1999), and extrapolation to ascend was suggested to correct the hover FM. The predicted hover FM shows a linear trend; however, the value of the FM falls between the experiments and the HELIX-I data. Overall, for the ascending rotor, the predicted rotor performance without trimming is in excellent agreement with experiments and HELIX-I data. Nevertheless, when the rotor trim condition is activated, a slight increase is observed in the rotor FM for the fast ascending case (Fig. 2d).

Figures 3a to 3d show the blade surface pressure coefficient, C_p, at 3 blade stations plotted for different ascend rates. From the comparison, the blade surface pressure is found to have the same stagnation values, but decreasing suction pressure, as rotor ascend rate increases. The distributions of spanwise blade lift

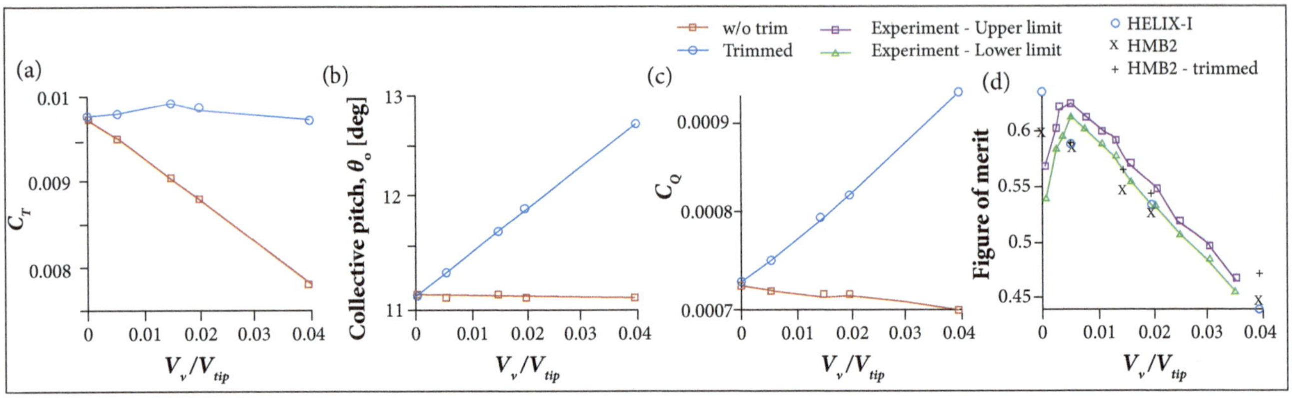

Figure 2. Performance of the Caradonna UH-1H rotor in hover and vertical ascent. (a) Thrust coefficient; (b) Collective pitch; (c) Torque coefficient; (d) Figure of merit.

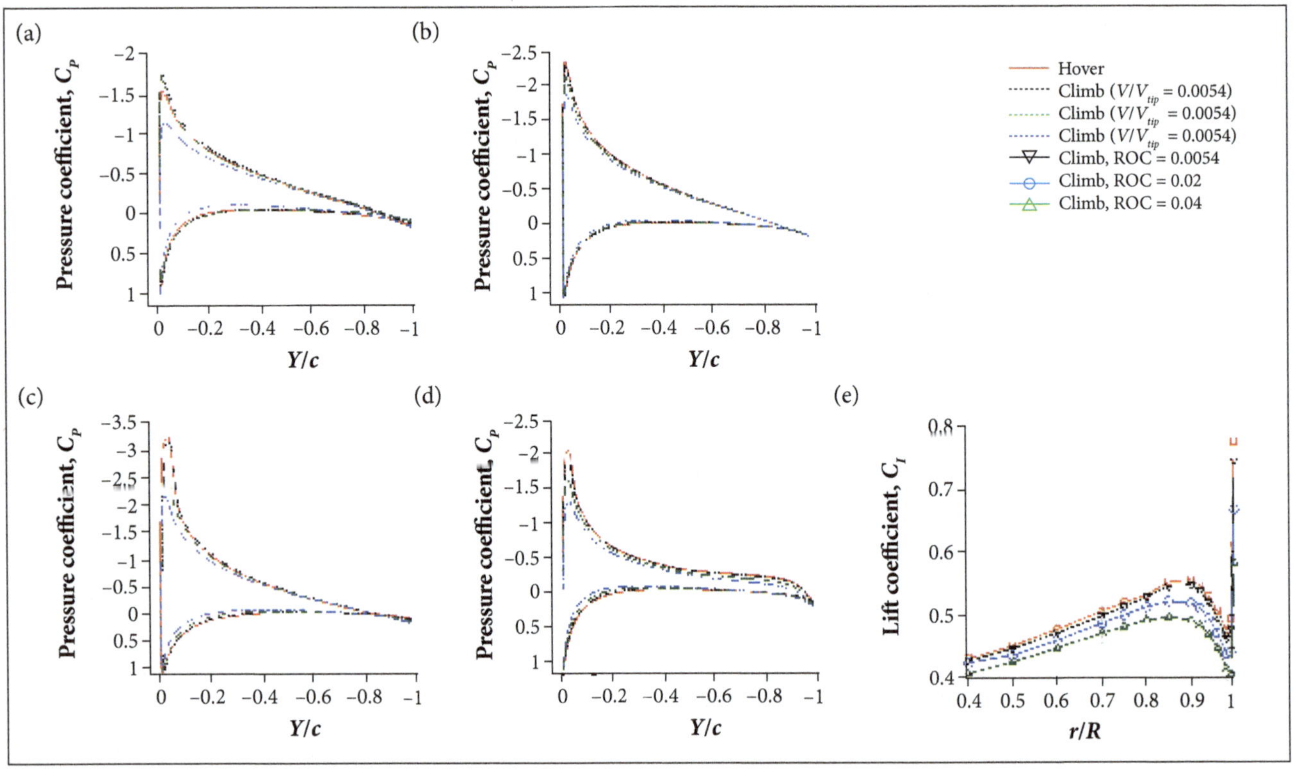

Figure 3. Comparison of CFD results of the Caradonna UH-1H blade surface pressure coefficients at different ascent rates. (a) r/R = 0.30; (b) r/R = 0.70; (c) r/R = 0.90; (d) r/R = 0.99; (e) Lift coefficient. ROC: Rate of climb.

coefficient, C_L, are presented in Fig. 3e and were obtained by integrating the C_p. From the figure, the spanwise C_L is found shifted vertically downwards as the normalised ascending velocity increases from hover to 0.04.

The vertical and axial rotor tip vortex trajectories obtained from CFD, and compared with the No. 1 blade of (Caradonna 1999) experiment are shown in Figs. 4a to 4b. It can be seen that, for low ascending speed, the calculated wake trajectories are in good agreement with the experimental data of Caradonna (1999). It can also be noted that the overall wake geometry is almost identical or independent of trim conditions. The comparison of the blade tip vortices for hovering and ascending flight cases represented by superimposed vorticity-cuts taken at $\Psi_w = 0°$ is depicted in Fig. 5, where, for the hovering case of $V_v/Vt_{ip} = 0$, the primary blade tip vortex of blade No. 1 after a complete revolution is predicted to be at 0.927 chord downstream the rotor plane. For a low ascending flight case of $V_v/Vt_{ip} = 0.04$, the primary blade tip vortex predicted by HMB is found to be displaced 2.634 chord downstream the rotor plane. However, due to the fast growth in the mesh spacing due to the exponential distribution used has resulted in a quick dissipation and dispersion of vorticity field.

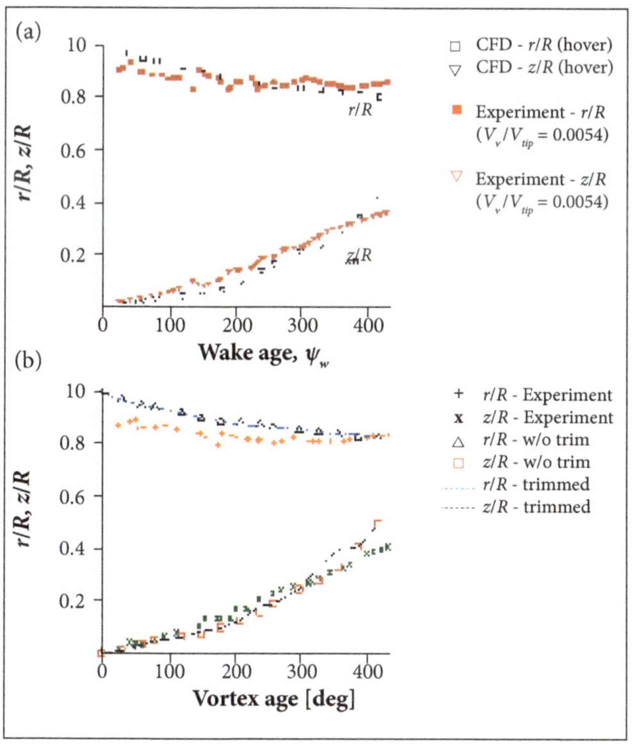

Figure 4. Normalised radial (r/R) and vertical (z/R) displacement of rotor tip vortex for $\theta_{0.7} = 11°$ and $M_{tip} = 0.5771$.(a) Hover; (b) $V_v/V_{tip} = 0.015$.

The distributions of vortex swirl velocity normalised with the rotor tip speed across the tip vortex cores for various wake ages are shown in Fig. 6. This information also shows how strong the axial velocity induced by the tip vortex is. The vortex swirl velocity of the tip vortex weakens as it evolves

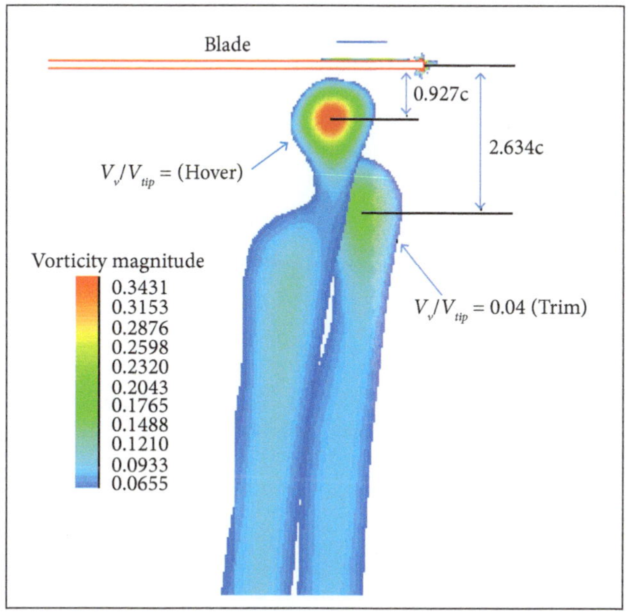

Figure 5. Superimposed vorticity cuts at $\Psi_w = 0°$.

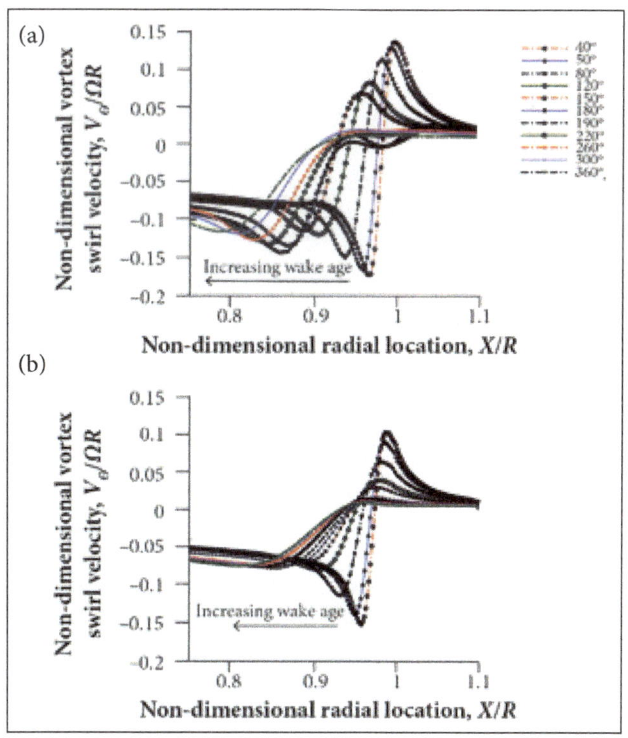

Figure 6. Distribution of vortex swirl velocity of the UH-1H rotor in ascending flight. (a) $V_v/V_{tip} = 0$ (Hover); (b) $V_v/V_{tip} = 0.04$.

but grows in core radius. Furthermore, for the same wake age considered, the vortex swirl velocity is found to be stronger in hover compared to the ascending flight. This quantity may be influenced by the dissipation of vorticity due to coarse mesh constructed below the rotor plane where an exponential point distribution was used.

The growth of the tip vortex core extracted for one rotor revolution is presented in Fig. 7a. It is found that, for the same mesh distribution, the core radius increases linearly with vortex age and grows faster in hovering flight. Similarly to the hovering cases, the tip vortex of an ascending rotor looks self-similar when the normalised swirl velocity of the core is plotted against the normalised radial distance (Fig. 7b).

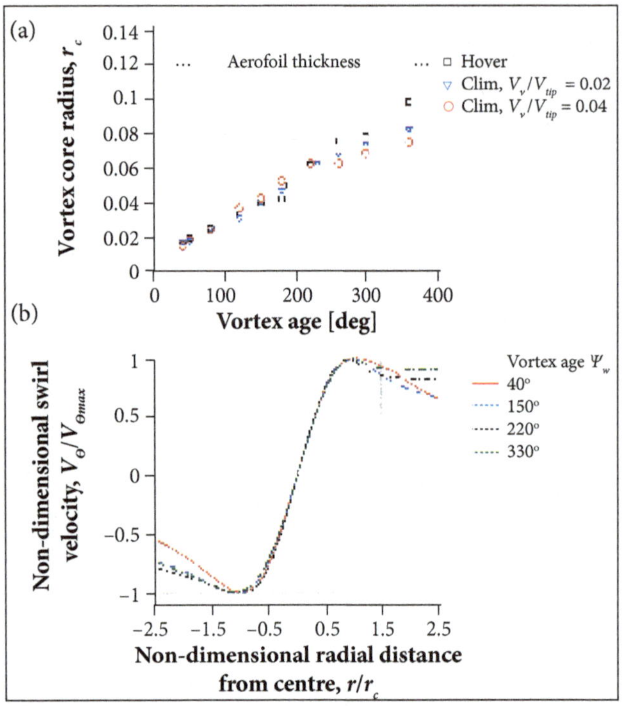

Figure 7. Predicted core growth and self-similarity of the UH-1H rotor tip vortices at different ascent rates. (a) Vortex core radius; (b) Vortex self-similarity.

DESCENDING WASHIZU ROTOR

Experiment for rotors in descending flight performed by Washizu et al. (1966) were referred in this study. In the experiment, the rotor was tested in a towing basin with the aim of eliminating the blockage effect that occurs in most wind tunnel experiments.

The experiments of the Washizu rotor in descending flight include hover, vortex-ring state ($V_v/Vt_{ip} = 0.054$, $V_v = 3.11$ m/s) and wind-mill brake state, WBS ($V_v/Vt_{ip} = 0.09$, $V_v = 5.184$ m/s), and these were computed using CFD. The test cases consider the blade tip Reynolds and Mach numbers fixed to a constant value of 1.306×10^6 and 0.1693, respectively. The rotor has 3 blades, and the blade collective pitch setting varied from 0°, 4.5° to 8°. The descending flight was simulated by increasing the descend speed while keeping the rotor tip speed constant. In CFD, the freestream far field boundary was specified for rotors in descending flight, and a Froude (source-sink) condition was used for the hovering flight case. The turbulence closure was achieved using the 2 equations of the *k-ω* model with transport of shear stress (SST) and baseline (BSL) variant. Figures 8a and 8b show the convergence history for rotor thrust and torque coefficients. It can be seen that rotor thrust and torque in WBS fluctuate even after 80,000 iteration steps. This behaviour is apparently caused by the highly separated flow developed on the upper blade surface. In descending flight, the rotor blades experience a flow coming from below the rotor plane.

The comparisons of the chordwise blade surface pressure at 3 blade stations (r/R = 0.3, 0.75, and 0.9) are presented in Fig. 9. The results suggest that the blade in VRS and WBS has large flow separation on the upper surface of the blade. This situation occurs up to r/R = 0.75. The chordwise plot of flow streamlines visualised at 3 blade stations are shown in

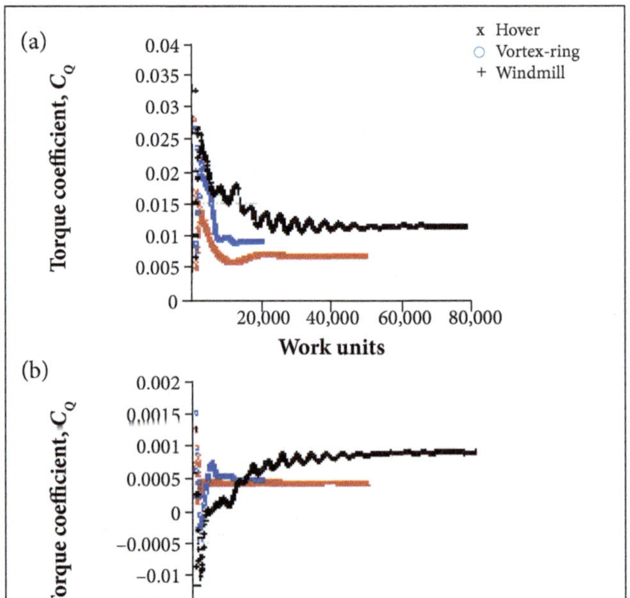

Figure 8. Convergence history plots for the Washizu rotor in descending flight ($\theta_{0.7}$ = 8°, M_{tip} = 0.1693). (a) Thrust history; (b) Torque history.

Figs. 10 and 11. As shown in Fig. 10, for the rotor in VRS, the blade at 20%R experiences leading edge separation and then growth to fully separated flow on the upper blade surface at 40%R. A much more severe flow separation can be verified for the blade in windmill brake as shown in Fig. 11, where almost 75%R of blade is separated.

For hovering rotor, the flow field enters the computational domain from all directions and is induced downstream below the rotor. In descending flight, however, the inflow velocity enters the domain from the boundary located below the rotor plane. In the VRS, where the transport of vorticity is approximately equal to the rate of descent, a large recirculation

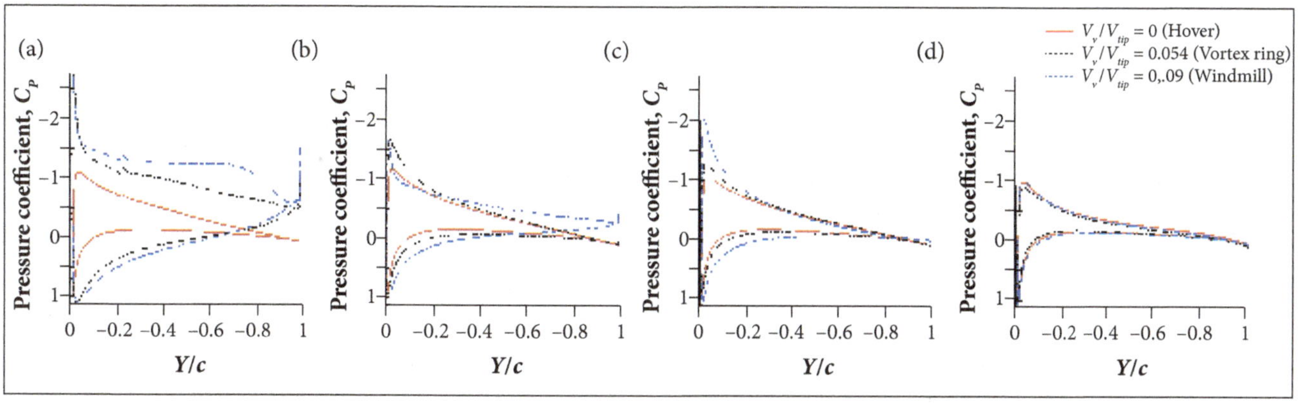

Figure 9. Comparison of CFD results of the Washizu rotor blade surface pressure coefficient at different descent rates ($\theta_{0.7}$ = 8°, M_{tip} = 0.1693). (a) r/R = 0.30; (b) r/R = 0.75; (c) r/R = 0.90; (d) r/R = 0.99.

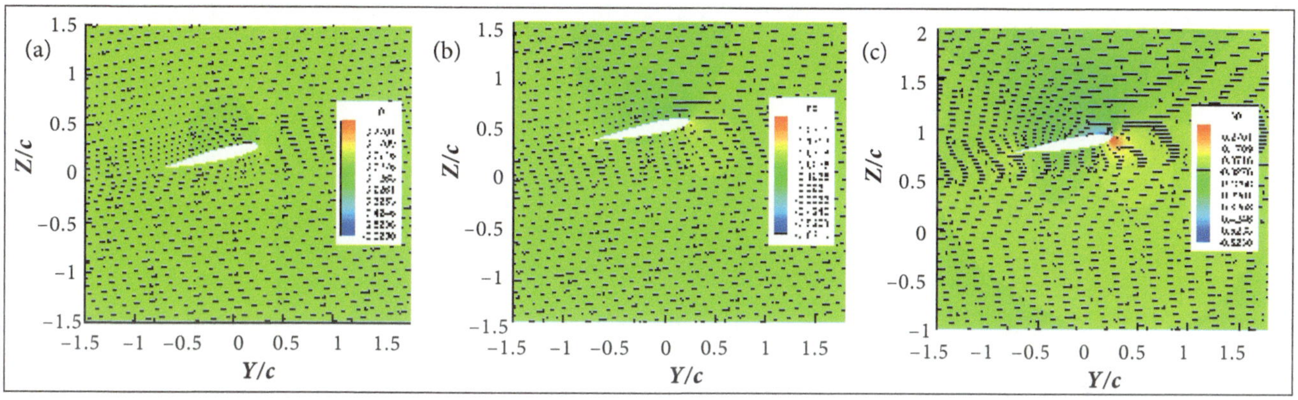

Figure 10. Streamlines and contour of pressure coefficient for rotor in VRS ($\theta_{0.75}$ = 8°, M_{tip} = 0.1693). (a) r/R = 0.20; (b) r/R = 0.40; (c) r/R = 0.70.

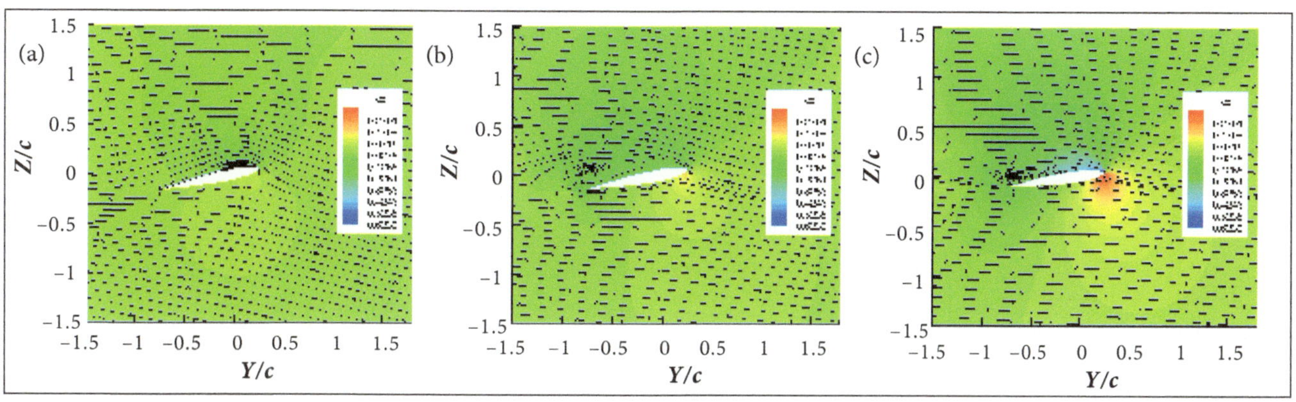

Figure 11. Streamlines and contour of pressure coefficient for rotor in WBS ($\theta_{0.75}$ = 8°, M_{tip} = 0.1693). (a) r/R = 0.20; (b) r/R = 0.40; (c) r/R = 0.70.

flow exists near the blade tip. Nevertheless, the mesh setting used was still unable to accurately capture the collection of tip vortices accumulated in this recirculation flow as found in most experiments. In the windmill brake state, where the rotor experiences a fast descent rate, the flow field continues to expand above the rotor plane. Iso-surfaces of Q-criterion coloured by pressure are depicted in Fig. 12. The Q-criterion is used here to visualise the highly concentrated vorticity regions by plotting the rotation-dominated strain in the flow. A large area of fully turbulence flow is seen to be developed around the rotor in WBS.

The comparisons of the rotor performance in terms of torque coefficient (C_Q) *versus* thrust coefficient (C_T) and FM *versus* C_T are given in Fig. 13. The effects of different turbulence models (k-ω BSL and SST) and trim conditions are also presented. The results suggest that, for low collective pitch settings, the CFD slightly overpredicts rotor torque for all test cases. Trimming the rotor improves the performance prediction compared to the untrimmed condition. Moreover, for a high collective pitch setting, a large offset of rotor torque was observed.

The FM of rotors in hover is plotted for different blade collective pitch settings in Fig. 13b. The CFD results show a well predicted FM for non-lifting rotor ($C_T = 0$), but this value is slightly below the experimental results for the 4.5° and 8° pitch settings. From these results, it can be concluded that the CFD method used can predict rotor performance in steep descent up to an acceptable accuracy.

Comparisons of the predicted rotor performance (trimmed condition) against momentum theory and experiments are shown in Fig. 14. As explained in the previous section, the rotor was trimmed to a specific thrust coefficient. The effect of trimming can be seen on the torque coefficient required to maintain the rotor descend speed for a particular rotor thrust and blade collective angles. There are discrepancies in torque prediction between CFD and the other 2 data. These could be due to potential errors in the experiments at high pitch angles and high rates of descent due to limited speed of the carriage, and the length of the track (Washizu *et al.* 1966).

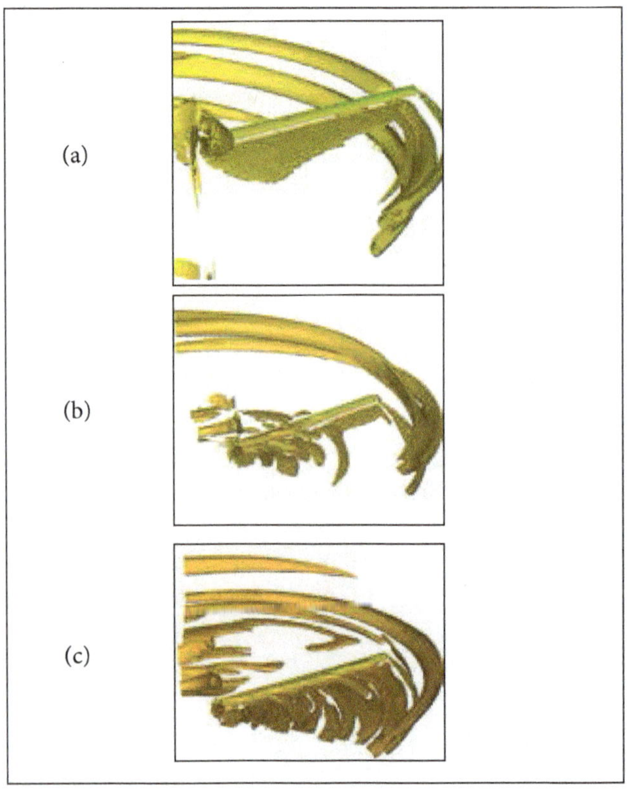

Figure 12. Vortical structures of rotors in descending flight visualised using iso-surface of Q-criterion coloured by pressure contour ($\theta_{0.75} = 8°$, $M_{tip} = 0.1693$). (a) $V_v/V_{tip} = 0$ (Hover); (b) $V_v/V_{tip} = 0.054$ (Vortex ring); (c) $V_v/V_{tip} = 0.09$ (Windmill brake).

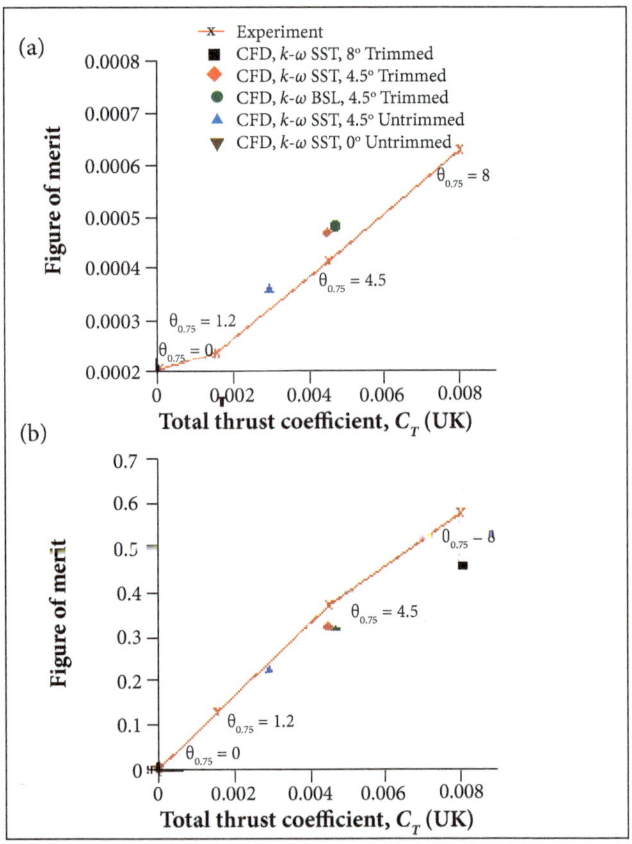

Figure 13. Performance of the Washizu rotor in hover. (a) C_T versus C_Q; (b) Figure of merit versus C_Q.

Furthermore, no detailed discussion on the hover and windmill test cases was given.

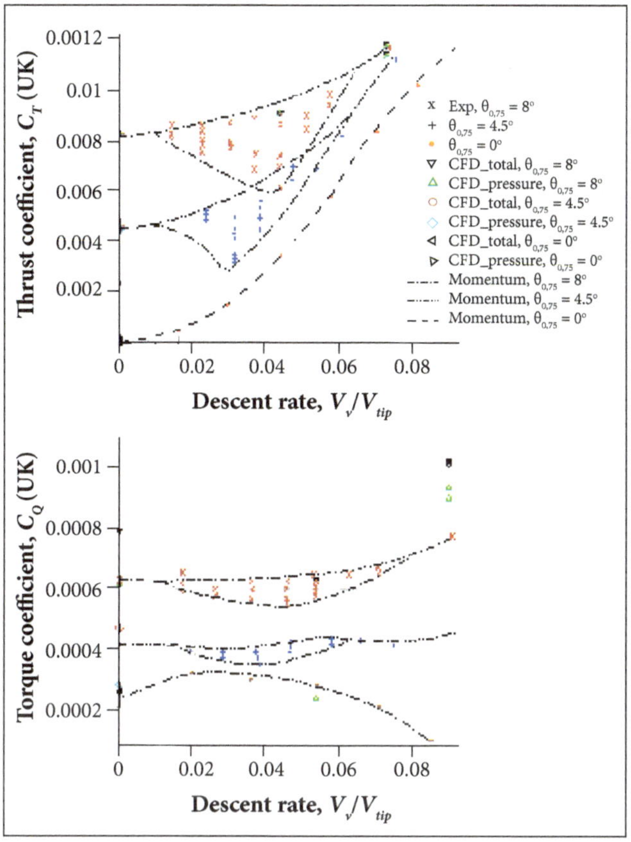

Figure 14. Comparison of the Washizu rotor performance plotted for different descent rates and collective pitch settings.

CONCLUSIONS

Accurate prediction of rotor blade aerodynamics in axial flight is a challenging task in computational fluid dynamics. This is due to the presence of large-scale vortical structures that interact with the blade near the tip and need to be accurately modelled. In this study, the performance of rotors in various axial flight conditions in vertical ascend, hover, and descent was computed using the Helicopter Multi-block (HMB) flow solver.

For rotors in hover and vertical ascend, the blade surface pressure, the integrated rotor performance, and the vortex wake trajectory were analysed using HMB and compared with the experimental data of the UH-1H rotor. It can be concluded that the HMB solver can predict well the blade aerodynamic performance in comparison to experimental and HELIX-I data. However, for low ascending rate, small discrepancies can be observed. This is expected due to the aperiodicity and recirculation of the flow field in the wind tunnel rather than CFD errors. Similar hover test cases were also employed for validation including the hovering Caradonna-Tung model rotor, with good success. For the same wake age considered, the distribution of the normalised vortex swirl velocity across the vortex core in climb was found to be weaker and smaller in radius than for the rotors in hover. Furthermore, the tip vortex showed more contraction in hover and fast descending than for ascending flight.

When the rotor descended into its own wake, a large flow separation developed on the upper surface of the blade extending up to 55%R. A more severe and chaotic flow field was observed for the rotor in windmill brake state where more than 75%R of the blade span experienced stall. The predictions of thrust and torque showed that the blade collective pitch had to be trimmed to the required thrust. The used of the baseline (BSL) and shear stress transport (SST) $k-\omega$ turbulence models showed little influence on the torque predictions.

Overall, the present work on rotors in axial flight has successfully validated the HMB solver using several rotor test data. The work presented, provides additional knowledge on rotor performance but there are still certain areas that are not well understood. Although CFD can be a good tool in predicting the basic flow physics on different axial flight regimes, future work and validation against new experiments are needed. This calls for experiments in various regimes of axial flight that should be the target of large international initiatives and should result in open databases of measurements.

ACKNOWLEDGEMENTS

The authors would like to acknowledge the Malaysian Ministry of Higher Education and Universiti Teknologi Malaysia for financial support for this research project.

AUTHOR'S CONTRIBUTION

Mohd NARN performed the numerical calculation and prepared the figures; Barakos G conceived the idea and co-wrote the main text. All authors discussed the results and commented on the manuscript.

REFERENCES

Ahlin GA, Brown RE (2005) Investigating the physics of rotor vortex ring state using the vorticity transport model. Proceedings of the 31st European Rotorcraft Forum; Florence, Italy.

Badcock K, Richards B, Woodgate M (2000) Elements of computational fluid dynamics on block structured grids using implicit solvers. Progr Aero Sci 36(5-6):351-392. doi: 10.1016/S0376-0421(00)00005-1

Bailly J (2010) A qualitative analysis of vortex ring state entry using full time marching unsteady wake model. Proceedings of the 34th European Rotorcraft Forum; Liverpool, UK.

Barakos G, Steijl R, Badcock K, Brocklehurst A (2005) Development of CFD capability for full helicopter engineering analysis. Proceedings of the 31st European Rotorcraft Forum; Florence, Italy.

Brand A, Drier M, Kisor R, Wood T (2011) The nature of vortex ring state. J Am Helicopter Soc 56:1-14. doi: 10.4050/JAHS.56.022001

Brinson P, Ellenrieder T (1998) The nature of vortex ring state. Proceedings of the 24th European Rotorcraft Forum; Marseille, France.

Brocklehurst A, Steijl R, Barakos G (2008) CFD for tail rotor design and evaluation. Proceedings of the 34th European Rotorcraft Forum; Liverpool, UK.

Caradonna FX (1999) Performance measurement and wake characteristics of a model rotor in axial flight. J Am Helicopter Soc 44(2):101-108. doi: 10.4050/JAHS.44.101

Caradonna FX, Tung C (1981) Experimental and analytical studies of a model helicopter rotor in hover. NASA-TM-81232.

Conlisk AT (2001) Modern helicopter rotor aerodynamics. Progr Aero Sci 37(5):419-476. doi: 10.1016/S0376-0421(01)00011-2

Drees JM, Hendal WP (1951) Airflow patterns in the neighbourhood of helicopter rotors: a description of some smoke tests carried out in a wind-tunnel at Amsterdam. Aircr Eng Aerosp Tech 23(4):107-111. doi: 10.1108/eb032021

Gagliardi A, Barakos GN (2008) Improving hover performance of low-twist rotors using trailing-edge flaps — a computational study. Proceedings of the 34th European Rotorcraft Forum; Liverpool, UK.

Green RB, Gillies EA, Brown RE (2005) The flow field around a rotor in axial descent. J Fluid Mech 534:237-261. doi: 10.1017/S0022112005004155

Johnson W (1980) Helicopter theory. Princeton: Princeton University Press.

Leishman JG (2006) Principles of helicopter aerodynamics. Cambridge: Cambridge University Press.

Line AJ, Brown RE (2004) Efficient high-resolution wake modelling using the vorticity transport equation. Proceedings of the 60th Forum of the American Helicopter Society; Baltimore, USA.

McAlister KW, Huang SS, Abrego AI (2001) A model rotor in axial flight. NASA/TM-2001-210925.

Menter FR (1994) Two-equation eddy-viscosity turbulence models for engineering applications. AIAA J 32(8):1598-1605. doi: 10.2514/3.12149

Nik Mohd NAR, Barakos GN (2010) Computational aerodynamics of hovering helicopter rotors. Proceedings of the 2010 RAeS Aerodynamics Conference; Bristol, United Kingdom.

Nik Mohd NAR, Barakos GN, Batrkov AS, Kusyumov AN, Nikolaev EI (2011) Computational aerodynamics of rotors in hover and vertical climb. Transactions of Academenergo 4:118-129.

Perry FJ, Chan WFY, Simon I, Brown RE, Ahlin GA, Khelifa BM, Newman SM (2007) Modelling the mean flow through a rotor in axial flight including vortex ring conditions. J Am Helicopter Soc 52(3):224-238. doi: 10.4050/JAHS.52.224

Srinivasan GR (1993) A free-wake Euler and Navier-Stokes CFD method and its application to helicopter rotors including dynamic stall. Mountain View: JAI Associates, Inc.; [accessed 2017 Jan 6]. http://oai.dtic.mil/oai/oai?verb=getRecord&metadataPrefix=html&identifier=ADA275416

Steijl R, Barakos G, Badcock K (2006) A framework for CFD analysis of helicopter rotors in hover and forward flight. Int J Numer Meth Fluid 51(8):819-847. doi: 10.1002/fld.1086

Wang SC (1990) Analytical approach to the induced flow of a helicopter rotor in vertical descent. J Am Helicopter Soc 35(1):92-97. doi: 10.4050/JAHS.35.92

Washizu K, Azuma A, Koo J, Oka T (1966) Experiments on a model helicopter rotor operating in the vortex ring state. J Aircraft 3(3):225-230. doi: 10.2514/3.43729

4

Management Process of a Frequency Response Flight Test for Rotorcraft Flying Qualities Evaluation

João Otávio Falcão Arantes Filho[1], Donizeti de Andrade[2]

ABSTRACT: This paper applies the frequency response methodology to characterize and analyze the flying qualities of longitudinal and lateral axes of a rotary-wing aircraft, AS355-F2. Using the results, it is possible to check the suitability of the aircraft in accordance with ADS-33E-PRF standard, whose flying qualities specifications criteria are based on parameters in the frequency domain. The key steps addressed in the study involve getting, by means of flight test data, the closed-loop dynamic responses including the design of the instrumentation and specification of the sensors to be used in the flight test campaign, the definition of the appropriate maneuvers characteristics for excitation of the aircraft, the planning and execution of the flight test to collect the data, and the proper data treatment, processing and analysis after the flight. After treatment of the collected data, single input-single output spectral analysis is performed. The results permit the analysis of the flying qualities characteristics, anticipation of the demands to which the pilot will be subjected during closed-loop evaluations and check of compliance with the aforementioned standard, within the range of consistent excitation frequencies for flight tests, setting the agility level of the test aircraft.

KEYWORDS: Flight characteristics, Helicopters, Flight tests, Frequency-domain analysis.

INTRODUCTION

The aircraft flying qualities, for both rotary and fixed-wing aircraft, determine the easiness and precision with which the pilot is able to perform the operational tasks that constitute the mission proposed for that air vehicle (Williams *et al.* 1995). Flying qualities evaluations aim to measure the pilot skill level required to operate the aircraft and are a method to verify the suitability of it for a given mission (Brasil 2003).

Flying qualities evaluations can be conducted in qualitative and quantitative ways. Qualitatively, they are assessed by testing in closed-loop form (pilot-in-the-loop), in which the pilot effectively acts on the flight controls to fulfill certain task.

There is a task requirement (desired or adequate performance to be reached), a pilot action in the flight control system to meet this requirement, the reaction of the aircraft according to its aerodynamic characteristics and an effectively achieved response. The pilot observes the aircraft response, determines the error between the actual and desired response and, if in fact the error exists, he (she) performs an additional control input to reduce or eliminate it. In the end, he (she) considers all actions applied on flight controls and the performance reached to determine the overall workload due to the compensation of the aircraft deficiencies (effort and attention required) added by that due to the properly execution of the task (Brasil 2003). Then he (she) employs the handling qualities rating (HQR) like the Cooper-Harper scale (Cooper and Harper Junior 1969). In addition, other factors influence the final rating given by the pilot, such as the interface with the cabin (displays, commands and controls), the environment (weather, visibility

1.Departamento de Ciência e Tecnologia Aeroespacial – Instituto de Pesquisas e Ensaios em Voo – Divisão de Ensaios em Voo – São José dos Campos/SP – Brazil.
2.Departamento de Ciência e Tecnologia Aeroespacial – Instituto Tecnológico de Aeronáutica – Departamento de Projetos – São José dos Campos/SP – Brazil.
Author for correspondence: João Otávio Falcão Arantes Filho | Departamento de Ciência e Tecnologia Aeroespacial – Instituto de Pesquisas e Ensaios em Voo – Divisão de Ensaios em Voo | Praça Marechal Eduardo Gomes, 50 | CEP: 12.228-901 | São José dos Campos/SP – Brazil | Email: jotaarantes@gmail.com

and turbulence), the requirements that define the mission and the stress (Brasil 2003). Figure 1 represents the block diagram of a closed-loop evaluation.

In the quantitative form, the characteristics of stability and control that directly affect the flying qualities are obtained. These tests evaluate the characteristics linking the flight controls to the aerodynamic response by means of the flight control systems. These tests are based on open-loop tests, carried with the pilot out of the control loop, in which the environment, the pilot or another device connected to the flight control system perform a predetermined control input. Then, the aircraft response is observed and measured (United States Naval Test Pilot School 1995). Objectively, these tests are responsible for system identification. Figure 2 illustrates the characteristic parameters of any rotary-wing aircraft response able to be obtained in an open-loop test, by a step control of lateral cyclic.

Aircraft system identification is the methodology by which a mathematical description of the dynamic characteristics of an air vehicle can be extracted from the flight data (Williams *et al.* 1995). It is a procedure that allows obtaining dynamic models from the aircraft response measured by flight tests specific control inputs. The inputs are applied to excite the dynamics of interest to which the model is employed, *e.g.* the one aiming at characterizing the flight mechanics of a complete aircraft, or a specific system, such as an actuator, engine or rotor of a rotary-wing aircraft. Typical applications of aircraft systems identification are: simulation models; development and validation of flight control systems; assessment of the flying qualities characteristics and check of compliance with the flying qualities requirements (Tischler and Remple 2012).

In the aircraft system identification field, a study of its historical development is done by Hamel and Jategaonkar (1996). This study briefly discusses from the estimation techniques used in the early 20th century to obtain flight stability derivatives to current methods, which were drastically changed by the data processing capability of digital computers. Special attention in that study is given to the "quad-M", four interrelated topics that should be thoroughly investigated and that are key to the success of a system identification: (a) maneuvers (control inputs) able to excite the dynamic modes; (b) selection of instrumentation system and filtering features for obtaining accurate measurements; (c) definition of the structure of the mathematical model (which depends on the type of vehicle to be investigated); and (d) quality of data analysis, based on the selection of the most appropriate method of identification.

The traditional techniques for flying qualities assessment involve, primarily, the time-domain analysis of the responses to certain control inputs, like steps, impulses, or doublets. The study focuses on the parameters such as rise time, time constant, time to double (or to decay by half) the amplitude, natural damped frequency and damping ratio. With the advent of augmented flight control systems in rotorcraft — with large bandwidth or

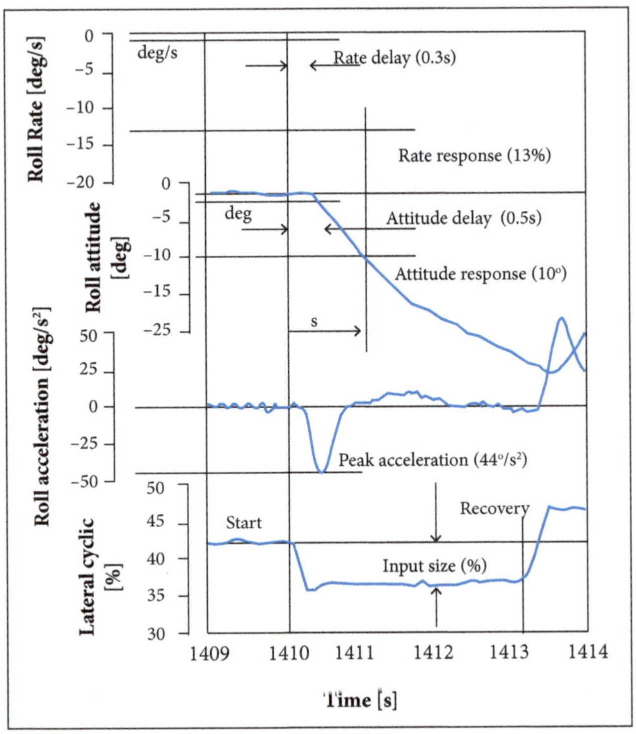

Figure 2. Characteristic parameters of a rotary-wing aircraft response — open-loop test (Cooke and Fitzpatrick 2002).

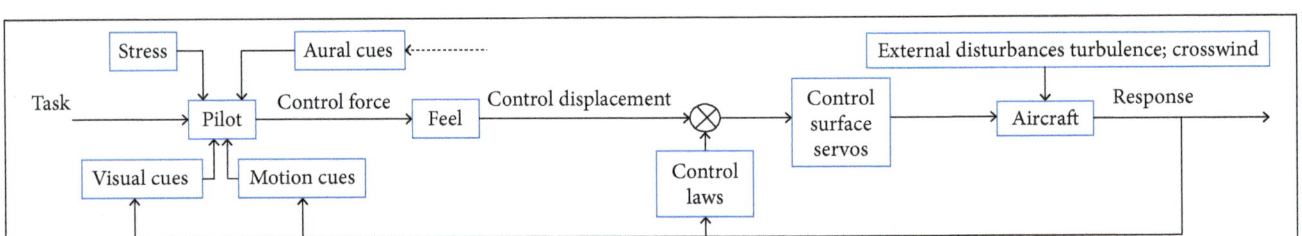

Figure 1. System pilot-aircraft (Cooper and Harper Junior 1969).

even full-authority fly-by-wire —, design, development, testing and evaluation of these systems require the use of more robust techniques of analysis. The use of frequency-domain analysis provides the required robustness (Williams *et al.* 1995).

The frequency response characterizes the dynamics of the input-output system part of a non-parametric model, without the need of describing a set of parameters of the model, such as the coefficients of a differential equation. For this reason, the method shows robustness for use in rotary-wing aircraft or other vehicles whose dynamics, compared to fixed-wing aircraft, features highest order, being unstable under certain conditions and carrying reduced signal-to-noise ratio of the data measured by the typical flight test instrumentation (Tischler and Remple 2012).

For rotorcraft frequency-domain flight test, aiming system identification, the frequency sweep is the most widely used control input to excite the aircraft dynamics. Its first application is given by Tischler *et al.* (1987). The research uses the Bell 214ST aircraft to demonstrate the frequency sweep excitation technique, in order to assist the upgrade program of the MIL-H-8501A (1961) standard — General Requirements for Helicopter Flying and Ground Handling Qualities — to characterize transient dynamic angular response of aircraft with augmented flight controls. Because the new combat helicopters required greater agility and maneuverability, the natural characteristics of stability could be "relaxed" by using more complex flight control systems, which were not actually considered in the previous flying qualities requirements. Based on these studies, the ADS-33E-PRF (2000) standard, whose objective is the flying qualities requirements of future generation of rotary-wing aircraft, has been established.

Either by robustness of the analysis or by compliance with the newest rotorcraft aeronautical requirements, frequency response flight tests are a fundamental step in the design or assessment of an aircraft, consuming time and resources when executed in a non-precise and non-objective way.

Thus, this paper aims to describe all the steps to apply the method of frequency response to characterize the dynamics of a rotary-wing aircraft and to analyze its flying qualities, compared to ADS-33E-PRF (2000) standard.

In order to address the key steps to obtain the rotorcraft frequency responses by means of flight test data, all the process of management to conceive a flying-qualities-frequency-domain flight test is divided into: (a) design of instrumentation and specification of the sensors to be used in the test campaign; (b) establishment of the appropriate maneuvers characteristics for excitation of the aircraft; (c) planning and execution of the flight test to collect all the data; (d) post-processing data to verify the consistency of the acquired information; (e) performance of the spectral analysis; and (f) confirmation of the flying qualities standard requirements with which frequency-domain data can be compared.

During each step description, to exemplify the proposed methodology, one employs records of frequency sweep excitation of the longitudinal and lateral axes, in level flight, obtained in flight test of an AS355-F2 helicopter, belonging by then to the Instituto de Pesquisas e Ensaios em Voo (IPEV). The tested aircraft is a light helicopter, single rotor with conventional tail rotor, designed and manufactured by AEROSPATIALE. It is equipped with two Allison 250 C-20F engines, with the total capacity for six people, including the crew. The aircraft has an irreversible flight control system, which is fully hydraulic-assisted, pitch and roll axes trim feel and trim actuator. The aircraft is equipped with a flight control system, SFIN 85 T31, composed by a stability augmentation system (SAS) and a flight director. A more detailed description is presented in its flight manual (Helibras 2005).

Additionally, one presents tools to support decision for validation, analysis and discussion of the results of a flight test campaign, as the consistency and angular reconstruction and the frequency-domain analysis codes, all of them built with MATLAB® environment.

MANAGEMENT PROCESS

FLIGHT TEST INSTRUMENTATION

Generally, the requirements for data acquisition and parameters needed to be acquired by flight test instrumentation (FTI) system are directly related to the purpose for which the data will be further processed. For verification of compliance with flying qualities requirements, which are based on the response of the aircraft to a control input, it is minimally necessary to perform measurements of the flight controls displacements (for rotorcraft: δ_{lon}, longitudinal cyclic; δ_{lat}, lateral cyclic; δ_{ped}, pedal; and δ_{col}, collective), rigid body attitudes and angular rates and linear vertical accelerations (Williams *et al.* 1995). In this case, as usual for this type of analytical and numerical analyses, the mechanical and hydraulic characteristics of the flight control systems are neglected and, when it does not happen, it yields highly non-linear models, usually difficult to identify. When SASs are active, they must also be monitored.

Primarily, for sensors design and specification, it is important to consider the existence of redundancies and the use of complementary sensors (Williams *et al.* 1995). In practice redundancy is rarely present due to the need for using extra channels for this purpose, additional cable routing, and effort with its evaluation and calibration, the use of additional sensors is common (Advisory Group for Aerospace Research and Development 1991). As an example, the angular attitudes, more reliable and useful for low-frequency responses, can be numerically differentiated to obtain angular rates, which can be also acquired, and have better responses in the higher frequency range (Williams *et al.* 1995). Furthermore, redundancy is one way to prevent the possibility of degradation of the sensors during the test and allows the reconstruction and compatibility checking (Advisory Group for Aerospace Research and Development 1991).

Another consideration, related to the helicopters dynamics, is the longitudinal and lateral linear acceleration sensors, which must be designed with sensitivity and accuracy for the typically low signal-to-noise ratio found (Williams *et al.* 1995), ensuring measurement at high frequencies. For these cases, often the acquired signals are exceed by standard vibration levels (in the example aircraft, around 0.02 g for lateral axis vibration), because these types of aircraft are not capable of producing large linear accelerations, other than the vertical axis (Advisory Group for Aerospace Research and Development 1991). However, oversensitivity can cause even more noise in the signal. For the sensors accuracy, the measuring range selection is a tradeoff between high resolution and saturation.

Challenges to accelerometers sensitivity and accuracy, however, are minimized due to the fact that in rotary-wing aircraft the reactions to control inputs are primarily angular rates (Advisory Group for Aerospace Research and Development 1991), despite the noise arising from vibration, which also affects these measurements. For this reason, while in most instrumentation projects the main concern is the accuracy of static measures, applying the methodology in the frequency-domain, priority is given to test dynamic accuracy in the frequency spectrum, especially to the angular rates signals (Williams *et al.* 1995).

In addition, for the method employed, the filtering characteristics, especially filter type and cutoff frequency, must be known (Williams *et al.* 1995). In FTI, the most common filters types are noise filtering or anti-aliasing (Tischler and Remple 2012). Primarily, the concern is due to the fact that filters having different cutoff frequencies for each channel relative to the control inputs and the response of the aircraft will cause additional spurious phase shifts. This occurs most often due to the fact that signals are conditioned taking into account the bandwidth of each sensor (Williams *et al.* 1995). So, to avoid the need for data correction during the analysis phase, all signals, whenever possible, are subjected to the same filtering (canceling each other when it obtained the Bode diagram for inputs and outputs), with cutoff frequency, at least, five times the maximum frequency of interest (Tischler and Remple 2012), ω_{max}, near 2 Hz (Tischler and Remple 2012). Ideally also the filter has to have a response flat in magnitude and minimum phase shift in the frequency spectrum (Williams *et al.* 1995).

Finally, the last characteristic that is designed for the FTI is the acquisition frequency sampling (ω_s). As practical interpretation of the Sampling Theorem (Nyquist 1928), the minimum sampling frequency to be used for an analog signal, sampled and bandlimited, which could be perfectly recovered from an infinite sequence of samples, is twice the higher frequency of the original signal (Nyquist 1928). Accordingly, the Nyquist frequency ($\omega_{Nyq} = 0{,}5\omega_s$) becomes the maximum frequency of interest. However, due to real characteristics of sensors noise and other sources of disturbances, such as atmospheric turbulence, the use of values close to ω_{Nyq} makes impossible an adequate identification at higher frequencies. Thus, commonly is employed the same value for all acquired signals, of at least five times the cutoff frequency of the selected filter (Tischler and Remple 2012). For this is, ideally, as explained, five times the maximum frequency of interest, one comes to a 25-fold factor in-between ω_s and ω_{max}. Logically, if the processing and recording units allow, it is recommended to use high sampling rates, which may be reduced in the post-processing involvements, using digital filters (Advisory Group for Aerospace Research and Development 1991).

The characteristics previously treated constitute the ideal condition for specification of flight test instrumentation for this purpose. This paper employs a flight test data record of a previous study, from Cruz (2009), and a no longer existing instrumentation project in IPEV. The information about the instrumentation system comes from the aforementioned reference.

The AS355-F2 test helicopter is equipped with an FTI system Aydin Vector PCU-816-I system, which provides 35 different parameters with ATD-800 digital recorder, also manufactured by Aydin Vector, with 4 GB of capacity.

For flight controls displacements, potentiometers (P/N D-23108) are installed in the flight control rods at the bottom of the aircraft, and the measurements are recorded and stored on FTI. In addition, flight controls displacement information is also presented in analog gauges installed on aircraft front panel (Cruz 2009).

Information of the longitudinal (θ) and lateral (ϕ) attitude are provided by a vertical gyro VG 208 C installed on the floor, on the right side of the aircraft. The magnetic heading (ψ) was acquired from the radio magnetic indicator (RMI) of the aircraft. Installed near the vertical gyro, a triaxial gyro provides the three angular rates (p, q, r). Accelerometers, JT21XX (manufactured by SFIN), one for each axis, mesure linear accelerations (a_x, a_y, and a_z) which are post-processing corrected by the relative position to the center of gravity (CG) of the aircraft, the angular attitude and linear velocities (u, v and w) of the rigid body (acquired from Differential Global Position System Z12 Astech, whose antenna is attached to the top of the vertical stabilizer).

Sampling acquisition for that FTI varies, depending on the parameter acquired. However, the presentation of the data is based on bigger sampling rate (60 Hz, for angular rates), repeating the information temporally prior of the acquisition, for parameters with a lower sampling rate.

Inevitably, as an FTI non-developed for this study or this flight test purpose, it is expected that the characteristics are not consistent with previously reported, defined as a possible reference. The data consistency verification techniques will be employed to confirm the suitability of the acquired data for the research objectives. Also, if inaccuracies in the results are a consequence of the instrumentation employed, they will be addressed in the results presentation. Furthermore, for the instrumentation design employed in this research, comercial packages are used. For them, the complete knowledge about the internal processing and filtering are unknown or not completely described in the documentation of the manufacturer of each item composing it.

IDENTIFICATION MANEUVERS

As mentioned, for the methodology employed in this research, it will be used a frequency sweep as flight controls inputs. Although they are not classified as optimized inputs, they have wide use in frequency-domain flight tests for security reasons, easiness, reliability and results accuracy (Tischler and Remple 2012).

For frequency sweep, whether generated by an automatic device, or by the pilot inputs, a flight control sinusoidal input is applied around a reference condition, starting at the lowest frequency, being increased smoothly and progressively, moving on to medium and reaching up high frequencies (Williams *et al.* 1995). One of the strengths of the frequency-domain methodology is robustness due to the fact that inputs, such as those mostly used by pilots, do not need to be strictly sinusoidal, perfectly symmetrical and have the same amplitude. In fact, irregularities in the control inputs are beneficial since they increase the richness of the power spectral density of the excitation (Tischler and Remple 2012). Thus, from excitations with these characteristics, it is possible to obtain a uniform distribution spectrum (*i.e.* constant power spectral density) in the entire frequency range of interest, ensuring that the excitation is persistent and with high coherence (the coherence function is an unbiased estimator of the accuracy and linearity of the nonparametric identification response).

Planning the execution of the maneuvers, two conditions must be met in order to assure the necessary feedback regulation to ensure that the closed-loop system response is bounded, regardless of their intrinsic dynamic feature to be stable or unstable: the existence of at least 3 s in the static condition "trimmed" at the beginning and end of data recording and the enforcement that they are delimited keeping the responses centered on this initial flight condition (Tischler and Remple 2012).

Regarding the amplitude, it is small, but allows the pilot to realize continuous control displacements, wide enough for effective response at medium and high frequencies, keeping them within the linear range of the response in attitude and angular rates. Generally, this is achieved with a sweep input typically in the range of ±10 to 20% of control inputs (what represents, for the example helicopter, ±1 to 2 in). The resulting aircraft angular response is typically in the range of, aproximately, ±0.1 to 0.3 rad, for angular attitude, and ±0.1 to 0.3 rad/s, for angular rate (Tischler and Remple 2012).

While maintaining the "trimmed" condition is essential for the reasons mentioned above, the responses in other coupled axes, than the effectively excited, are not artificially suppressed. Thus, intermittent and uncorrelated inputs on the non-excited flight controls are only allowed to contain large deviations from the reference condition (Tischler and Remple 2012). If any, this significant correlation will decrease the quality of the remaining axes identification, evidenced by a decrease in coherence and possibly mask or suppress important aspects of the aircraft dynamics. This fact also limits the use of SAS in their normal conditions of operation (standard gains) or to be active in all axes during the excitations.

The total duration of inputs and data recording are also another concern. From practical experiences (Tischler and Remple 2012), the total recording time is four to five times the maximum period of interest (T_{max}), required for precision conversion of time domain data to frequency-domain, in the whole range of frequencies of interest. Though this long test period, when compared to other classes of inputs, such as the optimized inputs, can be viewed as a negative aspect of frequency-domain methods, it ensures a high-quality frequency response result without detailing *a priori* knowledge of the dynamics of the test aircraft (Tischler and Remple 2012). Further, it has been shown that, for the same aircraft, frequency sweep inputs of the same time duration like optimized inputs give a comparable result (Tischler and Remple 2012).

Approximately 40% of the sweep interval is in the lower frequency inputs, the rest of recording was completed with the gradual increase in the frequency of the input (Williams *et al.* 1995). In the same flight condition, at least two - deally three - repetitions of these test runs are performed, allowing them to be concatenated, minimizing random error and increasing spectral richness of the excitation (Tischler and Remple 2012).

Moreover, the minimum and maximum excitation frequency are strictly planned and controlled in order to ensure accuracy and coherence of the frequency response. For flying qualities requirements compliance purposes, the pilot controls inputs are obtained in the frequency range from values lower than the frequency bandwidth (ω_{BW}), usually half of that, to values greater than twice the frequency in which the phase of the response in attitude is –180° (ω_{180}) (Tischler and Remple 2012). The bandwidth frequency is a measure of the quickness with which the aircraft responds (angular response) to a control input, and is the lowest closed-loop frequency which ensures at least 6 dB of gain margin (ω_{BWgain}) or 45° of phase margin ($\omega_{BWphase}$) of the frequency neutral stability (ω_{180}) (ADS-33E-PRF 2000).

Estimates of ω_{BW} and ω_{180} can be obtained analytically or by a response to a step input, in time domain, using the metric relations between the time domain and the frequency to second order systems. However, once one of the identification objectives is to properly obtaining these results, in most cases this information is not previously available. Then, typical or conservative values are initially used, which may be refined during the flight test campaign. Conservatively it is adopted for rotorcrafts without the augmentated systems, ω_{BW} equal to 1 rad/s and ω_{180}, 6 rad/s. Thus, the maximum and minimum frequency values of interest are 0.5 and 15 rad/s, respectively (Tischler and Remple 2012).

The maximum and minimum frequencies are also limited by other important factors that must be addressed in the planning phase. As for the low frequencies, the great difficulty in carrying out the long time pilots inputs is the. As for the low frequencies, the increased difficulty in carrying out the long period pilots inputs, due to the need to suppress the even more evident response in the other coupled axes, can be mastered with some practice sweeps on the ground and in flight (Tischler and Remple 2012). This topic will be addressed in the "Flight Test Planning" section.

For higher frequencies, and also taking into consideration the fact that pilots are responsible for the control inputs, the human muscle limits imposes restrictions on the ability to make flight control displacements changes as fast as necessary (Advisory Group for Aerospace Research and Development 1991). Generally, this limit is approximately 4 Hz (Tischler and Remple 2012).

Another attention in higher frequencies is the possibility of coalescence with some structural vibration modes. Usually natural frequencies are of the order of 12 rad/s, for the modes of rotor pylon, "soft" tail boom longitudinal, lateral, and torsional (Williams *et al.* 1995). On those conditions, care must be taken to avoid large amplitude inputs at the higher frequencies by using real-time telemetry of piloted inputs, aircraft responses, and key response variables. If structural excitation and identification is specifically desired, then the test should be conducted using automated sweeps, structural response instrumentation, and provision for automatic cutoff of the test if structural thresholds are reached (Tischler and Remple 2012). The inadvertent approach of these frequencies can result in rivets stress, structural damage or even catastrophic loss of the tail boom (Williams *et al.* 1995). Also, some hydraulic systems can show limit cycles due to frequency of the inputs.

FLIGHT TEST PLANNING

The first concern for all flight test activity, including employing frequency-domain methods, is safety. Historically, accidents during these types of flight tests occurred due to lack of research and knowledge of the system to be tested come with, among other factors, insufficient planning (Williams *et al.* 1995).

As in any other flight test activity, the doctrine of "step approach" is respected. Thus, it is prudent to progress gradually increasing up to maximum frequency of interest in order to identify any unknown aircraft dynamic mode or any other relevant characteristic (Williams *et al.* 1995). The risks can be mitigated not only by flight test technique, but also using accelerometers, strain gages or other sensors for monitoring the piloted inputs and key responses in real time, employing telemetry, as already mentioned (Tischler and Remple 2012).

When approaching the limits previously specified for the control input or response of the aircraft, and it is observed any unexpected behavior, with the necessary anticipation, the test run is interrupted. Aural alerts, visual or any other proprioceptive warning, to assist in identifying the proximity of such limits are extremely important (Williams *et al.* 1995). Moreover, recovery procedures, usually employing only the interruption of flight control inputs, maintain control of the aircraft (Williams *et al.* 1995) and, if necessary, carry an abnormal attitude recovery. The potential adverse behaviors of related systems are known and alerted during the mission briefing.

Not only for safety reasons but also to achieve satisfactory data quality, the frequency-domain test requires careful crew coordination. One technique for pilots frequency sweep is that the flight test engineer (FTE), or any other member of the crew, on board or on a real-time monitoring station, performs the timing of the first eight quarters of cycle (Tischler *et al.* 1987), focusing on quality of the minimum frequency to be tested and assisting the pilot to reference the flight control position as that time is announced. Passed the two periods of lower frequency, and started the gradual increase in frequency, the count is done in 10-second intervals, paying attention to the stopping criteria on maximum frequency (Tischler *et al.* 1987). Warnings of the current excitation frequency also assist in maintaining the attention to the limits.

Still aiming quality of the results, the execution of the flight control inputs is made around a reference flight condition and controlling the aircraft answers on coupled axes using non-correlated inputs to the primary excitation, but not completely eliminating them. To fulfill this task, the other flight controls are carried out in secondary and low-frequency inputs, preferably employing pulse-type (Williams *et al.* 1995). Psychomotor natural features or pilot inertia cause a coupling between the frequency sweep inputs and the actions performed in other controls (Williams *et al.* 1995). This is minimized by task division and cabin coordination in aircraft commanded by two pilots, one responsible for the excitation control inputs and the other to maintain the reference flight condition through the other remaining flight controls (Tischler *et al.* 1987), being the responsibility of each pilot in case of any abnormal behavior or emergency previously defined.

With the aircraft electrically and hydraulically external powered, whenever the flight control system allows, the frequency sweep inputs are trained and repeated on ground. This occasion is also used to make adjustments of crew coordination, to allow psychomotor pilots training for proper execution of the control inputs, to set phraseology and coordination between test pilot and FTE, and to reduce the need for training and repeat runs throughout the flight.

The planning phase of the flight test also includes the definition of the flight conditions to be flown. For flying qualities requirements applications, as objectified by this work, the flight conditions are restricted to hover and flight with some speed (Williams *et al.* 1995), respecting the requirements and standard settings to be used as reference. Since the aircraft response also depends on weight, balance and configuration of SAS (Williams *et al.* 1995), if any, the conditions of these parameters during the tests execution are previously defined.

The flight tests performed to obtain the data employed in this paper have a distinct goal and do not have been effectively made in order to apply frequency-domain methodology. For this reason, in some aspects, the technical guidelines previously addressed are not completely followed. It was not employed a device to monitor vibration or structural loads in real time, via telemetry, yielding a not-safely reaching of the higher frequencies, whose limit value or stop criterion was not specified.

The inadequacy of the employed flight test technique to the described recommendations causes consequently the distancing of data obtained to what is ideally required for rotary-wing aircraft flying qualities evaluation by means frequency response.

Thus Figs. 3 and 4 represent the three test runs of longitudinal (δ_{lon}) and lateral cyclic (δ_{lat}) control inputs, respectively, and the behavior of the other flight controls (pedal, δ_{ped}, and collective, δ_{col}) during the excitation. The flight test is performed from the initial flight condition of symmetrical level flight, 80 KIAS, 5,000 ft. The aircraft is weighing 2,200 kgf with rear CG (3.34 m).

As can be seen in Figs. 3 and 4, on the three runs of frequency sweep control inputs, along both axes, the excitation followed the previous recommendations: (a) starting and ending on the trimmed condition; (b) beginning the excitation with two periods in the lowest frequency of interest and, after, gradual increase of the excitation frequency; (c) control displacement less than ±2 in in all test runs for the longitudinal control, and ±3 in, for the lateral control.

Regarding the use of flight controls on non-excited axes, aiming the maintenance of flight conditions within small

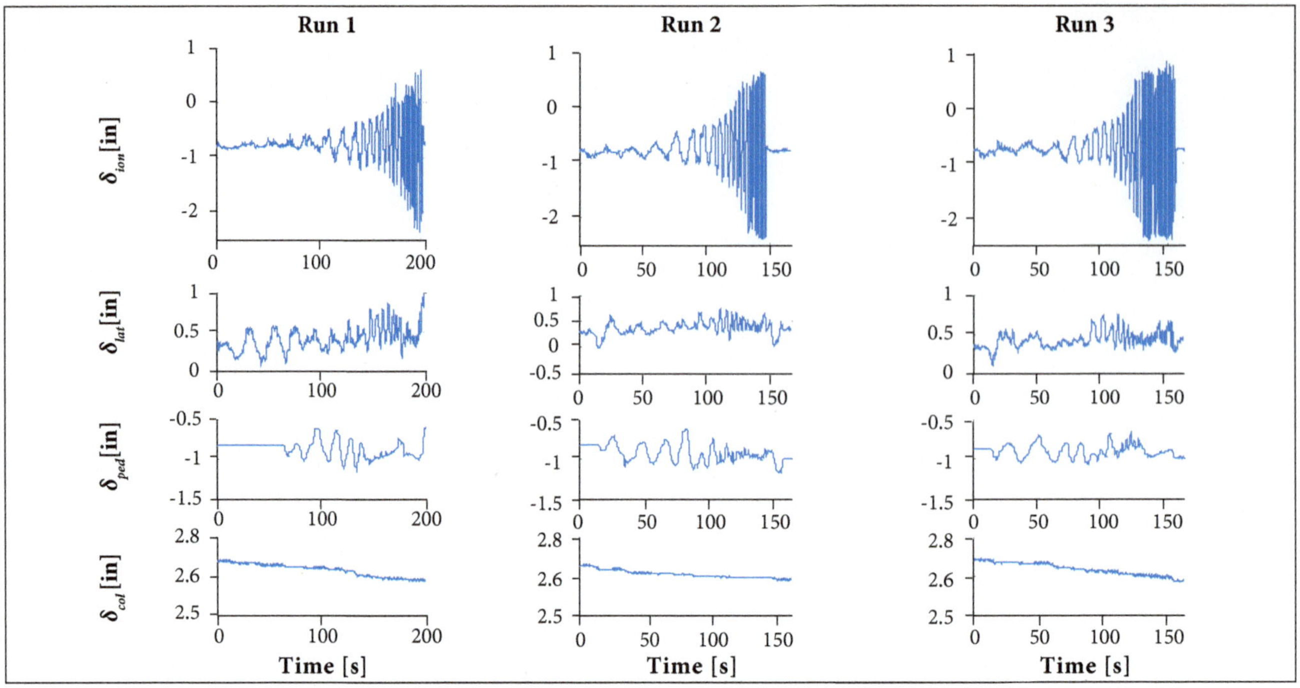

Figure 3. Longitudinal cyclic control inputs — level flight (80 KIAS).

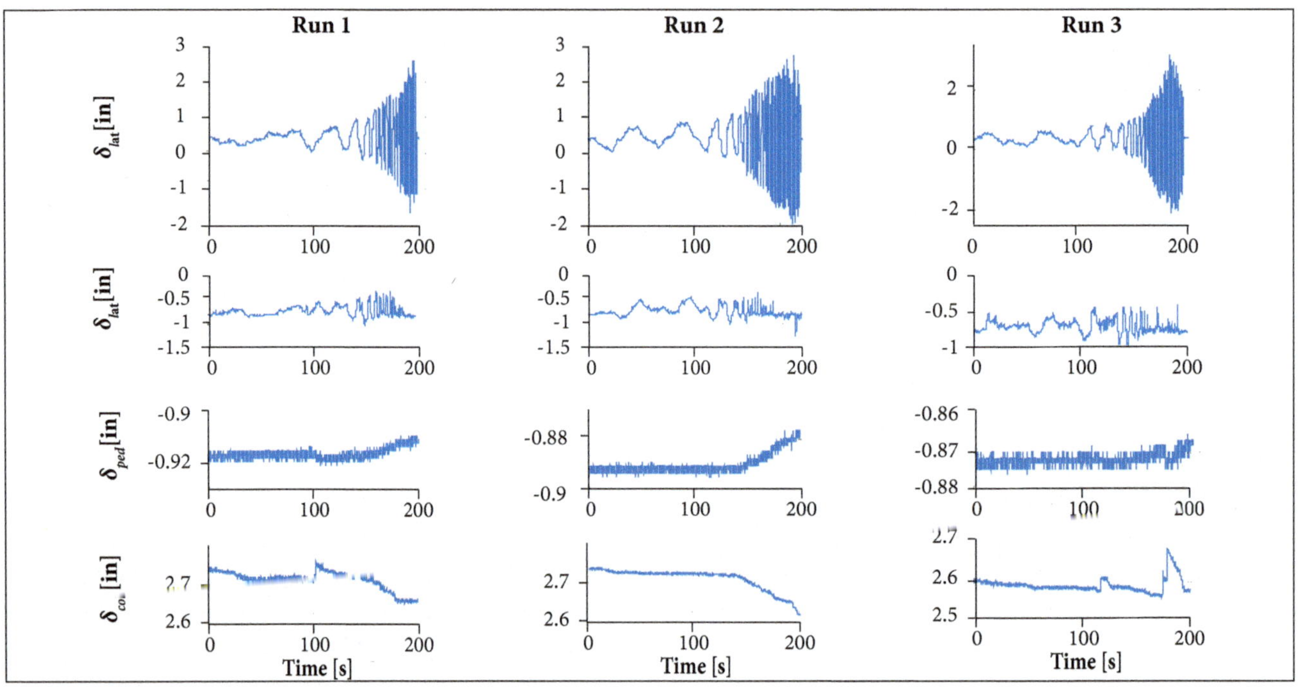

Figure 4. Lateral cyclic control inputs — level flight (80 KIAS).

amplitudes, can be observed that there is a certain regularity, correlated to the excitation, mainly between the longitudinal and lateral cyclic controls. Moreover, for the longitudinal cyclic control excitation one can also be observed a periodic variation of the pedal control.

Figures 5 and 6 show the responses to the excitation on longitudinal and lateral axis, respectively, to each of the control input performed, as a function of the respective angular attitude (θ, for longitudinal, and ϕ, for lateral axis) and their time rate (q, for longitudinal, and p, for lateral axis).

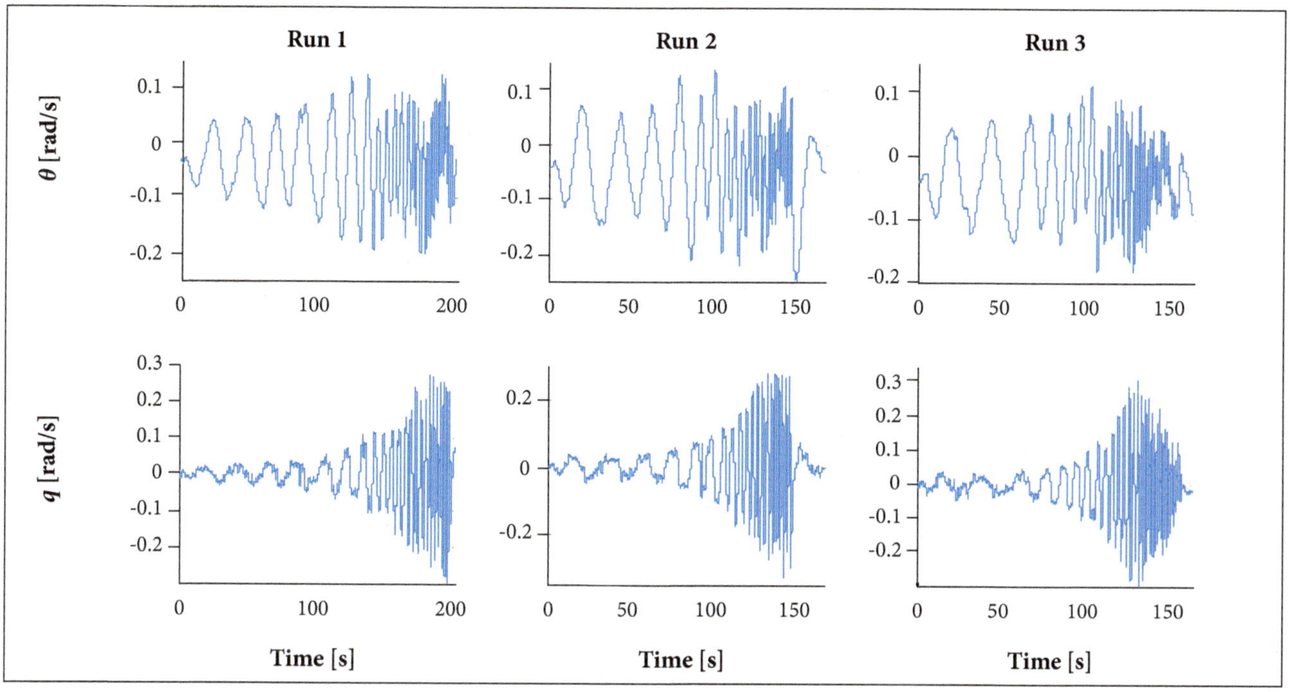

Figure 5. Longitudinal cyclic control input response — level flight (80 KIAS).

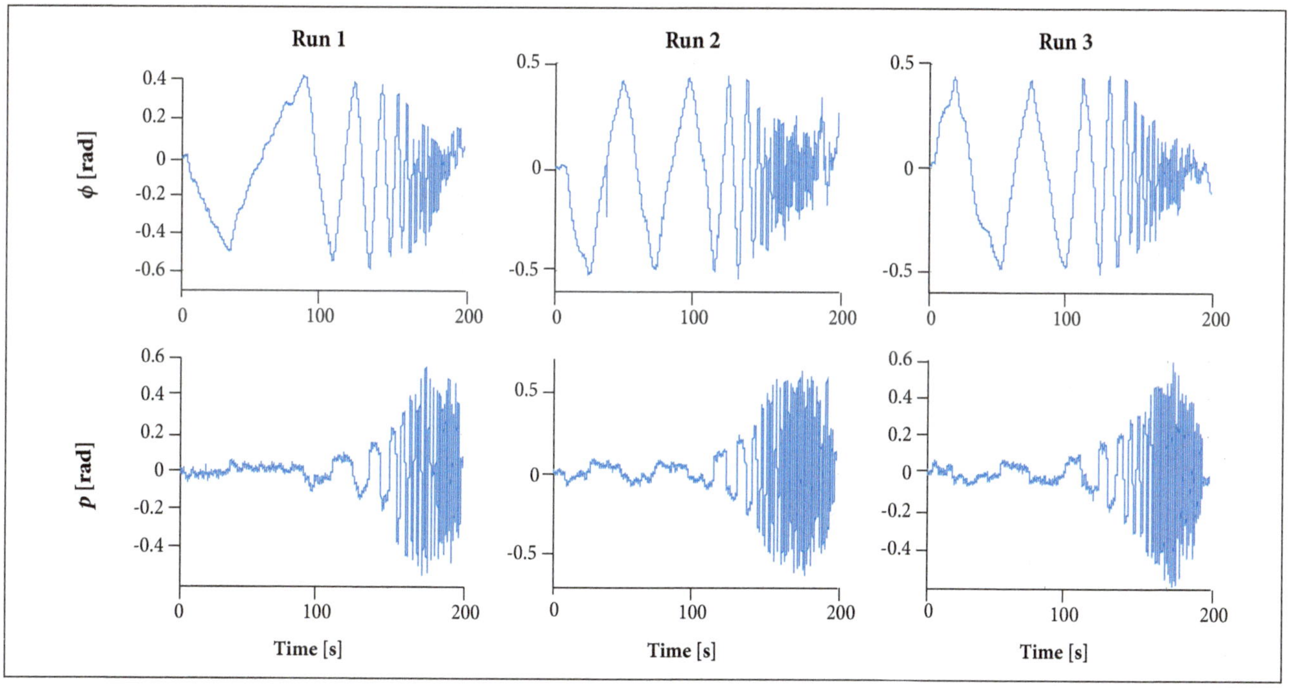

Figure 6. Lateral cyclic control input response — level flight (80 KIAS).

As can be observed in Figs. 5 and 6, the excitation causes similar characteristics in response to the inputs: (a) it starts and finishes in the trimmed condition (with some minor imperfections of "trimming"); (b) it responds with angular rate and attitude variations, respectively, less than ±0.2 rad/s or ±0.3 rad in all test runs for the longitudinal control and ±0.5 rad/s or ±0.5 rad for lateral control. Thus, just the response due to the longitudinal cyclic excitation can be considered within the acceptable limits of linearity, according to the typical and recommended values.

Higher amplitudes of lateral cyclic control lead hence to more pronounced responses. This result is justified by the lower inertia around the longitudinal axis, due to greater control sensitivity and dynamic characteristics of the tested aircraft, where more pronounced responses around the roll axis during dutch-roll excitation are in place.

In analyzing Figs. 3 to 6, one observes that for longitudinal cyclic excitation the average recording time is 176 s. That is at least eight times greater than the maximum excitation period, 20 s (corresponding to 0.31 rad/s for the minimum excitated frequency). These values are consistent with the previous recommendations. The maximum excitation frequency is 10.47 rad/s, a value that may limit reaching high coherence and prevent identification in the higher frequencies, as well as getting some results for comparison with ADS-33E-PRF (2000) standard. In addition, this value is lower than the typical employed in this type of evaluation.

For the lateral cyclic excitation, the average recording time is 215 s, at least five times greater than the maximum excited period, 41 s (corresponding to 0.15 rad/s for the minimum excited frequency). Again, these results are consistent with the recommendations imposed. The maximum frequency reached is 14 rad/s, larger than the longitudinal value.

The higher excitation frequency found for the lateral cyclic input is a consequence, again, of the lower inertia around the longitudinal axis, common in helicopter design, and the lowest pilot workload to maintain trimmed condition, that can be observed by means of the reduced use of controls on non-excited axes, allowing the pilot to focus only in the excitation.

An additional comment shall be made on the quality of sampled data related to the lateral cyclic control inputs (Fig. 4). Higher noise signals in the pedal and collective control displacements can be obtained as compared to signals observed from the longitudinal cyclic control inputs. Since the data were obtained in the same flight test campaign without any change on FTI during its course, this behavior can be justified due to a possible atmospheric turbulence during the test run or signal degradation for some interference due to excitation or other electromagnetic source. The absence of filters, for reasons already mentioned, justifies the fact that Fig. 4 represents the data without any treatment in post-processing.

DATA CONSISTENCY AND RELIABILITY

An important step when for frequency-domain techniques is data reduction. Systematic or random errors potentially present in the data, inevitably, contaminate subsequent analysis. The errors influence is particularly significant when aircraft states will be obtained by integration of observed or measured states. The suitable preparation of the data is thus crucial to obtain accurate system frequency response (Williams *et al.* 1995).

A first data-consistency analysis on flight test datais made seeking "hidden" errors, like the inversion of the sensors sign convention, scale factors and trend errors. Frequency-domain techniques are useful for their detection. The time delay and signal errors are detected by means of the response phase. During this first approach, some common DAS and processing unit errors and failures are quickly found, such as dropouts, spikes, saturation and large offsets (Advisory Group for Aerospace Research and Development 1991).

The observed dropouts and spikes, restricted to small sample intervals, are subsequently deleted. These intervals are then reconstructed by interpolation taking into account the nearest neighbor (or neighbor method). Although this procedure does not permit the accurate reproduction of lost data, this methodology allows making them more realistic when compared to the original with inconsistencies (Advisory Group for Aerospace Research and Development 1991).

In the frequency domain, the coherence function can be employed as metrics of quality of channels sampled since it measures the control inputs and the response relation, as the noise level. The noise is further analyzed by frequency-domain techniques to check the correct operation of each channel, detecting vibration or connector problems and also obtaining vital information to the most appropriate filter design.

For the duplicate or additional sensors, other data post-processing can be done with information from these sensors, in order to measure their quality. A special technique is called kinematics compatibility check. In that, the existing kinematic relationships between the different measured variables are used by applying them in several ways: from the comparison of two signals to full path reconstruction with a six-degree-of-freedom model which even may constitute a part of the identification process.

In case of simple comparison, one can be used frequency domain methodology tools, employing it in the signals of a gyro and a rate gyro, comparing the numerical integration of the angular rates to the attitude signals. Different error models are used.

For the data used in this research, besides making the initial inspection with consequent further exclusion of some samples and replacing them by means of interpolation by the neighbor method, the kinematic consistency is checked employing frequency domain techniques. Since the flying qualities requirements refer to angular responses as a function of flight controls displacement, the technique is restricted to checking the angular consistency between angular attitudes and rates, based on the model and methodology proposed by Tischler and Remple (2012).

Adopting an error model that includes scale factors in angular rates and attitudes and non-correlated noise only in the angular rate, the measured variables can be described as functions of these errors. Taking the Euler angle measured as input and angular rate as output, considering the small angles on the stabilized condition and small disturbances during identification maneuvers, from the linearized Euler equations, the frequency response is described as a ratio between the respective scale factors (K) and the data time delays, caused by filtering or distortion, represented in the effective time delay (τ). Values of K and τ can be determined by the fitting of the transfer function of the frequency response.

Though the scale factors from attitudes and rates could not be decoupled, analysis of kinematic compatibility provides values of K close to unity and zero effective time delays when they are consistent. If the values are different from these, for the error model adopted, the error source can be isolated conducting the same analysis in other redundant data sources (Tischler and Remple 2012).

Based on data acquired from the longitudinal and lateral cyclic control inputs, in level flight at 80 KIAS, the angular compatibility, in the frequency domain, between angular rates and attitudes, multiplied by an integrating element ($1/s$) for the longitudinal and lateral axes, respectively, are presented in Figs. 7 and 8.

As observed in Figs. 7 and 8, the integrator element ($1/s$) "cancels" the first-order characteristics of the attitude and angular rate relation, *i.e.* Bode diagrams featuring a constant response in magnitude and phase are also constant and equal to 0°. Fitting the frequency responses presented, it is obtained a K value close to unity (1.022, for longitudinal axis, and 1.045, for lateral axis) and practically zero effective time delay (0.058 s, for longitudinal axis, and 0.055 s, for lateral axis). Then one verifies the kinematic compatibility of the channels employed to flying qualities analysis, with a correct sign convention and the conversion to the units of measurement of interest held.

It must be emphasized that the results described are valid in the range in which the coherence function has no oscillatory behavior and its value is greater than 0.6, *i.e.* from 0.08 to 6.15 rad/s, for the longitudinal control input, and 0.03 to 8.49 rad/s, for the lateral control. This reference value for coherence function provides a signal-to-noise ratio and random error within acceptable levels (Tischler and Remple 2012). Further details and more emphasis on the interpretation of the results using the coherence function will be presented in the frequency-domain results.

In additionprimarily, the FTI employed does not have, as mentioned, the purpose of verification the flying qualities requirements employing frequency-domain techniques. Also,

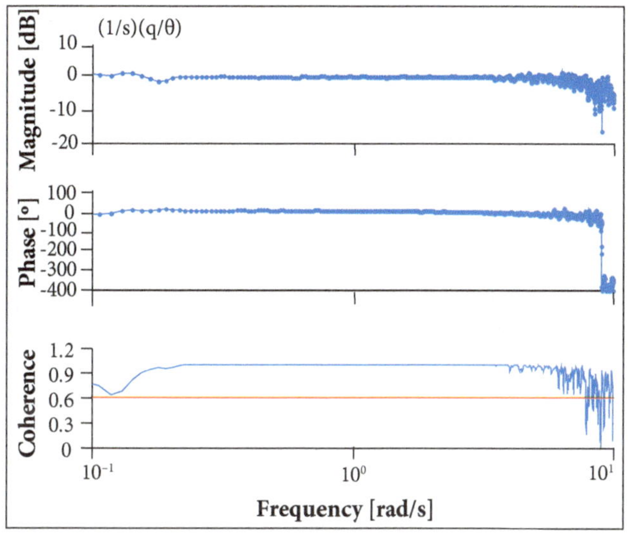

Figure 7. Frequency-domain angular consistency – longitudinal axis.

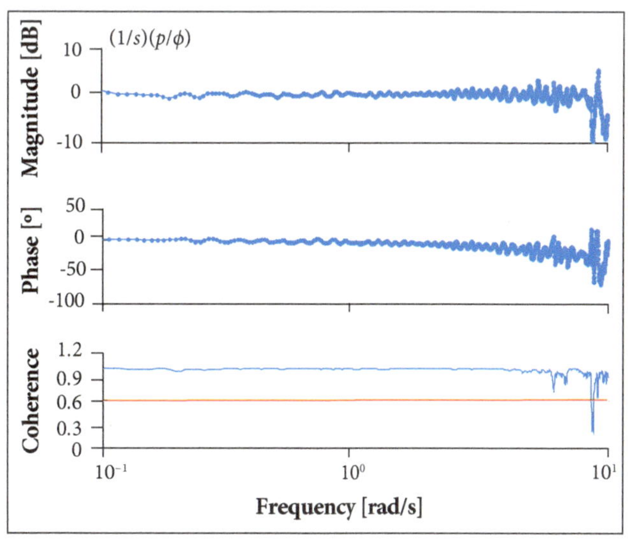

Figure 8. Frequency-domain angular consistency – lateral axis.

the detailed information of the dynamic characteristics of sensors and filters are not available, so the next stage of the data post-processing, referring to the filtering, cannot be applied.

During the initial inspection of the data one observes: an absence of too noise signals; no problem on the sampling rate; and lack of knowledge about sensors characteristics, in order to all of them be filtered as the same. Thus, given the results obtained in the visual inspection phase and angular consistency, for the purposes of this paper, the instrumentation characteristics and, consequently, the acquired data were considered appropriate without any correction in the sampling rate before the spectral analysis.

SPECTRAL ANALYSIS

From the study of dynamical systems, given an ideal physical system single input-single output (SISO) — physically realizable, characteristic parameters constant (time invariable), stable and linear (Bendat and Piersol 1980) — excited by a sinusoidal force function with frequency ($\omega = 2\pi f$), its properties lead to a particular and a general solution. After a period of time, the general solution, transient, is damped and remain, in steady state, the particular solution, whose response depends uniquely on the external force.

The response of this system is also characterized as a sine function with the same frequency (first harmonic) of the force function (input); the response has magnitude and phase relative to the input, modified by the intrinsic characteristics of the system and by means of the frequency input itself.

The analysis of the system frequency response, $H(f)$, of this simplified example that can be extended for any input and output, periodic, non-periodic (deterministic or random) or stationary random data, is the evaluation of the magnitude ratio of the input-output relation and the phase relation over a frequency spectrum. Once it is possible to extend the problem to physical systems with realistic features, $H(f)$ describes the system dynamic behavior, sometimes nonlinear, in terms of the best linear description of the input-output behavior, *i.e.* the first harmonic descriptive function, without imposing any prior knowledge of their internal structure, order, or equations of motion (Tischler and Remple 2012).

Mathematically, the frequency response, $H(f)$, is a complex function, characterized by its gain factor ($|H(f)|$) and phase delay ($\angle H(f)$), which relates the system output and input finite Fourier transform (Fourier coefficient, $Y(f)$ and $X(f)$, respectively), for each frequency f.

When one employs flight test data obtained from frequency-domain scope, the Fourier coefficients are obtained using the Discrete Fourier Transform (DFT), for the signals exist only in a finite-time interval, with a sampled data in a certain rate ($\omega_s = 2\pi/\Delta t$), for a recording with total duration of $N\Delta t$, where N is the number of discrete points of frequency (Bendat and Piersol 1980). They are calculated at points often equally distributed from the minimum resolution ($\omega_{min} = \Delta\omega = 1/N\Delta t$) to the sampling rate.

With the Fourier coefficients it is possible to determine other important one-sided spectral functions, also called spectral density functions: the input autospectrum (G_{xx}), or input power spectral density (PSD), which represents the excitation density of power as a function of frequency; the output autospectrum (G_{yy}), or output PSD, which represents the response power distribution as function of frequency; and cross spectrum (G_{xy}), or cross PSD, which represents the power transfer from input to output, as a function of frequency, which contains the magnitude and phase information from the Fourier coefficients (Tischler and Remple 2012).

The spectral functions are used directly to evaluate the frequency content of the excitation and the aircraft response for a particular identification maneuver. Its use in data analysis allows the selection of the most appropriate test runs to be concatenated, verification of turbulence levels, frequency sweep irregularities and finding the real uniformly excited frequency spectrum.

Using the simplified Fourier analysis, erroneous frequency information appears as side lobes of Fourier frequency of interest. This phenomenon, known as "leakage", permits the "leakage" of power at frequencies far from the central lobe and introduces significant abnormalities on spectral functions estimates, particularly for sinusoidal data samples (Bendat and Piersol 1980). To reduce it the technique employed is called windowing, with overlapping windows or periodograms, which produces smoother spectral estimated using the average of multiple segments of data (Tischler and Remple 2012) and reduces their random errors.

The window width is selected to optimize the quality of the spectral identification (Williams *et al.* 1995). Its frequency content is limited by its width, being the minimum identified frequency theoretically defined as the reciprocal of T_{win}. In practical applications, however, data are subjected to processing and measuring noise, nonlinearities and secondary control inputs, with satisfactory coherence from a frequency value which is twice the theoretical minimum. Thus, the selection of larger windows allows accurate identification of lower frequencies usually limited to the length of 50% T_{rec} (Tischler and Remple 2012).

On the other hand, the greater the number of windows, the larger the division of the total time interval, minimizing random error due to the larger sample for taking averages. In joining two or three recorded intervals of data one achieves a recording interval concatenated (T_P) of suitable length that allows the accuracy at low frequencies (function of T_{rec}) allied to the major capacity of segment division. It is generally recommended that Twin be less than 20% T_F (Tischler and Remple 2012).

Thus, selecting the width of the window involves a compromise between a smaller random error and a better identification of the lower frequency content (Williams *et al.* 1995). Even with an iterative process for its optimization, whose can show itself laborious, no single window size selection can produce a good result over the entire frequency range of interest. To obtain the most accurate identification estimates for a particular situation, the spectral calculations would have to be performed repeatedly, in an automatically matter, using several different window sizes, merging the results obtained. This technique, also called as composite windowing, is not employed in the present research.

Once the temporal history is divided in segments, data of each window segment are weighted by a window function ($w(t)$). Thus, it minimizes the "weight" of the beginning and end of each segment, avoiding the "leakage" caused by the addition of spurious dynamics during the data segmentation process (Tischler and Remple 2012). Several distinct functions are employed for this purpose as the half-sine or the Hanning window. Employing this method the spectral functions are now smoothly estimated and the frequency response function for SISO systems can be directly obtained.

Figures 9 and 10 represent, respectively, the longitudinal and lateral control inputs when concatenated, based on runs 1, 2 and 3 of Figs. 1 and 2.

Based on Fig. 9 one can observe that, for longitudinal cyclic control inputs, the inputs when concatenated provide an interval of 530 s. Likewise, based on Fig. 10, for lateral cyclic control inputs, the concatenated total time is 648 s. Thus, for this research, following the recommendations previously described herein, one selects: T_{win} of 53 s, for data on the excitation of longitudinal input, and T_{win} of 65 s, for the lateral input.

For non-linear systems analysis, as interpreted by Tischler and Remple (2012), the frequency response is descriptive, since it measures the output portion linearly related to the input. Thus, it is an equivalent linear model that minimizes the mean square difference

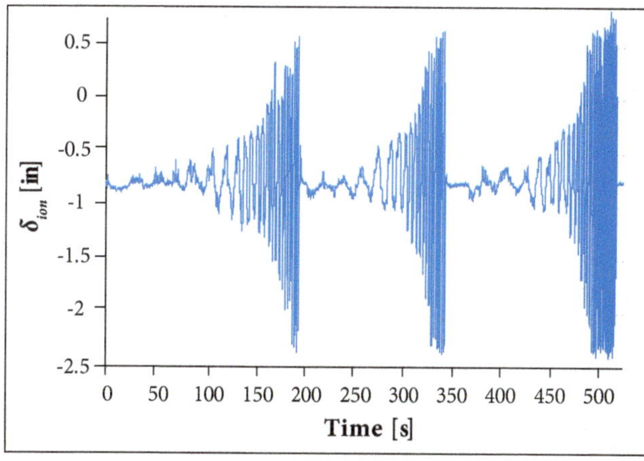

Figure 9. Concatenated longitudinal control inputs – level flight (80 KIAS).

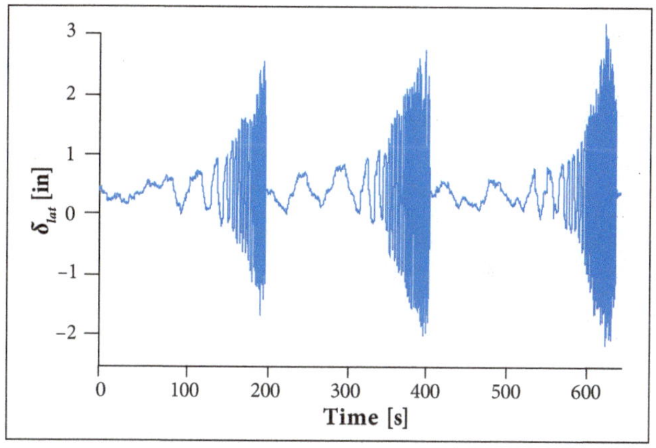

Figure 10. Concatenated lateral control inputs – level flight (80 KIAS).

between the current response signal and its approach by the first harmonic. The non-linear effects that are not characterized by the function are associated with higher order harmonics of the expansion of Fourier series. These harmonics appear as noise in the response measurement. Typically, the vast majority of nonlinear effects is filtered by the low-pass nature of the rotary-wing aircraft dynamics.

However, a way to measure the quality of this linear approximation by the first harmonic is the coherence function (γ^2_{xy}), defined as Eq. 1:

$$\hat{\gamma}^2_{xy}(f) = \frac{|\hat{G}_{xy}(f)|^2}{|\hat{G}_{xx}(f)||\hat{G}_{yy}(f)} \quad (1)$$

where: G_{xx} is the input PSD; G_{yy} is the output PSD; and G_{xy} is the cross PSD.

The function defined by Eq. 1 measures the portion of G_{yy} linearly attributable to G_{xx}, for a given frequency (Tischler and Remple 2012); its values range from 0 to 1, where the unit represents perfectly linear systems with noise free spectral estimates.

In practical terms, though, the coherence function exhibits values below unit, primarily due to three reasons. First, in certain systems, nonlinearity cannot be neglected or described simply by the first harmonic. This is the basis for the use of small perturbations and small excitation amplitudes, knowing that this is difficult to be met in rotary-wing aircraft, especially in hovering flight, when low-frequency control inputs are present.

A second common cause for the coherence reduction is the output or input noise on the measured signal (measurement noise). Though not contributing or not affecting directly the flight control input or the response of the aircraft, this type of noise contaminates the recorded signal and hence causes errors in the frequency response estimate (Williams *et al.* 1995). For this reason, once more, the importance of using high-quality sensors for flight control inputs which ideally free noise must be emphasized. Regardless of the correlation with the measured signal or with the own signal (output or input) in which the noise is present, always there are random errors (dispersion of the spectral estimate around an expected value) that are reflected in the coherence function. This error can be minimized and spectral accuracy significantly improved for a certain level of coherence, when the concatenation method is employed, for the normalized random error is inversely proportional to the average number of independent data intervals (Williams *et al.* 1995).

The third source of reduction of coherence is the existence of additional, secondary and not measured inputs in the system. They come from gusts, turbulence or other source of process noise, or secondary control inputs acting in not-excited axes arising from the pilot himself or some aircraft automatic control system function kept active, *e.g.* the SAS (Williams *et al.* 1995). Thus, to prevent the reduction of coherence due to those noise errors, the recommendations to realize frequency sweep tests in minimal wind conditions and low atmospheric turbulence must be implemented. Accordingly, recommendations involving: (a) action in the non-excited controls to maintain the flight condition around a trimmed reference; (b) amplitude; and (c) non-correlation of the entrances on non-excited controls to the proper excitation, are almost inevitable in helicopter flight test analyses.

STANDARD COMPLIANCE

Usually, the Bode diagrams are used to present the results of the flight data frequency response function, *i.e.* magnitude in dB ($\text{dB} = 20 \times \log_{10}|H(f)|$) and phase (*e.g.* in rad/s) *versus* frequency in a logarithmic scale. The properties and characteristics of the Bode diagrams allow easier treatment of the data, for they are additives for serial systems and able to present responses on the same graphic involving a wide frequency range, from the lowest to the highest.

The shape of the magnitude and phase curves on Bode diagrams, specifically the gradient of the magnitude curves and the phase shift, are indicative of the system structure: number of poles and/or zeros and cutoff frequencies. The existence of peaks in these figures are also interpreted as due to the natural system modes, and the frequency in which they occur, indicating an oscillatory response, allowing the estimates of its natural frequency and damping ratio.

Bode diagrams in Figs. 11 and 12 of the open-loop system's SISO relationship show magnitude, phase along with coherence for the longitudinal and lateral axes excitation, respectively. Since the attitudes and angular rates signals have different behavior and quality throughout the frequency range, the former being most used at low frequencies and the latter, for high frequencies of the spectrum. For this reason, the response in Bode diagrams employ the integration of the angular rate signal, what showed more wide frequency range in terms of best coherence. The angular consistency analysis between these two signals validates their use.

Figures 11 and 12 also show the 95% confidence levels of the frequency response, based on the normalized random error function, as given by Bendat and Piersol (1980).

The first important aspect to be noticed in Figs. 11 and 12 is the frequency range in which the data present coherence higher than the reference value (0.6), showing no oscillatory behavior or sudden drops. For the longitudinal axis excitation, it can be found in the range of 0.28 to 4.03 rad/s, while, for the lateral axis's, 0.1 to 14.90 rad/s. For frequencies not included in these ranges there are loss of coherence, be by the higher order dynamics of the aircraft and thus intrinsic to the system or due to the measurement or process noise from data acquisition system or in the data acquisition process. The latter is the main reason of the observed effects in this research.

Thus, in these frequency ranges in which the coherence is acceptable, one can ensure that an equivalent linear model (first order) can characterize the dominant dynamics modes of the tested aircraft. Furthermore, this equivalent model allows anticipating some model structure characteristics and dynamic modes (Tischler and Remple 2012).

Moreover, stability in closed-loop can be inferred by means of the Bode diagram using the gain margin (GM), defined as the permissible increase in closed-loop gain, before the system becomes unstable (Steven and Lewis 2003) and phase margin (PM), amount of phase that exceeds –180° (Steven and Lewis 2003). For piloted closed-loop systems, changes in the piloting gain affect stability, defining a corresponding phase delay (Williams *et al.* 1995). Because the gain margin can be interpreted as a measure of the pilot's allowable gain increase without it threatens stability. Thus, for flying qualities applications, the information from the Bode diagram is a measurement of pilot's workload. It allows designing the flight control systems.

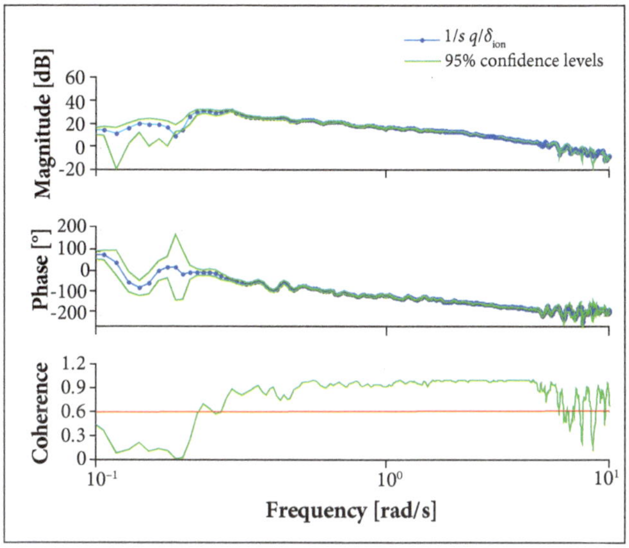

Figure 11. Bode diagram and coherence function for longitudinal cyclic input.

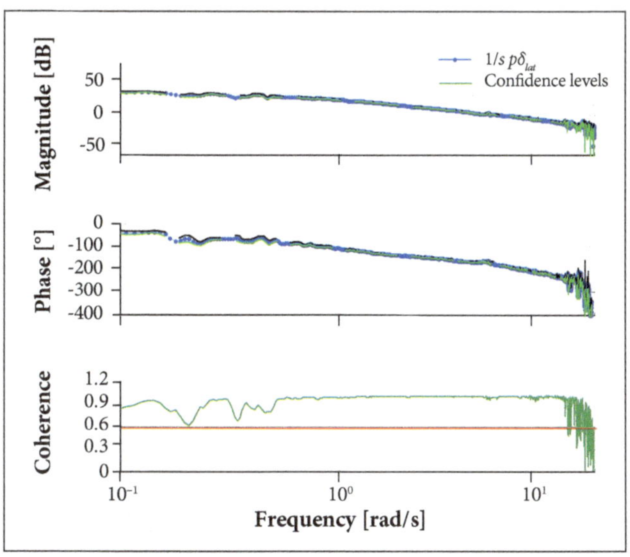

Figure 12. Bode diagram and coherence function for lateral cyclic input.

For lateral and longitudinal axis, the requirements to be check come from the ADS-33E-PRF (2000) standard, specifically paragraphs 3.4.1.1 Short-term response (bandwidth) and 3.4.6.1 Small-scale roll attitude response to control inputs (bandwidth), which employ frequency-domain parameters obtained from the Bode diagrams: bandwidth frequency (ω_{BW}), and phase delay (τ_p).

The bandwidth frequency is the lower closed-loop frequency of the angular attitude response to a pilot control input, which ensures, at least, 6 dB of gain margin (ω_{BWgain}) or 45° of phase margin ($\omega_{BWphase}$) from the frequency of neutral stability (ADS-33E-PRF 2000). Essentially, ω_{BW} is a measure of the quickness at which the aircraft responds to a control input by setting the maximum input frequency that result in a useful response, from magnitude and phase point of view. In addition, as defined in ADS-33E-PRF (2000), ω_{BW} is the maximum frequency that a pilot employing pure gain control strategy reaches a valid figure of merit, without affecting the stability. When an aircraft allows a pilot to employ, as control strategy, a pure gain element which does not require phase compensation, so-called synchronous control, the workload is minimized, reaching, for certain tasks, the bests HQR. For helicopters, greater workload tasks (also referred to as high-gain tasks) are directly affected, such as precision hover, pirouette, slope landing and offset running landing (Tischler and Remple 2012).

The phase delay is determined from the frequency at which the phase is equal to –180° (ω_{180}) and the value of the phase ($\varphi_{2\omega_{180}}$) measured in this double ($2\omega_{180}$), as Eq. 2:

$$\tau_p \equiv - \left[\frac{\varphi_{2\omega_{180}} + 180°}{57{,}3 \cdot (2\omega_{180})} \right] \tag{2}$$

As one can observe in Eq. 2, because there are data accuracy and coherence in this frequency range, the parameter is the slope of the phase curve at the point where the output delays 180° with respect to the input (neutral stability). As the pilot increases the gain in a certain task, he (she) approaches the frequency at which the aircraft responds out-of-phase with respect to the input. The pilot's natural reaction is to apply, mentally, a lead filter to compensate the phase lag. The success of this technique depends on the response predictability, measured by the phase lag: greater frequency changes will cause greater phase shifts, which turns response less predictable and more susceptible to pilot induced oscillations (PIO).

The ω_{BW} and the characteristic points to obtain τ_p (ω_{180}, $\phi 2\omega_{180}$ and $2\omega_{180}$), as defined in the ADS-33E-PRF (2000) standard, are shown in Fig. 13, which represents the Bode diagrams of the ratio between the angular attitude and the input, measured as flight control displacement (δ_s) or flight control load (F_s).

Comparison with the reference values of ADS-33E-PRF (2000) standard can be done based on linear interpolation of the Bode diagrams (Figs. 11 and 12) around the desired parameters.

Furthermore, the uncertainties of the presented results are based on the 95% confidence interval in magnitude and phase. In this case, it can be noticed that, once the normalized random error is a function of coherence, to γ^2_{xy} values close to unity, smaller errors and, consequently, smaller ranges of confidence are achieved, making them practically imperceptible in the shown curves.

For the longitudinal cyclic excitation, within the frequency range in which there is acceptable coherence, it is not possible to identify the phase of $2\omega_{180}$ (6.934 rad/s), thus preventing the calculation of the phase delay, applying Eq. 2. Table 1 summarizes the characteristic parameters found in the longitudinal response of the helicopter, at level flight (80 KIAS).

Even without the information regarding the phase delay, the bandwidth frequency already provides useful knowledge on pilot demands in carrying out tasks in closed-loop, applying HQR.

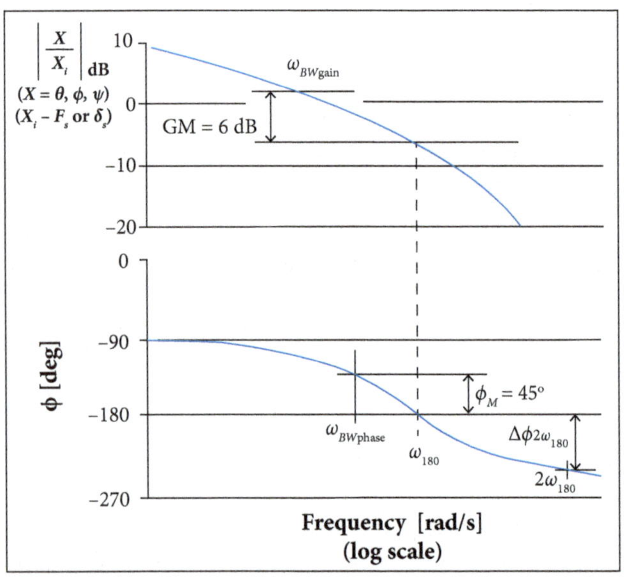

Figure 13. Definitions of ω_{BW} and τ_p (ADS-33E-PRF 2000).

Since this frequency is the maximum in which the pilot is able to close the loop by means of synchronous control and achieve adequate stability, its relationship to the cutoff frequency (ω_{co}), for a particular task, is a measure of the easiness with which the task is to be performed. Usually, tasks that require precise regulation associated with high pilot compensation gains result in increase of ω_{co}. If it exceeds ω_{BW}, it requires that the pilot imposes a phase delay control strategy to achieve the stability margin needed to close the control loop with adequate damping. Thus the bandwidth frequency is usually designed to be as high as possible, a requirement for an adequate flight control system.

For the lateral cyclic excitation, $2\omega_{180}$ (13.842 rad/s) lies within the frequency range in which the coherence is acceptable, allowing the definition of phase delay. Table 2 summarizes the ADS characteristic parameters for helicopter lateral response in level flight (80 KIAS).

Once obtained both parameters of interest, the ADS standard specifies the regions wherein the aircraft reaches a certain predicted level of flying qualities from the level 1, which is satisfactory in quickness and accuracy, not requiring modifications, to level 3, wherein the flying qualities are degraded, requiring improvements. The regions and hence the levels to be achieved, depends on the operating task to be performed

Table 1. AS355-F2 level flight (80 KIAS) longitudinal flying qualities ADS parameters.

Characteristic parameter	Frequency (rad/s)	Uncertanty (rad/s)
ω_{180}	3.467	±0.030
$\omega_{BWphase}$	1.612	±0.008
ω_{BWgain}	2.588	±0.023
ω_{BW}	1.612	±0.008

Table 2. AS355-F2 level flight (80 KIAS) lateral flying qualities ADS parameters.

Characteristic parameter	Value	Uncertanty
ω_{180}	6.921 rad/s	±0.012 rad/s
$\omega_{BWphase}$	2.013 rad/s	±0.003 rad/s
ω_{BWgain}	4.430 rad/s	±0.012 rad/s
ω_{BW}	2.013 rad/s	±0.003 rad/s
$\varphi_{2\omega_{180}}$	−231.04°	±1.1°
τ_p	0.013 s	±0.001 s

by a particular aircraft (Mission Task Elements, MTE) and the operating environment in which they will be executed, considering the visual acuity (Good Visual Environment, GVE or Degraded Visual Environment, DVE), weather condition (Visual Meteorological Conditions, VMC or instrument meteorological conditions, IMC), attention, among other factors.

The level flight pitch responses for the AS355-F2 aircraft are shown in Fig. 14 along with ADS-33E-PRF (2000) regions and flying qualities levels. To a wide range of phase delay values, results found permit to classify the aircraft as ADS level 1 only in tasks that did not demand great precision, in visual flight conditions and with all attention focused (Fig. 14b), mainly because of the low ω_{BW} obtained.

For the lateral response, involving the target acquisition and tracking maneuver (Fig. 15a), high gain task that requires precise regulation of the pilot in the control loop, the aircraft is classified as ADS level 2, meaning that its flight qualities, for tasks of this nature, exhibit deficiencies which require improvements (specifically, the high value of bandwidth frequency). The result is consistent with closed-loop pilots assessments at high gain tasks (e.g., lateral change in target acquisition), in which is commonly reported moderate workload and small amplitude and high frequency corrections in the lateral cyclic control, with occurrence of overshoots for reacquiring a target.

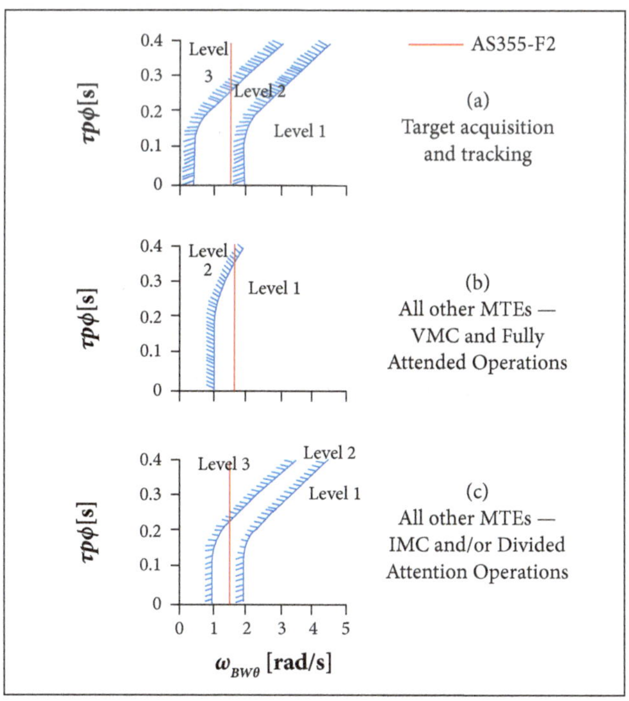

Figure 14. Flying qualities levels – AS355-F2 longitudinal response (ADS-33E-PRF 2000).

For all other tasks (Fig. 15b and c), whether in visual or instrument flight conditions, with full or partial attention, the aircraft is classified as ADS level 1, given the low phase delay value, ensuring predictability to the flight.

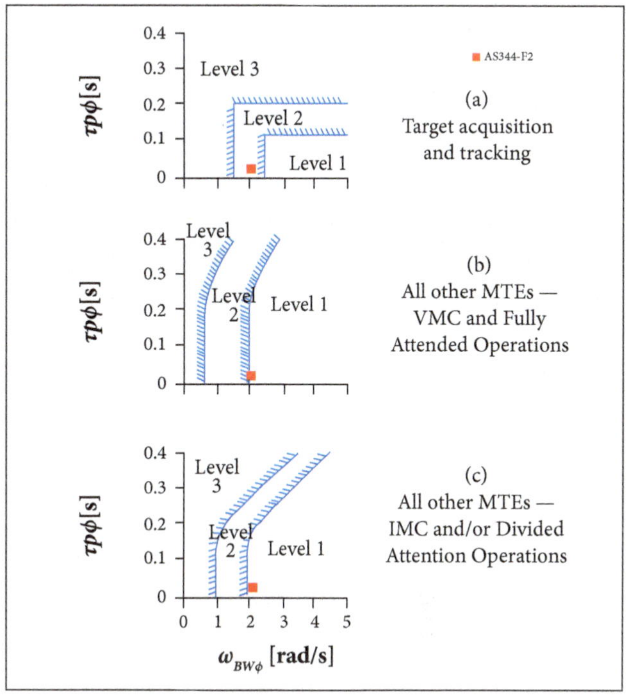

Figure 15. Flying qualities levels – AS355-F2 lateral response (ADS-33E-PRF 2000).

CONCLUSION

This work describes the key aspects of the management process to apply the frequency response methodology to characterize the dynamics of a rotary-wing aircraft, analyze its flying qualities and verify the adequacy of this aircraft to the ADS-33E-PRF (2000) standard.

Throughout the work, all the management process steps required for planning, acquisition and data analysis for frequency sweep flight tests are presented, and, when it is possible, exemplified with real flight test data from longitudinal and lateral frequency sweep, on level flight at 80 KIAS, of AS355-F2 helicopter.

It begins with the design of the instrumentation, defining the parameters that would be acquired, the need to acquire redundant signs and the sensors specification (their accuracy, sampling rates and filtering).

The second step is the definition of the appropriate maneuver characteristics for excitation. It is used sinusoidal inputs of longitudinal and lateral control frequency sweep. Additionally, one defines the optimal guidelines for the input execution as input and data recording duration, excited frequency ranges and discusses the safe aspects related to the high and low frequencies chosen.

It is also addressed the relevant factors for flight test planning to collect accurate data, from the basics of flight test doctrine to the safety factors, including division of tasks between the flight test crew.

Post-processing techniques are analyzed to proper treatment, consistency checking and accuracy verification of information acquired in flight, from simple visual inspection to application of frequency-domain techniques for compatibility analysis of acquired signals. Some of the techniques have been effectively applied to the data used as example. From kinematic consistency, it is shown that the attitude and angular rate channels, primarily used for flight qualities analysis, are consistent and showed the correct sign convention.

Finally, it is described the mathematical tools for SISO spectral analysis. It is shown the DFT implementation combined with windowing technique, which is applied to the time domain data of three test runs of longitudinal and lateral controls sinusoidal inputs concatenated, at 80 KIAS level flight. Bode diagrams along with coherence function are obtained for the complete frequency range. In the frequency range in which the coherence is acceptable one can assure good linear equivalent system describing the actual dynamic model of the aircraft by means of a non-parametric model.

The guidelines described during each management process phase permits to achieve high data quality aiming at the frequency-domain analysis of the flying qualities characteristics. The results extracted, within the range of frequencies consistently excited in flight tests, can anticipate the pilots demands during closed-loop evaluations and can be compared with the ADS-33E-PRF (2000), setting the agility level of the test aircraft.

Along the longitudinal axis, one verifies restrictions to the calculation of phase delay. Even without this information, the bandwidth frequency (1.612 rad/s) has provided useful knowledge of pilot demands in closed-loop tasks. For a wide range of phase delay values, the aircraft would be classified as ADS-33 level 1 only for tasks that did not demand tight precision, in visual flight conditions and with all attention focused, given the low ω_{BW} obtained.

For the lateral axis, both parameters of interest, phase delay and bandwidth frequency, are obtained. The aircraft has been classified as ADS-33 level 2 for target and acquisition tasks, because of its not so high bandwidth frequency (2.013 rad/s). This result is consistent with closed-loop evaluations for high gain tasks. For all other tasks, whether visual or instrument flight conditions, with partial or complete attention, the aircraft is considered level 1, part due to the high predictability consequence of the low-phase delay (0.013 s).

REFERENCES

Advisory Group for Aerospace Research and Development (1991) Rotorcraft system identification. Neuilly-sur-Seine: Advisory Group for Aerospace Research and Development. AGARD LS 178.

Bendat JS, Piersol AG (1980) Engineering applications of correlation and spectral analysis. New York: Wiley Interscience.

Brasil, Força Aérea Brasileira, Divisão de Formação em Ensaios em Voo (2003) QH-01: qualidades de pilotagem: teoria. São José dos Campos: Força Aérea Brasileira.

Cooke AK, Fitzpatrick WH (2002) Helicopter test and evaluation. Oxford: Blackwell Science.

Cooper GE, Harper Junior RP (1969) The use of pilot ratings in the evaluation of aircraft handling qualities. Moffett Field: NASA Ames Research Center. Technical Memorandum: NASA TM D-5133.

Cruz RV (2009) Desenvolvimento de um modelo dinâmico para simuladores de helicópteros (Master's thesis). São José dos Campos: Instituto Tecnológico de Aeronáutica.

Hamel PG, Jategaonkar RV (1996) Evolution of flight vehicle system identification. J Aircraft 33(1):9-28. doi: 10.2514/3.46898

Helibras (2005) Manual de voo do helicóptero AS355-F2. Itajubá: Helibras.

Nyquist H (1928) Certain topics in telegraph transmission theory. Transactions of the American Institute of Electrical Engineers 47(2):617-644. doi: 10.1109/T-AIEE.1928.5055024

Steven B, Lewis F (2003) Aircraft control and simulation. New York: Wiley Interscience.

Tischler MB, Fletcher JW, Diekmann VL, Williams RA, Cason RW (1987) Demonstration of frequency sweep testing technique using a Bell 214-ST Helicopter. Moffett Field: NASA Ames Research Center. Technical Memorandum: NASA TM 89422.

Tischler MB, Remple RK (2012) Aircraft and rotorcraft system-identification: engineering methods with flight test examples. Reston: AIAA. (AIAA Education Series).

United States Naval Test Pilot School (1995) Flight test manual: rotary wing stability and control. Patuxent River: United States Naval Test Pilot School. USNTPS-FTM No.: 107.

Williams JN, Ham JA, Tischler MB (1995) Flight test manual: rotorcraft frequency-domain flight testing. Edwards AFB: US Army Aviation Technical Test Center. AQTD Project No.: 93-14.

Detailed Numerical Study on the Aerodynamic Behavior of a Gurney Flapped Airfoil in the Ultra-low Reynolds Regime

Mahzad Khoshlessan[1], Seyed Mohammad Hossein Karimian[1]

ABSTRACT: The addition of Gurney flap changes the nature of flow around airfoil by producing asymmetric Von-Karman vortex in its wake. Most of the investigations on Gurney flapped airfoils have modeled the flow using a quasi-steady approach, resulting in time-averaged values with no information on the unsteady features of the flow. Among these, some investigations have shown that quasi-steady approach does a good job on predicting the aerodynamic coefficients and physics of flow. Previous studies on Gurney flap have shown that the calculated aerodynamic coefficients such as lift and drag coefficients from quasi-steady approach are in good agreement with the time averaged values of these quantities in time accurate computations. However, these investigations were conducted in regimes of medium to high Reynolds numbers where the flow is turbulent. Whether this is true for the regime of ultra-low Reynolds number is open to question. Therefore, it is deemed necessary to examine the previous investigations in the regime of ultra-low Reynolds numbers. The unsteady incompressible laminar flow over a Gurney flapped airfoil is investigated using three approaches; namely unsteady accurate, unsteady inaccurate, and quasi-steady. Overall, all the simulations showed that at ultra-low Reynolds numbers quasi-steady solution does not necessarily have the same correlation with the time averaged results over the unsteady accurate solution. In addition, it was observed that results of unsteady inaccurate approach with very small time steps can be used to predict time-averaged quantities fairly accurate with less computational cost.

KEYWORDS: Gurney flap, Ultra-low Reynolds number, NACA 0008 airfoil, Transient behavior, Incompressible flow, Laminar flow.

INTRODUCTION

Gurney flap (GF) is a tab of small height deployed at the trailing edge of the airfoil normal to the chord line at the pressure side. In the present paper, the effect of different numerical approaches on the aerodynamic behavior of a NACA 0008 airfoil equipped with a GF in laminar incompressible flow is investigated. GF is a simple device previously used on race cars, which is found to be useful in terms of providing the required downforce. Liebeck (1978) conducted an experimental study of the GF on a Newman airfoil in Douglas Long Beach low-speed wind-tunnel at Reynolds number based on cylinder diameter (Re_D) of 3×10^6. The results showed that the GF would provide higher value of maximum lift coefficient (C_L), increased lift at a specific angle of attack, and reduced drag at a specific C_L. The author hypothesized that the flow is composed of 2 counter-rotating vortices at the back side of the GF and a recirculating region upstream the flap. Later, this flow feature was also observed at low Reynolds number in a water tunnel by Neuhart and Pendergraft (1988).

Li *et al.* (2002), Myose *et al.* (1996), and Giguere *et al.* (1997) studied the effects of GF on the aerodynamic performance of various airfoils. Their results showed that the effect of GF is to substantially increase the maximum C_L and to reduce the stall angle. In addition, the zero-lift angle of attack becomes more negative with the increase in GF height. Overall, their results suggest that the effect of the GF is to increase the effective camber of the airfoil. Jeffrey *et al.* (2000) conducted an experimental study

1.Amirkabir University of Technology – Aerospace Engineering Department – Center of Excellence in Computational Aerospace Engineering – Tehran/Iran.

Author for correspondence: Mahzad Khoshlessan | Amirkabir University of Technology – Aerospace Engineering Department – Center of Excellence in Computational Aerospace Engineering | 424, Hafez Ave. | 15875-4413 – Tehran/Iran | Email: mahzadkhoshlessan@gmail.com over a single-element wing using laser Doppler anemometry along with pressure and force measurements. Based on the study performed by these authors, the time-averaged flow downstream a GF consists of 2 counter-rotating vortices, but the instantaneous flow structure actually consists of a wake of alternately shed vortices. Troolin *et al.* (2006) investigated the wake structure using time-resolved particle image velocimetry. They showed that 2 distinct vortex-shedding modes exist in the wake. The dominant mode resembles a Karman vortex street behind an asymmetric bluff body. The second mode is caused by the intermittent shedding of recirculating region upstream the flap, which becomes more coherent with the increase in the angle of attack. Manish *et al.* employed a compressible Reynolds-averaged Navier-Stokes (RANS) solver using Baldwin-Lomax turbulence model for the prediction of flow over NACA 0011 and NACA 4412 airfoils with different GF heights. All the computations in their study for NACA 4412 airfoil were performed for free-stream Mach number of 0.20 and chord Reynolds number of 2.0 million; those for NACA 0011 airfoil were carried out for Mach number of 0.14 and chord Reynolds number of 2.2 million.

They carried out their computations with the assumption that the flow is steady. The computed force coefficients had a good correlation with the experiment. But it was observed that the difference between numerical and experimental results increases with angle of attack and flap height. Ashby (1996) conducted experimental and computational studies to investigate the effects of lift-enhancing tabs on a multi-element airfoil. All of the computations were obtained using 2-D, incompressible RANS solver in the steady-state mode implementing the Spalart-Allmaras turbulence model. Good agreement was achieved between computed and experimental results, particularly for cases where no flow separation exists on the flap upper surface. However, poor agreement was obtained for configurations where significant flow separation exists over the flap upper surface. Overall, the computed results predicted all of the trends observed in the experimental data quite well.

Date and Turnock (2002) carried out a detailed computational study on the quasi-steady and periodic performance of NACA 0012 airfoil equipped with GFs with a flap height-to-chord ratio of 2 and 4% using the incompressible RANS solver and implementing the standard k-ε turbulence model at Reynolds number of 0.85×10^6. They showed that the time-averaged performance is identical to the performance predicted by the quasi-steady solution, showing the same correlation with the experimental data. According to their investigation, when periodic performance of airfoil is of secondary importance, the quasi-steady solution can be used to obtain estimates of airfoils, mean performance for practical purposes. According to the study performed by Date and Turnock (2002), quasi-steady approach has the same prediction of mean performance as time accurate computations.

Based on the experimental studies in the literature regarding Gurney-flapped airfoils, the addition of flap originates vortex shedding, and the flow becomes unsteady (Jeffrey *et al.* 2000; Troolin *et al.* 2006). As a result, all computations should be performed in unsteady mode. Nevertheless, many numerical computations assume a steady flow and use steady-state computations for the prediction of the flow (Singh *et al.* 2007; Ashby 1996; Jang *et al.* 1992; Ross *et al.* 1995). To the best of our knowledge, all of the studies in the literature are performed at medium-to-high Reynolds number flow regimes. However, less attention has been paid to the performance of GF at ultra-low Reynolds number (Re below 10,000) flow regime where the boundary layer, wake characteristics, and other flow features are different.

The regime of ultra-low Reynolds number has recently received special interest and has been the subject of many computational and experimental researches. According to the studies conducted in this flow regime, it has many applications in the analysis of aerodynamics at small physical scales (insect-sized aircraft), especially flight of Micro-Air Vehicles (MAVs). The choice of airfoil plays an important role in the design of aircraft suitable for low-speed aerodynamics. Also the payload, range, and endurance of the low-speed aircraft are limited by the performance of their high-lift systems (Storms and Jang 1994). Some investigations conducted a parameter study for analysis and design of airfoils suitable for ultra-low Reynolds numbers. Kunz (2003), in their computational study of ultra-low Reynolds numbers (between 1,000 and 6,000), developed an automated optimal design tool for 2-D airfoils at ultra-low Reynolds numbers and a hybrid method for the automated design and analysis of micro-rotors. Abdo (2004), in their study, presented an efficient numerical method for the incompressible flows past airfoils at low Reynolds numbers. Their analysis was based on a pseudo-time integration method using artificial compressibility to accurately solve the Navier-Stokes equations. They obtained solutions for airfoils at various incidences and very-low Reynolds numbers between 400 and 6,000. Furthermore, the generation of high lift can lead to steeper take-off ascent, increase in the climb lift-to-drag ratio, and fuel economy (Tejnil 1996). Therefore, the choice of an efficient high lift device would also be of great importance. GF is a good candidate to maintain high-lift coefficient required for approach and landing (Storms and Jang 1994) and is a mechanically-simple and cost-effective device in contrast with other complex multi-element high-lift ones, being used to control the flow in combination with other elements (Schatz *et al.* 2004). Thus, the motivation for this study was: (a) to model a Gurney-flapped airfoil at ultra-low Reynolds flow regime, since the understanding of aerodynamic behavior of the airfoil at this Reynolds number would be helpful for the design of low-speed aircraft; and (b) to investigate whether steady-state solution can still be used to predict mean performance of unsteady flow around Gurney-flapped airfoils at ultra-low Reynolds regime.

To this end, the Mass Corrected Interpolation Method (MCIM) algorithm of Alisadeghi and Karimian (2010) is employed for the solution of flow over a NACA 0008 Gurney-flapped airfoil. This method is one of the most robust ones for incompressible flow calculation that provides an efficient and accurate tool for solving practical engineering problems (Alisadeghi and Karimian 2010) and has been previously validated for internal flows (Alisadeghi and Karimian 2010), including Taylor problem, inviscid converging-diverging nozzle, and the lid-driven cavity, as well as for external flows (Khoshlessan *et al.* 2013; Khoshlessan 2013), including steady and unsteady flows over a circular cylinder section and unsteady flow over a NACA 0012 airfoil.

Three implicit time-marching approaches are implemented for the prediction of flow over a Gurney-flapped airfoil: (a) unsteady accurate, which refers to the solution in which linearization iterations are performed at each time step (Δt); (b) unsteady inaccurate, which refers to the solution in which no linearization iteration is performed at each Δt; and (c) quasi-steady, which refers to the solution in which the transient term is eliminated from the Incompressible Navier-Stokes (INS) equations, *i.e.* very-large time steps are used. Grid refinement and time-step independence studies are conducted to ensure the generation of accurate solution. Results obtained from these approaches are studied in detail and compared with each other. Moreover, the effect of linearization convergence criterion (e) on the time accurate solution and the sensitivity of all calculated parameters to the other values of e are investigated. As it will be seen, the present paper concludes that, in contrast to high Reynolds numbers flow, the quasi-steady solution cannot be used to correctly predict the mean aerodynamic behavior of a Gurney-flapped airfoil at ultra-low Reynolds numbers. It will be also shown that the unsteady inaccurate approach with very-small time steps can be used to fairly predict the time-averaged quantities accurately with 1/3 of computational cost needed for unsteady accurate calculations.

In addition, the governing equations are implicitly formulated and solved using direct methods. Two different direct solvers are examined here to study their effect on the required memory and speed up of the solution.

NUMERICAL ALGORITHM AND GOVERNING EQUATIONS

In this study, the MCIM (Alisadeghi and Karimian 2010) is employed for solving 2-D incompressible flows on structured grid systems. The present algorithm can be classified among the collocated grid schemes in control-volume-based methods in which 2 different velocity components at cell faces are defined to suppress the possible checkerboard problem in the solution domain. In the present approach, conservation equations are formed for each control volume throughout the solution domain. In addition, a fully-coupled formulation is used. The linear system of equations obtained in these algorithms is solved using a direct sparse solver. The transient term is modeled using a first-order backward difference in time. Convective term is linearized by lagging the stream-wise velocity and modeled so that it would cause the upstream value of velocity u_{up} to affect u_{ip} strongly, when convection is dominant. Also, the convective term is modeled using a first-order backward differencing in the upstream direction.

The computational domain is discretized into a number of quadrilateral elements, and all primitive variables are located at the vertices of these elements. A local non-orthogonal coordinate system (s,t) is defined inside each element. Medians within the element divide it into 4 sub-control volumes (SCVs), as shown in Fig 1a. Each SCV is associated with an elemental node at its vertex, and each node is surrounded by 4 SCVs that form a control volume for the given node. Sub-control surfaces (SCSs) are the representative of boundary surfaces of the SCVs fallen inside the element, and the integration points (i_p) are the representative of the midpoint of the SCSs. Diffusion fluxes and pressure terms at the integration points are related to the nodal variables using bi-linear interpolation. Mass and the convection fluxes at the i_p should be also related to the nodal values of velocity and pressure.

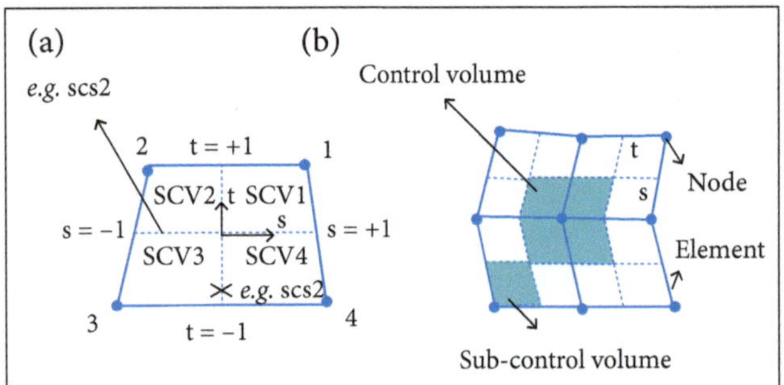

Figure 1. Definition of (a) element and (b) control volume and sub-control volume.

All primitive variables are located at the vertices of the elements and should be evaluated at the i_p to determine the flux at each SCS. Using the linear interpolation, any variable and its gradients within the element can be calculated. More details of the method by which these fluxes are defined at the i_p and the formation of the system of linearized equations and time-marching scheme can be found in Karimian and Schneider (1995, 1994) and Karimian (1994). Governing equations are the INS equations, and its integral form for SCV2 in Fig. 1, for instance, is given by:

$$\int_{SCV2} \frac{\partial Q}{\partial t} dV + \int_{SCS1\&2} (E ds_x + F ds_y) = \int_{SCS1\&2} (G ds_x + H ds_y), \tag{1}$$

where: dV represents volume differential; ds_x and ds_y represent components of the outward vector normal to the surface; SCS denotes the inner sub-control surface associated with SCV2; Q, E, F, G, and H are defined as follows:

$$Q = \begin{pmatrix} 0 \\ \rho u \\ \rho v \end{pmatrix}, \quad E = \begin{pmatrix} (\rho u) \\ (\rho u)u + p \\ (\rho u)v \end{pmatrix}, \quad F = \begin{pmatrix} (\rho v) \\ (\rho v)u \\ (\rho v)v + p \end{pmatrix},$$

$$G = \begin{pmatrix} 0 \\ \tau_{xx} \\ \tau_{xy} \end{pmatrix}, \quad H = \begin{pmatrix} 0 \\ \tau_{xy} \\ \tau_{yy} \end{pmatrix} \tag{2}$$

$$\tau_{xx} = 2\mu \frac{\partial u}{\partial x}, \quad \tau_{yy} = 2\mu \frac{\partial v}{\partial y}, \quad \tau_{xy} = \tau_{yx} = \mu\left(\frac{\partial u}{\partial y} + \frac{\partial v}{\partial x}\right)$$

where: ρ is the density; u is the convected stream-wise velocity; v is the convected transverse velocity; μ is the dynamic viscosity.

The volume integral of the transient term is modeled using a first-order backward difference in time. Any integral argument, *i.e. E, F, G,* and *H*, is approximated by its average value over the SCS, which is evaluated at its midpoint location (i_p). For diffusion flux vectors *G* and *H*, bi-linear interpolation is used to directly evaluate the components of stress tensor (Karimian and Schneider 1995, 1994; Karimian 1994). For convective flux vectors *E* and *F*, pressure is evaluated using bi-linear interpolation, and the momentum fluxes are linearized with respect to mass fluxes (ρu) and (ρv). Velocity components u and v in mass fluxes are called integration point mass-conserving velocities, being denoted by a hat, *i.e.* $\rho\hat{u}$ and $\rho\hat{v}$. Other values of u and v in the momentum fluxes, which are convected by the mass fluxes through the control-volume surface, are called integration point convected velocities. Mass-conserving and convected velocities are cell-face velocities. Details of the modeling of the cell-face velocities are represented in Alisadeghi and Karimian (2010) in detail. It should be noted that, since Navier-Stokes equations are non-linear by nature, and in order to have a linear system of equations, convection terms in momentum equations are linearized by lagging the stream-wise velocity. This means that the mass fluxes (ρu and ρv) in convection term are calculated from previous iteration, which introduces an error to the solution. Therefore, it is necessary to perform the linearization iteration at each Δt in order to reduce the linearization error to the lowest possible amount. The ε used to stop the linearization iteration process is given by Eqs. 3 and 4, in which z_j is the jth unknown variable. For all of the solutions over a circular cylinder shown here, ε is set equal to 0.1%.

$$\left(\frac{\sum_{j=1}^{3} \sum_{i=1}^{N} \left| \frac{(z_j)_i - (z_j)_i^{o}}{Z_j} \right|^2}{3 \times N} \right)^{0.5} < \varepsilon \tag{3}$$

where:

$$Z_j = \left(\frac{\sum_{i=1}^{N} \left| (z_j)_i \right|^2}{N} \right)^{0.5} \tag{4}$$

Three approaches are implemented for all the simulations performed in the present study. Comparison between these approaches are made for the case of Gurney-flapped airfoil. These approaches are as follows:

- Unsteady accurate solution (UAS), in which linearization iterations are performed at each Δt. The ε used to stop linearization iteration process is defined by Eqs. 3 and 4.
- Unsteady inaccurate solution (UIS), in which no linearization iteration is performed at each Δt.
- Quasi-steady solution (QSS), in which the transient term is eliminated from the INS equations and is considered equivalent to UIS, with $\Delta t \rightarrow \infty$.

BOUNDARY CONDITIONS

Velocity inlet boundary condition is specified at the inflow boundary, *i.e.* $u = U_{\infty}cos\alpha$ and $v = U_{\infty}sin\alpha$, where U_{∞} is the free stream velocity. Pressure outlet boundary condition is specified at the outflow boundary, *i.e.* $\partial v/\partial s = 0$ and $p_{outflow} = P_{\infty}$. No-slip boundary condition is employed at all walls, *i.e.* velocities are set equal to 0 on all boundary nodes. GF is considered as a single flat plate without thickness; on its both sides, no-slip boundary condition is applied.

IMPLEMENTATION OF BAND SOLVER AND DIRECT SPARSE SOLVER

As mentioned before, in the present numerical study, the governing equations are implicitly formulated, which leads to the formation of a system of linearized equations that should be solved at each Δt (UAS approach). Since the number of unknown variables is typically very large, solving the resulted coefficient matrix has a significant impact on computational time. To reduce the amount of required memory and to speed up the convergence rate in the iterative process for the solution of the system of equations, the following tasks and their effects are examined:

- Implementation of band solver.
- Implementation of a direct sparse solver.

The details about the band solver are not given here and can be found in Khoshlessan *et al.* (2013) and Khoshlessan (2013). Only Sparse solver is discussed, as it is used in all simulations. It should also be noted that all the simulations are executed in serial mode in the present study.

DIRECT SPARSE SOLVER

Note that the main part of the present study is dedicated to unsteady flows. Therefore, very-fine grids are required to predict time evolution of the wake and the flow periodicity. As a result, the number of unknown variables will be very large, which results in a very large coefficient matrix. This leads to a noticeable increase in both CPU time and required memory, especially when extremely intensive computations are necessary to be carried out with small time steps. To handle such cases, a direct sparse solver called BLOCKPACK (Karimian 1994) is implemented instead of the band solver to solve the renumbered band matrix. In order to test the efficiency of the sparse solver, the case of unsteady flow over a circular cylinder at $Re_D = 100$ is examined. Table 1 shows the comparison of CPU time taken to complete 7,500 time steps to reach the solution at $t = 7.5$ s using band solver and direct sparse solver.

As illustrated in Table 1, the average time needed for each linearization iteration is reduced to 20 s for sparse solver as compared to 9 min for band solver. Thus, the amount of CPU time taken to complete 7,500 time steps using sparse solver is 1/27 of that of the band solver (41.67/1,125). As can be seen, the implementation of direct sparse solver has led to significant speed up in convergence rate and saving in computational time.

ALGORITHM VALIDATION

For the validation of the present algorithm, 2 test cases are studied. In both, cases results are compared against other numerical and experimental results available in the literature (Berthelsen and Faltinsen 2008; Russel and Wang 2003; Linnick and Fasel 2005; Herfjord 1996; Calhoun 2002; Medjroubi 2011; Coutanceau and Bouard 1977; Tritton 1959). All the simulations in validation section are performed using the UAS approach. The first test case is steady flow over a circular cylinder at $Re_D = 40$. Grid refinement study is conducted to ensure that there is enough near wall resolution, and the obtained results are accurate. The numerical grid used for the simulation is an

Table 1. Comparison of CPU time taken to reach periodic state of unsteady flow over a circular cylinder at $Re_D = 100$ with $\Delta t = 10^{-3}$ s using band solver and direct sparse solver.

Solver	Average time needed for each linearization iteration	Average number of linearization iterations in each Δt	CPU Time (h) needed to complete 7,500 time steps to reach the solution at $t = 7.5$ s
Band	9 min	1	1,125
Sparse	20 s	1	41.67

O-type grid topology, composed of quadrilateral elements. The cylinder is located at $(x, y) = (D/2, 0)$, where D is the cylinder diameter. The far-field boundary is located at 20 chords (C) away from the center of the cylinder in all computations. In this case, grid refinement study is performed based on 2 independent radial (N_r) and angular (N_θ) directions, and the sensitivity of the calculated drag force with change in grid resolution is studied. Three different mesh counts with sizes of 220 × 40, 220 × 60, and 220 × 80, in N_r, and with sizes of 180 × 60, 220 × 60, and 260 × 60, in N_θ, are used for this purpose. Drag coefficient (C_D) calculated from the steady-state solution and the corresponding relative error (E) on these grids are reported in Tables 2 and 3, which are defined as:

$$\Delta_{j,j-1} = \left|C_{D_j} - C_{D_{j-1}}\right| \tag{5}$$

$$E_{i,j-1} = \Delta_{i,j-1}/C_{D_j} \tag{6}$$

Table 2. Angular grid independence study at Re_D= 40 (N_θ = 220).

Grid	N_r	C_D	$E_{i,i-1}$ (%)
1	40	1.5513	--
2	60	1.5505	0.0516
3	80	1.5502	0.0194

Table 3. Angular grid independence study at Re_D= 40. (N_r = 60)

Grid	N_θ	C_D	$E_{i,i-1}$ (%)
1	180	1.5505	--
2	220	1.5483	0.1421
3	260	1.5478	0.0323

In Tables 2 and 3, indices 1, 2, and 3 refer to large, medium, and fine grids used for grid independence study. In Table 2, E_{21} and E_{32} of C_D are 0.05 and 0.02, respectively, which are very low. The grid refinement study provides details of fine mesh (220 × 80) compared to medium (220 × 60) and coarse (220 × 40) meshes. Mesh refinement study performed on the coarse, medium, and fine grids on CD shows very little variation when increasing mesh size, leaning towards their limiting trends. Therefore, 60 nodes in radial direction would be the best choice here. The first grid node should be located at 0.00044D away from the cylinder surface for the mesh to be able to capture details of the flow field with this number of grid points. With the same argument, one can conclude that, based on grid refinement study for fine (260 × 60), medium (220 × 60), and coarse (180 × 60) meshes shown in Table 3, the medium grid (220 × 60) is enough tin in the angular direction. Therefore, the best grid to obtain grid independent results in this case is the medium one.

In the next step, the present results are compared with others reported in the literature (Berthelsen and Faltinsen 2008; Russel and Wang 2003; Linnick and Fasel 2005; Herfjord 1996; Calhoun 2002; Medjroubi 2011; Coutanceau and Bouard 1977; Tritton 1959). Drag force and geometrical parameters of the steady flow structure around a circular cylinder, defined in Fig. 2 and calculated on the 220 × 60 grid, are shown in Table 4, where they are compared with other experimental and numerical

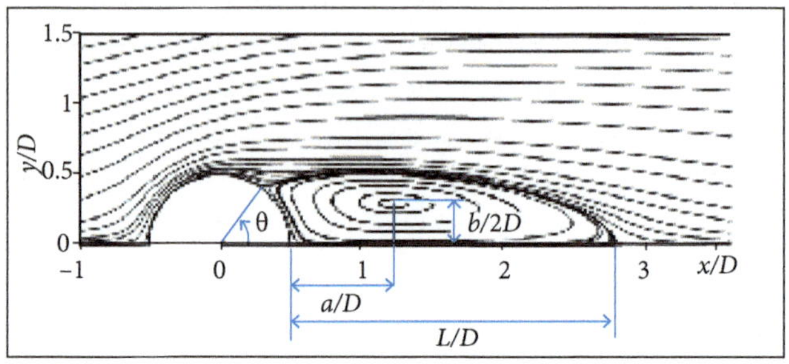

Figure 2. Nomenclature of the geometrical parameters for the flow over a circular cylinder at Re_D = 40 (Berthelsen and Faltinsen 2008).

Table 4. Comparison of geometrical parameters for the flow over a motionless cylinder at Re_D = 40.

Reference	L/D	a/D	b/D	θ(°)	C_D
Berthelsen and Faltinsen (2008)	2.29	0.72	0.6	53.9	1.59
Russel and Wang (2003)	2.29	--	--	53.1	1.6
Linnick and Fasel (2005)	2.28	0.72	0.6	53.6	1.54
Herfjord (1996)	2.25	0.71	0.6	51.2	1.6
Calhoun (2002)	2.18	--	--	54.2	1.62
Medjroubi (2011)	2.21	0.7	0.59	53.8	1.6
Coutanceau and Bouard* (1977)	2.13	0.76	0.59	53.5	--
Tritton* (1959)	--	--	--	--	1.57
Average of other numerical results	2.25	0.7125	0.5975	53.3	1.592
Error of the average of other numerical results with respect to experiment (%)	5.6	6.6	1.3	0.4	1.4
Present study	2.25	0.72	0.6	53.8	1.55
Present study with respect to experiment (%)	5.6	5.3	1.7	0.56	1.3

*Experimental results.

results (Berthelsen and Faltinsen 2008; Russel and Wang 2003; Linnick and Fasel 2005; Herfjord 1996; Calhoun 2002; Medjroubi 2011; Coutanceau and Bouard 1977; Tritton 1959). In order to compare the efficiency of the present method concerning other numerical methods, the average of error in other numerical results is calculated. Thus, the error of the average of other numerical results as well as the error of the present data, with respect to the experiment (%), are compared. As can be seen, the relative error of the present results for L/D, a/D, b/D, θ/D, and C_D are 5.6; 5.3; 1.7; 0.56 and 1.3, respectively, with respect to the experiment, where L is the length of the recirculation region; a is the length of the center of recirculation region; b is the width of the center of recirculation region; θ is the separation angle. These errors are less than or about the same amount as compared to the relative error in the average of other numerical results, *i.e.* 5.6; 6.6; 1.3; 0.4 and 1.4 with respect to experiment. Thus, it can be concluded that the present algorithm predicts the details of the steady external flows over 2-D bodies very well as compared to other numerical results.

A second test case is performed to validate the present algorithm for the solution of unsteady flow that, over a circular cylinder at $Re_D = 100$, at which Von-Karman street is produced in its wake, is used for this purpose. The same grid of 220 × 60 points is adopted for the solution of the unsteady flow based on the grid study performed in previous section. This case is solved with different time steps to ensure Δt independent solution. During the transient regime, solution linearization iterations are performed at each Δt (refer to Eqs. 3 and 4 for definition). Solution is continued until the fully periodic state is achieved. Strouhal number (St), mean drag coefficient ($C_{D,m}$), amplitude drag coefficient ($C_{D,a}$), and amplitude lift coefficient ($C_{L,a}$) are extracted from the numerical results obtained from the present solution. Time steps of 10^{-3}; 10^{-4}, and 5×10^{-5}s are considered here, and the corresponding results are shown in Table 5. The change in time step from 10^{-3} to 10^{-4} s has altered St, $C_{D,m}$, $C_{D,a}$, and $C_{L,a}$ by 1.93; 0.44; 3.25 and 1.12%, respectively, while a decreasing time step from 10^{-4} to 5×10^{-5} s has changed St, $C_{D,m}$, $C_{D,a}$, and $C_{L,a}$ by only 0.175; 0.004; 0.165 and 0.35%, respectively. Relative error in the change of parameters listed in Table 5 from time step of 10^{-4} to 5×10^{-5} s are less than 1%. If a relative error of less than 1% is fine enough, then the time step of $\Delta t = 10^{-4}$ s is an appropriate value to accurately capture unsteady flow field around a circular cylinder at $Re_D = 100$.

The results are compared with those of the other numerical or experimental studies (Russel and Wang 2003; Calhoun 2002; Ding *et al.* 2007; Liu *et al.* 1998; Xu and Wang 2006; Williamson 1989) available in the literature in Table 6. In order to compare the efficiency of the present method as compared to other numerical methods, the average of error in other numerical results is calculated. The maximum relative error of the other numerical results with respect to their average (%) and the error of present data with respect the average of other numerical results

Table 6. Unsteady flow past circular cylinder at Re_D=100.

Reference	St	$C_{D,m}$	$C_{D,a}$	$C_{L,a}$
Ding *et al.* (2007)	0.166	1.356	0.01	0.287
Russel and Wang (2003)	0.169	1.38	0.007	0.3
Liu *et al.* (1998)	0.165	1.35	0.012	0.339
Calhoun (2002)	0.175	1.33	0.014	0.298
Xu and Wang (2006)	0.171	1.42	0.013	0.34
Williamson** (1989)	0.164	--	--	--
Average of other numerical results	0.1692	1.3672	0.0112	0.3128
Max. relative error in the above results with respect to average listed above (%)	3.42	3.8	37.5	8.69
Error of the average of numerical results with respect to experiment (%)	3.2	--	--	--
Present study (Δt=10-4s)	0.1711	1.37776	0.0123	0.2868
Error* (%)	1.12	0.77	9.8	8.3
Present study with respect to experiment (%)	4.3	--	--	--

*This is error of the results of the present study (Δt = 10^{-4} s) with respect to the average of results listed above; **Experimental study.

Table 5. Time step independence study for unsteady flow past circular cylinder at Re_D= 100.

Case	Present study	St	$E_{i,i-1}$(%)	$C_{D,m}$	$E_{i,i-1}$ (%)	$C_{D,a}$	$E_{i,i-1}$(%)	$C_{L,a}$	$E_{i,i-1}$ (%)
1	($\Delta t = 10^{-3}$ s)	0.1678	--	1.372	--	0.0127	--	0.2900	--
2	($\Delta t = 10^{-4}$s)	0.171	1.93	1.378	0.44	0.0123	3.25	0.2868	1.12
3	($\Delta t = 5 \times 10^{-5}$s)	0.171	0.175	1.378	0.0	0.0121	0.165	0.2858	0.35

are compared. As can be seen in Table 6, the relative errors of the present results for *St*, $C_{D,m}$, $C_{D,a}$, and $C_{L,a}$ with respect to the average of other numerical results (Russel and Wang 2003; Calhoun 2002; Ding *et al.* 2007; Liu *et al.* 1998; Xu and Wang 2006; Williamson 1989) (1.12; 0.77; 9.8 and 8.3) are less than the maximum relative error in other numerical results (Russel and Wang 2003; Calhoun 2002; Ding *et al.* 2007; Liu *et al.* 1998; Xu and Wang 2006; Williamson 1989) and their average values (3.42; 3.48; 37.5 and 8.69).

Again, from the results obtained here, one can conclude that the present algorithm predicts excellently the details of external unsteady flows over 2-D bodies as well. The relative error of *St* for the present results and the average of other numerical results (Russel and Wang 2003; Calhoun 2002; Ding *et al.* 2007; Liu *et al.* 1998; Xu and Wang 2006; Williamson 1989) are 4.3 and 3.2, respectively with respect to experimental data (Table 6). The comparison presented in Table 6 shows that the present algorithm does a good job in predicting the *St* and aerodynamic coefficients for unsteady flow over a circular cylinder.

RESULTS

In this section, the MCIM algorithm of Alisadeghi and Karimian (2010) validated before for external flows over 2-D bodies is used for the analysis of laminar flow over a NACA 0008 airfoil fitted with GF. All of the simulations presented herein are performed by solving the INS equations with the assumption that the flow is laminar over the entire domain. Steady solutions are obtained when aerodynamic forces converge to their final value. Unsteady solutions are obtained after several periodic cycles, where the force coefficient time series reach a periodic state. Grid and time refinement studies are performed to ensure independent solutions. Grid refinement study is first performed in order to find proper number of grid points in different directions of the computational grid. Discussions on grid refinement study are given in the following subsection.

Numerical grid used for the simulation is a structured C-type grid topology. Gurney flapped NACA 0008 airfoil is located at $(x,y) = (C/2,0)$, and the far-field boundary is located at $20C$ away from the center of the airfoil in all directions. Uniform flow over the airfoil is from left to the right in the x-direction, and the entire domain is initialized using the undisturbed uniform flow at the far-field inlet boundary condition.

GRID REFINEMENT STUDY

Grid refinement study is conducted on a Gurney flapped airfoil with a flap height of $H_f = 4\%C$, at Reynolds number based on chord length (Re_C) of 2,000 and angle of attack $\alpha = 4°$. Grid resolution is changed in different regions of computational domain independently, *i.e.* inside the flap region (N_f), in the radial direction (N_η), along the chord (N_ξ), and in the wake (N_ζ) to ensure enough grid points in each direction. This will be of great importance to capture physics of the flow accurately. Figure 3 shows the schematic diagram of the computational grid around a Gurney flapped airfoil.

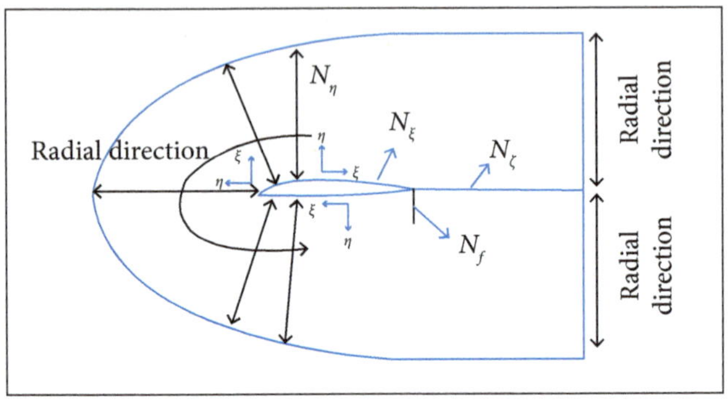

Figure 3. Schematic diagram of the computational grid.

It is important to note that the computational time of the unsteady accurate solution approach is much higher than that of the quasi-steady approach, due to the linearization iterations, which should be performed at each Δt in unsteady accurate solution approach. Therefore, quasi-steady approach is used in all simulations carried out in this section to avoid the extremely-intensive computational burden due to time accurate computations. Thus, the quasi-steady values of lift and drag coefficients are the parameters which their changes due to the grid refinement will be monitored in this study. In each test case, the number of grid points is changed in one direction, while the number of grid points in the other directions is kept constant. Grid refinement study is of great importance in order to reach an independent and accurate solution. Calculated force coefficients along with their relative errors computed on 12 grids with different resolutions are considered next. Note that, in the subsequent sections, the error of the lift coefficient (C_L) is calculated similarly to error of C_D, as defined in Eqs. 3 and 4.

Flap Grid Independence Study

As seen in Table 7, the relative error E_{21} of C_D and C_L is 0.3694 and 1.1894%, respectively, from the coarse grid to the medium one, and E_{32} of C_D and C_L is 0.1353 and 0.0744%, respectively, from the medium grid to the finest one. From the coarsest grid to the finest one, the solution changes, and relative errors of both C_D and C_L are reduced, leaning toward their limiting trend. Therefore, N_f=30 would be the best choice here as the baseline resolution for all subsequent computations.

Table 7. Flap Grid Independence Study with H_f/C = 4% at Re_C = 2,000 and α = 4°.

Grid	N_f	$N_\eta = 70, N_\xi = 100, N_\zeta = 80$			
		Lift data		Drag data	
		C_L	$E_{i,i\text{-}1}$	C_D	$E_{i,i\text{-}1}$
1	20	0.5445	--	0.08092	--
2	30	0.5381	1.1894	0.08122	0.3694
3	40	0.5377	0.0744	0.08133	0.1353

Radial Grid Independence Study

As seen in Table 8, relative errors of C_D and C_L are very small. The relative error E_{32} of C_D and C_L is approximately 0.012 and 0.074% from the medium grid to the finest one, and E_{21} of C_D and C_L is approximately 0.037 and 0.17% from the coarse grid to the medium one. Based on these data, one can conclude that C_D and C_L have reached their limiting values. Therefore, N_η = 70 would be the best choice here as the baseline resolution for all subsequent computations.

Table 8. Radial grid independence study with H_f/C = 4% at Re_C = 2,000, and α = 4°.

Grid	N_η	$N_f = 30, N_\xi = 100, N_\zeta = 80$			
		Lift data		Drag data	
		C_L	$E_{i,i\text{-}1}$	C_D	$E_{i,i\text{-}1}$
1	60	0.539	--	0.08119	--
2	70	0.5381	0.1673	0.08122	0.0369
3	80	0.5377	0.0744	0.08123	0.0123

Chord Wise Grid Independence Study

As can be seen in Table 9, the relative error E_{32} of C_L is approximately 0.1485% from the medium grid to the finest one, and E_{21} of C_L is approximately 0.1858% from the coarse grid to the medium one, which shows that C_L has reached its limiting value. Regarding the relative error of C_D, we believe that, although E_{32} (0.0002) is larger than E_{21} (3×10^{-5}), both are small enough to accept that C_D has also reached to its limiting value, as can be seen in Table 9. Therefore, selection of the medium grid with N_ξ = 100 is an appropriate choice in this case.

Table 9. Chord wise grid independence study with H_f/C = 4% at Re_C = 2,000, and α = 4°.

Grid	N_ξ	$N_f = 30, N_\eta = 70, N_\zeta = 80$			
		Lift data		Drag data	
		C_L	$E_{i,i\text{-}1}$	C_D	$E_{i,i\text{-}1}$
1	80	0.5371	--	0.08119	--
2	100	0.5381	0.1858	0.08122	0.0369
3	120	0.5389	0.1485	0.08142	0.2456

Wake Grid Independence Study

As seen in Table 10, the relative errors of C_D and C_L are very small. The relative error E_{32} of C_D and C_L is approximately 0 and 0.0186% from the medium grid to the finest one, and E_{21} of C_D and C_L is approximately 0.0123 and 0.0186% from the coarse grid to the medium one. Based on these data, one can conclude that C_D and C_L have reached their limiting values. Therefore, the selection of the medium grid with N_ζ = 90 is an appropriate choice in this case.

Finally, based on the error reduction achieved through grid refinements, a grid with N_f= 30, N_η = 70, N_ξ= 100, N_ζ= 90 is selected for all the subsequent solutions of Gurney flapped airfoil. This grid is based on the medium cases which yield lift and drag coefficients with very small errors (within 0.2%) from the finest grids. Figure 4a shows the C-grid topology used for the computations, and Fig. 4b presents details of numerical grid used for the simulations in the flap region.

Table 10. Wake grid independence study with H_f/C = 4% at Re_C = 2,000, and α = 4°.

Grid	N_ζ	$N_f = 30, N_\eta = 100, N_\xi = 80$			
		Lift data		Drag data	
		C_L	$E_{i,i\text{-}1}$	C_D	$E_{i,i\text{-}1}$
1	80	0.5381	--	0.08122	--
2	90	0.5382	0.0186	0.08121	0.0123
3	100	0.5383	0.0186	0.08121	0

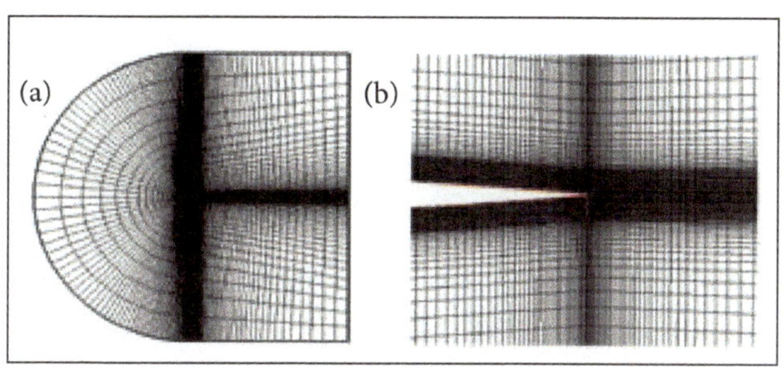

Figure 4. Details of computational grid (a) around the section and (b) inside the flap region.

SIMULATIONS OF FIXED GURNEY FLAPPED AIRFOIL

In the present section, the performance of the quasi-steady approach in comparison with unsteady accurate approach will be evaluated in the regime of ultra-low Reynolds numbers. Different approaches for the prediction of flow over a Gurney flapped airfoil will be investigated and compared with each other.

Simulations are carried out for a NACA 0008 airfoil with a flap to chord ratio of $H_f/C = 4\%$ at angles of attack of 3°, 4°, and 7° and at Reynolds number of 2,000 using the 3 approaches introduced in the section "Implementation of band solver and direct sparse solver".

Parameters under consideration in the present study are *St*, mean lift coefficient ($C_{L,m}$), $C_{D,m}$, $C_{L,a}$, and $C_{D,a}$, reported in Tables 11 to 13. All the calculated parameters are obtained from C_L and C_D time series, once the computed flow field has converged to a periodic solution. The linearization convergence criterion used to stop linearization iteration process for all the calculations summarized in Tables 11 to 13 is $\varepsilon = 0.1\%$. To ensure a Δt independent solution for UAS approaches at each angle of attack, variation of *St*, $C_{L,m}$, $C_{D,m}$, $C_{L,a}$, and $C_{D,a}$ is monitored for a variety of time steps to achieve accurate transient solutions for the prediction of unsteady flow structure and aerodynamics of a Gurney flapped airfoil.

TIME STEP INDEPENDENCE STUDY

Tables 11 to 13 illustrate the sensitivity of the calculated quantities using UAS approach to different time steps ranging from $\Delta t = 10^{-3}$ s to $\Delta t = 10^{-5}$ s. As seen in these tables, $C_{L,m}$, $C_{D,m}$, and *St* extracted from the numerical results calculated using UAS approach show a monotone and asymptotic convergence leaning toward their limiting trend as Δt decreases from $\Delta t = 10^{-3}$ s to $\Delta t = 10^{-5}$ s. The relative errors in $C_{L,m}$, $C_{D,m}$, and *St* from $\Delta t = 10^{-3}$ s to $\Delta t = 10^{-5}$ s are reduced from 2.59 to 1.37%, 0.33 to 0.15%, and 78.9 to 1.53% at $\alpha = 3°$; 8.5 to 1.16%, 1.33 to 0.14%, and 16.7 to 1.17% at $\alpha = 4°$; and 19.03 to 0.74%, 7.75 to 0.23%, and 10.7 to 0.17% at $\alpha = 7°$, respectively. The E_{43} of $C_{L,m}$, $C_{D,m}$, and *St* is 1.37; 0.15 and 1.57% at $\alpha = 3°$; 1.16; 0.14 and 1.17% at $\alpha = 4°$; and 0.74; 0.23 and 0.17% at $\alpha = 7°$, respectively, which are fine enough and in the same order of magnitude.

Moreover, amplitude lift and drag coefficients extracted from the numerical results of UAS approach also show a monotone and asymptotic convergence as time step decreases from

Table 11. Unsteady accurate solution at $Re_C = 2{,}000$, $\alpha = 3°$ with $H_f/C = 4\%$, $\varepsilon = 0.1\%$.

Case	Δt (s)	$C_{L,m}$	$E_{j,j\text{-}1}$	$C_{L,a} \times 10^2$	$E_{j,j\text{-}1}$	*St*	$E_{j,j\text{-}1}$	$C_{D,m}$	$E_{j,j\text{-}1}$	$C_{D,a} \times 10^3$	$E_{j,j\text{-}1}$
4	0.00001	0.50659	1.37	0.3215	29.9	1.570	1.53	0.07858	0.15	0.5056	29.6
3	0.00005	0.49963	1.59	0.2253	41.2	1.546	2	0.07846	0.19	0.3561	41.7
2	0.0001	0.49170	2.59	0.1325	99.5	1.515	78.9	0.07832	0.33	0.2075	99.5
1	0.001	0.47897	--	0.00069	--	0.319	--	0.07806	--	0.001068	--

Table 12. Unsteady accurate solution at $Re_C = 2{,}000$, $\alpha = 4°$ with $H_f/C = 4\%$, $\varepsilon = 0.1\%$.

Case	Δt (s)	$C_{L,m}$	$E_{j,j\text{-}1}$	$C_{L,a} \times 10^2$	$E_{j,j\text{-}1}$	*St*	$E_{j,j\text{-}1}$	$C_{D,m}$	$E_{j,j\text{-}1}$	$C_{D,a} \times 10^3$	$E_{j,j\text{-}1}$
4	0.00001	0.60256	1.16	0.6978	15.3	1.5448	1.17	0.08256	0.14	1.2289	15.2
3	0.00005	0.59558	1.20	0.5909	16.9	1.5267	1.75	0.082442	0.16	1.0421	16.3
2	0.0001	0.58844	8.5	0.4908	98.8	1.5000	16.7	0.082306	1.33	0.8721	98.9
1	0.001	0.53822	--	0.00565	--	1.2500	--	0.081215	--	0.0096	--

Table 13. Unsteady accurate solution at $Re_C = 2{,}000$, $\alpha = 7°$ with $H_f/C = 4\%$, $\varepsilon = 0.1\%$.

Case	Δt (s)	$C_{L,m}$	$E_{j,j\text{-}1}$	$C_{L,a} \times 10^2$	$E_{j,j\text{-}1}$	*St*	$E_{j,j\text{-}1}$	$C_{D,m}$	$E_{j,j\text{-}1}$	$C_{D,a} \times 10^3$	$E_{j,j\text{-}1}$
4	0.00001	0.849753	0.74	0.0830	3.37	1.163	0.17	0.114126	0.23	0.02001	3.7
3	0.00005	0.843436	0.95	0.0802	3.9	1.161	0.26	0.113859	0.29	0.01927	4.8
2	0.0001	0.835436	19.03	0.0771	77.04	1.158	10.7	0.113528	7.75	0.01835	79.3
1	0.001	0.676484	--	0.0177	--	1.034	--	0.104730	--	0.00380	--

All errors are in percent.

$\Delta t = 10^{-3}$ s to $\Delta t = 10^{-5}$ s. The relative errors in $C_{L,a}$ and $C_{D,a}$ from $\Delta t = 10^{-3}$ s to $\Delta t = 10^{-5}$ s is reduced from 99.9 to 29.5% and 99.5 to 29.6% at $\alpha = 3°$; 98.8 to 15.3% and 98.9 to 15.2% at $\alpha = 4°$; and 77.04 to 3.37% and 79.3 to 3.7% at $\alpha = 7°$, respectively. However, the downward trend is dependent on the angle of attack. For example, E_{43} of $C_{L,a}$ and $C_{D,a}$, is 3.37 and 3.7% at $\alpha = 7°$; 15.3 and 15.2% at $\alpha = 4°$; 29.9 and 29.6% at $\alpha = 3°$, respectively. This is due to the fact that values of amplitude force coefficients are 1- or 2-order of magnitude lower than the mean values, *i.e.* $C_{L,a} << C_{L,m}$ and $C_{D,a} << C_{D,m}$. Moreover, the reason for which the relative errors of $C_{L,a}$ and $C_{D,a}$ at $\alpha = 3°$ are higher than those of $\alpha = 4°$ and the relative errors of $C_{L,a}$ and $C_{D,a}$ at $\alpha = 4°$ are higher than those of $\alpha = 7°$ is that $C_{L,a\ \text{at}\ \alpha=3°} << C_{L,a\ \text{at}\ \alpha=4°} << C_{L,a\ \text{at}\ \alpha=7°}$ and $C_{D,a\ \text{at}\ \alpha=3°} << C_{D,a\ \text{at}\ \alpha=4°} << C_{D,a\ \text{at}\ \alpha=7°}$.

Based on the data obtained in the Δt independence study shown in Tables 11 to 13, it was decided that a value of 10^{-5} s would be the proper time-step size for the time accurate computations in this Reynolds number. This Δt provides a compromise between accuracy and CPU time. Time steps smaller than $\Delta t = 10^{-5}$ s would noticeably increase the computational time.

Table 14 shows the comparison of CPU time estimated using UAS approach to reach periodic state with $\Delta t = 10^{-4}$ s at $Re_C = 2{,}000$, with $H_f/C = 4\%$, and $\varepsilon = 0.1\%$ at different angles of attack. The CPU time taken to reach a periodic state is 350, 275, and 50 for 3°, 4°, and 7°, respectively. As can be seen, there is a noticeable difference between the CPU time taken to reach a periodic state, at different angles of attack. The computational time at low angles of attack is much higher than that of the large angles of attack. Note that the unsteadiness and amplitude of the oscillation increases as the angle of attack increases. Thus, at higher angles of attack, the oscillations dominate uniform flow coming from the far-field boundary in a shorter span of time as compared to lower angles of attack. As a result, for small angles of attack at which the level of unsteadiness is low, forming the vortices and establishing periodic oscillations will take more time as compared to higher angles of attack.

OBSERVATIONS

One of the main purposes of the present study is to evaluate the possibility of using QSS approach instead of time accurate computations for the prediction of performance. Many numerical studies used the QSS approach for the prediction of flow over a Gurney flapped airfoil with the assumption that QSS approach has the capability to predict the flow characteristics and the time averaged behavior resulting from time accurate computations with a very good accuracy. Table 15 shows mean lift and drag coefficient errors of QSS approach, *i.e.* with $\Delta t = \infty$, with respect to UAS approach with $\Delta t = 10^{-5}$ s. As shown in Table 15, QSS approach has predicted the mean lift and drag coefficients with errors of 5.5 and 0.67%, 10.6 and 1.67% and 35.9 and 14.83%, at $\alpha = 3°$, 4° and 7°, respectively. As can be seen, the amount of calculated errors is not negligible, and, therefore, QSS approach cannot be used to predict mean lift and drag coefficients with the same accuracy as UAS approach does. In addition, relative errors increase with the angle of attack. For example, the relative error in mean lift and drag coefficients has increased from 5.5 to 35.9% and 0.67 to 14.83%, respectively, by increasing angle of attack from 3° to 7°. This is because the QSS approach introduces the average of flow properties instead of details of flow transient, while UAS approach presents instantaneous flow structure. At higher angles of attack, the flow structure is more complicated, and the vortices which are the main cause of unsteadiness are larger. As a result, the error of QSS approach would be higher with respect to UAS approach due to the fact that QSS approach ignores the capture of these vortices of flow.

Table 15. Relative error of quasi-steady approach with respect to unsteady accurate solution ($Re_C = 2{,}000$, $H_f/C = 4\%$, $\varepsilon = 0.1\%$).

		E* [%]		
		$\alpha = 3°$	$\alpha = 4°$	$\alpha = 7°$
$\Delta t = \infty$	$C_{L,m}$	5.5	10.6	35.9
	$C_{D,m}$	0.67	1.67	14.83

*This is the error of mean values of force coefficients from QSS approach with respect to UAS approach at $\Delta t = 10^{-5}$ s.

Table 14. Comparison of CPU time taken to reach periodic state using UAS approach with $\Delta t = 10^{-4}$ s for unsteady flow over a Gurney flapped NACA 0008 airfoil at $Re_C = 2{,}000$, with $H_f/C = 4\%$, and $\varepsilon = 0.1\%$.

α [°]	Average time needed for each linearization iteration [s]	Average number of linearization iterations at each Δt	Overall number of time steps to reach the periodic state	CPU time [h]
3	20	3	21,000	350
4	20	3	16,500	275
7	20	1	9,000	50

In addition, the performance of UIS approach in comparison with that of the UAS approach is also studied. Table 16 shows relative errors of calculated $C_{L,m}$, $C_{D,m}$, $C_{L,a}$, $C_{D,a}$, and St by UIS approach for 3 time steps of 10^{-3}, 10^{-5}, and 5×10^{-5} s with respect to those calculated by UAS approach with $\Delta t = 10^{-5}$ s. Navier-Stokes equations are non-linear by nature, and, in order to have a linear system of equations, convection terms in momentum equations are linearized with respect to mass fluxes ρu and ρv. Therefore, it is necessary to perform the linearization iteration at each Δt in order to reduce the linearization error to the lowest possible amount. As a result, UIS approach in which no linearization iteration is performed does not have the same accuracy as the UAS one.

Table 16 illustrates that, for small time steps, there would be no need to perform linearization iterations. This is reasonable since, when the Δt gets smaller, the error in flow properties between 2 successive time steps becomes lower, and, therefore, the linearization error decreases. For example, with $\Delta t = 5 \times 10^{-5}$ s, relative errors of calculated $C_{L,m}$, $C_{L,a}$, $C_{D,m}$, $C_{D,a}$, and St by UIS approach with respect to those calculated by UAS approach are 0.29; 2.02; 0.03; 2.23 and 1.02; 0.29, 0.01, 0.04, 0.2, and 0.89; and 0.31; 1.3; 0.25; 1.4 and 0.79, at $\alpha = 3°$, 4° and 7°, respectively. As can be seen, all errors are less than or about 2%.

Table 16. Errors of unsteady inaccurate solution with respect to unsteady accurate solution (R_{eC} = 2,000, H_f/C = 4%, ε = 0.1%).

α (°)		E* (%)		
		Δt = 0.001	Δt = 0.0001	Δt = 0.00005
3	$C_{L,m}$	5.3	0.65	0.29
	$C_{L,a}$	80.7	5.1	2.02
	$C_{D,m}$	0.63	0.09	0.03
	$C_{D,a}$	81.01	5.4	2.23
	st	20.4	2.5	1.02
4	$C_{L,m}$	5.3	0.65	0.29
	$C_{L,a}$	15.2	0.54	0.01
	$C_{D,m}$	0.75	0.08	0.04
	$C_{D,a}$	15.4	0.41	0.2
	st	19.1	2.4	0.89
7	$C_{L,m}$	8.6	0.72	0.31
	$C_{L,a}$	5.1	2.5	1.3
	$C_{D,m}$	4.8	0.58	0.25
	$C_{D,a}$	4.4	3.4	1.4
	st	16.8	1.92	0.79

*This is the error with respect to $\Delta t = 10^{5}$ s results from unsteady accurate solution.

As observed in Table 16, relative errors decrease as Δt decreases for all parameters studied. For example, relative errors of calculated $C_{L,a}$ by UIS approach with respect to those calculated by UAS approach reduce from 80.7 to 2.02, 15.2 to 0.01, and 5.1 to 1.3 with the decrease in Δt from 1×10^{-3} s to 5×10^{-5} s at $\alpha = 3°$, 4° and 7°, respectively. This shows the limiting trend of all parameters with the decrease in Δt from 1×10^{-3} s to 5×10^{-5} s. Overall, the present results show that UIS approach at small time steps in the order of $\Delta t = 5 \times 10^{-5}$ s or lower is comparable with UAS approach and capable of predicting aerodynamic coefficients with a reasonable accuracy.

Table 17 shows the comparison of CPU time estimated for the solution to reach its periodic nature using UAS and UIS approaches for the case of unsteady flow over a Gurney flapped NACA 0008 airfoil at Re_C = 2,000, $\alpha = 7°$, and H_f/C = 4%.

As illustrated in Table 17, the amount of CPU time taken to solve the flow using the UIS approach (100) is approximately 1/3 of that of UAS approach (340). As can be seen, UIS approach at small time steps is capable of producing time-averaged data with less CPU time than UAS approach. Therefore, using UIS approach with small time steps, such as $\Delta t = 5 \times 10^{-5}$ s, can lead to savings in computational cost.

EFFECT OF LINEARIZATION CONVERGENCE CRITERION ON THE SOLUTION

The linearization convergence criterion used to stop linearization iteration process was 0.1% in all the simulations presented in the previous sections. In order to ensure that the linearization convergence criterion used to stop linearization iteration process is small enough, simulations of UAS approach are carried out for 3 angles of attack of α = 3°, 4°, and 7° with

Table 17. Comparison of CPU time taken to reach periodic state using UAS (ε = 0.1%, with and $\Delta t = 10^{-5}$ s) and UIS ($\Delta t = 5 \times 10^{-5}$ s) for unsteady flow over a Gurney flapped NACA 0008 airfoil at Re_C = 2,000, α = 7° with H_f/C = 4%.

Aproach	Average time needed for each linearization iteration (s)	Average number of linearization iterations in each Δt	Overall number of time steps to reach the periodic state	CPU Time (h)
UIS	20	1	18,000	100
UAS	20	1	61,000	340

linearization convergence criterion values of ε = 0.1, 0.01, and 0.001% at $\Delta t = 10^{-5}$ s. The errors indicated in Fig. 5 are defined as:

$$\Delta_{j,j-1} = \left|C_{L,m_j} - C_{L,m_{j-1}}\right| \tag{7}$$

$$E_{i,j-1} = \Delta_{i,j-1}/C_{L,m_j} \tag{8}$$

where: indices 1; 2 and 3 refer to the linearization convergence criterion used for each simulation, *i.e.* ε = 0.1; 0.01 and 0.001%. Note that errors of *St*, $C_{L,a}$, $C_{D,m}$, and $C_{D,a}$ are calculated similarly to the error of $C_{L,m}$, as defined in Eqs. 7 and 8. Figure 5a shows sensitivity of mean lift and drag coefficients calculated from UAS approach to convergence criterion. The relative errors of E_{21} of $C_{L,m}$ and $C_{D,m}$ for 3 angles of attack α = 3°, 4°, and 7° are 0.19; 0.19 and 0.06% as well as 0.03; 0.012 and 0.009%, respectively. The relative errors of E_{32} of $C_{L,m}$ and $C_{D,m}$ are all very small and approximately zero. As can be seen, mean lift and drag coefficients have negligible changes with linearization convergence criterion and, overall, from ε = 0.1% to ε = 0.01%, their changes are less than 0.2%.

Figure 5b shows sensitivity of Strouhal number calculated from UAS approach to convergence criterion. The relative error of E_{21} of *St* for three angles of attack of α = 3°, 4°, 7°, is 0.13; 0.05, and 0.08%, respectively. The relative errors of E_{32} are 0 for all angles of attack. As observed, the *St* also has negligible changes with linearization convergence criterion and, overall, its changes are less than 0.2%, similarly to what has been predicted for mean lift and drag coefficients.

Figure 5c shows sensitivity of amplitude lift and drag coefficients calculated from UAS approach to convergence criterion. The relative errors of E_{21} of $C_{L,a}$ and $C_{D,a}$ for 3 angles of attack of α = 3°, 4°, 7°, are 6.2; 3.32; 0.85% and 5.84; 3.31; 0.81%, respectively. The relative errors of E_{32} of $C_{L,a}$ and $C_{D,a}$ are all very small and approximately zero. As seen amplitude lift and drag coefficients have different behavior from mean lift and drag coefficients and Strouhal number. The changes in relative errors of $C_{L,a}$ and $C_{D,a}$ from ε = 0.1% to ε = 0.01% (E_{21}) are noticeable and dependent on the angle of attack. As seen in Fig. 5c, the relative errors of E_{21} of $C_{L,a}$ and $C_{D,a}$ decreases when the angle of attack increases. Indeed, amplitude lift and drag coefficients at low angles of attack (α = 3°, 4°) are subjected to major changes, *i.e.* 6.2 and 3.32%; and 5.84 and 3.31%, respectively, than higher angles of attack (α = 7°), which is 0.85 and 0.81%, respectively. This is expected since $C_{L,a\ \text{at}\ \alpha=3°} << C_{L,a\ \text{at}\ \alpha=4°} << C_{L,a\ \text{at}\ \alpha=7°}$ and $C_{D,a\ \text{at}\ \alpha=3°} << C_{D,a\ \text{at}\ \alpha=4°} << C_{D,a\ \text{at}\ \alpha=7°}$.

The tests carried out on the effect of linearization convergence criterion ε show that the relative errors of E21 of $C_{L,m}$, $C_{D,m}$ and St undergo minor changes and are all less than 0.2%. However relative errors of E_{21} of $C_{L,a}$ and $C_{D,a}$ have considerable changes, and their changes are dependent on the angle of attack. Overall the relative errors of E_{32} are approximately zero for all parameters at all angles of attack. All in all, it can be concluded that if error of less than 7% is fine enough, then, linearization convergence criterion of ε=0.1% is also an appropriate value to accurately capture unsteady flow field around a Gurney-flapped airfoil at Re_C = 2,000, with H_f/C = 4% and $\Delta t = 10^{-5}$ s. Nevertheless, $\Delta t = 10^{-5}$ s and ε = 0.01% are proper to be used in the computations when transient behavior of the airfoil and amplitude force coefficients are of primary interest.

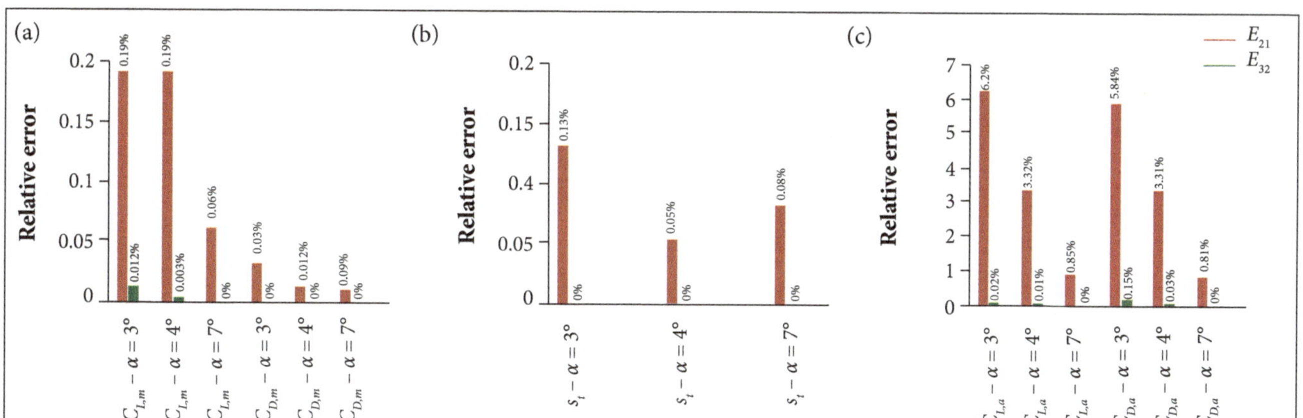

Figure 5. Linearization convergence criterion independence study at $\Delta t = 10^{-5}$ s using UAS for unsteady flow over a Gurney flapped NACA 0008 airfoil at Re_C = 2,000, with H_f = 4%C, and α = 3°, 4°, and 7°. (a) Mean values of lift and drag coefficients; (b) Strouhal number; (c) Amplitude of lift and drag coefficients.

CONCLUSION

A detailed study into the aerodynamic behavior of a Gurney flapped NACA 0008 airfoil with a flap height of $H_f/C = 4\%$ at angles of attack of 3°, 4°, and 7°, and at ultra-low Reynolds number of $Re_C = 2{,}000$ was conducted using a control volume based finite-element collocated scheme. In order to reduce the amount of required memory and to speed up the convergence rate two solver namely band solver and direct sparse solver were examined and their performances were compared with each other. The results showed that the amount of CPU time taken to solve the flow using the sparse solver is 1/27th of that of the band solver. Hence, the sparse solver with node renumbering was used for all computations throughout the present research.

The Δt independence study carried out here demonstrated the importance of proper time-step size selection. All the parameter showed a monotone and asymptotic convergence as Δt decreased. Δt independence study showed that, when the value of 10^{-5} s was employed for the calculations, $C_{L,m}$, $C_{D,m}$ and St were predicted with a very good accuracy and the calculated relative errors were very small. However, the relative errors of $C_{L,a}$ and $C_{D,a}$ were higher in comparison with those of $C_{L,m}$, $C_{D,m}$ and St, depending on the angle of attack. Overall the obtained accuracy with $\Delta t = 10^{-5}$ s is satisfactory providing a compromise between accuracy and CPU time.

Our investigation in this paper showed that for a Gurney flapped airfoil QSS approach is not able to predict the time averaged behavior resulting from UAS approach. The predicted QSS mean parameters showed remarkable discrepancy from those calculated by UAS approach at the regime of ultra-low Reynolds numbers. This is in contrast to what was reported in the literature for Gurney flapped airfoil at higher Reynolds numbers where the flow is turbulent.

Moreover, the present study revealed that results obtained from UIS approach at sufficiently small time steps, were comparable with the results of UAS approach. UIS approach was capable of producing time accurate results with a reasonable accuracy and less CPU time than computationally intensive time accurate simulations.

Finally, simulations of UAS approach at the Δt of 10^{-5} s on the effects of linearization convergence criterion demonstrated that $C_{L,m}$, $C_{D,m}$ and St are not affected by the linearization convergence criterion. However, $C_{L,a}$ and $C_{D,a}$ changed with linearization convergence criterion, and their changes were dependent on the angle of attack. Overall, if error of less than 7% is fine enough, then, linearization convergence criterion of $\varepsilon = 0.1\%$ is a proper value to capture unsteady flow field around a Gurney flapped airfoil. Nevertheless, $\varepsilon = 0.01\%$ is the proper one to be used in the computations especially when transient behavior of the airfoil and amplitude force coefficients are of primary interest.

AUTHOR'S CONTRIBUTION

Khoshlessa M elaborated the idea, performed all the simulations, and prepared all the data and figures. Karimian SMH is the principal investigator of this research, who critically revised the article for important intellectual content and gave the final approval of the version to be published.

REFERENCES

Abdo M (2004) Theoretical and computational analysis of airfoils in steady and unsteady flows (PhD thesis). Montreal: McGill University.

Alisadeghi H, Karimian SMH (2010) Different modelings of cell-face velocities and their effects on the pressure-velocity coupling, accuracy and convergence of solution. Int J Numer Meth Fluid 65(8):969-988. doi: 10.1002/fld2224

Ashby D (1996) Experimental and computational investigation of lift-enhancing tabs on multi-element airfoil. NASA-CR-201482.

Berthelsen PA, Faltinsen OM (2008) A local directional ghost cell approach for incompressible viscous flow problems with irregular boundaries. J Comput Phys 227(9):4354-4397. doi: 10.1016/j.jcp.2007.12.022

Calhoun D (2002) A Cartesian grid method for solving the two-dimensional stream-function vorticity equations in irregular regions. J Comput Phys 176:231-275, doi: 10.1006/jcph.2001.6970

Coutanceau M, Bouard R (1977) Experimental determination of the main features of the viscous flow in the wake of a circular cylinder in uniform translation. Part 1. Steady flow. J Fluid Mech 79(2):231-256.

Date JC, Turnock SR (2002) Computational evaluation of the periodic performance of a NACA 0012 fitted with a Gurney flap. J Fluid Eng 124:227-234.

Ding H, Shu C, Yeo KS, Xu D (2007) Numerical simulation of flows around two circular cylinders by mesh-free least square-based finite difference methods. International Journal for Numerical Methods in Fluids 53:305-332.

Giguere P, Dumas G, Lemay J (1997) Gurney flap scaling for optimum lift-to-drag ratio. AIAA J 35(12):1888-1890.

Herfjord K (1996) A study of two-dimensional separated flow by a combination of the finite element method and Navier-Stokes equations (PhD thesis). Trondheim: Norwegian Institute of Technology.

Jang CS, Ross JC, Cummings RM (1992) Computational evaluation of an airfoil with a Gurney flap. Proceedings of the 10th AIAA Applied Aerodynamics Conference; Palo Alto, USA.

Jeffrey DRM, Zhang X, Hurst DW (2000) Aerodynamics of Gurney flaps on a single-element high-lift wing. J Aircraft 37(2):295-301. doi: 10.2514/2.2593

Karimian SMH (1994) Pressure-based control-volume finite-element method for flow at all speeds (PhD thesis) Waterloo: University of Waterloo.

Karimian SMH, Schneider GE (1994) Numerical solution of two-dimensional incompressible Navier-Stokes equations: treatment of velocity-pressure coupling. Proceedings of the AIAA 25th Fluid Dynamics Conference; Colorado Springs, USA.

Karimian SMH, Schneider GE (1995) Pressure-based control-volume finite element method for flow at all speeds. AIAA J 33(9):1611-1618. doi: 10.2514/3.12700.

Khoshlessan M (2013) Numerical investigation of unsteady flow over a Gurney flapped airfoil with plunge oscillation (Master's thesis). Tehran: Amirkabir University of Technology.

Khoshlessan M, Karimian SMH, Daemi N (2013) Evaluation of a control-volume based finite-element collocated scheme for the solution of external steady and unsteady incompressible flows at low Reynolds numbers. Proceedings of the 11th International Conference of Numerical Analysis and Applied Mathematics; Rhodes, Greece.

Kunz PJ (2003) Aerodynamics and design for ultra-low Reynolds number flight (PhD thesis). Stanford: Stanford University.

Li Y, Wang J, Zhang P (2002) Effects of Gurney flaps on a NACA 0012 airfoil. Flow Turbulence and Combustion 68(1):27-39.

Liebeck RH (1978) Design of subsonic airfoils for high lift. J Aircraft 15(9):547-561. doi: 10.2514/3.58406

Linnick MN, Fasel HF (2005) A high-order immersed interface method for simulating unsteady compressible flows on irregular domains. J Comput Phys 204(1):157-192. doi: 10.1016/j.jcp.2004.09.017

Liu C, Zheng X, Sung CH (1998) Preconditioned multigrid methods for unsteady incompressible flows. J Comput Phys 139:35-57. doi: 10.2514/6.1997-445

Manish KS, Dhanalakshmi K, Chakrabartty SK (2007) Navier-Stokes analysis of airfoils with Gurney flap. J Aircraft 44(5):1487-1493.

Medjroubi W (2011) Numerical simulation of dynamic stall for heaving airfoils using adaptive mesh techniques (PhD thesis). Oldenburg: Carl von Ossietzky University.

Myose R, Heron I, Papadakis M (1996) The effect of Gurney flaps on a NACA 0011 airfoil. AIAA Paper 1996-0059.

Neuhart DH, Pendergraft Jr OC (1988) A water tunnel study of Gurney flaps. NASA TM-4071.

Ross JC, Storms BL, Carrannanto PG (1995) Lift-enhancing tabs on multi-element airfoils. J Aircraft 32(3):649-655. doi: 10.2514/3.46769

Russel D, Wang ZJ (2003) A Cartesian grid method for modeling multiple moving objects in 2D incompressible viscous flow. J Comput Phys 191:177-205.

Schatz M, Gunther B, Thiele F (2004) Computational modeling of the unsteady wake behind Gurney-flaps. Proceedings of the 2nd AIAA Flow Control Conference, Fluid Dynamics and Co-located Conferences; Portland, USA.

Storms LB, Jang SC (1994) Lift enhancement of an airfoil using a Gurney flap and vortex generators. J Aircraft 31(3):542-547.

Tejnil E (1996) Computational investigation of the low speed S1223 airfoil with and without a Gurney flap (master's thesis). San Jose: San Jose State University.

Tritton DJ (1959) Experiments on the flow past a circular cylinder at low Reynolds numbers. J Fluid Mech 6:547-567.

Troolin DR, Longmire EK, Lai WT (2006) Time resolved PIV analysis of flow over a NACA0015 airfoil with Gurney flap. Exp Fluid 41(2):241-254.

Williamson CHK (1989) Oblique and parallel modes of vortex shedding in the wake of a circular cylinder at low Reynolds numbers. J Fluid Mech 206:579-627.

Xu S, Wang ZJ (2006) An immersed interface method for simulating the interaction of a fluid with moving boundaries. J Comput Phys 216:454-493. doi: 10.1016/j.jcp.2005.12.016

GPS-Dependent Systems: Vulnerabilities to Electromagnetic Attacks

Lester de Abreu Faria[1], Caio Augusto de Melo Silvestre[1], Marcelino Aparecido Feitosa Correia[1]

ABSTRACT: The GPS is a satellite navigation system that provides location and time information. Such a system currently supports critical applications for military, civil and commercial users worldwide and is accessible to any operator by using a single GPS receiver. However, although being too dependent on GPS signals, just a few of these applications present some kind of countermeasure to electromagnetic attacks, showing a high level of vulnerability to intentional attacks. In this paper, we pose questions and situations related to security and vulnerability of different kinds of platforms/vectors and systems which directly afflict the situational awareness of operators. Experiments were made with different general GPS receivers as a function of distance and incidence angle, showing that they fail to work even at low jamming powers (−25 and 0 dBm at 10 m). More complex GPS systems, such as aeronautical receivers, were also tested, losing completely the tracking at −30 dBm, when a 0° (levelled) and 10 m far electromagnetic jamming signal is incident on its antenna. A specific open source, free software (JammPy) allows extending these experimental results, providing a roadmap and estimating how much jamming power is necessary to cause damage to these systems.

KEYWORDS: Aerospace, Electronic warfare, GPS, Jammer, GPS-dependent systems.

INTRODUCTION

The Global Positioning System (GPS) is a satellite navigation system that provides location and time information in all weather conditions and at any position where there is a line of sight to 4 or more GPS satellites (The Library of Congress 2011). Such a system supports critical applications for military, civil and commercial users worldwide, having the US government as its main sponsor. It shows itself accessible to any operator by using a single GPS receiver.

Currently, a series of applications are developed and supported primarily by this "tool", becoming highly dependent of it. They are called GPS-dependent systems, which include different tracking, encryption and navigation systems. GPS applications, such as transfer time, traffic signal timing and synchronization of cell phone base stations use the cheap and highly-accurate GPS infrastructure. However, although being essentially dependent on the GPS signals, just a few of them present some kind of countermeasure to electromagnetic attacks (jamming), becoming highly vulnerable to them.

Considering the fundamentals governing the operation of such a system, it is relatively simple to interfere with GPS signals, being either the military or a civilian. The effectiveness of interference (jamming effectiveness), depending on the distance and power of the equipment, is also quite simple to calculate, making the dimensioning of a jammer system an easy and accessible task for the majority of the population. In general, jammers can be used against any radio communication signals, whether being GPS signals, Wi-Fi signals and/or communication between mobiles. Despite being illegal in different countries, projects teaching how to build this kind of jammers are easily

1. Departamento de Ciência e Tecnologia Aeroespacial – Instituto Tecnológico de Aeronáutica – Divisão de Engenharia Eletrônica – São José dos Campos/SP – Brazil.

Author for correspondence: Lester de Abreu Faria | Departamento de Ciência e Tecnologia Aeroespacial – Instituto Tecnológico de Aeronáutica – Divisão de Engenharia Eletrônica | Praça Marechal Eduardo Gomes, 50 – Vila das Acácias | CEP: 12.228-900 – São José dos Campos/SP – Brazil | Email: lester@ita.br

found on the Internet. This is where the vulnerability of "GPS-dependent" devices comes from.

In a recent past, GPS vulnerability has been demonstrated in different episodes. In 2010, 2011 and 2012, South Korea suffered many blockages originated in North Korea. As a countermeasure to such interference, an integration of GPS with Enhanced Long Range Navigation (eLoran) was used (Lee 2013). Another incident was demonstrated in December 2011, when Iran surprised the world imposing a controlled landing to a drone RQ-170 Sentinel. It is supposed that the communication link between the control station and the aircraft was jammed, and then the GPS receiver was subjected to spoofing (Petersonm 2011).

THEORETICAL CONCEPTS

Jamming is defined as "the emission of radio frequency with enough power and with the needed features to avoid, in a given area, the receivers to track GPS signals" (Oonincx and van der Wal 2014). The low power of GPS signals (around −160 dBm on the Earth's surface) allows the jammer to be effective even at low power levels, making relatively simple and cheap to be realized. Various efficient jammers are available and cost about $ 1,000, being capable of generating 100 W of power (Carroll 2003). This power is sufficient to provide a high operation distance, as will be seen next.

Meaconing refers to the process of interception and retransmission of a signal. Considering GPS signals, a device receives the signal, amplifying and retransmitting it. The target receiver receives the original signal and the retransmitted one. Once the malignant signal arrives with higher power, the target processes this signal as the true one, thereby providing information of positioning, navigation and time corresponding to the information of the malicious transmitter (Landry *et al.* 1998).

Finally, the spoofing occurs when a false PNT information is inserted into the target receiver, which is unable to identify the attack. Although this technique is more complex, it constitutes a real threat to the current military and civil operations.

The Friis' Transmission Equation gives the power received by an antenna, under idealized conditions, given another antenna some distance away and transmitting a known amount of power. It serves as a good approximation for estimating signal levels through free space, considering the main factors that influence the propagation. It is considered just a simplified equation because it does not consider the atmospheric attenuation (or the path loss) and eventual influences of clouds and rain. It can be written as:

$$P_r = \frac{P_t A_{ef}}{4\pi(R)^2} = \frac{P_t \lambda^2}{(4\pi R)^2} = G_t G_r \frac{\lambda^2}{(4\pi R)^2} \qquad (1)$$

or, in dB:

$$P_r = P_t + G_t + G_r + 20\log_{10}\frac{\lambda}{4\pi R} \qquad (2)$$

where: G_t and G_r are the gains (with respect to an isotropic radiator) of the transmitting and receiving antennas, respectively; P_t is the transmitted power; A_{ef} is the effective area of the receiving antenna; R is the distance between antennas; λ is wavelength of the carrier.

Thus, based on Eq. 1, it can be seen that, not considering path loss, the power density that reaches the GPS receiver is inversely proportional to the square of the distance (R) of the jammer antenna to the GPS antenna. If a more complex propagation equation must be considered, Eq. 2 becomes:

$$P_r = P_t + G_t(\theta_t,\phi_t) + G_r(\theta_r,\phi_r) + \left(\frac{\lambda}{4\pi R}\right)^2 \ldots$$
$$\ldots\left(1-|\Gamma_t|^2\right)\left(1-|\Gamma_r|^2\right)\left|a_t . a_r^*\right|^2 e^{-\alpha R} \qquad (3)$$

where: $G_t(\theta_t,\phi_t)$ is the gain of the transmitting antenna in the direction (θ_t,ϕ_t); $G_r(\theta_r,\phi_r)$ is the gain of the receiving antenna in the direction (θ_r,ϕ_r); Γ_t and Γ_r are the reflection coefficients of the transmitting and receiving antennas; a_t and a_r are the polarization vectors of the transmitting and receiving antennas; α is the absorption coefficient of the intervening medium.

TARGET APPLICATIONS (SCOPE)

Although being originally a military project, the GPS is currently considered a dual-use technology, having become a widely used tool in trade, scientific applications, monitoring and surveillance.

Many civil applications use one or more of the 3 basic components of GPS (absolute location, relative movement and transfer time). Among these, some applications may be highlighted:

1. Autonomous vehicles, Unmanned Aerial Vehicles (UAVs) or drones: location and routes.
2. Cell-phones: clock synchronization allows the transfer of time, being fundamental to the synchronization of its spreading codes with other base stations (± 10 s) (National Institute of Standards and Technology 2011).
3. Navigation: designation of targets as well as troop and supply coordination (Sinha 2003).
4. Tracking targets: various weapons systems use GPS to track potential ground and aerial targets before setting them as hostiles. Such weapons systems transfer the coordinates of targets for high-performance guided weapons.
5. Missiles and projectiles guidance: GPS allows accurate pointing and guidance of different military weapons, including ICBMs, cruise missiles, precision-guided munitions and artillery projectiles (GlobalSecurity.org 2007).

Among the previously mentioned applications, the items 1, 2 and 5 may highly affect complex and high added-value military systems.

DRONES

Also known as (UAVs) or Remotely Piloted Aircraft (RPA), the flight can be remotely controlled by a pilot on the ground in real time or a route where the vehicle will fly can be defined in advance, based on GPS coordinates. There are a variety of shapes, sizes, configurations and features, depending on the target applications of the drone. They have 2 communication systems (How to kill... Date unknown):

- Radio line-of-sight: in the military C-band of 500 – 1,000 MHz, which is vulnerable and can be jammed with a simple radio.
- Satellite communications: in K_u band from 10.95 to 14.5 GHz, where the satellite may be jammed (uplink-band: from 13.75 to 14.5 GHz; downlink band: from 10.95 to 12.75 GHz).

The satellite communication system generally uses the same technology used by civil applications, except the encrypted systems. Once it is possible to jam both communication links, as suggested earlier, the operator is unable to control the aircraft, flying until colliding or the fuel is over. Radio frequencies in the C-band, in particular, are used during takeoff and landing, letting the drone to be an easy target.

JOINT DIRECT ATTACK MUNITION

Joint Direct Attack Munition (JDAM) is a guidance kit that enables the conversion of unguided bombs or "dumb" bombs into "smart" munitions. JDAM-equipped bombs are guided by an integrated inertial guidance system coupled to a GPS receiver, giving them a published range of up to 15 nautical miles (28 km). These can range from 500 pounds (227 kg) to 2,000 pounds (907 kg) (Hansen 2006). When installed on a bomb, they are called Guided Bomb Unit (GBU). The key components of the system consist of a tail section with aerodynamic control surfaces, a (body) strake kit as well as a combined inertial guidance system and GPS guidance control unit.

SAFE COMMUNICATIONS NETWORK

The communications network stations need to be synchronized to function properly. The more precise the synchronization, the smallest is the loss of information or the presence of noise. In modern synchronization systems, the time information coming from GNSS is generally used due to its accuracy and reliability.

Some systems automatically synchronize with the GPS-time, as it is the case in many civil systems. However, other systems allow the user to enter an arbitrary time information, which becomes more effective against GPS interference, if used correctly.

MATERIALS AND METHODOLOGY

Firstly, in this research, electromagnetic jamming was realized onto general receivers, such as automotive GPS and mobile phones. These experiments focus on a validation of the method and of the experimental setup, besides assessing the robustness of different receivers to jammer.

Such experiments were carried out outdoor and with the receivers vertically positioned. Distances from the radiating antenna of 10 m, with incidence angles of −10°; 0° and +10°, were measured. Finally, a jamming test at 20 m and 0° was also conducted. The power of the radiated signal was gradually increased at intervals of 3 min to allow a stabilization of the signals and the acquisition time of the receivers.

In all cases, signals generated by the jammer compete with the ones from the GPS satellites, reproducing the real environment. The experimental setup is composed of (Fig. 1):

- Signal Generator Agilent E8257D.
- DHR antenna 0118.

- SMA coaxial cable.

Two different receivers were tested:

- Cell phone Samsung Galaxy S3, with the "GPS Test" App, in order to monitor the available satellites, parameters of the coordinates and the date/time group.
- Automotive receiver (Foston).

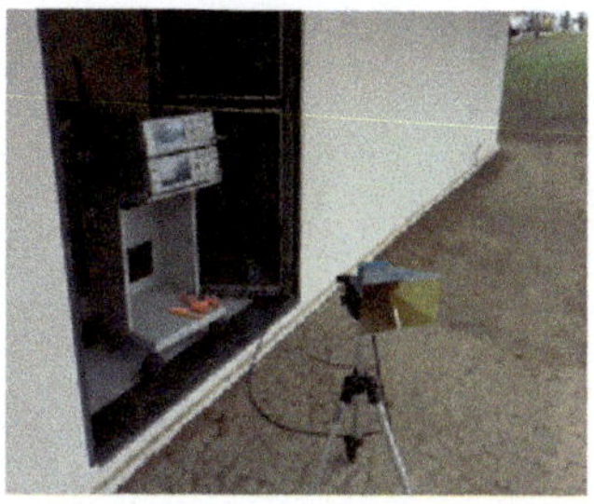

Figure 1. Experimental setup for general GPS receivers jamming.

Once assessed all possible situations for general GPS receivers, the next step was to build an experimental setup for assessing aeronautical GPS receivers. For safety reasons, data relating to aircraft and used equipment are not detailed in this paper.

The aircraft was placed outside the flight hangars where a direct line of sight from the jamming antenna to the receiving GPS antenna and to the GPS satellites would be possible. Two different GPS systems could be assessed: a stand-alone and a more complex integrated one, composed of GPS/inertial/radio-altimeter. Both work independently (including different antennas) and provide good conditions to evaluate the efficiency of the jamming system.

The integrated GPS system provides 3 independent solutions (Inertial Navigation System pure — INS, GPS pure and GPS/INS combined) simultaneously. It is possible to continuously monitor the performance of each one of the solutions, calculating a figure of merit (FOM) associated with the expected error. Such features were used as a database for this research.

An initial alignment was carried out in both systems and, after a full alignment, the initial coordinates reported by the aircraft system were presented: lat 23°13.34'S; long 45°51.96'W, as well as the airfield altitude (São José dos Campos Airport - SBSJ) of 2,040 ft, coincident in both systems.

The experimental setup, the used equipment and the jammer power procedures were the same as those mentioned for general GPS receivers tests. The receiving and transmitting antennas were positioned at a distance of 10 m and at angles of 0° and –6° in respect to the antenna of the aircraft, as shown in Fig. 2, where the type of aircraft is merely illustrative.

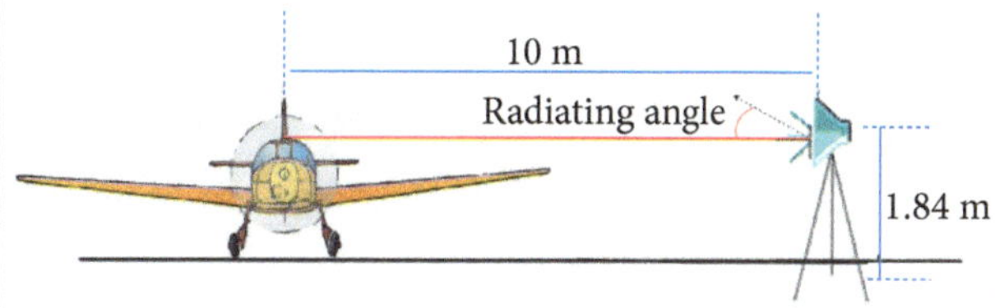

Figure 2. Experimental setup for aeronautical GPS receivers jamming.

Jamming was performed using narrowband signals without modulation, interfering in the C/A signal of the L1 carrier (1,575.42 MHz). For a better comparison of the effects of the jamming, the stand-alone GPS position was referenced to the non-directional (radio) beacon of São José dos Campos (NDB SJC) on the results of the first test and to the airfield of Lagoa Santa (SBLS) in the second test. These benchmarks are at 1 and 240 NM of the actual position of the tested equipment, respectively.

EXPERIMENTAL RESULTS

The first tests focused on general GPS receivers, where an automotive GPS receiver and a cell phone were assessed.

As could be confirmed in the first experiment, both receivers (Samsung S3 cell phone and Foston automotive receiver) are vulnerable at fairly low powers, when jammed by an antenna placed 10 m far and at an angle of 0°.

The automotive GPS receiver (Foston) proved to be more sensitive, once at –20 dBm it stopped receiving signals from all available satellites. It clearly demonstrates a high vulnerability and no countermeasures of any kind for such interference. Besides, since GPS systems need at least 4 GPS signals to work properly, it is possible to conclude that, at –25 dBm, it was not reliable anymore, because, at this jammer power, only one satellite signal could be received.

The Samsung S3 receiver showed to be more robust, acquiring 5 satellites up to –5 dBm and, therefore, generating a reliable location information. However, with an interference of 0 dBm, it also succumbed, losing all signals.

In the second part of the experiment, the number of visible satellites (in line of sight with the GPS receiver) increased for both receivers. Increasing the distance between the jamming antenna and the GPS receiver to 10 m far, it could be perceived that the Foston receiver needed approximately more 10 dB to cancel all the received GPS signals, agreeing with the expected behavior related to Eq. 1.

On the other hand, Samsung S3 cell phone presented an unexpected behaviour, since all the GPS signals were cancelled with a power lower than that used when the jamming antenna was 10 m far from the GPS antenna. This can be easily explained because the weather conditions noticeably deteriorated during the experiment, which has attenuated the signal from the GPS satellites, letting them weaker and then easier to be cancelled by the jammer. It can be supported by Eq. 3, in which the parameter α refers to this kind of signal loss. Thus, although at the beginning of the experiment there were more visible satellites, their signal arrived to the ground considerably weaker, making them easier to be cancelled by a jammer.

In the third experiment, it could be verified the sensitivity of the receiver to jamming with angles of incidence different from 0°, 10 m far, being the case in a real attack coming from the ground or from a flying vector. Firstly, it was simulated a jammer illuminating the receiver from the ground, at an angle of −10°.This change did not considerably affect the Foston receiver, which had its signal cancelled at the same radiated power (maybe some small differences of power between −25 and −20 dBm). This suggests that the antenna radiation diagram of this receiver presents a similar sensitivity to radiated signals coming from the ground or from leveled jamming antennas.

On the other hand, the Samsung S3 cell phone needed 5 dBm instead of 0 dBm to have the signal cancelled. This result suggests that this antenna radiation pattern is more directed to the upper hemisphere, presenting a large attenuation for the lower one.

Therefore, depending on the radiation pattern of the antenna, installed in the target vector, more or less power will be necessary to cancel the GPS signal. It is always necessary to evaluate the target and the antenna before going to an actual jamming procedure.

Finally, in the fourth experiment, a jammer illuminating the GPS receiver from the upper hemisphere was simulated at an angle of 10°, 10 m far. The Foston receiver showed the same radiation power profile for the cancelling of the GPS signals than the ones seen in the previously experiments, showing an antenna radiation pattern very isotropic in all directions from −10° to 10°.

Since the cell phone Samsung S3 showed the extinction of the signal with a very low radiated power (−20 instead of 0 dBm), it can be concluded that the sensitivity of this equipment is much higher for the upper hemisphere. A simple comparison of both radiation patterns leads to the scheme in Fig. 3, where the left radiation pattern suggests the one from Foston automotive GPS and the right one leads to the Samsung S3 radiation pattern (vertical profile).

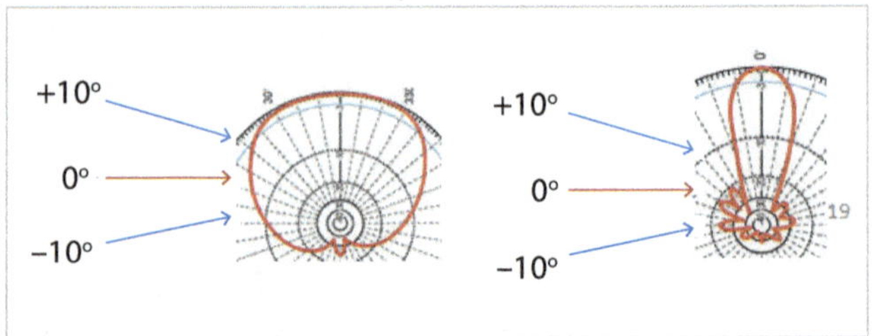

Figure 3. Suggested profiles of the radiation pattern of the tested devices.

After testing the general application receivers, some tests were performed in aeronautical GPS devices, including a stand-alone GPS and an integrated solution (GPS/inertial/radio-altimeter — GPS+IMU-RALT).

In a first test, an incidence angle of −6° was used between the jamming antenna and the aircraft antenna, while, in a second test, an angle of 0° between them was used. The results can be seen in Tables 1 and 2.

In Tables 1 and 2, it is possible to evaluate, for different jamming powers, the threshold power, the integrated GPS solution error (GPS solution) and the stand-alone GPS error.

Concerning specifically to the stand-alone GPS system, it showed to be highly vulnerable, once, in both tests, even using very low jammer power, the system lost the detection, showing very high errors before that.

In the first test, it can be seen that, at −50 dBm of jamming power, the receiver already indicates a wrong position (based on SJC benchmark), although keeping the initial position. This error gradually increases with the increasing in the jamming power.

In the second test, when the jamming signal started to be radiated, the receiver indicated that the aircraft was moving with a ground speed (GS) of 50 kt. The position indication did not change too much when compared to the first test indications. Figure 4 shows the displacement of the coordinates of the GPS relatively to the jamming power.

At the early stages of the experiment, the increasing power of the jamming signal apparently did not influence significantly the position indications. However, as the power increases above −85.9 dBm (which may be considered a threshold power for an incidence angle of 0°), considerable errors could be perceived. Figure 4d shows the final position indicated by the system, with a power of −40 dBm. From −30 dBm on, the receiver was unable to provide any coordinates information.

Table 1. First test results (angle of −6° between jammer and GPS antenna).

Jammer power (dBm)	Power at the GPS antenna (dBm)	Integrated GPS solution error (m)	Stand-alone GPS (present position) distance from the NDB SJC
-60	- 115.96	25	1 NM — Initial position
-50	- 105.96	1,000	2.6 NM — Initial position
-40	- 95.96	75	3.6 NM — Initial position
-30	- 85.96	500	4.8 NM — lat 23°09.81′S; long 45°54.63′W
-20	- 75.96	500	8.0 NM — lat 23°08.42′S; long 45°59.61′W
-10	- 65.96	500	No present position (lost coordinates)

Table 2. Second test results (angle of −0° between jammer and GPS antenna).

Jammer power (dBm)	Power at the GPS antenna (dBm)	Integrated GPS solution error (m)	Stand-alone GPS (present position) distance from the NDB SJC
-60	- 115.9	50	240 NM — lat 23°13.04′S; long 45°51.92′W — GS 50 kt
-50	- 105.9	50	232 NM — lat 23°13.00′S; long 45°51.92′W — GS 50 kt
-40	- 95.9	700	240 NM — lat 23°12.15′S; long 45°51.93′W — GS 50 kt
-30	- 85.9	200	No present position (lost coordinates)
-20	- 75.9	500	No present position (lost coordinates)
-10	- 65.9	500	No present position (lost coordinates)

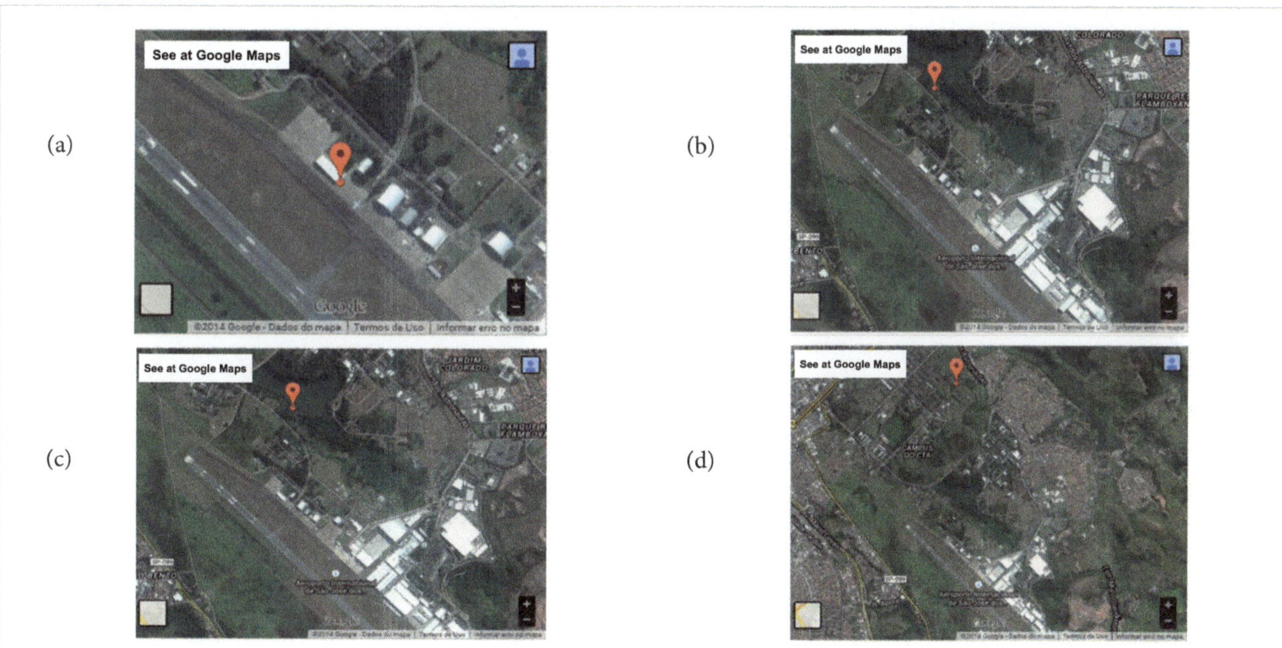

Figure 4. Indicated GPS position. (a) Actual; (b) −60 dBm; (c) −50 dBm; (d) −40 dBm, for 0° of incidence angle between the jammer and the GPS antennas.

Improving the robustness of the GPS system and testing the integrated solution (which is used in various actual systems, as the ones previously mentioned in this article), the same tests were conducted. In this case, the base parameter for the evaluation was the FOM, highlighting that the GPS position solution mode was active. This parameter increased as the jamming signal increased. Error spikes (outliers) could be seen during the tests, although no explanation was found untill the final version of this article.

Some additional tests are important to evaluate these outliers, but the preliminary results are those presented in Tables 1 and 2.

Concerning the variation of the incidence angle, the error showed to be different for both tests. Figure 5 shows the average calculated error (integrated GPS solution) for different jamming signal levels.

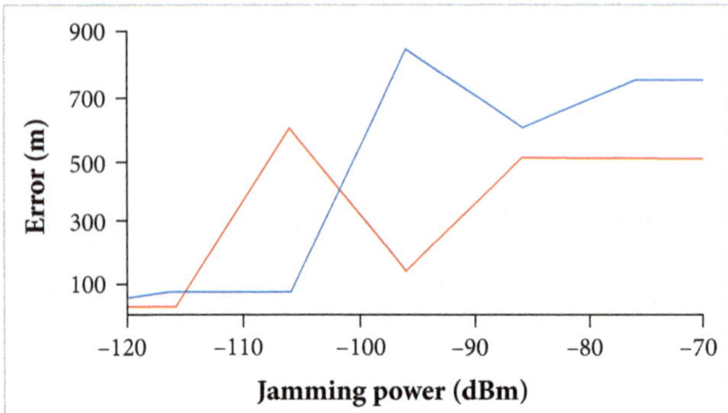

Figure 5. Average error for the GPSA integrated solution. Results for an incidence angle of −6° (red); Results for 0° (blue).

Based on Fig. 5, it is possible to conclude that, with an incidence angle of −6°, the error increases rapidly for lower jamming powers and then stabilizes at 500 m. For an incidence angle of 0°, the error remains lower than that up to −100 dBm, approximately, and then stabilizes around 750 m. These results suggest that, even using complex GPS systems, considerable errors may be inputted in real systems.

SIMULATION RESULTS

Based on the previously presented results, a very simple analysis was performed in order to estimate the necessary power to damage the control system of GPS-dependent devices. For that, it was considered a flying vector (drones, JDAM, aircrafts or safe communications network) with a specific altitude (in feet) and distance (in meters) from the jamming antenna. The used threshold power was −30 dBm, found experimentally in the previous experiments. This threshold power can be considered the worst case, because, with less than −30 dBm, the GPS systems already present a loss. It was adopted just to have an idea of a jammer efficiency. The atmospheric transmittance was considered 100%; the gains of the jammer antenna and the GPS antenna are the same as those evaluated for the aeronautical GPS receiver. The results are based on the Eq. 2 of this article. Figure 6 depicts the 3-D solution for this kind of interference.

In Fig. 6, it is possible to see the necessary jamming power for a complete extinction of the GPS signal on the receiving antennas, where Jamming power levels below 1 W are presented in green, while those above 5 W are shown in red. Powers between these values are presented in blue. All calculi were performed in an open source and free software called JammPy, specially developed for this research. Based on the threshold power (in dBm) and the distance (in meters) at which this power is efficient, it is possible to generate 3-D graphs, like the one presented in Fig. 6, for any range of altitudes and distances, allowing to evaluate and estimate the necessary jamming power for very specific targets and flight profiles.

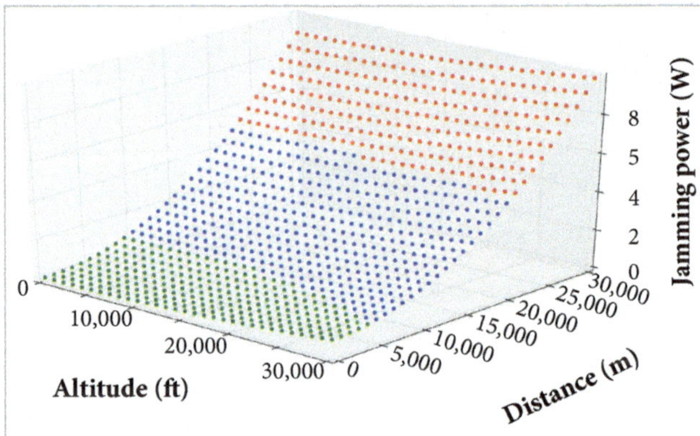

Figure 6. Necessary jamming power for a complete extinction of the GPS signal in aeronautical vectors, considering the distance and the altitude from the jamming antenna.

From the graph, one can conclude that, even for vectors at a high altitude (as high as 30,000 ft) and in a high distance from the jamming antenna (30 km), it is possible to have a successful interference with relatively low jamming powers (less than 10 W). These results show that GPS jamming is a very dangerous threat to any kind of GPS-dependent system.

CONCLUSIONS

GPS supports critical applications for military, civil and commercial users worldwide. Just a few of these applications present some kind of countermeasure to electromagnetic attacks, showing a high level of vulnerability to intentional attacks. Experimental results of jamming on general receivers, including cell phones and automotive GPS, show that all of them present a high vulnerability to this kind of interference, losing the satellite GPS signal at a fairly low jamming power. At −25 dBm, no more satellites can be seen by the automotive GPS under an attack 10 m far from it. The same situation occurs with a cell phone GPS, but at 0 dBm. The experiments were repeated at different incidence angles and distances, showing variable behaviors for the receivers and suggesting that, depending on the target, different powers and strategies of interference must be applied.

Concerning more complex systems, such as the aeronautical ones, it was possible to test 2 different and independent systems: a stand-alone GPS receiver and an integrated solution, composed of a GPS+IMU-RALT. Both GPS systems were jammed and lost their signals at a relatively low jamming power, showing a high vulnerability to electromagnetic attack. Different distances and incidence angles were tested showing that the stand-alone GPS is more vulnerable to jamming, although the integrated solution also presents a high error at low jamming power.

Thus, based on the experimental results, a software was developed (JammPy) in order to dimension the necessary power to jam a flight vector. The threshold power and distance at which the experiment was run are the input values for the calculus of the jamming power to interfere with a vector, as a function of distance and altitude. These conclusions allow to generate a danger area (map) where any kind of vector can be jammed with a specific jamming device (power generator and antenna). This can be used to develop doctrine to attack or to dimension a safe zone for our vectors.

Finally, it must be highlighted that the presented results lead to a high concern for the most part of countries that use GPS-dependent systems embedded on their terrestrial and aerial vectors, because less than 1% of defense communications, nowadays, has some kind of countermeasure to any type of electromagnetic attack, even the most simple ones (How to kill… Date unknown; Thompson 2010; ABC 2000).

AUTHOR'S CONTRIBUTION

Faria LA conceived the idea and wrote the main text, providing the simulation. Silvestre CAM and Correa MAP performed the experiments and co-wrote the main text. All 3 authors discussed the results and commented on the manuscript.

REFERENCES

ABC (2000) Backyard satellite jammers concern US Airforce; [accessed 2015 Feb 15]. http://www.abc.net.au/science/news/space/SpaceRepublish_120537.htm

Carroll JV (2003) Vulnerability assessment of the U.S. transportation infrastructure that relies on the global positioning system. J Navigation 56(2):185-193. doi: 10.1017/S0373463303002273

GlobalSecurity.org (2007) XM982 Excalibur precision guided extended range artillery; [accessed 2015 Mar 19]. www.globalsecurity.org/military/systems/munitions/m982-155.htm

Hansen R (2006) JDAM continues to be warfighter's weapon of choice; [accessed 2015 Feb 15]. www.archive.today/k4VX

How to kill UAVs (Date unknown); [accessed 2015 Mar 19]. http://privat.bahnhof.se/wb907234/killuav.htm

Landry RJ, Calmettes V, Bousquet M (1998) Impact of interference on a generic GPS receiver and assessment of mitigation techniques. Proceedings of the 5th International Symposium on Spread Spectrum Techniques and Applications; Sun City, South Africa.

Lee SJ (2013) GNSS vulnerability issues in Korea. South Korea: Chungnam National University.

National Institute of Standards and Technology (2011) Commom view GPS time transfer; [2015 Feb 22]. http://web.archive.org/web/20121028043917/http:/tf.nist.gov/time/commonviewgps.htm

Oonincx PJ, van der Wal AJ, editors (2014) Optimal deployment of military systems: technologies for military missions in the next decade. The Hague: T.M.C. Asser Press.

Petersonm S (2011) Iran hacked RQ-170 GPS—fooled into autopilot landing in Iran; [accessed 2014 Sept 17]. www.uasvision.com/2011/12/16/iran-hacke-rq-170-gps-fooled-in-autopilot-landing-in-iran/

Sinha V (2003) Soldiers take digital assistants to war; [2015 Feb 22]. https://gcn.com/articles/2003/07/24/soldiers-take-digital-assistants-to-war.aspx

The Library of Congress (2011) What is a GPS? How does it work?; [accessed 2015 Feb 24]. http://www.loc.gov/rr/scitech/mysteries/global.html

Thompson LB (2010) Lack of protected satellite communications could mean defeat for joint force in future war; [accessed 2015 Feb 22]. http://lexingtoninstitute.org/lack-of-protected-satellite-communications-could-mean-defeat-for-joint-force-in-future-war/?a=1&c=1171

7

Structural and Morphological Characteristics of Polyaniline Synthesized in Pilot Scale

Maria Alice Carvalho Mazzeu[1,2], Lohana Komorek Faria[3,4], Andreza de Moura Cardoso[3], Adriana Medeiros Gama[3], Maurício Ribeiro Baldan[5], Emerson Sarmento Gonçalves[1,3]

ABSTRACT: Polyanilines have many applications in Aerospace, especially in their doped form. Studies on their synthesis in a pilot scale can contribute to obtain products with desirable characteristics for such applications. The present study reports the chemical oxidative synthesis of polyaniline in pilot scale and different reaction times in order to determine if there are variations in the polyaniline structure, morphology and conductivity due to these synthesis conditions. It is very common to analyze these data for polymers obtained through bench scale. However, several parameters change the properties of final material in major scales, such as thermal, mechanic and diffusive variables. Therefore, the reaction time is the only variable into the 9 syntheses carried out, and polyaniline is obtained in a doped form, being dedoped with ammonium hydroxide and redoped with dodecylbenzenesulphonic acid. The doped and redoped samples were characterized by their molecular structure, thermal behavior, crystallinity and morphology. The electrical conductivity of redoped samples was determined. Some differences in the structure and morphology of doped and dedoped forms, identifying the doping structures, were reported. This paper aims to present the relationship between changes on structure and morphology of doped and undoped polyaniline obtained by the mentioned experiments. Furthermore, some addicts on conductivity are carried out. It was possible to contribute in order to obtain a more conductive polyaniline in pilot scale.

KEYWORDS: Polyaniline, Crystallinity, Morphology, Raman spectroscopy, Electrical conductivity.

INTRODUCTION

Since the observation of the remarkably high electrical conductivity of a halogen-treated polyacetylene, a number of other conjugated polymers have shown transition from an insulating to a highly-conductive state, such as polyaniline (PAni), poly (phenylenevinylene), polypyrrole, and polythiophene. Because of their unique electrical properties covering the whole insulator-semiconductor-metal range, unusual conducting mechanism as well as controllable chemical and electrochemical properties, conducting polymers show not only great potential in a range of applications, but also great contribution to the fundamental materials science research (Luo *et al.* 2011).

Among conducting polymers, PAni has received wide-spread attention because of its outstanding properties including simple and reversible doping-dedoping chemistry, stable electrical conduction mechanisms, high environmental stability and ease of synthesis (MacDiarmid 2001; Fratoddi *et al.* 2015).

Nevertheless, PAni exhibits poor physical and mechanical properties and is insoluble in common solvents (Panah *et al.* 2012). To improve solubility and induce fusibility of the stiff chain of this polymer, a doping procedure with functionalized protonic acids such dodecylbenzenesulfonic acid (DBSA) can be used (Jaymand 2013). The DBSA enhances the solubility and the stability of the PAni, and the complex salt exhibits a good conductivity. However, the emulsion needs a high DBSA/PAni molar ratio because some molecules of DBSAs act as a dopant

1.Departamento de Ciência e Tecnologia Aeroespacial – Instituto Tecnológico de Aeronáutica – Programa de Pós-Gradução em Ciência e Tecnologias Espaciais – São José dos Campos/SP – Brazil. **2.**Departamento de Ciência e Tecnologia Aeroespacial – Instituto de Fomento e Coordenação Industrial – Divisão de Certificação de Sistema de Gestão – São José dos Campos/SP – Brazil. **3.**Departamento de Ciência e Tecnologia Aeroespacial – Instituto de Aeronáutica e Espaço – Divisão de Materiais – São José dos Campos/SP – Brazil. **4.**Universidade do Vale do Paraíba – Faculdade de Engenharia, Arquitetura e Urbanismo – São José dos Campos/SP – Brazil. **5.**Instituto Nacional de Pesquisas Espaciais – Laboratório Associado de Sensores – São José dos Campos/SP – Brazil.

Author for correspondence: Maria Alice Carvalho Mazzeu | Departamento de Ciência e Tecnologia Aeroespacial – Instituto de Fomento e Coordenação Industrial – Divisão de Certificação de Sistema de Gestão | Praça Marechal Eduardo Gomes, 50 – Vila das Acácias | CEP: 12.228-900 – São José dos Campos/SP – Brazil | Email: aie.mzz@hotmail.com

of PAni, and the others act as a surfactant to keep the stability of the reactive medium (Kohut-Svelko *et al.* 2005).

PAni has potential utilization in a large number of applications: rechargeable batteries, sensors, electronic devices, light-emitting diodes, conducting paints and glues, gas-separation membranes, coatings etc. (Ibrahim 2013). Moreover, many studies have been conducted regarding application of PAni in the aerospace sector, and some of these applications are cited next.

The lifetime and efficiency of electronic devices and electrical equipment can be increased through effective electromagnetic interference (EMI) shielding. Conducting polymers such as doped polyaniline or polypyrrole are promising materials to replace or supplement typical metals for EMI shielding applications. Conducting polymers have merits such as lightweight, physical flexibility, and a tunable shielding response (Joo and Lee 2000).

The stealth technology makes use of geometrical means and material engineering in order to develop low reflection and high absorption structures. The most suitable way of reducing the reflection is by covering the aircraft with Radar Absorbing Materials (RAM) as of paint or polymeric sheet (Franchitto *et al.* 2007). Conducting polymers have been applied as RAM due to the possibility of variation in their conductivity with the frequency of the incident radiation (Faez *et al.* 2000).

Aircrafts routinely operate in atmospheric environments that, over time, may affect their structural integrity. Despite using appropriate materials and process for protection, recurrent exposure to chlorides, pollution, temperature gradients, and moisture provide the necessary electrochemical conditions for the development and profusion of corrosion in aircraft structures (Rinaldi *et al.* 2012). There are several new technologies now under consideration for aluminum alloys protection including conductive polymers as primers without Cr-based metal pre-treatments (Bierwagen 2001). For magnesium and its alloys, the conducting polyaniline-blended organic coatings have been found to protect the pinholes due to the passivation ability of the conductive polymer (Sathiyanarayanan *et al.* 2008).

Since commercial airliners are struck by lightning quite often during their operation, there is a need to elaborate new materials for lightning strike protection combining high conductivities with lightweight. Conducting polymers are thought to be one of the most promising compounds that can be used for this application. From a variety of commercially available conducting polymers, PAni fits the criteria of the best chemical stability, appropriate conductivity, and density (Krukiewicz and Katunin 2016).

These applications require PAni at industrial scale, and the optimization of manufacturing processes is essential for this purpose. Since pilot scale influences the hydrodynamics of polymerizations system (Roichman *et al.* 1999), this kind of experiment is important for evaluating amendments in the process.

In this study, each synthesis was performed on a pilot scale, with variation of reaction time for every synthesis, keeping the other parameters unchanged. The PAni salt first obtained was dedoped, and the PAni in a base form, non-conductive (PAni-B) was redoped with DBSA, when PAni in a salt form, conductive (PAni-DBSA) was achieved. The effects of synthesis conditions on the structural and morphological characteristics of PAni-B and PAni-DBSA are investigated by Raman spectroscopy, thermogravimetric analysis (TGA), X-ray diffractometry (XRD), and scanning electron microscopy (SEM). Electrical conductivity was determined for redoped samples.

METHODOLOGY

Synthesis of PAni-B in different reaction times: 9 syntheses of polyaniline powder were obtained by the usual chemical oxidative polymerization of aniline with ammonium peroxydisulfate (APS). For each synthesis, aniline (1.0 M) and APS (1.9 M) solutions have been prepared separately, in aqueous solution containing 1.0 M hydrochloric acid and 1.0 M sodium chloride; the synthesis was performed in a 5-L glass reactor with glass rods and Teflon naval type propellers, for agitation. The reactor temperature was maintained at -5 ± 1°C.

Aniline solution was added to the reactor under constant stirring, and then APS solution was added dropwise. The reactions started with the oxidant addition and stopped 10; 20; 35; 50; 65; 80; 95; 110 and 170 min after the beginning. Fifty minutes were necessary for the complete addition of APS so, in the first 3 syntheses, it was not completely added yet. The polyaniline salts obtained were filtered and washed with approximately 10 L of distilled water until the washing liquid was colorless and the pH was around 6.

Dedoping process: PAni-B was prepared by dedoping of the powder in a 1.0-M ammonium hydroxide solution, being held under agitation by 12 h. The material was filtered again, washed, and the power (dark brown color) was dried in a conventional oven at 60 °C until constant weight. Figure 1 represents the chemical structure of PAni-B, according to Galiani *et al.* (2007).

Redoping process: PAni-B (5 g) was added to an aqueous solution containing 9 g of DBSA and left under constant stirring for 24 h at room temperature. The doped PAni-DBSA was washed with distilled water and ethyl alcohol, filtered, and the green powder was drying under vacuum at 60 °C until constant weight. The chemical structure of PAni-DBSA is represented in Fig. 2 (Galiani *et al.* 2007).

To confirm the chemical structures of PAni-B and PAni-DBSA, including doping, it was used a Raman spectrometer Labram HR Evolution, with Argon ion laser at 514 nm. A thermogravimetric analyzer model 7HT (PerkinElmer®) was used to study moisture content and thermal behavior of the polymers. The crystallinity study was performed by an X-ray diffractometer Panalytical, model XPert PRO, with a rotating anode X-ray generator working at $5° \leq 2q \leq 90°$, with Cu Kα monochromatic radiation of 0.154 nm; to study the surface morphology, it was employed a Zeiss Leo 440 SEM. AC electrical conductivity measurements were carried out by dielectric impedance spectroscopy (DIS), with a 2-terminal probe arrangement, by using a potentiostat-galvanostat AUTOLAB PGSTAT302 with impedance module.

Samples were used in powdered form, without dilution for all techniques, except for conductivity measurements, when the powder was compacted up to 3 MPa to obtain the pellets.

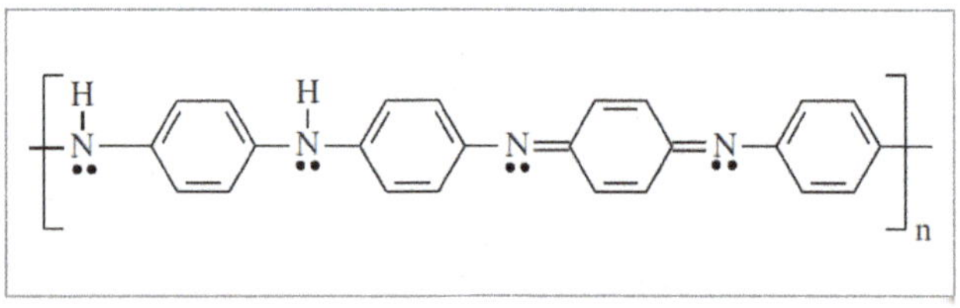

Figure 1. Chemical structure of PAni-B.

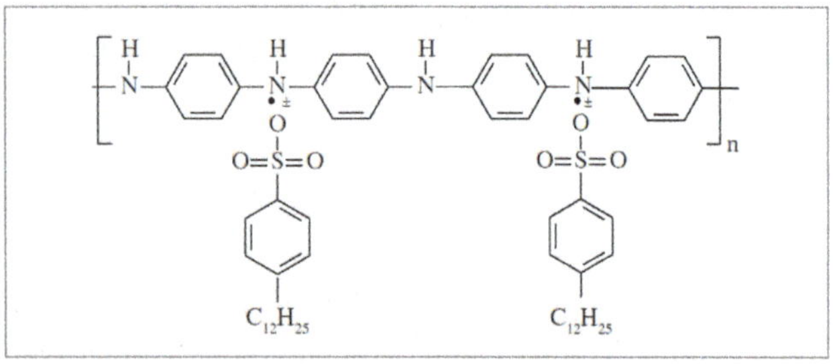

Figure 2. Chemical structure of PAni-DBSA.

RESULTS AND DISCUSSION

RAMAN SPECTROSCOPY

Raman spectroscopy has demonstrated to be a useful analytical tool to investigate conducting polymers (Engert *et al.* 1994). Raman spectra of PAni-B and PAni-DBSA, at 110 min reaction time, are presented in Fig. 3.

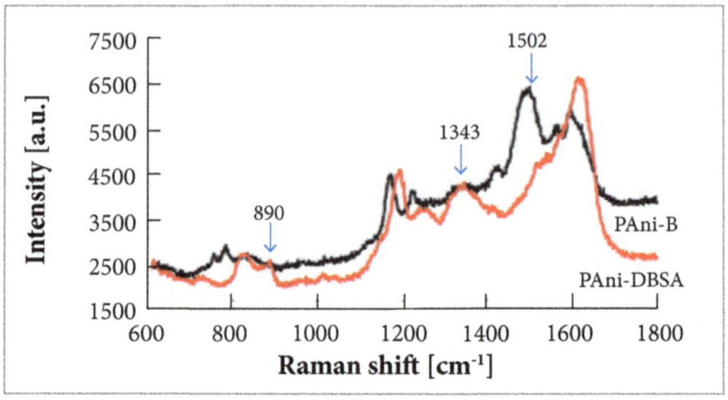

Figure 3. Raman spectra of PAni-B and PAni-DBSA (l0 = 514 nm).

PAni-B spectra presented characteristic bands at 1170 (C-H bending); 1206 (C-N stretching); 1502 (C=N stretching); 1560 (N-H bending) and 1598 cm^{-1} (C=C stretching) (Mažeikienė *et al.* 2007; Cirić-Marjanović *et al.* 2008; Moravková *et al.* 2012). PAni-DBSA spectra, in turn, presented characteristic bands at 1190 (C-H bending); 1,260 (C-N stretching); 1,343 (C-N^+ stretching); 1,530 (C-C stretching); and 1620 cm^{-1} (C-H stretching) (Nobrega *et al.* 2012; Ibrahim 2013; Shakoor and Rizvi 2010).

The conductive nature of PAni-DBSA can be associated to the band located at 890 cm^{-1} (arrow in Fig. 3), assigned to deformation C-N of secondary amine next to aromatic ring in the polaronic form C-N^+-C, and to the band at 1343 cm^{-1} (arrow in Fig. 3) that is assigned to vibration of a delocalized polaron structure (Belaabed *et al.* 2010). The band at 1502 cm^{-1} (arrow in Fig. 3), which is assigned to C=N, is very intense for PAni-B but significantly decreases to PAni-DBSA, due to the association between PAni and DBSA. Figures 1 and 2 illustrate what happens with the band assigned to C=N, where one can observe changes in the chemical structure due to the doping process.

THERMOGRAVIMETRIC ANALYSIS

The knowledge of PAni's thermal stability is important for its use in many practical applications. The thermal degradation of PAni is shown by thermogravimetry (Gul *et al.* 2013).

TGA results for PAni-B, PAni-DBSA, and DBSA are shown in Fig. 4, and the differential thermogravimetry curves (DTG), calculated from TGA, are shown in Fig. 5.

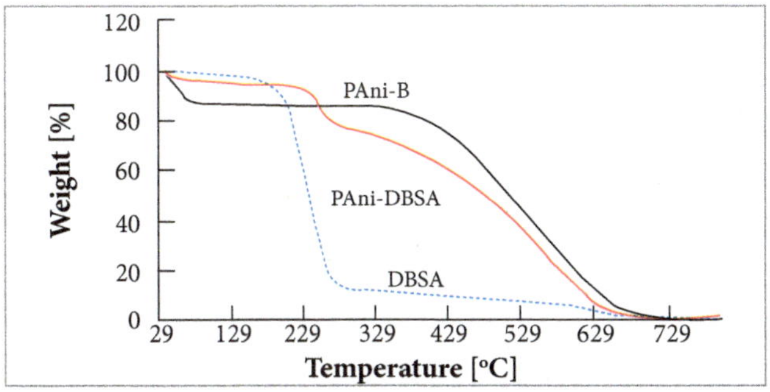

Figure 4. Thermogravimetric curves of PAni-B, PAni-DBSA, and pure DBSA.

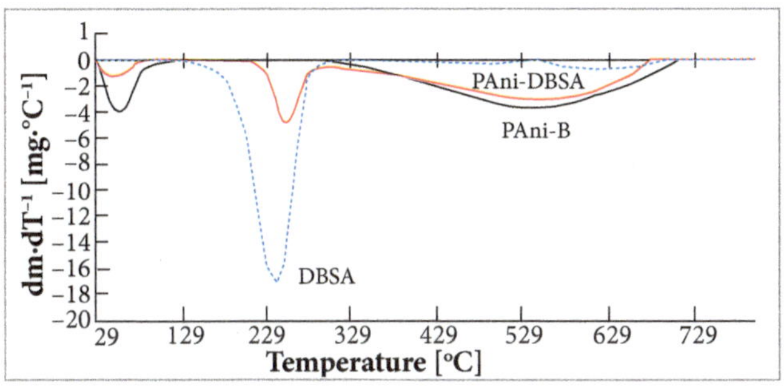

Figure 5. DTG curves of PAni-B, PAni-DBSA, and pure DBSA.

The temperature ranges mentioned next refer to measurements carried out at room temperature and atmospheric air.

PAni-B presents 2 stages of weight loss. The first weight loss occurs in the range of ~ 30 to 90 °C and is related to the loss of physically absorbed water plus moisture evaporation and, perhaps, outgassing of unknown small molecules (Zeng andKo 1998; Nithyaprakash *et al.* 2014). The second weight loss, starting at ~ 330 °C and finishing at ~ 700 °C, is attributed to complex and irreversible processes of decomposition of the polymeric chain (Nobrega *et al.* 2012).

PAni-DBSA also presents a first weight loss in the range of ~ 30 to 80 °C and a second one at ~ 130 to 290 °C, attributed to breakdown of the interaction between the PAni chain and the DBSA dopant ($NH^{+}....SO_3^{-}$), with decomposition of DBSA. The decomposition temperature is higher than that of pure DBSA. A third weight loss that starts at ~ 300 °C and is complete at ~ 680 °C indicates a structural decomposition of the PAni backbone (Babazadeh 2009). Weight losses obtained from thermogravimetric curves of PAni-B and PAni-DBSA are shown in Table 1.

Table 1. Weight losses results from PAni-DBSA and PAni-B.

Event	Weight losses (%)	
	PAni-DBSA	PAni-B
Water + moisture evaporation	3.0	7.8
DBSA decomposition	12.9	------
PAni decomposition	70.3	59.8
Residual weight	13.8	32.4

Thermal behavior change is observed when comparing PAni associated with DBSA molecules. The degradation temperature is higher than that of pure DBSA, and PAni decomposition begins before than PAni-B, probably due to PAni-DBSA interaction. This interaction affects the weight losses too, and higher values are obtained for PAni decomposition, probably due the chain is more vulnerable when associated with DBSA. The results of thermal degradation have shown that PAni-DBSA could be used for applications where high temperature is necessary.

ELECTRICAL CONDUCTIVITY

The 2-probe and 4-probe techniques are commonly used to characterize almost all semiconductor products available in the world today. These techniques have paved the way in discovering semiconductor materials (Ashraf 2014; Barsoukov and MacDonald 2005).

Two-probe is the simplest method for measuring resistivity and it is useful when the sample has large resistance. It is used mainly for the measurement of high-impedance materials where the impedance of cables is not significant (Barsoukov *et al.* 2005).

Considering that conductivity is the inverse of resistivity, in this study the conductivity was obtained by utilizing the 2-probe method to measure the electrical resistivity of PAni-DBSA, using Eq. 1 (Abed *et al.* 2014):

$$R = \rho\ L/A \qquad (1)$$

where: R is the electrical resistance; ρ is the electrical resistivity; L is the length of the material in the direction of the current; A is the cross-sectional area.

The conductivity s (S/cm) was calculated according to Eq. 2:

$$\sigma = 1/\rho \qquad (2)$$

The electrical conductivity for some reaction times is presented in Table 2.

Table 2. Electrical conductivity of PAni-DBSA.

Reaction time (min)	50	80	95	110	170
Electrical conductivity ($S \cdot cm^{-1}$)	1.8×10^{-5}	4.6×10^{-6}	4.0×10^{-7}	4.0×10^{-4}	8.1×10^{-4}
Error ($S \cdot cm^{-1}$)	4.22×10^{-7}	2.08×10^{-6}	6.54×10^{-8}	5.14×10^{-5}	3.61×10^{-5}

The sample preparation may interfere with the results since the sample is pressed. However, these results confirm that conductive structures are present at PAni-DBSA, as Raman spectroscopy has shown.

The obtained PAni-DBSA is a semiconductor, and a 170-min reaction time shows the highest electrical conductivity result.

For applications, PAni-DBSA is normally used as a blend with others materials which serve as a support. Such materials may be rubber, paint, silicone or other polymers and they may influence electrical conductivity too.

X-RAY DIFFRACTOMETRY ANALYSIS

XRD is used for the determination of the crystallinity degree of polymeric compounds (Araujo *et al.* 2013). Crystalline orientation of conducting polymer is very interesting, because a more highly-ordered system could display a metallic property such as conductive state (Vivekanandan *et al.* 2011), but the crystallinity of polyaniline and the intensity of peaks depend on the synthesis conditions (Pouget *et al.* 1991; Laska and Widlarz 2005). Polyanilines are semi-crystalline in nature and a 2-phase system. The phase in which the polymer chains are parallel and ordered in close packed array is the crystallites region, while the phase where the chains are not ordered and do not have parallel alignment is the amorphous region (Bhadra and Khastgir 2008). The faster grown of the polymer results in disorder orientation of polymer chains and amorphous polymer structure (Kumar *et al.* 2013).

Figure 6 shows XRD patterns of PAni-B and PAni-DBSA for a reaction time of 110 min. The diffraction patterns of PAni-DBSA agree with Soares *et al.* (2006), in relation to redoped method of preparation, and the diffraction patterns of PAni-B agree with Zilberman *et al.* (1997). The crystalline phase can be identified in both XRD spectra; however, PAni-DBSA shows 2 peaks at 2θ = 20 and 25° while PAni-B shows 4 peaks at 2θ = 10; 15; 20 and 25°, which is an indicative of more ordered structures than PAni-DBSA.

Besides the preparation methods, other factors may contribute to differences in crystallinity of PAni-B and PAni-DBSA, such as variation in inter-chain hydrogen bonding and electrostatic interaction at these times, polymer morphology, as well as speed of polymer formation.

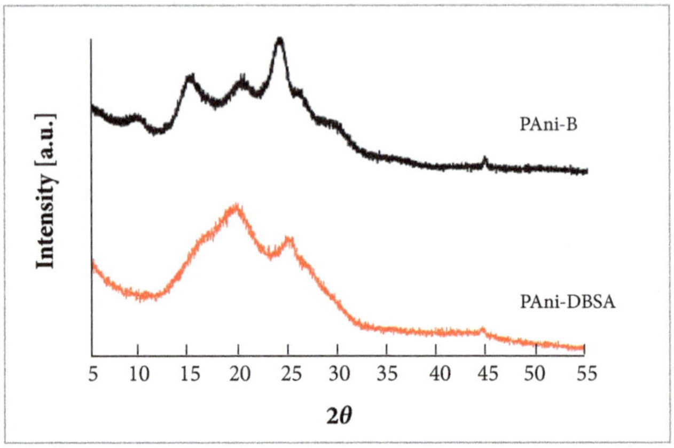

Figure 6. XRD pattern of PAni-B and PAni-DBSA.

The proportion between amorphous phase contributions in overall structure (*R*) of PAni-B and PAni-DBSA, in each reaction time, was calculated using the ratio between minimum and heights of peaks, as described in Saravanan *et al.* (2006) and Manjunath *et al.* (1973).

For PAni-DBSA pattern, *R* was calculated using Eq. 3, with minimum and heights of the peaks at 2θ = 20 and 25°.

$$R = 2m_1/h_1 + h_2 \quad (3)$$

where: m_1 represents the heights of minima between 2 peaks; h_1 and h_2 are the heights of these peaks from the base.

For PAni-B spectrum, *R* was calculated using Eq. 4, with minimum and height of the sharp peaks at 2θ = 15; 20; and 25°.

$$R = m_1 + 2m_2/h_1 + h_2 + h_3 \quad (4)$$

where: m_1 and m_2 are the heights of minima between 2 peaks; h_1, h_2, and h_3 are the heights of peaks from the base.

Crystallinity indexes were given by 1-*R*, and the results for PAni-B in each reaction time are shown in Fig. 7.

Crystallinity index for PAni-B showed variation against the reaction time, with some specific higher values like 50 and 170 min. Since each reaction is unique, the changes in the crystallinity index seem to be more related to speed of formation of the polymer than to reaction time.

Crystallinity index results for PAni-DBSA in each reaction time are shown in Fig. 8 and present very slight variation. The main factors that can influence these results are the method of preparation (dedoping-redoping) and the fact that DBSA is a bulky dopant. The previously synthetized material (PAni-B) appears to have little influence on the crystallinity of the redoped PAni (PAni-DBSA).

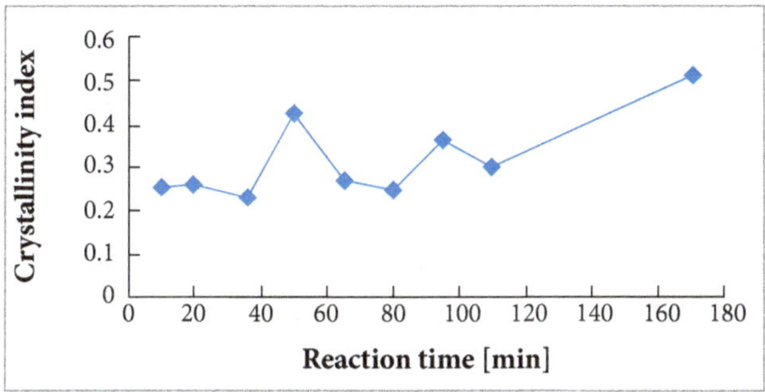

Figure 7. Crystallinity index variation with reaction time for PAni-B.

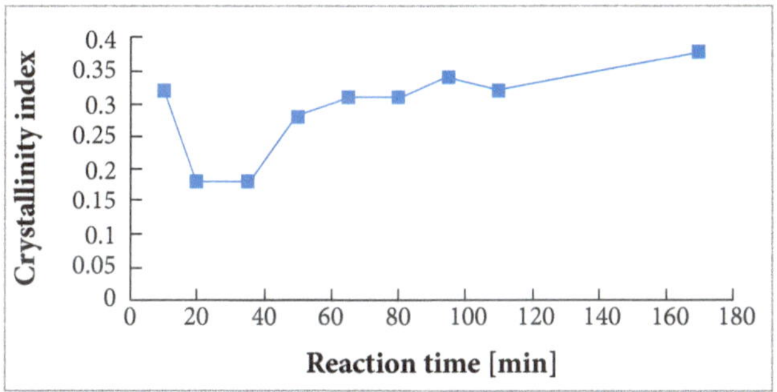

Figure 8. Crystallinity index variation with reaction time for PAni-DBSA.

SCANNING ELECTRON MICROSCOPY

The formation of PAni morphology is determined by the synthesis conditions (Abdolahi *et al.* 2012). The granular morphology of PAni powders is most common in PAni prepared by precipitation polymerization when using strong oxidants and high aniline concentrations under strongly-acidic conditions, at pH < 2.5. Globular structure has often been used as a synonym of the granular type (Stejskal *et al.* 2010).

Polyaniline prepared in this way is highly aggregated, and rapid sedimentation from solution is generally observed. However, careful examination reveals that a small quantity of nanofibers also appear in the product. Apparently, these nanofibers are formed without any extra structural directing agents, suggesting that nanofibers may be a natural morphology of polyaniline (Huang and Kaner 2006).

Most relevant morphological characteristics of PAni-B and PAni-DBSA are shown in Fig. 9. It is rather hard to distinguish the individual particles or grains from the images because the particles are aggregated. Therefore, it is difficult to identify the size and shape of the grains.

The structures more discernible at PAni-B (Fig. 9a, obtained at 110 min) that appear at any reaction time are tiny coral reefs closely covered with globules (or granules). In our synthesis condition, it is expected that the formed spheres or fibers tend to agglomerate, forming micro-particles. At reaction time of 50 and 170 min (Fig. 9b, arrow, obtained at 170 min), it is possible to identify several structures that appear to be fiber by linking agglomerates structures. These structures seem to be more related with agglomeration phenomena during the synthesis than with time of reaction, and it could cause deviations in analysis by XRD.

PAni-DBSA showed agglomerates of globular structures, similarly to PAni-B, but with forms that are more spherical and with some flatter multilayer structure (Fig. 9c, obtained at 110 min). PAni-DBSA also showed some shapeless areas, ascribed to the high concentration of dopant (Fig. 9d, arrow, obtained at 170 min).

After PAni-B is redoped and PAni-DBSA is obtained, the reaction time seems to have no influence on morphology.

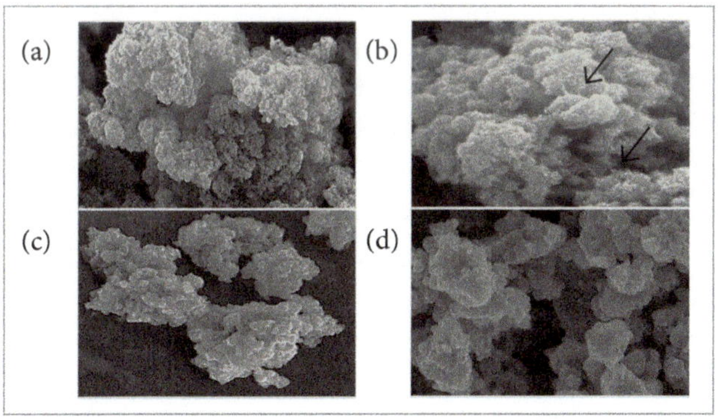

Figure 9. Most relevant morphological characteristics. (a, b) PAni-B; (c, d) PAni-DBSA. Increase of 5,000 X.

CONCLUSION

Getting to know the influence of pilot scale and reaction time on structural and morphological characteristics of PAni-B and PAni-DBSA can lead to important changes on structural properties related to electric characteristics.

In this study, polyaniline obtained in base and doped form was characterized as to identify significant changes in structure and morphology because synthesis has been carried out in a pilot scale and different reaction times.

Comparing both PAni forms, it is possible to identify the doping structure for PAni-DBSA by Raman spectroscopy due to the increase of the 890 and 1,343 cm^{-1} bands, assigned to polaronic structures, and the decrease of the 1,502 cm^{-1} band, indicating break of C=N bond.

The thermal behavior of PAni-B and PAni-DBSA shows that it is possible to distinguish components of the obtained polymers. However, some changes in temperature indicate that the combination of PAni and DBSA can change degradation temperature, increasing the onset temperature of DBSA degradation and reducing the onset temperature of PAni degradation. These changes affect weight losses percentage too, increasing weight losses percentage of PAni-DBSA.

Since changes in Raman spectra and TGA results of PAni-B and PAni-DBSA due to reaction time were not found, this variation seems to have little influence on polymer structure after doping.

The electrical conductivity is higher when the synthesis gets more time, indicating that reaction time can give some positive influence on conductivity.

The XRD result shows differences in crystalline peaks of PAni-B and PAni-DBSA that can be attributed mainly to the redoping process.

When the morphology shows the presence of fibers in the polymeric structure (reaction time of 50 and 170 min), the crystallinity index of PAni-B increases, which might indicate a correlation between these factors.

Whereas the formation of crystals on a pilot scale may change because of effects caused by water flow, the speed of polymerization can affect the formation of crystals too, causing random effects in different syntheses results and changing crystallinity index and morphology of PAni-B. However, these effects do not seem to be completely transferred to the doped form, probably due to many washing processes and association of PAni with a bulky molecule such as DBSA.

Evaluating the set of techniques, it is possible to assert that very important changes occur from PAni-B to PAni-DBSA. These changes allow getting a conductive behavior, as observed by Raman peaks and DIS. However, sensitivity *versus* reaction time is explained in detail by the last technique, in which it was possible to verify changes on 3 magnitude orders. It can be noticed that, for a pilot reactor system, it is required a time of 50 min for APS addition and 2 h more to process reaction mixing so that more conductive PAni is obtained.

ACKNOWLEDGEMENTS

Thanks to the Instituto de Aeronáutica e Espaço of the Departamento de Ciência e Tecnologia Aeroespacial (DCTA/IAE) that, through the Physical Chemistry Characterization Laboratory, Materials Division (AMR), provided the necessary structure and materials for this study.

AUTHOR'S CONTRIBUTION

Mazzeu MAC and Gonçalves ES conceived the idea and co-wrote the main text; Baldan MR and Gama AM discussed the results; Faria LK and Cardoso AM performed the experiments. All authors discussed the results and commented on the manuscript.

REFERENCES

Abdolahi A, Hamzah E, Ibrahim Z, Hashim S (2012) Synthesis of uniform polyaniline nanofibers through interfacial polymerization. Materials 5(8):1487-1494. doi: 10.3390/ma5081487

Abed MY, Youssif MA, Aziz HA, Shenashen MA (2014) Synthesis and enhancing electrical properties of PAni and PPA composites. Egyptian Journal of Petroleum 23(3):271-277. doi: 10.1016/j.ejpe.2014.08.003

Araujo JR, Adamo CB, Robertis E, Kuznetsov AY, Archanjo BS, Fragneaud B, Achete CA, De Paoli MA (2013) Crystallinity, oxidation states and morphology of polyaniline coated curauá fibers in polyamide-6 composites. Compos Sci Tech 88:106-112. doi: 101016/jcompscitech.2013.08.039

Ashraf A (2014) Electrical conductivity measurement of polyaniline and its composites; [accessed 2016 Jun 14]. https://www.Researchgate.net/publication/271840942_Electrical_Conductivity_measurement_of_polyaniline_and_its_composites

Babazadeh M (2009) Aqueous dispersions of DBSA-doped polyaniline:

one-pot preparation, characterization and properties study. J Appl Polymer Sci 113(6):3980-3984. doi: 10.1002/app.30460

Barsoukov E, MacDonald JR (2005) Impedance spectroscopy theory, experiment, and applications. 2nd Edition. New Jersey: John Wiley & Sons.

Belaabed B, Lamouri S, Naar N, Bourson P, Hamady SOS (2010) Polyaniline-doped benzene sulfonic acid/epoxy resin composites: structural, morphological, thermal and dielectric behaviors. Polymer J 42:546-554. doi: 10.1038/pj.2010.41

Bhadra S, Khastgir D (2008) Determination of crystal structure of polyaniline and substituted polyanilines through powder X-ray diffraction analysis. Polymer Test 27(7):851-857. doi: 101016/j.polymertesting.2008.07.002

Bierwagen G (2001) Next generation of aircraft coatings systems. J Coating Tech 73(915):45-52. doi: 10.1007/BF02730030

Ćirić-Marjanović G, Trchová M, Stejskal J (2008) The chemical oxidative polymerization of aniline in water: Raman spectroscopy. J Raman Spectros 39(10):1375-1387. doi: 10.1002/jrs.2007

Engert C, Umapathy S, Kiefer W, Hamaguchi H (1994) Dynamic structure of charge carrier in polyaniline by near-infrared excited resonance Raman spectroscopy. Chem Phys Lett 218(1-2):87-92. doi: 10.1016/0009-2614(93)E1468-V

Faez R, Rezende MC, Martin IM, Paoli MA (2000) Polímeros condutores intrínsecos e seu potencial em blindagem de radiações eletromagnéticas. Polímeros 19(3):130-137. doi: 10.1590/S0104-14282000000300009

Franchitto M, Orlando AJF, Faez R, Rezende MC, Martin IM (2007) Measurements of reactivity and complex permittivities of radar absorbing materials based on conducting polymers. Proceedings of the 2007 Progress in Electromagnetics Research Symposium; Beijing, China.

Fratoddi I, Venditti I, Cametti C, Russo MV (2015) Chemiresistive polyaniline-based gas sensors: a mini review. Sensor Actuator B Chem 220:534-548. doi: 10.1016/j.snb.2015.05.107

Galiani PD, Malmonge JA, Santos DP, Malmonge LF (2007) Compósitos de borracha natural com polianilina. Polímeros 7(2):93-97. doi: 10.1590/S0104-14282007000200007

Gul S, Shahb AA, Bilal S (2013) Synthesis and characterization of processable polyaniline salts. J Phys Conf 439(2013):012002. doi: 10.1088/1742-6596/439/1/012002

Huang J, Kaner RB (2006) The intrinsic nanofibrillar morphology of polyaniline. Chem Commun (Camb) (4):367-376. doi: 10.1039/B510956f

Ibrahim KA (2013) Synthesis and characterization of polyaniline and poly(aniline-co-o-nitroaniline) using vibrational spectroscopy. Arabian Journal of Chemistry. doi: 10.1016/j.arabjc.2013.10.010. In press.

Jaymand M (2013) Recent progress in chemical modification of polyaniline. Prog Polymer Sci 38(9):1287-1306. doi: 10.1016/j.progpolymsci.2013.05.015

Joo J, Lee CY (2000) High frequency electromagnetic interference shielding response of mixtures and multilayer films based on conducting polymers. J Appl Phys 88(1):513-518. doi: 10.1063/1.373688

Kohut-Svelko N, Reynaud S, François J (2005) Synthesis and characterization of polyaniline prepared in the presence of nonionic surfactants in an aqueous dispersion. Synthetic Met 150(2):107-114. doi: 10.1016/j.synthmet.2004.12.022

Krukiewicz K, Katunin A (2016) The effect of reaction medium on the conductivity and morphology of polyaniline doped with camphorsulfonic acid. Synthetic Met 214:45-49. doi: 10.1016/j.synthmet.2016.01.017

Kumar SA, Bhandari H, Sharma C, Khatoon F, Dhawan SK (2013) A new smart coating of polyaniline-SiO2 composite for protection of mild steel against corrosion in strong acidic medium. Polymer Int 62(8):1192-1201. doi: 10.1002/pi.4406

Laska J, Widlarz J (2005) Spectroscopic and structural characterization of low molecular weight fractions of polyaniline. Polymer 46(5):1485-1495. doi: 10.1016/j.polymer.2004.12.008

Luo C, Peng H, Zhang L, Lu G-L, Wang Y, Travas-Sejdic J (2011) Formation of nano-microstructures of polyaniline and its derivatives. Macromolecules 44(17):6899-6907. doi: 10.1021/ma201350m

MacDiarmid AG (2001) "Synthetic Metals": a novel role for organic polymers (Nobel Lecture). Angew Chem Int Ed 40(14):2581-2590. doi: 10.1002/1521-3773(20010716)40:14<2581::AID-ANIE2581>3.0.CO;2-2

Manjunath BR, Venkataraman A, Stephen T (1973) The effect of moisture present in polymers on their X-ray diffraction patterns. J Appl Polymer Sci 17(4):1091-1099. doi: 10.1002/app.1973.070170407

Mažeikienė R, Tomkutė V, Kuodis Z, Niaura G, Malinauskas A (2007) Raman spectroelectrochemical study of polyaniline and sulfonated polyaniline in solutions of different pH. Vib Spectros 44(2):201-208. doi: 101016/j.vibspec.2006.09.005

Moravková Z, Trchová M, Exnerová M, Stejskal J (2012) The carbonization of thin polyaniline films. Thin Solid Films 520(19):6088-6094. doi: 101016/j.tsf.2012.05.067

Nithyaprakash D, Chandrasekaran J, Punithaveni B, Sasikumar L (2014) The study of optical, dielectric and optoelectronic properties of dodecylbenzene sulfonic acid doped polyaniline. Optik 125:5343-5347. doi: 101016/j.ijleo.2014.06.057

Nobrega MM, Silva HB, Constantino VRL, Temperini MLA (2012) Spectroscopic study on the structural differences of thermally induced cross-linking segments in emeraldine salt and base forms of polyaniline. J Phys Chem B 116(48):14191-14200. doi: 101021/jp309484e

Panah NB, Payehghadr M, Danaee I, Nourkojouri H, Sharbatdaran M (2012) Investigation of corrosion performance of epoxy coating containing polyaniline nanoparticles. Iran Polym J 21(11):747-754. doi: 10.1007/s13726-012-0080-8

Pouget JP, Józefowicz ME, Epstein AJ, Tang X, MacDiarmid AG (1991) X-ray structure of polyaniline. Macromolecules 24(3):779-789. doi: 101021/ma003a022

Rinaldi G, Huber T, McIntosh H, Lebrun L (2012) Corrosion sensor development for condition-based maintenance of aircraft. International Journal of Aerospace Engineering (2012):Article ID 684024, 11 pages. doi: 10.1155/2012/684024

Roichman Y, Titelman GI, Silverstein MS, SiegmannA, Narkis M (1999) Polyaniline synthesis: influence of powder morphology on conductivity of solution cast blends with polystyrene. Synthetic Met 98(3):201-209. doi: 10.1016/S0379-6779(98)00190-8

Sathiyanarayanan S, Azim SS, Venkatachari G (2008). Corrosion protection of magnesium alloy ZM21by polyaniline-blended coatings. J Coat Technol Res 5(4):471-477. doi: 10.1007/s11998-008-9097-5

Saravanan S, Mathai CJ, Anantharaman MR, Venkatachalam S, Prabhakaran PV (2006) Investigation on the electrical and structural properties of polyaniline doped with camphor sulphonic acid. J Phys Chem Solid 67(7):1496-1501. doi: 101016/j-jpcs.2006.01.100

Shakoor A, Rizvi TZ (2010) Raman spectroscopy of conducting poly (methylmethacrylate)/polyanilinedodecylbenzenesulfonate blends. J Raman Spectros 41(2):237-240. doi: 10.1002/jrs.2414

Stejskal J, Sapurina I, Trchov M (2010) Polyaniline nanostructures and the role of aniline oligomers in their formation. Progr Polymer Sci 35(12):1420-1481. doi: 10.1016/j.progpolymsci.2010.07.006

Soares B, Leyva ME, Barra GMO, Khastgir D (2006) Dielectric behavior of polyaniline synthesized by different techniques. Eur Polymer J 42(3):676-686. doi: 101016/eurpolymj.2005.08.013

Vivekanandan J, Ponnusamy V, Mahudeswaran A, Vijayanand PS (2011) Synthesis, characterization and conductivity study of polyaniline prepared by chemical oxidative and electrochemical methods. Arch Appl Sci Res 3(6):147-153.

Zeng X-R, Ko T-M (1998) Structure and properties of chemically reduced polyanilines. Polymer 39(5):1187-1195. doi: 10.1016/S0032-3861(97)00381-9

Zilberman M, Titelman GI, Siegmann A, Haba Y, Narkis M, Alperstein D (1997) Conductive blends of thermally dodecylbenzene sulfonic acid-doped polyaniline with thermoplastic polymers. J Appl Polymer Sci 66(2):243-253. doi: 10.1002/(SICI)1097-4628(19971010)66:2<243::AID-APP5>3.0.CO;2-W

8

Electromagnetic Properties of Multifunctional Composites Based on Glass Fiber Prepreg and Ni/Carbon Fiber Veil

Daniel Consoli Silveira[1], Newton Gomes[2], Mirabel Cerqueira Rezende[2,3], Edson Cocchieri Botelho[1]

ABSTRACT: Multifunctional composites combine structural and other physicochemical properties, with major applications in aeronautical, space, telecommunication, automotive, and medical areas. This research evaluates electromagnetic properties of multifunctional composites based on glass fiber woven fabric pre-impregnated with epoxy resin laminated together carbon fiber non-woven veil metalized with Ni. In this way, searching for possible application as radar absorbing structures or electromagnetic interference shielding structures. The scattering parameters, in the frequency range of 8.2 to 12.4 GHz, show that the epoxy resin/glass fiber prepreg allows the transmission of the electromagnetic waves through its microstructure, independently of the glass fiber reinforcement orientation (98% transmission, S_{24} = −0.09 dB). However, the carbon fiber/Ni veil shows highly reflector behavior (91% reflection, S_{22} = −0.43 dB). Energy dispersive spectroscopy of the veil, before and after nitric acid attacks, confirmed the Ni coating removal from the carbon fiber surface. Still, the scattering parameters show reflector behavior (77% reflection, S_{22} = −1.13 dB), attributed to the electrical conductivity of carbon fibers. Multifunctional composites based on glass fiber/epoxy/carbon fiber/Ni veil laminates were processed by hot compression molding. The scattering parameters show that the laminates do not behave as good radar absorbing structures. Nevertheless, the laminates present promising results for application as light weight and low thickness structural composites with electromagnetic interference shielding effectiveness (91.4% reflection for 0.36 mm thickness and 100% for ~ 1.1 mm) for buildings, aircraft, and space components.

KEYWORDS: Composite structures, Microwave absorption, Electromagnetic shielding, Glass fiber reinforced plastics, Carbon fiber reinforced plastics.

INTRODUCTION

Many technological advances in modern society have been achieved since the introduction of advanced and multifunctional materials and structures (Gibson 2012). The development of new multifunctional materials involves the combination of structural requirements with other properties not related to mechanical strength (Gibson 2010; Silveira 2015). Therefore, materials with structural properties may be combined — such as tensile, flexural, shear strength, and fatigue resistance — with non-structural materials possessing active properties, for example, electrical and thermal conductivity, self-healing ability, capability to interact with electromagnetic waves, biodegradability, among others (Chin and Lee 2007; Gibson 2012). The multifunctional composite materials with ability to interact with electromagnetic waves are categorized into 2 major classes: the radar absorbing structures (RAS) and the electromagnetic interference (EMI) shielding materials. Both classes may be developed with the ability to withstand mechanical loads, thereby having structural properties, combined with the inherent property of interacting with the electric and/or magnetic wave fields (Lee *et al.* 2016; Thomassin *et al.* 2013).

RAS act attenuating the electromagnetic wave energy via Joule effect, i.e., electrons, nucleus, permanent and/or induced dipoles of material are forced to move, inducing electric current, friction at molecular level, and, consequently, the Joule effect (Soethe *et al.* 2011). If material employed presents impossibility for free movement at the molecular level, the wave may be also attenuated by its reorientation, entering in contrary phase with the incident electromagnetic field. In this case we have the

1.Universidade Estadual Paulista Júlio de Mesquita Filho – Faculdade de Engenharia de Guaratinguetá – Departamento de Materiais e Tecnologia – Guaratinguetá/SP – Brazil. **2.**Departamento de Ciência e Tecnologia Aeroespacial – Instituto Tecnológico de Aeronáutica – Laboratório de Guerra Eletrônica – São José dos Campos/SP – Brazil. **3.**Universidade Federal de São Paulo – Instituto de Ciência e Tecnologia – Curso de Engenharia de Materiais – São José dos Campos/SP – Brazil.

Author for correspondence: Daniel Consoli Silveira | Universidade Estadual Paulista Júlio de Mesquita Filho – Faculdade de Engenharia de Guaratinguetá – Departamento de Materiais e Tecnologia | Avenida Dr. Ariberto Pereira da Cunha, 333 – Portal das Colinas | CEP: 12.516-410 – Guaratinguetá/SP – Brazil | Email: danielsilveira.tech@gmail.com

cancelling of wave (Jacob *et al.* 1995; Zhang *et al* 2015). The depreciation of wave energy by Joule effect inside the material or by phase cancelling reduces the target detection, since the wave is not reflected or is reflected with less intensity to the microwave source. In this way, this class of material assists military and government institutions by controlling spurious electromagnetic radiation, staying under necessary limits and reaching accordance with their respective laws (Silveira 2016).

Valuable RAS have been developed using multi-layered structures (Lee and Yang 2014; Lee *et al.* 2016) or employing frequency selective surfaces (FSS), wherein periodic geometric patterns are incorporated in conductive material (Wan *et al.* 2015; Lee *et al.* 2016). Generally FSS structures have reduced thickness in comparison with multi-layered RAS (Silveira 2016). Also, there are numerous applications of these materials in electronic industries, telecommunication, computers, wind power generation, medical sector, among others (Jang *et al.* 2014; Qin and Brosseau 2012). Some examples of successful application of absorbing materials are found in aviation field. In this area aircrafts equipped with stealth technology may be mentioned, such as the F-117A one, since this is the first aircraft that holds low detection at low altitude. Going further, the B-2 bomber, the F/A-22, and the F-35 aircraft family may be mentioned (Jenn 2005; Nangia and Palmer 2005).

Multi-layered RAS are based on materials responsible for dielectric losses (Lee *et al.* 2016) and/or magnetic losses (Xu *et al.* 2016). Several investigations on multi-layered dielectric structures have been reported in the literature, e.g., epoxy resin composites loaded with C-based materials, such as granular graphite, fullerenes and carbon fibers (Micheli *et al.* 2010), ferrite composites doped with dielectric lossy material, such as $Ba_{0.7}La_{0.3}Co_2Fe_{16}O_{27}$ ferrites (Shen *et al.* 2006), sandwich constructions based on carbon nanotube (CNT) combined with polymethacrylimide (PMI) and carbon/epoxy composite (Choi *et al.* 2012). Multi-layer panels with metallic back plane and dielectric layers are strategically positioned to a quarter of microwave length (λ/4) from the microwave incidence surface — the last employed for a considerable time and known as Jaumann absorbing structure (Hyde *et al.* 2014). Alternatively, FSS RAS have 3 different and relevant intrinsic features: the effective bandwidth, resonant frequency, and reflection loss (RL), all being dependent on structure characteristics (Lee *et al.* 2016; Zang *et al.* 2015). Also, the investigation on the FSS layer material is essential for the determination of the behavior of the FSS RAS (Silveira 2016; Xu *et al.* 2016), prioritizing materials with low reflection and transmission of microwave and maximized wave energy absorption.

Another important class of multifunctional materials, with ability to interact with microwaves, includes those that promote protection against EMI, which is a critical problem nowadays due to the development and spread of electronic information and its employment in several electronic devices (Lan *et al.* 2014). EMI structures have their shielding effectiveness (SE) mostly based on the reflection of the electromagnetic wave, with considerable metallic behavior (Seo *et al.* 2004), microwave energy absorption can also be observed on a smaller scale (Mishra *et al.* 2016). The total shielding effectiveness ($SE_T = SE_R + SE_A$) of a material is composed by SE due to reflection (SE_R) and SE due to absorption (SE_A). They can be calculated using scattering parameters results from electromagnetic characterization in a vector network analyzer equipped with wave guide and low-loss cables and connectors (Mishra *et al.* 2016), such as those applied in this research. Going further, the total shielding effectiveness ($SE_T = 10 \log P_t/P_i$) can be obtained by the logarithmic ratio between incoming power (P_i) from emitting port connected at a wave guide and transmitted power through material (P_t). The result is a negative quantity ($P_t < P_i$) expressed in decibels (dB) as the standard unit (Cao *et al.* 2010; Mishra *et al.* 2016).

In most cases, polymeric matrix materials are transparent to microwaves, thus requiring the addition of conductive fillers into the polymer to improve conductivity and reflectivity (Chen *et al.* 2013; Deng *et al.* 2014). An important consideration lays in the comparison of electrical conductivity among conductive particles dispersed into polymeric matrices and fibers and their respective fabrics combined as layer into polymeric matrices. The proximity between metallic particles dispersed in a polymeric matrix is critical for the formation of conductive pathways (Chan *et al.* 2011; Tang *et al.* 2013). Fiber surfaces, mainly carbon fiber (CF), which presents intrinsic conductivity (Gallo and Thostenson 2015; Tzeng and Chang 2001), ideally ensure the formation of conductive pathways, which in turn can help to optimize the microwave reflection and EMI SE.

Even though good EMI shielding is obtained by adding more CFs in the matrix, higher fiber volume fractions lead to difficulties in the production process with associated high costs. The aforementioned difficulties can be minimized or completely solved by coating the CF surface with thin layers of metal or conducting polymers (Tzeng and Chang 2001; Mishra *et al.* 2016). Mishra *et al.* (2016) presented the development of composites based on polyaniline (PANI) doped with β-naphthalene sulfonic acid (β-NSA), produced via chemical oxidative polymerization and with CF as filler. SE was calculated using scattering parameters (Par_S) results acquired from vector network analyzer in the frequency range from 8.2 to 12.4 GHz. A maximum of −31 dB was obtained for EMI SE (Mishra *et al.* 2016).

Ni is one of the main metals applied for metallization processes of CF, with cementation and electroless deposition employed as main metallization processes (Chung 2012; Sunitha *et al.* 2016). The strategic positioning of the metalized fabric layer into the composite material may be tailored to control the EMI phenomenon (Silveira 2016). Balaraju *et al.* (2016) carried out extensive research into electroless plating of Ni-based polyalloys over CF, *e.g.*, Ni-P, Ni-Cu-P, and Ni-W-Cu-P, deposited via alkaline citrate bath. The EMI SE results for the quaternary coating presented important values, mainly based on absorption (SE_A) and equal to −37 dB. The effect of electroless plating of Ni-P-Cu alloys onto CF and combined with polyether ether ketone composites (CF/PEEK) was also studied by Su *et al.* (2014). The resulted material presented EMI SE over −76 dB, providing CF/PEEK composite, with Ni-P-Cu coating over fibers, potential for different applications satisfying several requirements of EMI (Su *et al.* 2014).

The objective of this study was to process and characterize multifunctional composites, based on epoxy resin/glass fiber (GF) woven fabric prepreg and CF/Ni veil, a non-woven material, evaluating their behavior as RAS or EMI shielding structure in the microwave band (8.2 to 12.4 GHz). In order to evaluate the application of the processed composites as RAS, the CF/Ni veil was treated with HNO_3 acid for the Ni removal from the CF surface.

MATERIALS AND METHODS

The laminates prepared in this study used glass fiber/epoxy resin prepreg and a CF veil metalized with Ni (CF/Ni veil). The prepreg, code HexPly® F155, was based on GF fabric, plain weave style, impregnated with epoxy resin (cured at 120 °C), from Hexcel Co. The conductive layer was based on CF veil bonded with a polyester resin and metalized with Ni. This material is produced by Advanced Fiber Nonwovens, a group of Hollingsworth & Vose Company, code 8000826. This veil is characterized by weight of 25.4 g/m^2, thickness of 0.18 mm, and surface electrical resistance equal to 0.25 DC Ω/square. This unity relates the specific electrical resistance in 2 dimensions for materials with uniform and reduced thickness (Bugnet *et al.* 2001; Joffe and Lock 2010).

Aiming to act upon the microwave reflection and absorption properties of the CF/Ni veil, acid treatments were performed on the veil surface. This procedure intended to remove or to reduce, in a controlled manner, the Ni content on the CF surface. For this purpose, samples of CF/Ni veil (7.0 cm × 7.0 cm) were immersed, separately, in aqueous solution of HNO_3 5.0 mol/L for 2 different times, 45 and 120 min, with moderate manual agitation every 5 min. This procedure resulted in 2 new samples of veil, one attacked in acid solution for 45 min (named V_1) and other treated during 120 min (named V_2). After the acid attacks, the veil samples were washed, being shaken carefully in deionized water and dried in an oven at 60 °C for 270 min.

Laminates based on GF/F155-epoxy resin prepreg have been used routinely in aeronautical industry, with applications in primary and secondary structures, due to their high modulus (Paiva *et al.* 2005). Table 1 shows the families of laminates processed by hot-compression molding in the present study. The processing was performed at a heating rate of 7.6 °C/min up to 115 °C, with an isothermal at this temperature for 100 min. The applied pressure was 0.64 MPa. The used sequence of layers is depicted in Table 1, in whose caption it is possible to identify each composite layer. The laminate composite $P/V_0/P_2$ in Family 1, for example, indicates a prepreg layer (P) followed by a unique veil layer as received (V_0) and 2 prepreg layers (P_2). The composite shown in Family 7 (Table 1) is formed by the combination of 2 laminates, $P_3/V_1/P_4$ and $P/V_0/P_2$. So, during the electromagnetic evaluation, the microwave may reach the composite surface formed by 3 cured prepreg layers (P_3), and, presuming that the microwave travels across the complete thickness of the material, it sequentially encounters the V_1 layer and then 5 layers ($P_4 + P$) of the cured prepreg plus 1 layer of V_0 and finally 2 cured prepreg layers (P_2). The contrary can also occur, with the wave reaching the opposite side of this composite, i.e., reaching P_2, travelling across the laminate in the opposite direction.

Table 1. Multilayer composites electromagnetically characterized.

Families	Laminates
1	$P/V_0/P_2$
2	$P_3/V_0/P_4$
3	$P/V_1/P_2$
4	$P_3/V_1/P_4$
5	$P/V_2/P_2$
6	$P_3/V_2/P_4$
7	$P_3/V_1/P_4 + P/V_0/P_2$
8	$P_2/V_2/P + P/V_1/P_2 + P/V_0/P_2$

P_n: n layers of cured prepreg; V_0: 1 layer of original veil as received; V_1: 1 layer of veil treated with acid solution during 45 min; V_2: 1 layer of veil treated with acid solution during 120 min; "/": Interfacial region between different layers.

The scanning electron microscopy (SEM) analyses were performed in a ZEISS, EVO series LS15 model, with a microprobe INCAx-act from Oxford-Instruments, resolution of 133 eV at 5.9 keV, for the elemental mapping of chemical composition via energy dispersive spectroscopy (EDS). This accessory was used to evaluate the Ni removal from the CF surface of the treated veil. SEM images were obtained by back scattered electron detector (CZBSD) with 20 kV electron accelerating voltage, working distance (WD) of 8.5 mm, without metallization of the samples.

The electromagnetic characterization of the CF/Ni veil, epoxy resin/GF prepreg, and the processed composites (Table 1) were performed using a vector network analyzer (VNA) from Agilent Technologies, model PNA-L N5230C, with 4 ports for signal generation/capture, frequency generator capacity between 300 kHz and 20 GHz, low-loss cables/connectors, and rectangular waveguide, also from Agilent Technologies. The scattering parameters measurements were performed using 2 of the 4 ports available in the VNA, in the frequency range of 8.2 to 12.4 GHz.

Equations 1 and 2 are related with S_{ii} (microwave reflection on material) and S_{ji} (microwave transmission through the material), respectively (Karmel *et al.* 1998; Nicolson and Ross 1970).

$$S_{ii} = \frac{\text{reflected power at port } i}{\text{incident power from port } i} = \left.\frac{V_i^-}{V_i^+}\right|_{V_j^+ = null} \quad (1)$$

$$S_{ji} = \frac{\text{transmitted power at port } j}{\text{incident power from port } i} = \left.\frac{V_j^-}{V_i^+}\right|_{V_j^+ = null} \quad (2)$$

Equation 1 relates the total energy generated at port i (V_i^+) of VNA, *i.e.*, microwave energy emitted from port i and incident on the material located at the VNA rectangular sample-holder, with the portion of energy reflected back from the material to the same port (V_i^-). Thus the total energy measured in dB refers effectively to the reflected wave on the material surface, with no energy emitted from port j (V_j^+ = null). In contrast, Eq. 2 relates the total energy generated at port i (V_i^+) of VNA with the total energy that crosses the material thickness, finally achieving the port j (V_j^-) located beyond the material. Also, in this case, the generation of microwave energy from port j is turned off (V_j^+ = null). Equation 2 measures the microwave energy that propagates through the entire thickness of the material. The portion of energy not received at port j refers to the attenuated energy, due to the reflected energy on the material surface returning to the emitting port (S_{ii}) and due to the portion of absorbed energy by the material microstructure (E_a) and the dissipated energy (E_d) in the process, the last being considered null.

Equation 3 allows the conversion of S_{ii} and S_{ji} from dB to percentage (%) of signal attenuation or gain (Lee 1991):

$$\text{Signal gain or loss } (\%) = 100.\left[1 - 10^{\left(\frac{-dB}{10}\right)}\right] \quad (3)$$

Equation 4 supports the understanding of the relationship between transmittance and reflectivity, more specifically the transmission coefficient (T) and the reflection coefficient (Γ). According to this equation, the highest impedance matching between the veil (Z_L) and the medium of wave propagation (Z_0) the nearest to the unit (1.0) is the ratio between the impedances. Substituting this value ($Z_L = Z_0$) in Eq. 4, we have the sum of the squares of Γ and T equal to 1.0, i.e., the reflection decreases as the transmission through the material increases (Bayrakdar 2011; Liao 1990).

$$1 - \Gamma^2 = \frac{Z_0}{Z_L} T^2, \quad (4)$$

where: Z_0 is the impedance of the medium of wave propagation; Z_L is the intrinsic impedance of the material under research.

RESULTS AND DISCUSSION

ELECTROMAGNETIC CHARACTERIZATION

The electromagnetic behavior of epoxy resin/GF prepreg was measured in 2 different GF orientations, in order to verify the possible influence of the arrangement of GF fibers in the interaction with microwaves in the X-band (8.2 to 12.4 GHz). Figure 1 displays the assessed GF orientations.

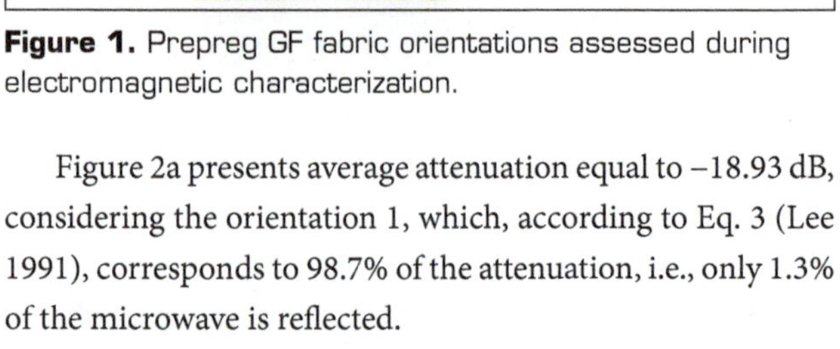

Figure 1. Prepreg GF fabric orientations assessed during electromagnetic characterization.

Figure 2a presents average attenuation equal to −18.93 dB, considering the orientation 1, which, according to Eq. 3 (Lee 1991), corresponds to 98.7% of the attenuation, i.e., only 1.3% of the microwave is reflected.

The same wave-material interaction behavior was identified for the epoxy resin/GF fabric prepreg with orientation 2. As confirmed by S_{22} parameter presented in Fig. 2b, the material is highly transparent to the microwaves in the X-band, presenting average attenuation of −18.78 dB. This value indicates an attenuation of 98.7%, and only 1.3% of the signal returns to port 2.

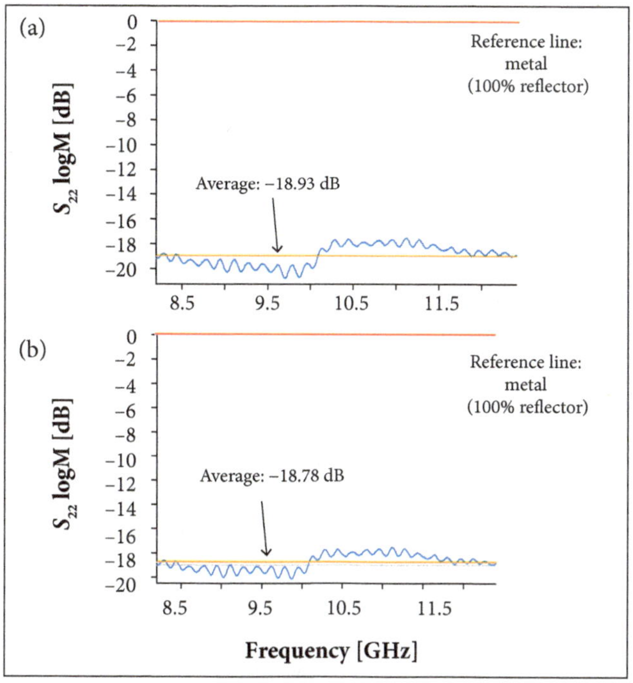

Figure 2. S_{22} curves versus frequency for the epoxy/GF prepreg with GF orientation (a) 1 and (b) 2.

The epoxy resin/GF prepreg transmission (S_{24} and S_{42}) was also measured, and the results confirm the low reflectivity of the material, with average attenuation values through the material of only −0.09 dB, which, according to Eq. 3, indicates that 98% of the microwave is transmitted through the prepreg, independently of the incident wave side on the material (S_{24} is equal S_{42}). The parameter S_{44} was also measured (−18.1 dB for orientations 1 and 2). These results are similar to those shown in Fig. 2. Therefore, the epoxy resin/GF prepreg can be applied during manual lamination process on both sides and at any of the 2 fibers orientations evaluated in this paper, with no impact on the material-microwave interaction. Based on the scattering parameters of the epoxy resin/GF prepreg, it is verified a low reflector behavior (S_{22} equal to S_{44} and highly attenuated) and a high transmission of microwaves through the laminate (S_{24} and S_{42} near to 0 dB of attenuation). Considering the well-known resistance of this material in deteriorative conditions, such as ultraviolet radiation (UV) exposure (Zainuddin *et al.* 2011), moisture absorption (Botelho *et al.* 2008), and thermal degradation (Silveira 2016), the studied prepreg shows potential application as outer layer of EMI structures.

Figure 3 illustrates the electromagnetic properties of the CF/Ni veil, as received, in the frequency range of 8.2 to 12.4 GHz. Taking the average attenuation value of only −0.43 dB for S_{22} as basis, the material in question presents reflector behavior, and 90.7% of wave energy is reflected on the CF/Ni veil surface. This reflector behavior is more pronounced in frequencies below 10.3 GHz as can be seen in Fig. 3a.

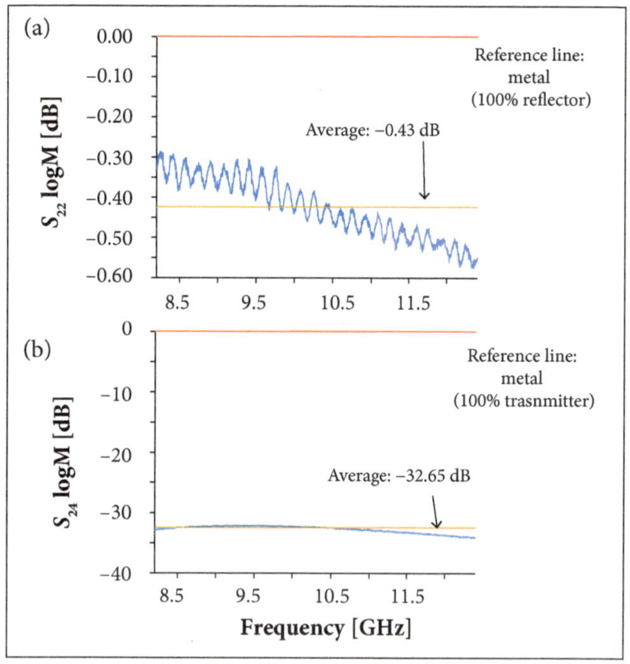

Figure 3. Curves of (a) S_{22} and (b) S_{24} versus frequency of the CF/Ni veil, as received.

Figure 3b presents the S_{24} parameter. The average value of −32.7 dB corresponds to 99.95% of attenuation, i.e., the signal emitted from port 4 does not cross the CF/Ni veil towards port 2. Correlating this result with the S_{22} data, it is clear that the signal is practically all reflected on the veil surface.

The S_{44} and S_{42} parameters were also measured: −0.46 and −32.5 dB average values, respectively. Comparing these values with those presented in Fig. 3, the reflection and transmission present similar behavior independently of the incidence side of the electromagnetic wave on the CF/Ni veil, for frequencies ranging from 8.2 to 12.4 GHz. Based on this information, the veil lamination process can be performed with no worries related with the lamination side previously to the hot-compression molding process.

ENERGY DISPERSIVE SPECTROSCOPY

Once identified the high reflectivity of the CF/Ni veil, as received, acid attacks were conducted, aiming the removal of the Ni layer from the CF surface. This procedure was an attempt to decrease the reflector behavior of the veil, which, in turn, could lead the material to a condition of greater absorption or transmission of electromagnetic waves in the X-band, providing field of application in RAS. Figure 4 shows representative SEM/EDS chemical mapping in 3 different regions of the CF/Ni veil as received, representative of the veil in its totality. Figure 4a exhibits non-uniform region formed with the presence of deposition and accumulation of Ni (Spectrum 1). Figure 4b presents exposed CF (Spectrum 2), and Fig. 4c is representative of a region with uniform and continuous Ni coating (Spectrum 3).

Table 2 shows the average content, in mass and atomic percentages, of Ni and other chemical elements related to each of the 3 different assessed regions from Fig. 4. The values in Table 2 and the characteristic peaks in Fig. 4a make evident the presence of Ni, indicating high percentage in mass for this metal in the veil, as received. Spectrum from Fig. 4b is related to the black region, with high content (79.2%) of carbon element, where the CF surface is partially exposed. This aspect is attributed to the mechanical removal of Ni layer due to friction between fibers, during the veil handling. This, in turn, is the root cause for the deposition and accumulation of Ni as pointed in Spectrum 1. The chemical content indicated in the spectrum of Fig. 4c, as expected, shows high content of Ni, 72.75% in mass. The EDS spectra also show the presence of O, probably related to the presence of oxides (Burakowski and Rezende 2001) and/or adsorbed gases (An *et al.* 2013).

At sequence, Fig. 5 presents SEM images and results of chemical mapping via EDS in 3 different regions of the CF/Ni veil after the HNO_3 acid attacks.

Table 3 shows the average Ni content determined by SEM/EDS in the veil exposed to the acid attack during 45 min (V_1). Two of the spectra, specifically the Spectra 2 and 3, did not detected the presence of Ni, and the spectrum related to the region pointed in Fig. 5a detected a small percentage of Ni content (0.15% in mass). In addition, it need to be mentioned that the Ni content of the veil attacked in HNO_3 solution during 120 min (V_2) was also surveyed, with no presence of Ni or only traces (average ~ 0.2% in mass and 0.05% atomic). These results show that the acid attacks in different exposure times were effective in the Ni removal from the CF surface.

The EDS technique of CF/Ni veil after acid attack also detected traces of other elements; among these contaminants, Cl, Si, and Fe were identified. The presence of Fe is attributed to the use of a Fe-C steel support to keep the veil surface suspended during the drying process in the oven, at 60 °C for 270 min. In addition,

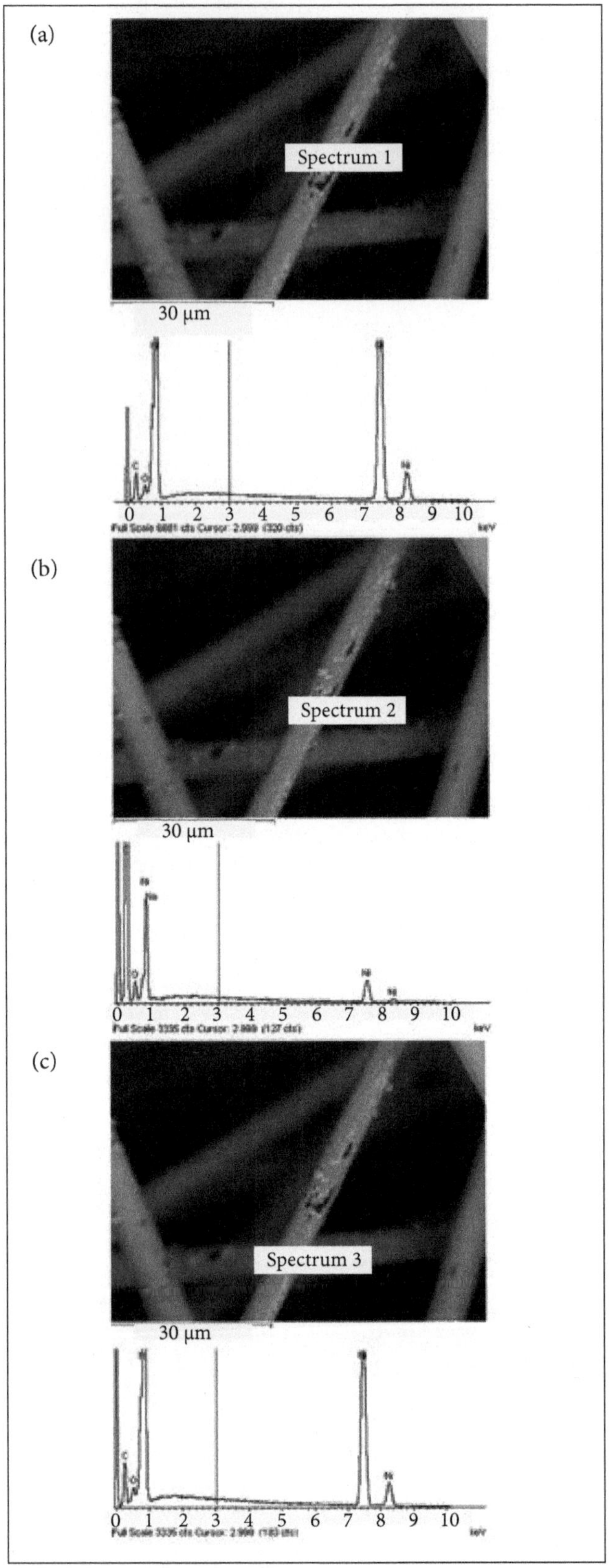

Figure 4. SEM/EDS analyses of the CF/Ni veil as received. (a) Spectrum 1, (b) Spectrum 2, and (c) Spectrum 3.

the comparison of O contents in Tables 2 and 3 shows an increase in this element after the acid attack. This fact can be explained by oxidation processes during acid attack and high temperatures during drying process in the oven, which may add O element on the surface and/or into the samples (An *et al.* 2013). The high carbon content determined is expected. Firstly, because the precursor material of the veil is CF; secondly, due to the presence of polyester polymer, used as binder for the CFs in the veil, as can be observed in SEM images (Spectra 1 and 3 from Fig. 5), which has the C as the main chemical element forming its macromolecules.

The 2 veil samples exposed to acid attacks had their scattering parameters measured. Figure 6a is representative of the S_{22} curves versus frequency for veils V_1 and V_2. This figure shows that the S_{22} curve for the veils submitted to the acid attack moved downward in relation to the original veil, represented by the black curve in Fig. 6 (−0.43 dB, ~ 91% of reflectivity), therefore indicating the drop in the reflectivity of the veils exposed to the acid treatments. Specifically for the veil sample exposed to the acid attack during 45 min (V_1), the average S_{22} is −1.13 dB, in other words, an average reflectivity of 77%. Thus, it was found

Table 2. Chemical contents in mass and atomic percentages determined by EDS of the CF/Ni veil as received.

Element	% Mass			% Atomic		
	Figure 4a*	Figure 4b	Figure 4c	Figure 4a	Figure 4b	Figure 4c
C	16.7	79.2	24.5	47.7	88.5	59.16
O	2.2	10.9	2.75	4.8	9.15	4.97
Ni	81.1	9.6	72.75	47.5	2.2	35.87
Na	--	0.3	--	--	0.15	--

*Regions indicated in Fig. 4.

Table 3. Chemical contents in mass and atomic percentages determined by EDS of the CF/Ni veil after the acid attack.

Element	% Mass			% Atomic		
	Figure 5a*	Figure 5b	Figure 5c	Figure 5a	Figure 5b	Figure 5c
C	63.07	84.19	71.8	69.8	88.1	77.8
O	35.87	14.47	26.43	29.8	11.4	21.5
S	0.91	0.52	1.77	0.37	0.21	0.70
Ni	0.15	--	--	0.03	--	--
Si	--	0.19	--	--	0.08	--
Cl	--	0.46	--	--	0.16	--
Fe	--	0.17	--	--	0.05	--

*Regions indicated in Fig. 5.

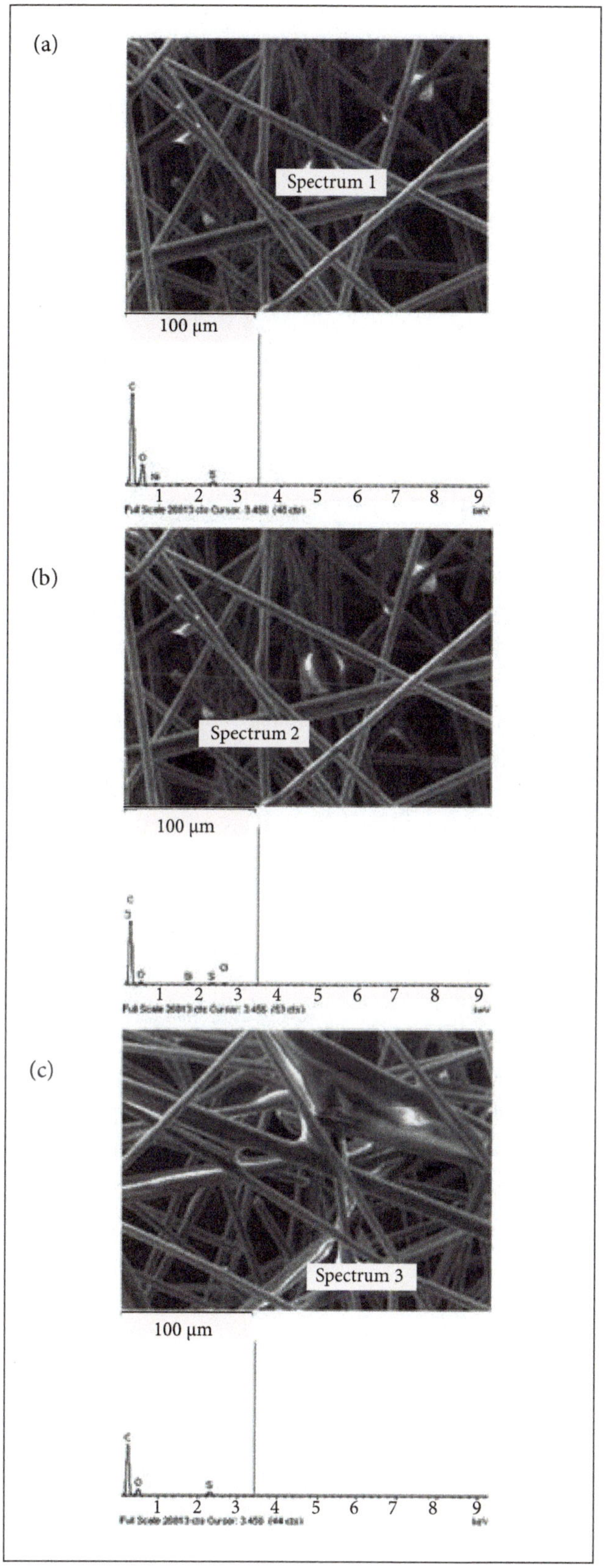

Figure 5. SEM/EDS analyses of the CF/Ni veil after acid attack. (a) Spectrum 1, (b) Spectrum 2, and (c) Spectrum 3.

a maximum reflectivity reduction equal to 14% in comparison with the original CF/Ni veil, as a consequence of the Ni removal. For the veil sample exposed to the acid attack during 120 min (V_2), the average S_{22} is −0.56 dB, corresponding to an average reflectivity of ~ 88%.

The S_{42} transmittance parameters for veils V_0, V_1, and V_2 were also measured, and the results are presented in Fig. 6b. The S_{42} values show that the transmittance of the veils increases from V_0, through V_2, and upward to V_1. Therefore veils after acid attacks present lower attenuation of microwaves through its microstructure. This behavior was expected based on S_{22} results and Eq. 4.

Now the study is focused on understanding the electromagnetic behavior of the epoxy resin/GF/CF/Ni laminates when impinged by microwaves. According to the literature, when a wave reaches a structure, wave displacements around its axis of propagation may occur, followed by multiple reflections into the laminate. This phenomenon can contribute to phase cancellation between the incident and the reflected wave, leading to the signal attenuation and approaching the material to the necessary conditions for application in RAS (Zhang *et al.* 2015).

Figures 7 present the S_{22} and S_{44} curves versus frequency for the laminates presented in Table 1. In these figures it is observed that the $P/V_2/P_2$ laminate presents the lowest value of attenuation (S_{44} ~ 0 dB), featuring no attenuation and reflectivity close to 100%. The greatest average attenuation occurs for the $P/V_1/P_2$ laminate (S_{22} = −0.39 dB) when the electromagnetic waves are emitted from port 2, i.e., composite side with 1 prepreg layer and V_1 layer exposed to the incident wave. Thus, the lowest average reflectivity among all these laminates is equal to 91.4%. Also, for the laminate $P_3/V_0/P_4$, it is observed an attenuation peak (S_{22} = −0.82 dB) of electromagnetic wave with frequency equal to 9.7 GHz, i.e., 82.8% of reflectivity at this specific frequency.

From the investigation and based on the electromagnetic characterization results, e.g., S_{22} and S_{44}, it was found that the processed laminates do not present desired properties for potential development of RAS. However, the processed laminates combine low thickness (between 0.36 mm for $P/V_1/P_2$ and 1.08 mm for $P_2/V_2/P + P/V_1/P_2 + P/V_0/P_2$), light weight, structural properties, and, as important as the other mentioned properties, significant reflector behavior (low attenuation for S_{ii} parameters, average reflection between 91.4 and 100% of incident microwave in the X-band). In this case, the studied materials present potential for application as multifunctional composites in EMI shielding of buildings, aircraft, and space components.

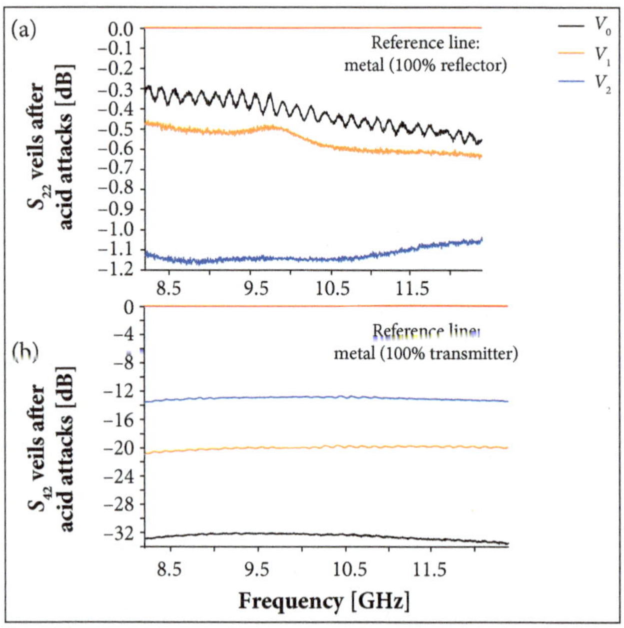

Figure 6. S_{22} (a) and S_{42} (b) curves versus frequency for the CF/Ni veil after acid attacks.

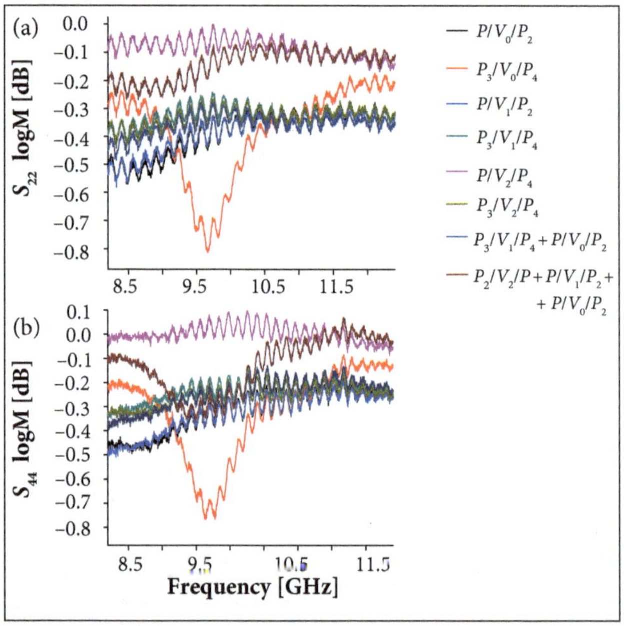

Figure 7. S_{22} (a) and S_{44} (b) parameters versus frequency for GF/epoxy resin/CF/Ni laminates.

CONCLUSIONS

The results presented in this research allow concluding that the epoxy resin/GF fabric prepreg can be applied as outer layer of RAS, since this material allows the transmission, through its microstructure, of 98.7% of electromagnetic waves in the frequency range of 8.2 to 12.4 GHz. The CF/Ni veil used in this research is predominantly reflective (~ 91%). Acid treatments of CF/Ni veil were effective in the Ni removal from the CF surface, as confirmed via EDS, with 13.6% of maximum decrease in reflector behavior of the CF veil subjected to acid treatment in 5.0 mol/L HNO_3 solution during 45 min. Hence, even after the complete removal of Ni metallized layer, the CF veil still presented low attenuation for S_{ii} parameters, and it is not adequate for application in RAS, when combined with the epoxy resin/GF fabric prepreg. On the other hand, the results indicate compatibility for processing light-weight and low-thickness reflective multifunctional composites based on epoxy resin/GF/CF/Ni, designed for EMI shielding applications, for example, in buildings, communication, automation, bio-medicine, and aerospace areas. Considering that the studied composites were processed based on continuous GF fabric that presents good mechanical properties, and CF/Ni veil with reflector behavior for microwaves in X-band, multifunctional composites with intrinsic structural and EMI properties have been developed.

ACKNOWLEDGEMENTS

The authors are grateful to the Conselho Nacional de Desenvolvimento Científico e Tecnológico (CNPq; Proc.: 303287/2013-6, 303559/2012-8), Coordenação de Aperfeiçoamento de Pessoal de Nível Superior/Programa Professor Visitante Nacional Sênior (CAPES/PVNS), Laboratório de Guerra Eletrônica (LAB-GE) at the Instituto Tecnológico de Aeronáutica (ITA), and Laboratório de Análise de Imagens de Materiais (LAIMat) at the Faculdade de Engenharia of the Universidade Estadual Paulista "Júlio de Mesquita Filho" (UNESP), Guaratinguetá Campus, Brazil.

AUTHOR'S CONTRIBUTION

All authors conceived the idea as well as contributed to design creation, manufacturing, characterizations, data acquisition, results analysis, and comments on the manuscript.

REFERENCES

An H, Feng B, Su S (2013) Effect of monolithic structure on CO_2 adsorption performance of activated carbon fiber-phenolic resin composite: A simulation study. Fuel 103:80-86. doi: 10.1016/j.fuel.2011.06.076

Balaraju JN, Radhakrishnan P, Ezhilselvi V, Kumar AA, Chen Z, Surendran KP (2016) Studies on electroless nickel polyalloy coatings over carbon fibers/CFRP composites. Surf Coating Tech 302:389-397. doi: 10.1016/j.surfcoat.2016.06.040

Bayrakdar H (2011) Complex permittivity, complex permeability and microwave absorption properties of ferrite-paraffin polymer composites. J Magn Magn Mater 323(14):1882-1885. doi: 10.1016/j.jmmm.2011.02.030

Botelho EC, Rezende MC, Pardini LC (2008) Hygrotermal effects evaluation using the iosipescu shear test for glare laminates. J Braz Soc Mech Sci Eng 30(3):213-220. doi: 10.1590/S1678-58782008000300006

Bugnet B, Costa M, Doniat D, inventors; Porous structures having a pre-metallization conductive polymer coating and method of manufacture. 2001 Sept 18. United States Patent US 6290832 B1.

Burakowski L, Rezende MC (2001) Modificação da rugosidade de fibras de carbono por método químico para aplicação em compósitos poliméricos. Polímeros 11(2):51-57. doi: 10.1590/S0104-14282001000200006

Cao M-S, Song W-L, Hou Z-L, Wen B, Yuan J (2010) The effects of temperature and frequency on the dielectric properties, electromagnetic interference shielding and microwave-absorption of short carbon fiber/silica composites. Carbon 48(3):788-796. doi: 10.1016/j.carbon.2009.10.028

Chan KL, Mariatti M, Lockman Z, Slim LC (2011) Effects of the size and filler loading on the properties of copper- and silver-nanoparticle-filled epoxy composites. J Appl Polymer Sci 121(6):3145-3152. doi: 10.1002/app.33798

Chen Z, Xu C, Ma C, Ren W, Cheng H-M (2013) Lightweight and flexible graphene foam composites for high-performance electromagnetic interference shielding. Adv Mater 25(9):1296-1300. doi: 10.1002/adma.201204196

Chin WS, Lee DG (2007) Development of the composite RAS (radar absorbing structure) for the X-band frequency range. Compos Struct 77(4):457-465. doi: 10.1016/j.compstruct.2005.07.021

Choi I, Kim JG, Seo IIS, Lee DG (2012) Radar absorbing sandwich construction composed of CNT, PMI foam and carbon/epoxy composite. Compos Struct 94(9):3002-3008. doi: 10.1016/j.compstruct.2012.04.009

Chung DDL (2012) Carbon materials for structural self-sensing, electromagnetic shielding and thermal interfacing. Carbon 50(9):1972-2012. doi: 10.1016/j.carbon.2012.01.031

Deng H, Lin L, Ji M, Zhang S, Yang M, Fu Q (2014) Progress on the morphological control of conductive network in conductive polymer composites and the use as electroactive multifunctional materials. Progr Polymer Sci 39(4):627-655. doi: 10.1016/j.progpolymsci.2013.07.007

Gallo GJ, Thostenson ET (2015) Electrical characterization and modeling of carbon nanotube and carbon fiber self-sensing composites for enhanced sensing of microcracks. Materials Today Communications 3:17-26. doi: 10.1016/j.mtcomm.2015.01.009

Gibson RF (2010) A review of recent research on mechanics of multifunctional composite materials and structures. Compos Struct 92(12):2793-2810. doi: 10.1016/j.compstruct.2010.05.003

Gibson RF (2012) Principles of composite material mechanics. 3rd ed. Boca Raton: Taylor & Francis.

Hyde MW, Bogle AE, Havrilla MJ (2014) Nondestructive characterization of Salisbury screen and Jaumann absorbers using a clamped rectangular waveguide geometry. Measurement 53:83-90. doi: 10.1016/j.measurement.2014.03.025

Jacob J, Chia LHL, Boey FYC (1995) Thermal and non-thermal interaction of microwave radiation with materials. J Mater Sci 30(21):5321-5327. doi: 10.1007/BF00351541

Jang H-K, Choi W-H, Kim C-G, Kim J-B, Lim D-W (2014) Manufacture and characterization of stealth wind turbine blade with periodic pattern surface for reducing radar interference. Compos Part B-Eng 56:178-183. doi: 10.1016/j.compositesb.2013.08.043

Jenn DC (2005) Radar and laser cross section engineering. 2nd ed. Reston: AIAA (Education Series).

Joffe EB, Lock K-S (2010) Grounds for grounding: a circuit-to-system handbook. IEEE Electromagnetic Compatibility Society. New Jersey: John Wiley & Sons.

Karmel PR, Colef GD, Camisa RL (1998) Introduction to electromagnetic and microwave engineering. New York: Wiley.

Lan M, Cai J, Zhang D, Yuan L, Xu Y (2014) Electromagnetic shielding effectiveness and mechanical property of polymer-matrix composites containing metallized conductive porous flake-shaped diatomite. Compos Part B-Eng 67:132-137. doi: 10.1016/j.compositesb.2014.06.029

Lee C-S, Yang C-L (2014) Thickness and permittivity measurement in multi-layered dielectric structures using complementary split-ring resonators. IEEE Sensor J 14(3):695-700. doi: 10.1109/JSEN.2013.2285918

Lee S-E, Lee W-J, Oh K-S, Kim C-G (2016) Broadband all fiber-reinforced composite radar absorbing structure integrated by inductive frequency selective carbon fiber fabric and carbon-nanotube-loaded glass fabric. Carbon 107:564-572. doi: 10.1016/j.carbon.2016.06.005

Lee SM (1991) International encyclopedia of composites. New York: VCH Publishers. Vol. 4.

Liao SY (1990) Microwave devices & circuits. 3rd ed. Upper Saddle River: Prentice-Hall.

Micheli D, Apollo C, Pastore R, Marchetti M (2010) X-Band microwave characterization of carbon-based nanocomposite material, absorption capability comparison and RAS design simulation. Compos Sci Tech 70(2):400-409. doi: 10.1016/j.compscitech.2009.11.015

Mishra M, Singh AP, Gupta V, Chandra A (2016) Tunable EMI shielding effectiveness using new exotic carbon: Polymer composites. J Alloy Comp 688:399-403. doi: 10.1016/j.jallcom.2016.07.190

Nangia RK, Palmer ME (2005) A comparative study of UCAV type wing planforms-aero performance and stability considerations. Paper presented at: 23rd AIAA Applied Aerodynamics Conference; Toronto, Canada.

Nicolson AM, Ross GF (1970) Measurement of the intrinsic properties of materials by time-domain techniques. IEEE T Instrum Meas 19(4):377-382. doi: 10.1109/TIM.1970.4313932

Paiva JMF, Mayer S, Rezende MC (2005) Evaluation of mechanical properties of four different carbon/epoxy composites used in aeronautical field. Mater Res 8(1):91-97. doi: 10.1590/S1516-14392005000100016

Qin F, Brosseau C (2012) A review and analysis of microwave absorption in polymer composites filled with carbonaceous particles. J Appl Phys 111(061301). doi: 10.1063/1.3688435

Seo IIS, Chin WS, Lee DG (2004) Characterization of electromagnetic properties of polymeric composite materials with free space method. Compos Struct 66(1-4):533-542. doi: 10.1016/j.compstruct.2004.04.076

Shen G, Xu Z, Li Y (2006) Absorbing properties and structural design of microwave absorbers based on W-type La-doped ferrite and carbon fiber composites. J Magn Magn Mater 301(2):325-330. doi: 10.1016/j.jmmm.2005.07.007

Silveira DC (2016) Obtenção e caracterização de estruturas absorvedoras de micro-ondas baseadas em laminado de fibra laminado de fibra de vidro/resina epoxy/véu de C/Ni (Master's thesis). Guaratinguetá: Universidade Estadual Paulista "Júlio de Mesquita Filho".

Silveira DC (2015) Processing and characterization of structural and electromagnetic microwave absorber multifunctional composite based on GF/epoxy/CF/Ni. Proceedings of the 2ª Jornada de Iniciação Científica da Pós-Graduação, UNESP, Faculdade de Engenharia de Guaratinguetá [accessed 2016 April 04]. https://www.researchgate.net/publication/282671863_Processing_and_characterization_of_structural_and_electromagnetic_microwave_absorber_multifunctional_composite_based_on_GFEpoxyCFNi

Soethe VL, Nohara EL, Fontana LC, Rezende MC (2011) Radar absorbing materials based on titanium thin film obtained by sputtering technique. J Aerosp Technol Manag 3(3):279-286. doi: 10.5028/jatm.2011.03030511

Su Y, Zhou B, Liu L, Lian J, Li G (2014) Electromagnetic shielding and corrosion resistance of electroless Ni-P and Ni-P-Cu coatings on polymer/carbon fiber composites. Polymer Compos 36(5):923-930. doi: 10.1002/pc.23012

Sunitha JN, Rajesh CS, Rai SK (2016) Electromagnetic interference shielding effectiveness and electrical conductivity of Ni coated PCABS/PPS composites with reinforcement of carbon fibre. Polymer Polymer Compos 24(1):57-63.

Tang L-C, Wan Y-J, Peng K, Pei Y-B, Wu L-B, Chen L-M, Shu L-J, Jiang J-X, Lai G-Q (2013) Fracture toughness and electrical conductivity of epoxy composites filled with carbon nanotubes and spherical particles. Compos Part A-Appl S 45:95-101. doi: 10.1016/j.compositesa.2012.09.012

Thomassin J-M, Jérôme C, Pardoen T, Bailly C, Huynen I, Detrembleur C (2013) Polymer/carbon based composites as electromagnetic interference (EMI) shielding materials. Mater Sci Eng R Rep 74(7):211-232. doi: 10.1016/j.mser.2013.06.001

Tzeng S-S, Chang F-Y (2001) EMI shielding effectiveness of metal-coated carbon fiber-reinforced ABS composites. Mater Sci Eng 302(2):258-267. doi: 10.1016/S0921-5093(00)01824-4

Wan D, Bie S-W, Zhou J, Xu H, Xu Y, Jiang J (2015) A thin and broadband microwave absorber based on magnetic sheets and resistive FSS. Progress in Electromagnetics Research C 56:93-100. doi: 10.2528/PIERC14122203

Xu H, Bie S, Xu Y, Yuan W, Chen Q, Jiang J (2016) Broad bandwidth of thin composite radar absorbing structures embedded with frequency selective surfaces. Compos Part A-Appl S 80:111-117. doi: 10.1016/j.compositesa.2015.10.019

Zainuddin S, Hosur MV, Barua R, Kumar Ashok, Jeelani S (2011) Effects of ultraviolet radiation and condensation on static and dynamic compression behavior of neat and nanoclay infused epoxy/glass composites. J Compos Mater 45(18):1901-1918. doi: 10.1177/0021998310394693

Zang Y, Xia S, Li L, Ren G, Chen Q, Quan H, Wu Q (2015) Microwave absorption enhancement of rectangular activated carbon fibers screen composites. Compos Part B-Eng 77:371-378. doi: 10.1016/j.compositesb.2015.03.059

Zhang J, Xiao P, Zhou W, Hong W, Luo H (2015) Preparation and microwave absorbing properties of carbon fibers/epoxy composites with grid structure. J Mater Sci Mater Electron 26(2):651-658. doi: 10.1007/s10854-014-2445-6

9

Design Analysis of Solar-Powered Unmanned Aerial Vehicle

Karthik Reddy Buchireddy Sri[1], Poondla Aneesh[1], Kiran Bhanu[1], M Natarajan[1]

ABSTRACT: One of the main problems in micro unmanned aerial vehicles is endurance or flight time since the general domain aircraft uses conventional fuel, which is a pollutant, has a limited life and is costly. Then, there is a huge demand for using an unlimited non-exhaustible source of energy. Solar energy is one of the unlimited available renewable energy which can be used to increase the endurance of unmanned aerial vehicle without adding significant mass or increasing the size of the fuel system. This paper aims to encourage research on renewable energy sources for aviation considering the basic challenges for a solar-powered aircraft: geographical area of operation, energy collection and storage, payload and design parameters. Hence, a plane is designed for 2 kg, including payload, and is analyzed in various aspects. Besides, the design is optimized starting from airfoil to complete structure for better performance.

KEYWORDS: Solar cell, Unmanned aerial vehicle, Renewable energy, Endurance, Conceptual design.

INTRODUCTION

In today's world, there are more than 11,000 unmanned aerial vehicles (UAVs) in service (or planned for future services) by the Military for various purposes (Kumar *et al.* 2015).These vehicles can fly remotely as well as autonomously. In spite of their usage in various applications, they lack in performance because of power restrictions, which means they either have to land to refuel or to depend on another UAV to complete the task.

Nowadays the ability to fly without using conventional fossil fuels is primarily focused in recent years, both in application point of view and scientific field since the major concerns are an increase in global warming and a decrease in natural resources. Since then, the use of electric aircraft has been widespread but here the crucial issue is their high power consumption when compared with their limited energy storage capability, which leads to an endurance that can rarely exceed half an hour. Increasing the size of the battery or incorporating more batteries increases the weight of the plane, which affects the flight time of the UAV as it is inversely proportional. So, now, the ability to fly an aircraft to increase the endurance limit has been a key issue in both UAVs and civilian aviation.

One of the possibilities to increase the flight time is by using unlimited solar energy through solar cells. These cells, by connecting them with an electronic circuit, can provide sufficient power for motor and the electronics and, if in excess, it can also be stored in the battery; here the battery can be used as a buffer when flying in darkness or under clouds. Hence, the possible solution to enhance the endurance is by using solar-powered aircraft driven by electric-based propulsion systems in which the power is supplied continuously throughout the

1.VIT University – School of Mechanical Engineering Mechanical Department – Vellore/Tamil Nadu – India.

Author for correspondence: Karthik Reddy Buchireddy Sri | VIT University – School of Mechanical Engineering Mechanical Department | Near Katpadi Road | Vellore – Tamil Nadu 632014 – India | Email: bskarthikreddy@gmail.com day by the unlimited solar energy, which can eliminate fuel and also solve the limited energy storage capability problem.

This paper studies the complete design of a solar-powered UAV, which is mostly designed for low-altitude long-endurance (LALE) applications.

The main principle is to make use of available unlimited solar energy by converting it into electricity through solar cells. When sunlight strikes the solar cell, the cell creates electrons and holes as charge carriers and, when a circuit is made, the free electrons pass through a certain load in order to recombine with holes and, in this way, the current is generated. Here, by arranging solar cells in series on the top of the wing and then wrapping the entire wing with transparent material for the safety of solar cells during flight, cells are arranged in series to get the required voltage in order to safely charge the 3S battery; from there, the battery power is supplied to the motor for throttling during constant level flight. In this study, the aircraft was assumed to be a glider, which means it also stores the energy in the battery during the gliding period. In this way, both flying the plane by using solar energy alone and storing the energy in the battery in order to extend the flight time can be achieved.

DESIGN METHODOLOGY

The design process of the entire aircraft is composed of two phases: conceptual and preliminary. In conceptual design, the basic configuration arrangement, size, weight, and performance are calculated. Then, designing the CAD modeling of individual parts in order to meet the required targets and optimizing it are done in preliminary design. The design requirements are used to guide and evaluate the development of the overall aircraft configuration. There are two simple balances:

- Weight balance: the lift force has to be equal to the weight of all the elements constituting the airplane.
- Energy balance: the energy that is collected during a day from the solar panel has to be equal to or higher than the electrical energy needed by the airplane during its level flight.

MISSION SPECIFICATIONS

One of the main aims of this paper is to check whether the power available from the Sun is required to power an aircraft, which, in turn, helps in increasing the endurance of the aircraft from a couple of minutes to hours. Here the gross weight of the plane (2 kg) is kept constant (Table 1). There was no range requirement since the goal was simply to remain flying, not to reach any destination.

Table 1. Mission specifications.

Parameter	SI units
Gross weight	2 kg
Payload	0.5 kg
Altitude	30 – 50 m
Average air density	1.22 kg·m^{-3}
Clearness factor	0.7 (1 = clear sky)
Take-off distance	None (hand toss launch)

ESTIMATION OF THE NUMBER OF SOLAR CELLS REQUIRED AND ITS ARRANGEMENT ON THE WING

For this design, 3S battery has been chosen for two reasons: using more lithium polymer (LiPo) cells in series (> 3S) requires more voltage to initiate the charging process, leading to an increase in solar cells and wingspan, which, in turn, increases the weight of the model. On the other hand, if less LiPo cells in series (2S) are used, it will be difficult to supply the required power to achieve a good climb rate.

As for charging the 3S battery, it requires a constant safe charging voltage of about 12.4 V. This voltage has to be supplied by the solar cells; those selected for this design were Sun power C-60 photovoltaic. The Sun power cells were more efficient than most silicon-based ones. The 22% efficiency of the C-60 (Table 2)was significantly better than typical silicon solar cells with efficiencies around 15% (Kederoglu *et al.* 2012).

According to Table 2, each solar cell gives out 0.57 V, which means 22 solar cells are required to meet the targeted voltage; here, two more cells are added for safe side. These 24 cells should be connected in series to achieve the required voltage.

Since the number of cells is fixed, a minimum area of the wing is calculated. Two possible combinations in arranging

Table 2. Specifications of sun power C-60 solar cell.

Parameter	SI unit
Mass of Solar cell	0.008 kg
Length and width	0.125 × 0.125 m
Area of single solar panel	0.0150 m^2
Efficiency of solar cell	22%
Rated voltage	0.57 V
Rated current	5.37 A

Source: sunpowercorp.com

these cells in series (Type 1 and Type 2) over the wing can be seen in Fig. 1.

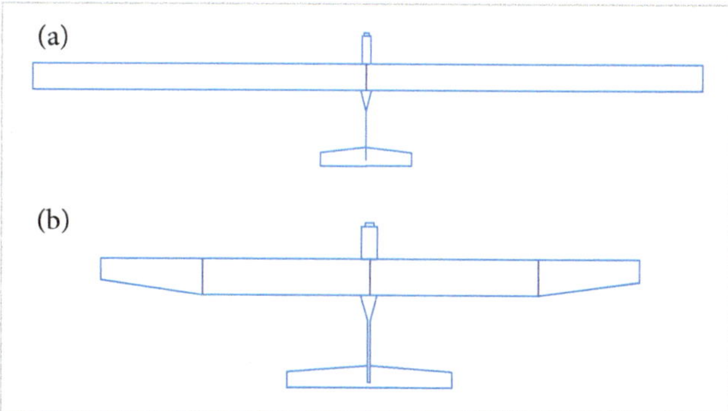

Figure 1. (a) Single row of cells — high-aspect-ratio wing (Type 1); (b) Two rows of cells — medium-aspect-ratio wing (Type 2).

The problems in Fig. 1a are:

- The wing is two long, which requires a tough structure, increasing the weight and cost.
- Controlling will be difficult as it does not have ailerons.
- While connecting the circuit with tapping wire, the 2 ends of the wire are too far, resulting in loss of power efficiency.

In Fig. 1b, all these problems are solved as the solar cells are placed only on the mid wing; the end wings are for stability with a polyhedral angle of about 7° on both sides, which also includes ailerons for easy control. In Fig. 2, it is observed that, in Type 1, more tapping wire is used in order to close the circuit, which means more loss while transferring the current. In Type 2, this problem is nullified as we can close the circuit with the usage of very less wire.

As each solar cell has 0.125 m in length, for 12 solar cells in a row, a wingspan (middle portion) of 1.5 m is required. However, apart from this length, extra space is required for the soldering purpose. Besides, at the ends, at least 2 – 3 cm should be left because there will be abnormal forces at the junction. Then, a central wingspan of 1.63 m was chosen by considering

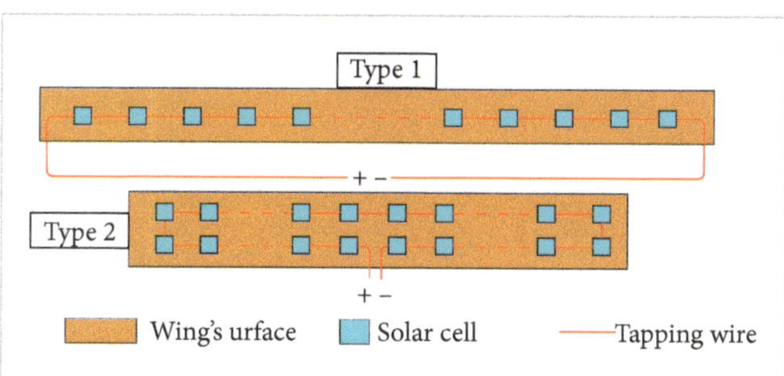

Figure 2. Arrangement and connection of solar cells on the wing.

all parameters. The ends of the wing are tapered from a root chord of 30 cm to a tip one of 20 cm over a length of 50 cm. The aspect ratio of the entire wing is 9.36.

AIRFOIL SELECTION

After finalizing the wing design, the airfoil is selected based on the following requirements: high lift coefficient; low drag coefficient; less camber for cell placement. Here, the 4 airfoils are analyzed, which are taken from previous designs (Bruscoli 2011). The approach to find the airfoil lift and drag is shown in Fig. 3, where the input parameters are aspect ratio (AR) and wingspan (b). The airfoils can be seen in Figs. 4.

For these 4 airfoils, XFLR analysis has been done to select the airfoil with the best performance characteristics. The analysis is done for 30-cm chord length and by keeping the 4-degree angle of attack for all airfoils.

In Fig 5, various graphs have been plotted by comparing the 4 airfoils for the 4-degree angle of attack (α), and the important parameters lift (C_L) and drag (C_D) coefficients are shown in Table 3. Hence, according to this table, Airfoil-4 WE3.55 is chosen for its best performance as it has high C_L/C_D ratio, which is the same airfoil designed for Sky-Sailor (Noth 2008).

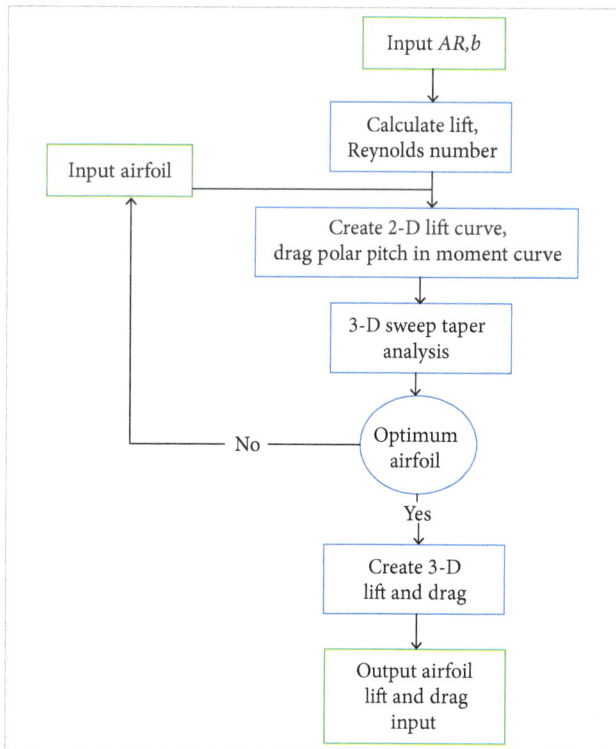

Figure 3. Flowchart to find airfoil lift and drag.

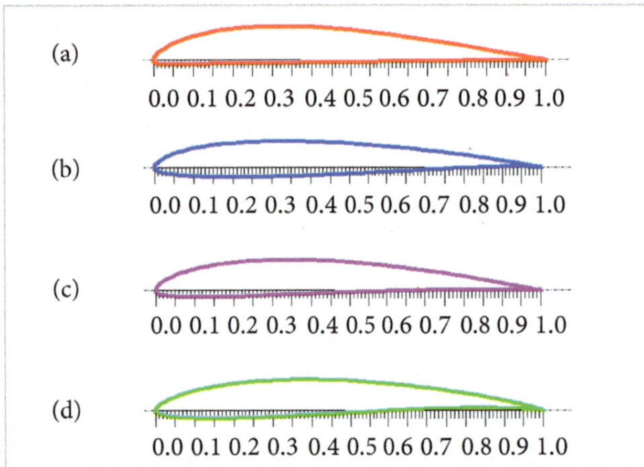

Figure 4. (a) Aquila 9.3%; (b) Medium S9000 (9%); (c) S9037 (9%); (d) WE3.55.

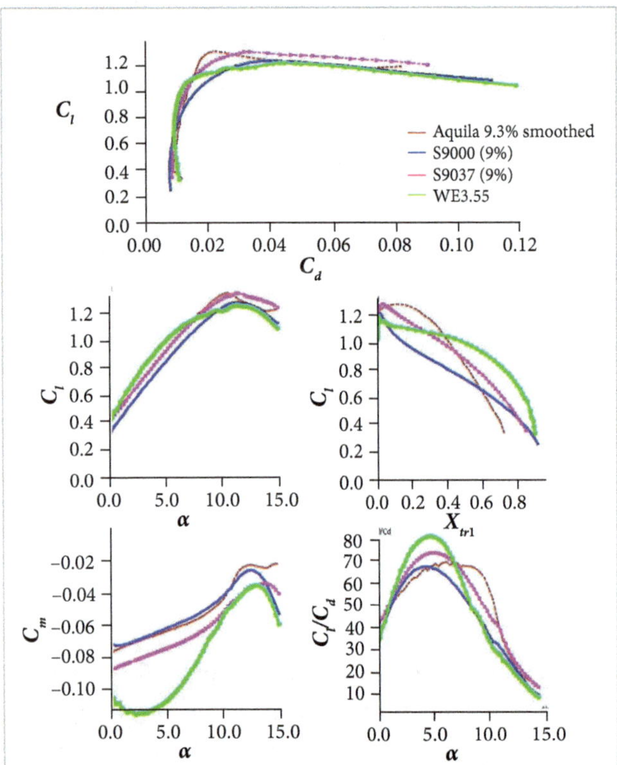

Figure 5. Comparison of airfoil performance.

Table 3. Comparison of the 4 selected airfoils.

Parameter	Aquila 9.3%	Medium S9000 (9%)	S9037 (9%)	WE3.55
C_L	0.83	0.75	0.83	**0.91**
C_D	**0.013**	0.01127	0.116	0.0115
C_L/C_D	64	66.5	71	**78.5**

Circuit Diagram

From the circuit diagram shown in Fig. 6, the energy received from the Sun is collected by the solar panel, which converts it into electricity and then there is the maximum power point tracker (MPPT), which helps in tracking the maximum power of any solar cell and provides it at all times. In MPPT, there are 3 terminals: one is connected to the solar panel; the other, to load (motor); and the last one, to the battery. Then, in level flight, the MPPT sends power directly to the motor from the solar cells and, when gliding, as the motor does not require power, the battery starts charging. If an excess of power is required, during climbing or when the solar intensity is low, the battery supplies the required power, which is stored into the motor. Thus, from the battery, the energy goes to the motor, which rotates the propeller and, between them, there is an electronic speed control (ESC) regulating the speed. Besides the present battery, there is an additional small pencil battery to control the avionics; the reason to separate it is, even if there is any problem in the main circuit or else, any damage in the cells still controlling the plane can be done as the control system's battery is not connected to the main circuit, which helps in safe landing.

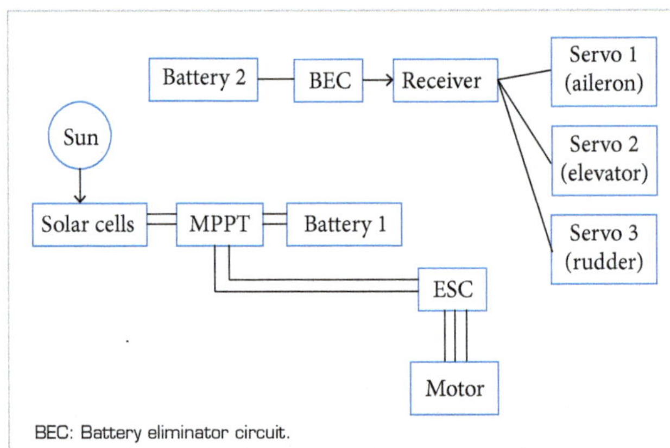

Figure 6. Circuit diagrams for entire plane.

POWER REQUIRED FOR LEVEL FLIGHT

At steady level flight, the lift force generated by the wing exactly compensates for the weight, and the propeller thrust, for the drag force (Simons 1994).

$$W = mg = L = \frac{\rho}{2} C_L S V^2 \tag{1}$$

where: m is the mass of the aircraft (kg); L is the lift force; ρ is the air density (kg·m^{-3}); S is the surface area of the wing (m^2); V is the cruise velocity of the aircraft (m·s^{-1}).

The axial thrust is calculated as:

$$T= D= \frac{\rho}{2} C_D S V^2 \quad (2)$$

where: D is the drag force.

The induced drag is obtained as:

$$C_{Di} = \frac{C_L{}^2}{\pi * e * AR} \; ; C_D = c_d + C_{Di} \quad (3)$$

where: c_d is the coefficient of drag of airfoil.

From Bandel *et al.* (2015), *e* means Oswald's efficiency, equal to $1.78 \times (1 - 0.045 \times AR^{0.68}) - 0.64$. The power required for level flight is:

$$P_{req} = T \times V \quad (4)$$

The required parameters for calculating the level flight are shown in Table 4. As the mass of the plane was initially fixed as 2 kg, the total weight of the plane becomes 19.6 N. From the above two equations, the cruise velocity of the plane that can be obtained as the lift coefficient for the airfoil is 0.91, and the surface area of the wing can be calculated as the chord equal to 30 cm and the length of 2.63 m, which is 0.8 m^2. Hence, the velocity of the plane at cruise is 7.5585 m·s^{-2}. The total drag is the sum of 3 drags: one caused by the wing c_d, which is around 0.2479 N; the other is the induced drag, which accounts for 0.7855 N due to very high lift; and the last one is due to the body (fuselage), which is taken 1.7 times the airfoil drag. Then, the total drag for the entire plane is 1.4646 N. From the total drag force and the cruise velocity, one can find the power required for a cruise: $P_{req} = T \times V = 10.13$ W.

In order to get this much power, more power should be extracted from batteries as there are electronics in the circuit for which there will be losses during conversion and efficiencies involved must be included: propeller efficiency of 80%, motor efficiency of 80%, ESC efficiency of 70% and the battery discharge efficiency of 90%, so that it is obtained a total of 40.38%. Here, extra 2 W are added to other losses. Hence, the total power required from the batteries is: (Power required for cruise/Battery to thrust efficiency) + P_{other} = 27.11 W.

The other circuit was from the battery to solar cells in which solar encapsulation was 90%; solar cells, 22%; camber efficiency, 90%; MPPT efficiency, 90%; and battery charging efficiency, 90%, totaling 14.435%. In order to get this power, the required solar irradiance is: Total power required from batteries/ (Sunlight to battery efficiency × Area of Solar cells) = 451.23 W·m^{-2}. Hence, 451.23 W·m^{-2} of solar irradiance is required in order to have a level flight. Now, whether this much amount of solar irradiance is available from the Sun has to be determined.

Table 4. Required parameters for calculating level flight.

Parameters	SI units
Gross weight	2 kg
Gravity	9.81 m·s^{-2}
Airfoil lift coefficient	0.91
Airfoil chord	0.3 m
Surface area of wing	0.7389 m^2
Airfoil drag coefficient	0.0115
Aspect ratio	9.36
Oswald's efficiency	0.773

POWER AVAILABLE FROM SOLAR ENERGY

Here, the power available from solar energy is calculated for April 1st, 2016 since April and May represent the hottest seasons in which maximum solar radiation occurs ($n = 92$, as the numeric sequence starts with 1 for January 1st).

Then, the latitude (ϕ) of VIT Vellore, India, for the calculation, is 12.9692°. From this, the declination angle (δ) is calculated (Eq. 5), which represents the angular displacement of the Sun's rays at the north (or south) of the Equator and is the angle made by the Sun and the Earth with its projection on the equatorial plane.

$$\delta = 23.45 \sin\left(360 \frac{284+n}{365}\right) \quad (5)$$

By using latitude and declination angle, the hour angle (ω_s) can be found, which is the angle through which the Earth must turn to bring the meridian at a point directly in line with the Sun's rays. It is equivalent to 15° per hour. From Jashnani *et al.* (2013):

$$\omega_s = \cos^{-1}(\tan\phi \times \tan\delta) \quad (6)$$

Thus, daily average irradiance (H_o) is:

$$H_o = \frac{24}{\pi} I_{sc} \times 3600\left(1 + 0.033\left(\cos\frac{360n}{365}\right)\right) \left(\omega_s.\sin\phi.\sin\delta + \cos\phi.\cos\delta.\sin\omega_s\right) \text{ kJ/m}^2\cdot\text{day} \quad (7)$$

where: I_{sc} is the extra-terrestrial irradiance.

The day length or maximum sunshine hours can be found using:

$$S_{max} = \frac{2}{15}\cos^{-1}(-\tan\phi.\tan\delta) \quad (8)$$

which was for 12:13. Then, monthly average daily global radiation (H) can be calculated for an altitude of 30 m and estimated using:

$$\frac{H}{H_0} = a + b\left(\frac{S}{Smax}\right) \quad (9)$$

where: $a = -0.309 + 0.539.\cos\phi - 0.0693E + 0.29\ (S/S_{max})$; $b = 1.527 - 1.027.\cos\phi + 0.0926E - 0.359\ (S/S_{max})$; S is the average sunshine hour per day; E is the elevation of the location above mean sea level (200 m — Vellore).

At each hour of the day (h), the instantaneous hour angle for a particular hour can be estimated as (in degrees):

$$\omega = (15h - 180) \quad (10)$$

The hourly global solar irradiance can be achieved (in $kJ{\cdot}m^{-2}{\cdot}day^{-1}$) by using:

$$I_g = \left[\frac{\pi}{24}(c + d.\cos\omega)\frac{\cos\omega - \cos w_s}{\sin w_s - \frac{\pi w_s}{180}\cos w_s}\right] \times H \quad (11)$$

where: $c = 0.409 + 0.516 \times \sin(\omega_s - 60)$ and $d = 0.6609 - 0.4767 \times \sin(\omega_s - 60)$.

The calculated solar parametric values are shown in Table 5 and, by calculating the instantaneous hour angle for a particular hour, global solar irradiance can be found for each solar hour; by taking the clear sky factor (N_{sky}) as 0.7, the readings in Table 6 are obtained for the entire day.

Table 5. Calculated solar parametric values.

Parameter	SI units
Declination angle	4.4139°
Hour angle	91.0186°
Daily average irradiance	37,537 $kJ{\cdot}m^{-2}{\cdot}day^{-1}$
Day length	12.13 h
Monthly average daily radiation	24,266 $kJ{\cdot}m^{-2}{\cdot}day^{-1}$

From the graph shown in Fig. 7 and the data in Table 6, it is observed that, by using solar power alone, one is able to fly the plane for up to 5 h. This flight time can be increased as, during gliding, some amount of energy is stored in the battery.

Table 6. Estimated global solar radiation on April 1st, 2016 *versus* power required for level flight.

Duration of the day (h)	Global radiation ($W{\cdot}m^{-2}$)	Estimated radiation N_{sky} = 0.7 ($W{\cdot}m^{-2}$)	Power for level flight ($W{\cdot}m^{-2}$)
1 – 5	0	0	451.23
6	10.27	7.19	451.23
7	185.72	130.0	451.23
8	392.16	274.51	451.23
9	602.52	421.76	451.23
10	784.69	549.28	451.23
11	908.41	635.89	451.23
12	952.22	666.55	451.23
13	908.41	635.89	451.23
14	784.69	549.28	451.23
15	602.52	421.76	451.23
16	392.16	274.51	451.23
17	185.72	130.0	451.23
18	10.27	7.19	451.23
19 – 24	0	0	451.23

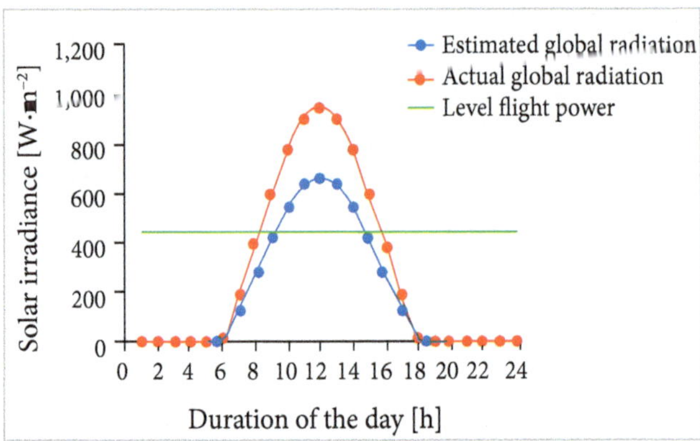

Figure 7. Duration of the day (April 1st, 2016) *versus* hourly global solar radiation — level flight power.

PRELIMINARY DESIGN

DESIGN OF THE WING

Wing Configuration

With respect to the angle made with the fuselage, from Fig. 8, the central part of the wing is perpendicular to the fuselage (straight) because it contains the solar cells and, in order to get maximum efficiency, these cells should be parallel to the ground; this kind of wing has high rigidity but poor roll stability. Hence, in order to have a good counter roll and lift, the ends of the main wing are extended by keeping a positive dihedral angle of 7°, which makes an obtuse angle with the main wing.

Regarding the position of the wing on the fuselage, a high wing is selected, which means it is situated over the fuselage; this kind of design helps in having high ground clearance, less interference drag, better field performance at low speeds and maximum space for having a payload.

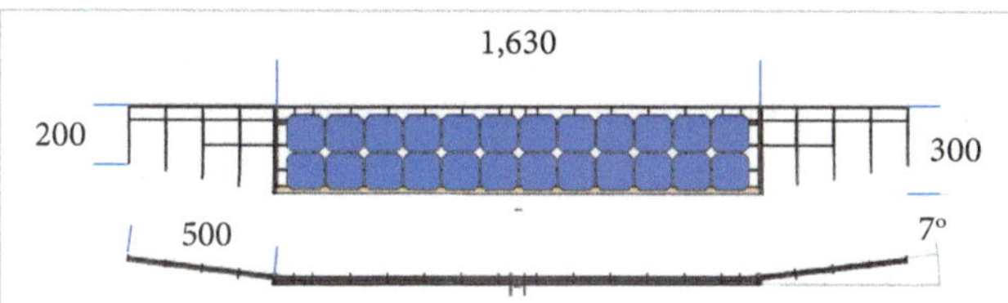

Figure 8. A draft model of the entire wing assembly.

Taper Ratio

It is the ratio (λ) between the tip and the root chord of the wings. A trapezoidal shaped ($0 < \lambda < 1$) is selected for the ends of the main wing. Wing tapering is used to improve lift distribution characteristics, which will also result in better lateral control since the mass moment of inertia of the wings around the longitudinal axis will be less. Here, $\lambda = 0.66$.

$$\lambda = \frac{C_t}{C_r} \tag{12}$$

where: C_t is the tip chord; C_r is the root chord.

Structure

In the central part of the wing, the construction consists of spars, ribs, and the skin. As the central part of the wing is 0.0163 m, which has solar cells integrated into it, the main priority is to achieve very small deflection. Then, 3 spars have been included, which are made of a carbon fiber (zoltac) material; two of them are 6 × 4 mm and the middle one is 4 × 2 mm. The spars will handle the tension and compression forces on the wing and are attached to the fuselage so they can transfer the forces to it; here, the forces will be spread over the entire construction. The ribs (made of lightweight balsa wood) are intended to give the wing the desired airfoil shape and also to stabilize the construction; they provide rigidity and strength to the wing by transferring the forces to the spars. The skin is also a part of the construction to resist the forces on the wing, which covers only the top surface with 1 mm of balsa sheet in order to make a non-conductive flat surface for the integration of solar cells.

The ends of the wing are tapered from a root chord of 30 cm to a tip chord of 20 cm over a length of 50 cm. This part of the wings includes ailerons in order to have stability and control over the plane; In these parts, two carbon fiber spars are kept in order to give support and help in integrating the tapered parts to the main wing, since, at the junction of two parts of the wing, large forces will act. In Fig. 12, all dimensions are in mm.

DESIGN OF THE TAIL

The airfoil selected for the horizontal and vertical tail is the NACA 0009 (Fig. 9) because it has a symmetrical profile in nature and a maximum lift coefficient of 0.8. For the horizontal tail, the mean chord length is 0.205 m and, for the vertical one, the mean chord length is 0.25 m (AirfoilTools.com 2016).

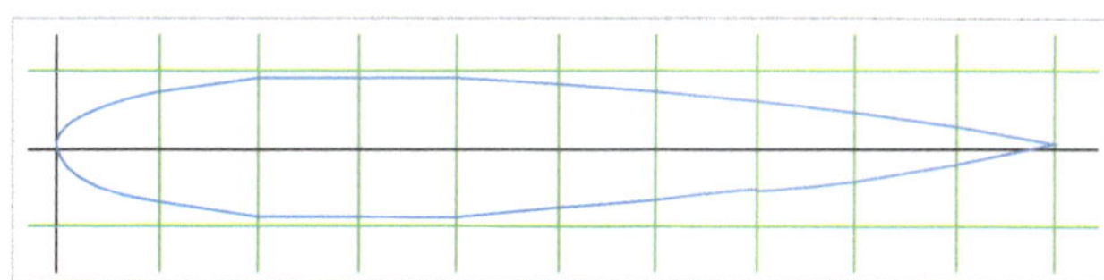

Figure 9. NACA 0009 airfoil shape.

Horizontal Tail

The dominant factors which control the aircraft pitch stability are gravity center (GC) position and horizontal tail size and position. The effectiveness of horizontal tail is measured by horizontal tail volume coefficient (V_h):

$$V_h = \frac{S_h * L_h}{S * c} \tag{13}$$

where: S_h is the horizontal tail area; L_h is the distance from the tail's aerodynamic center (more of which later) to the aircraft's GC; S is the wing area; c is the mean aerodynamic chord.

After several iterations, the horizontal stabilizer parameters are shown in Table 7.

Table 7. Horizontal tail configuration.

Horizontal stabilizer parameter	SI units
Area (S_h)	0.14 m^2
Moment arm (L_h)	0.96 m
Volume coefficient (V_h)	0.64
Span (b_h)	0.7 m
Mean chord (c_h)	0.205 m
Taper ratio (λ_h)	1.0
Aspect ratio (AR_h)	3.41
Elevator ratio (S_r/S_h)	0.3

Vertical Tail

The role of vertical tail is to control the yaw movement by providing yaw damping. The effectiveness of vertical tail is measured by the vertical tail coefficient (V_v):

$$V_v = \frac{S_V * L_V}{S * b} \tag{14}$$

where: S_v is the vertical tail area; L_v is the distance from the vertical tail's aerodynamic center to the aircraft's GC; b is the wingspan.

After several iterations, the vertical stabilizer parameters are shown in Table 8. In Fig. 10, all dimensions are in mm.

FUSELAGE

This part is designed after a number of iterations in order to improve the lift-to-drag ratio. The trails shown in Table 9 are analyzed in fluent software as shown in Fig. 11, respectively, for a Reynolds number of 150,000 and by using SST k-ω viscous model. The boundary conditions are given in Table 10.

Table 8. Vertical tail configuration.

Vertical stabilizer parameters	SI units
Area (S_v)	0.075 m^2
Moment arm (L_v)	0.85 m
Volume coefficient (V_v)	0.0328
Span (b_v)	0.3 m
Mean chord (c_v)	0.25 m
Aspect ratio (AR_v)	2.4
Taper ratio (λ_l)	1.0
Rudder ratio (S_r/S_v)	0.3

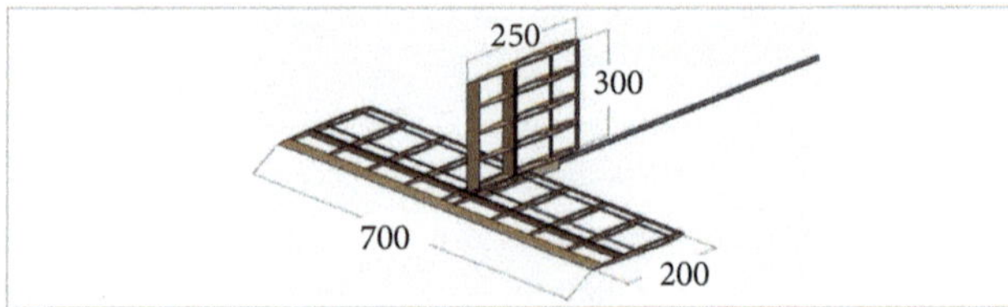

Figure 10. A draft model of the entire tail assembly.

Table 9. Comparison of L/D for different trails.

Trail number	Design modification	L/D
1	Basic cuboidal	15.244
2	Tapered front and back	15.498
3	Lightly streamlined	15.72
4	Final design	15.81

Table 10. Boundary conditions.

Edge	Type of boundary	Boundary condition
Inlet	Velocity	8 m·s^{-1} (x direction)
Outlet	Pressure	Gauge pressure = 0 Pa
Wall	Stationary	Velocity = 0 m·s^{-1}

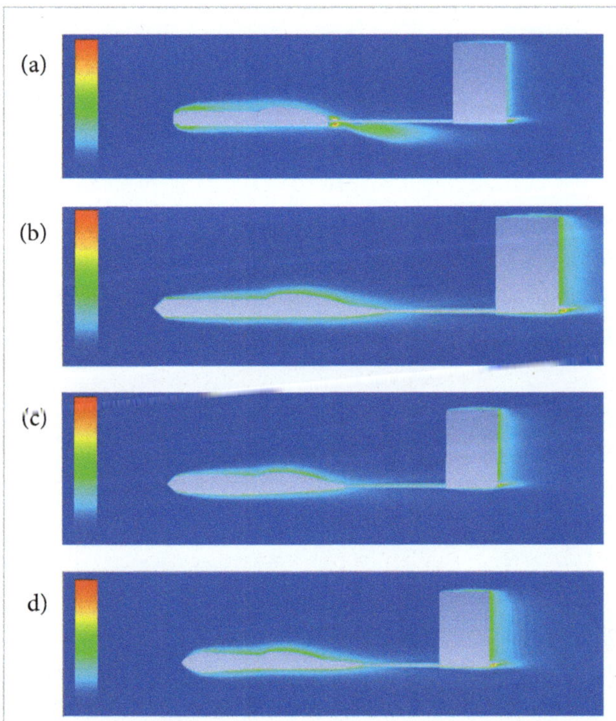

Figure 11. (a) Basic cuboidal (Trail 1); (b) Tapered ends (Trail 2); (c) Streamlined (Trail 3); (d) Final design (Trail 4).

The final design is selected, having a lift force of 18.06 N and a drag one of 1.142 N. The primary function of the fuselage is to carry the electronics and payload. Hence, a length of 0.65 m and 0.82 m in width and height are obtained for the design. The propulsion system components can be mounted in the front of the fuselage, and the electronics and payload may be shifted back or fronted in such a way to adjust the center of gravity position of the plane, which is at a distance of 9 cm from the leading edge of the wing. The tail boom is connected to the back of the fuselage, which is integrated to the main wing through carbon fiber rods. In Fig. 12, all dimensions are in meters. Finally, all parts are assembled forming the complete plane, as shown in Fig. 13, in which all the dimensions are in mm.

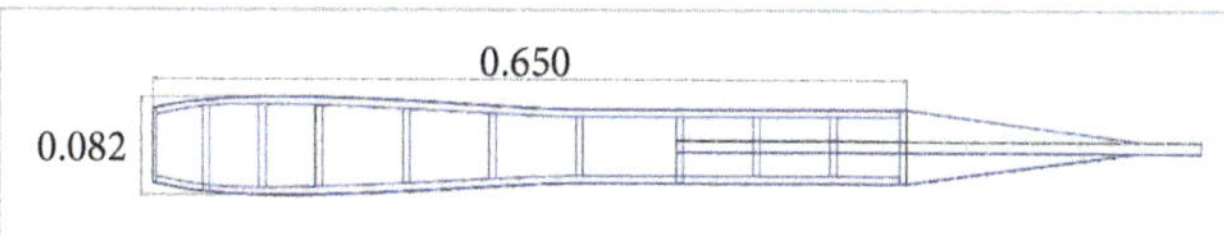

Figure 12. A draft model of the fuselage.

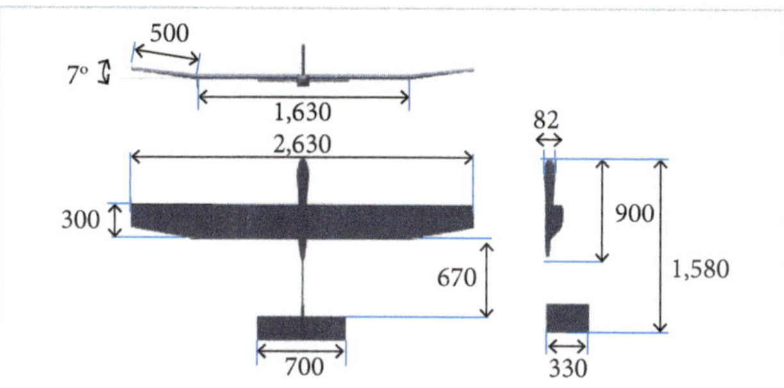

Figure 13. Entire plane draft model.

STRUCTURAL ANALYSIS

The two primary structures for the design are the wing spar and the tail boom. Both of these are analyzed in detail so they could be lightweight but strong enough to support the design loads. The wing spar has to carry 3 primary loads: the wing bending load produced by lift, the transverse shear load caused by wing bending, and the torque load produced by the pitching moment of the wing airfoil. The tail boom has to carry the bending loads produced by the horizontal and vertical stabilizers.

The wing spar and tail boom are made of carbon fiber material which can support very high loads before they fail. For aerodynamic surfaces, even a small amount of deformation could potentially upset the stability or control of the airplane. Calculating the structural deformation using an analytical method would have been impractical. Instead, the potential approach is to use finite element analysis to determine how the structures are deformed.

A software has been used to find the deflection and stress using finite element method. In order to validate it, we took a problem which can be solvable by the analytical approach and compared it with the software results.

For the validation, a problem (Bansal 2010) has been solved both manually and in the software; the deviation was found to be less than 5%. Then, for further complex analyses, the software is used to find the results. The input properties for balsa wood (MakeitFrom.com 2016) and carbon fiber (GoodWinds Composites 2016) have been taken for all the analyses.

In the wing deformation, the mid wing excluding the tapered parts is analyzed since mostly the lift force is acted only on the middle section of the wing. Then, a load of 4 kg is applied to the entire wing, taking a factor of safety equal to 2 and keeping the central position of about 10 cm constant since that portion is fixed to the fuselage, as shown in Fig 14. A deflection of only 1.4 cm is observed.

The tail boom has to resist to bending loads, keeping the empennage surfaces at the proper aerodynamic incidence with the wing. If the tail boom flexed too much, control effectiveness would be reduced. Bending load is the greatest on the tail boom when the empennage surfaces operate at a high lift coefficient and the aircraft is flying at a high speed. The bending load on the tail boom could be treated as a point load acting at the end of the tail boom and its critical value can be calculated as:

$$P_{crit} = (0.5)\,\rho V^2 S_h C_{Lmax} \tag{15}$$

The critical bending load characteristics for the tail boom are shown in Table 11. A bending load that we got from the calculation of 4.4 N is applied on one end of the tail, keeping the other end constant. A deflection of 0.8 cm is observed.

The fuselage has to carry the payload and electronics, which can weight more than 1 kg. Hence, a force of 30 N, which is

Table 11. Critical bending load characteristics for tail boom.

Parameter	SI unit
Velocity (V)	8 m·s^{-1}
Air density (ρ)	1.227 kg·m^{-3}
Horizontal stabilizer area (S_h)	0.14 m^2
Max tail coefficient (C_{Lmax})	0.8
Critical bending load (P_{crit})	4.4 N
Tail boom (L)	0.6 m
Modulus of elasticity (E)	131 Gpa

3 kg (more than the entire plane weight), is applied to the front section by keeping the end section fixed, as shown in Fig 15. A deformation of 0.279 cm is observed.

Concerning the stress analysis, the fuselage is analyzed to find out whether the stress is below the permissible range or not for a load of 30 N. Besides, according to the analysis, as shown in Fig. 16, a von Mises stress of 4.4 MPa is observed, which is within the permissible value of 9.6 MPa.

In the middle section of the wing, more stress is acted on the carbon fiber rods than that of balsa ribs and skin for a force of 40 N, as shown in Fig. 17. Nevertheless, the stress which is acted on carbon rods is negligible when compared to its permissible stress. The red portion is only observed on the carbon fiber rods, whereas the balsa wood portion is blue in a color which is the minimum stress. Hence, all the parts are under the safer limit even under a factor of safety equal to 2.

RESULTS AND CONCLUSION

In this paper, the suitable airfoil has been chosen by analyzing different airfoils from previous designs for a high

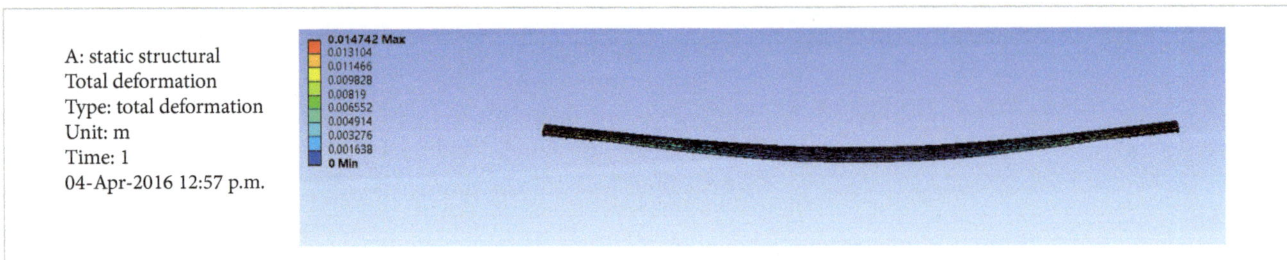

Figure 14. Deflection of mid-wing.

Figure 15. Deflection of the fuselage.

Figure 16. Von Mises stress analysis of the fuselage.

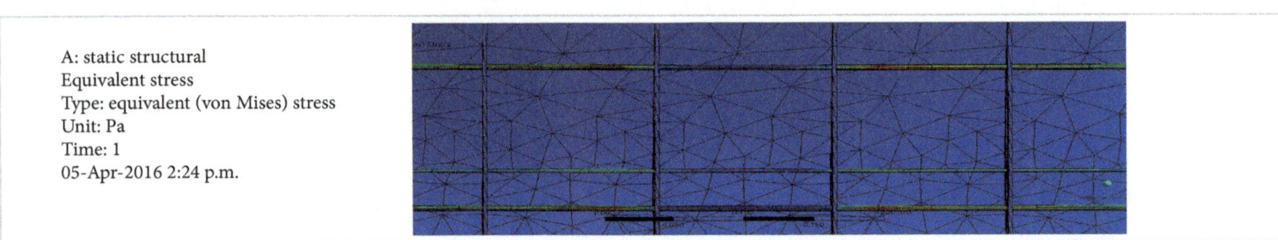

Figure 17. An enlarged view of stress region near the middle portion of the mid-wing.

lift-to-drag ratio. It was presented a way to calculate a number of solar cells required for giving enough power to the motor lift the 2-kg plane, which can extend for heavier ones. It was performed detailed energy and power analysis to check whether the design would be feasible or not. Then, all the components were designed, considering various factors for better performance, and analyzed for deflection and stress to keep the design under the safer limit. This gives a basic foundation in order to fabricate a solar-powered UAV.

ACKNOWLEDGEMENTS

We wish to express our profound gratitude and indebtedness to Prof. M. Natarajan, for his inspiring guidance, and Prof. Arivazhagan N, for granting permissions to use the required equipment during the course of this research. We would also like to show our gratitude to Team Vimaanas, VIT University, which provided insight and expertise that greatly assisted the research.

AUTHOR'S CONTRIBUTION

Aneesh P was responsible for conceptual study and design. Data acquisition, analysis and drafting were carried out by Reddy BSK. Bhanu K and Natarajan M performed solar cells experimentation and interpretation of data. All authors discussed the results and commented on the manuscript.

REFERENCES

AirfoilTools.com (2016) Airfoil tools; [accessed 2016 Aug 11]. http://airfoiltools.com/index

Bandel A (2015) Aviation studies design build and fly minor. Amsterdam: University of Applied Sciences.

Bansal RK (2010) A textbook of the strength of materials. Bangalore: Laxmi Publications.

Bruscoli S (2011) Airfoil optimization for a solar powered aircraft (PhD thesis). Pisa: Università degli Studi di Pisa.

GoodWinds Composites (2016) Carbon rods; [accessed 2016 Aug 11]. Available: http://gwcomposites.com/carbon-rods/

Jashnani S, Nada TR, Ishfaq M, Khamker A, Shaholia P (2013) Sizing and preliminary hardware testing of solar-powered UAV. The Egyptian Journal of Remote Sensing and Space Sciences 16(2):189-198. doi: 10.1016/j.ejrs.2013.05.002

Kederoglu OG, Yanbastioglu S, Saraç B, Kocaoğlan MH (2012) Solar powered aircraft: PV Power System Definition & Solution Development Project; [accessed 2016 Aug 11]. http://www.academia.edu/7515954/Solar_Powered_Aircraft_Project

Kumar G, Sepat S, Bansal S (2015) Review paper of solar-powered UAV. International Journal of Scientific & Engineering Research 6(2):41-44.

MakeitFrom.com (2016) Balsa; [accessed 2016 Aug 11]. http://www.makeitfrom.com/material-properties/Balsa/

Noth A (2008) Design of solar powered airplanes for continuous flight (PhD thesis). Zürich: ETH Zürich.

Simons M (1994) Model aircraft aerodynamics. 3rd edition. Hemel Hempstead: Argus.

10

Design and Manufacture of a Solar-Powered Unmanned Aerial Vehicle for Civilian Surveillance Missions

Nelson Javier Pedraza Betancourth[1], Julio Enoc Parra Villamarin[1], John Jairo Vaca Rios[1], Pedro David Bravo-Mosquera[2], Hernán Darío Cerón-Muñoz[2]

ABSTRACT: In order to promote the development of renewable energy and take advantage of the new technologies for the benefit of sustainability, both the design and the manufacture methodologies of an experimental solar-powered unmanned aerial vehicle for civilian surveillance applications are presented. Throughout the document, it is provided the historical process around the development of the aircraft. Therefore, in the first part, it is shown the aerodynamic design, which includes the 2-D and 3-D analyses of the wing platform using numerical and experimental methods, the analytical design of the empennage configuration, and the main characteristics of the performance analysis. In addition, major systems and components that characterize the aircraft are described, such as the photovoltaic solar cells configuration as well as the electronics and control system into the unmanned aerial vehicle. Lastly, the modeling for the weights distribution of the components was carried out in a preliminary test using CAD tools. Thus, it was obtained a suitable process for the manufacture of the unmanned aerial vehicle, considering that the purpose of the aircraft is to be as light and aerodynamic as possible to accomplish the mission for which it was created.

KEYWORDS: Unmanned aerial vehicle, Aerodynamic design, Performance analysis, Photovoltaic solar cells, Manufacture processes.

INTRODUCTION

In the quest to find energy systems less harmful to the environment and more efficient than the currents, the processes of design and manufacture of a unmanned aerial vehicle (UAV) that operates with photovoltaic solar energy are described in this paper. The application of alternative energy sources in UAVs has established one of the broader challenges in its design and implementation, as specified by Ramanathan *et al.* (2007), promoting ideas through the use of solar cells (Lubkowski *et al.* 2010), liquid fuel cells (Fürrutter and Meyer 2009), lithium-polymer (LiPo) batteries (Song and James 2000), among others. These energy sources can be a clean way to supply the power needs of the UAVs, such as the power plant system, the telemetry systems and the avionics. According to Herwitz *et al.* (2003) and MacCready *et al.* (1983), the application and use of new energy management systems are joined by the innovative and clean technology trends, which encourage the use of new energy resources in order to reduce the impact of the pollutants as the fossil fuels derived from the coal and petroleum.

Based on the experience of Austin (2010) and Wilkins *et al.* (2009), the determining factors for the use of alternative types of energy in UAVs are the mission and the size for which the vehicles are designed; these factors are reflected in the weight, which is one of the most important aspects when calculating the model. Therefore, it has been established, according to survey data, that hybrid energy configurations exist linking various types of energy; among the different alternatives are: the combination of hydrogen fuel cells (LH_2) with Diesel

1.Fundación Universitaria Los Libertadores – Facultad de Ingeniería Aeronáutica – Bogotá – Colombia. 2.Universidade de São Paulo – Escola de Engenharia de São Carlos – Departamento de Engenharia Aeronáutica – São Carlos/SP – Brazil.

Author for correspondence: Julio Enoc Parra Villamarin | Fundación Universitaria los Libertadores – Facultad de Ingeniería - Ingeniería Aeronáutica | Carrera 16 # 63 A - 68 – Bogotá – Colombia | Email: jeparrav@libertadores.edu.co

fuels or gas turbines (Nickol *et al.* 2007) and electric engines with solar cells (Brandt and Gilliam 1995; Romeo *et al.* 2007). Consequently, one can opt for the tendency to use renewable fuels. The efficient use of clean energy is attached to an important electronic development. Therefore, the adaptation of systemic mathematical algorithms from microcontrollers increases the energy efficiency and distributes the load capacity to all electronic components, establishing a system of energy management in the UAVs (Baoan *et al.* 2009).

According to data and models obtained by Nickol *et al.* (2007), Noth (2008) and Panagiotou *et al.* (2016), some analyzes and preliminary experiments have indicated that hybrid propulsion systems with solar energy offer excellent autonomy due to their energy storage capacity and continuous recharge. However, they require a large wing surface to the location of the solar panels. Nowadays, photovoltaic cells have been the largest source of alternative energy studied and applied. Therefore, there are references about 40 manned and unmanned experimental aircraft that take advantage of its features, performing missions in civil and military scenarios as: the Sky Sailor, the Helios and the Solar Impulse, that have successfully used photovoltaic solar energy as a source of power for their operation (Nickol *et al.* 2007; Sun *et al.* 2015).

The conceptual design of the UAV was carried out taking into account the optimal performance of the amount of harnessing solar energy achieved by the UAVs (Cestino 2006; Youngblood *et al.* 1984), which depends on the following environmental and technical factors: seasons of the year, due to the amount of sunlight that is exposed on the Earth's surface; time of the day, as a result of the sunlight intensity increase from sunrise until noon; the sun irradiance, exploitable in some periods of the day; efficiency of the photovoltaic cell, in which typical polycrystalline cell efficiencies are between 13 and 18%; and the surface area of the solar cells, directly proportional to the amount of electric power supplied. Therefore, a larger wing surface was necessary to get better results from electrical power generation.

In this regard, a rigorous aerodynamic study was performed in order to select the best airfoil and the best wing configuration that meet the requirements of the aircraft. Thus, the wing was designed with a large aspect ratio (AR) of 17, aiming to obtain the characteristics of a sailplane wing; at the wingtip, there is a taper ratio (λ) of 0.375 for the sake of reduction of the wing tip stall effects. Thereby, the amount of 90 cells was distributed on the upper surface of the wing and the horizontal empennage, which are intended to power the motor and the subsystems of the aircraft. Due to the wingspan (b) of 6 m and the maximum chord (c) of 0.4 m, it was decided that the wings of the UAV load internally all the structural elements and the systems for the operation, simplifying the use of a fuselage and reducing the weight of the vehicle. Figure 1 shows the main dimensions of the wing platform of the UAV.

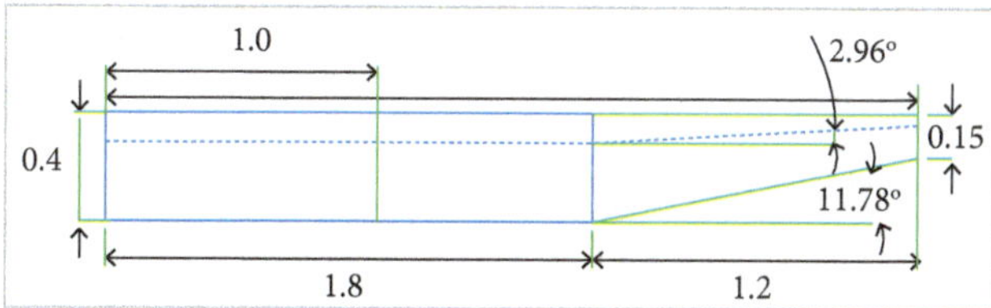

Figure 1. Wing dimensions in meters — UAV-SOLVENDUS full scale.

UAV DESCRIPTION

Currently, UAVs have become a very important device for the aviation and the industrial development, presenting ideal characteristics for recognition of inaccessible areas on missions than the traditional aircraft. In Colombia, UAVs designed and manufactured in different countries have been used in order to carry out different tasks in agriculture, aerial photography, security and other areas. However, in the last years, new academic research in UAVs have been proposed in the Colombian aeronautical sector, mainly in partnerships with Brazilian universities (Bravo-Mosquera *et al.* 2015; Herrera *et al.* 2014; Martínez and Cerón 2013; Parra *et al.* 2013).

Throughout this article, the methodological framework for the design of a surveillance UAV integrating the use of alternative energies and lightweight structures is presented. The UAV called SOLVENDUS (which means "worthy of being released") is a project that has been developed by the Fundación Universitaria Los Libertadores, Colombia, in cooperation with the Escola de Engenharia de São Carlos, Universidade de São Paulo, Brazil, in the period of 2010 – 2015. The project was divided into 3 phases.

In the first one (2010), the establishment of the mission criteria was conducted. Therefore, a market study focused on the integration of alternative energy in aircrafts was developed. This gave way to consolidate the conceptual design processes of the UAV-SOLVENDUS.

At the second stage (2011 – 2012), the conceptual design of the UAV-SOLVENDUS was initiated, taking as reference the methodological proposals by Anderson Jr (1999), Raymer (1992) and Roskam (1997) in relation to different configurations for fixed wing aircraft. Thereby, the characterization of the UAV-SOLVENDUS mission in its different phases of flight as well as the selection of the different mechanical and electronic elements for the operation were established.

In the third phase (2013 – 2015), the complete development of the preliminary and detail design, including manufacture, was executed. In this regard, the aerodynamic design was performed using analytical, numerical and experimental methods in order to validate each stage calculated, aiming to analyze how would be the aerodynamic behavior of the UAV performing the established mission.

The solar cells configuration on the aircraft and the arrangement that best fits the needs of the electrical system were calculated. The upper surface of the airfoils should be treated as a polygonal line connected by several short line sections, as shown in Fig. 2. In this way, it is possible to arrange the photovoltaic cells without considerable deformations. This type of configuration can also be seen in Cestino (2006), Gao *et al.* (2014) and Romeo *et al.* (2007). On the other hand, new methodologies to optimize the construction of the aircraft were implemented following the global trend by applying manufacturing processes with composite materials. In this context, the material selection was based on the ASTM D3039. Nowadays, flight tests are being developed.

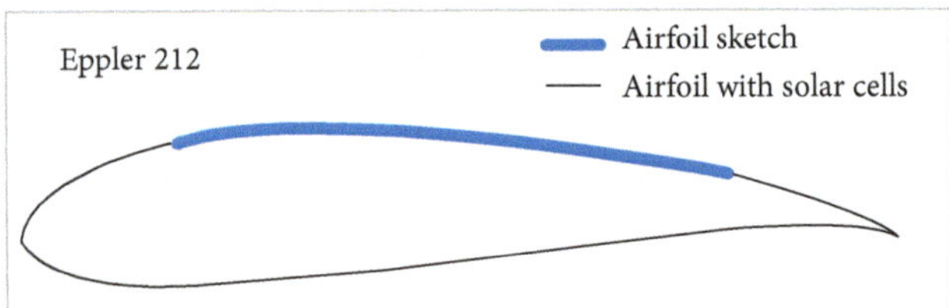

Figure 2. Configuration of solar panels on the selected wing airfoil.

CONCEPTUAL DESIGN

The conceptual design was based on an extensive parametric study on the characteristics of UAVs at the same category, aiming to interpret and analyze the main design features (geometric, aerodynamic and performance) that make up an aircraft with solar panels (Arai *et al.* 2014; Marta and Gamboa 2014). The particular requirements of the UAV-SOLVENDUS were established taking into account the first estimation of the maximum take-off weight (MTOW) and the desired power for flight. In this regard, most of the UAVs that operate with solar energy use battery systems as a supplement to its functionality. As a result, LiPo rechargeable batteries have adapted in flight, providing a continuous-cyclical storage, which supplies the necessary autonomy to carry out powerful flights, taking advantage of the daylight and storing energy for night missions.

The maximum efficiency offered by the solar panels is only able to electrically supply the electronic control and navigation systems. However, to supply the electrical engines, auxiliary batteries charged previously in the ground station were required. These reasons were the base to design a lightweight aircraft in order to reduce the energy consumption. According to Raymer (1992), the first estimation of the MTOW of an aircraft is supported by the following equation:

$$\text{MTOW} = W_{fuel} + W_{crew} + W_{payload} + W_{empty} \quad (1)$$

where: W means weight.

Analyzing the elements of the UAV-SOLVENDUS, the fuel weight is zero because the UAV does not operate with combustion engines; on the other hand, the crew and payload weights are irrelevant. Therefore, the MTOW of the UAV-SOLVENDUS is characterized by the components of the empty weight, which, by definition, are: Aircraft structure, Engines, Landing gear and Avionics. In this respect, the weight of the aircraft depended strictly on the manufacturing weight. Therefore, the use of materials with a high strength/weight ratio was necessary. This relation is the main characteristic of materials with high mechanical properties and low densities. Thus, composite materials were used to manufacture the experimental prototype of the UAV-SOLVENDUS, reducing the energy consumption of the power plant and increasing the performance of the aircraft.

From this perspective, the flight path and the mission profile studies for the UAV-SOLVENDUS were developed considering the following requirements and constraints:

- Low power required (aerodynamic efficiency and low structural weight).
- Low flight velocities.
- Use of solar energy through photovoltaic cells.
- Wing area sufficient for the distribution of cells.
- Mechanical, chemical or mixed fastening for the photovoltaic cells.
- Electric engines.
- Integration of autonomous navigation systems.

The mission profile of the UAV-SOLVENDUS is shown in Fig. 3. The UAV-SOLVENDUS moves at low flight regime, given the nature of its mission and the selection of its power plant system; this should maintain a high lift coefficient (C_L) for the fulfillment of its operation. Therefore, airfoils that provide high lift coefficient at low speed flight were evaluated (Herrera *et al.* 2014).

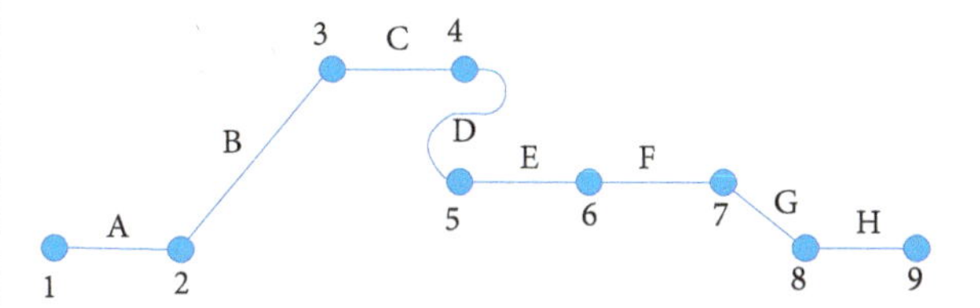

Transition points — 1: Engine ignition; 2: Take-off; 3: Operational altitude point; 4: Descent point for surveillance; 5: Start surveillance; 6: End surveillance; 7: Start descent; 8: Landing; 9: Stop engine. Flight stage — A: Take-off run; B: Climb; C: Cruise I; D: First descent; E: Surveillance; F: Cruise II; G: Runway approach; H: Landing.

Figure 3. Mission profile of the UAV-SOLVENDUS.

PRELIMINARY DESIGN

AERODYNAMIC STUDY

Initially, different airfoils were studied in order to determine the appropriate selection that meets the best features for the aircraft mission. Therefore, the best airfoil was chosen based on the "Stall" behavior and the drag polar curve. The airfoil selected was the Eppler 212, which presents a satisfactory aerodynamic performance at low Reynolds number. Its geometry allows attaching the solar panels. Two-dimensional computational analysis using a free-software program was carried out, resolving the velocity field and the pressure distribution around the airfoil. These results are suitable for maneuverability use on the UAV-SOLVENDUS. Further information is reported in Herrera *et al.* (2014).

THREE-DIMENSIONAL ANALYTIC DEVELOPMENT

A free software called (Tornado® 2001) was used for the preliminary analysis of the wing, which is a vortex lattice method. Figure 4 shows the lift distribution *versus* wingspan at an angle of attack equal to 4° and geometric torsion in the wingtips. The twist angles analyzed were 0°, −2° and −4°, being negative when the leading edge was lowered relative to the trailing edge. Although distributions were similar, the twist angle equal to −4° was selected aiming to prevent the wing stalling on the wingtip region. The results show a lift force (L) of 121.61 N.

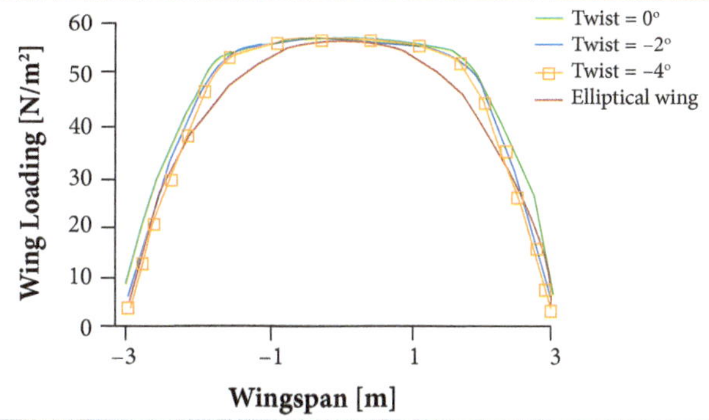

Figure 4. Comparison of lift distribution on the wing.

COMPUTATIONAL ANALYSIS

For this stage, the ANSYS®-FLUENT code (ANSYS® 2011) was used in the modeling of airflow over the wing (scale 1:3). The computational domain used was large enough to avoid the fluid restriction in the simulations. It was discretized in both space and time through an appropriate mesh, where the Navier-Stokes equations were solved. In this regard, the distance to the nearest wall was set at 1×10^{-4}, resulting in a $Y^+ \approx 5$. The number of elements generated was 3.8×10^6, using an unstructured mesh (tetrahedral). The mesh generated for the wing model is shown in Fig. 5.

All the simulations were performed with the International Standard Atmosphere (ISA) parameters at the atmospheric conditions of São Carlos, São Paulo, Brazil, where the experiments were executed. The variation of the angle of attack was from −4° to 18° at a Reynolds number of 2.05×10^5. The Shear Stress Transport (SST) turbulence model was chosen by its most common applications in the aerospace field related with limited flows by walls and simulations at low Reynolds numbers (ANSYS® 2011; Kontogiannis and Ekaterinaris 2013).

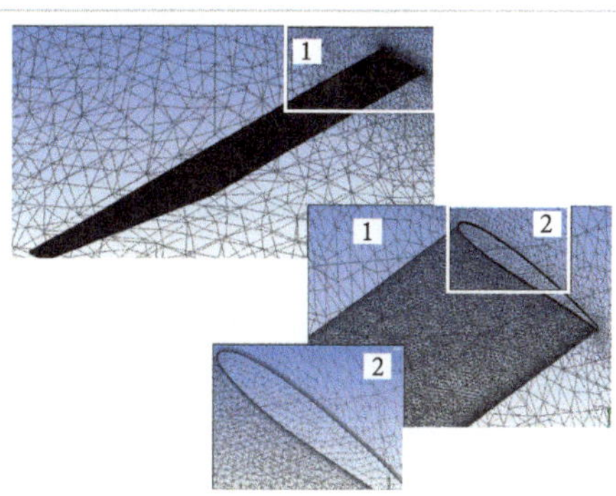

Figure 5. Unstructured mesh and grid details for the wing simulation.

In addition to the good numerical approximations of the model's aerodynamic qualities compared with the experiments, the Computational Fluid Dynamics (CFD) simulations allowed to understand the behavior of the streamlines in each condition. In this connection, Figs. 6a and 6b show the streamlines at angles of attack equal to 4° and 10°, respectively. As can be observed, at 10° there is a cross flow and fluid detachment compared with the simulations at 4°.

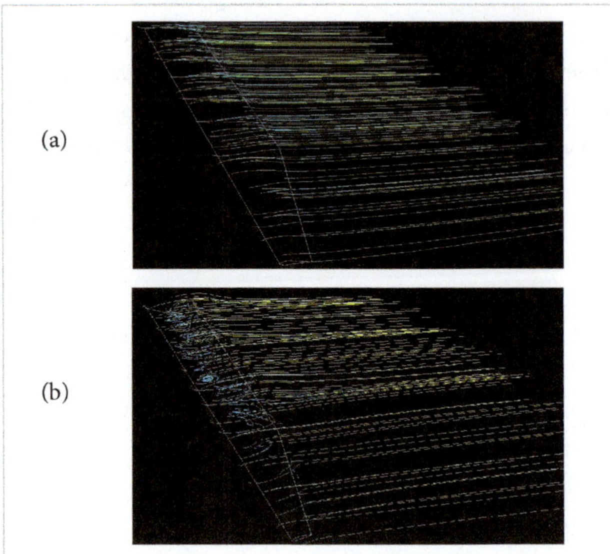

Figure 6. Streamlines at (a) 4°; (b) 10°.

EXPERIMENTAL AERODYNAMIC MODEL

Taking into account the dimensions of the test section of the wind tunnel (LAE, Fig. 7) (1.3 × 1.7 × 3 m) located at the Escola de Engenharia de São Carlos, Universidade de São Paulo (Catalano 2001), the wing model was built at a 1:3 scale. The experiments were executed at 25 m·s^{-1}. For the manufacture of the model, the ribs were made in medium density fiberboard (MDF), with thickness of 6.3 mm, the coating was made of polyvinyl chloride (PVC), with thickness of 0.5 mm, and plastic tubes were built at 25% chord on the scale for the pressure taps. The experimental model used was a semi-span wing of 1 m, 0.05 m tip chord and 0.13 m in root chord, as well as taper ratio of 0.375, having a constant section of 0.60 m and a tip section of 0.40 m.

Figure 7. Wing model at the wing tunnel

ACQUISITION AND ANALYSIS RESULTS

Figure 8 presents the comparison of the numerical results obtained by the ANSYS®-FLUENT code with the experimental data. In Fig. 8a, the numerical and experimental results at low angles of attack were very close, being obtained a numerical lift coefficient at zero angle of attack equal to 0.22 (for the experimental case, the angle was equal to 0.2). However, a divergence in the result begins to exist from 8°. From

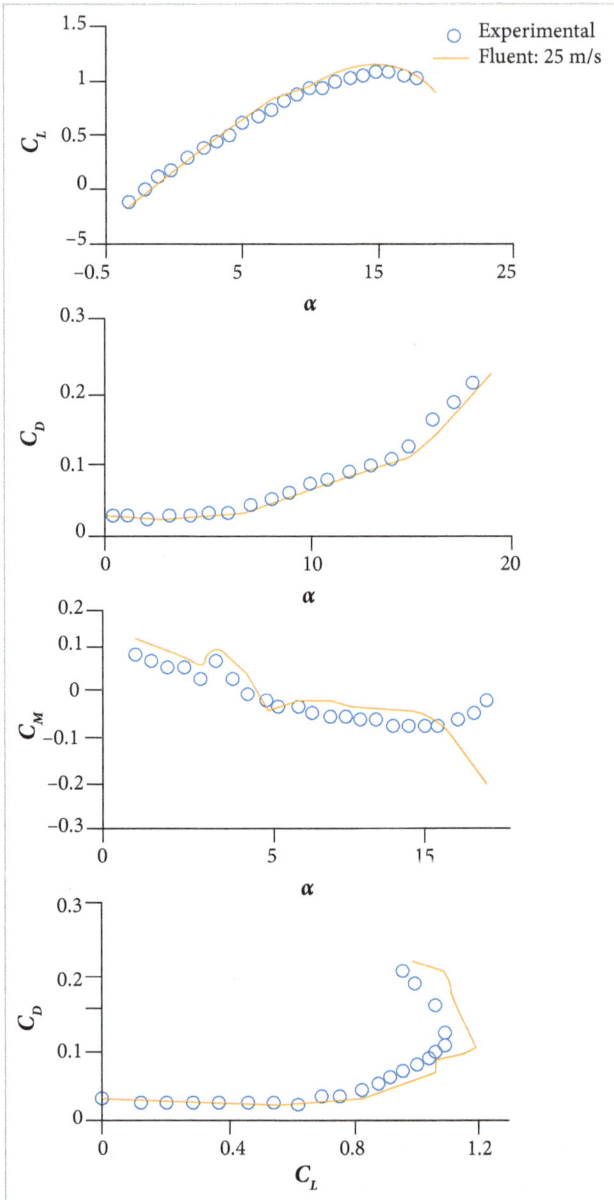

Figure 8. Aerodynamic curves — wing model. Scale 1:3.

an angle of attack of 10°, the inclination of the curve changes and the growth rate of the lift coefficient decreases. As could be observed in Fig. 6b, the presence of cross flow suggests that the wing is beginning to stall. Furthermore, an excellent approxi mation of the experimental and numerical results regarding the drag coefficient (C_D) was obtained, as can be seen in Fig. 8b, where the drag coefficient at zero angle of attack was practically set at 0.03. In Fig. 8c, both experimental and computational moment coefficients around the aerodynamic center were negative, confirming that the wing has a tendency to contribute to the stability of the aircraft. Figure 8d shows the drag polar curve, in which it was observed that the maximum aerodynamic efficiency (*L/D*) was set at 4° for the experimental results and at 5° for the numerical ones. Therefore, this angle of incidence was selected to optimize the mission profile, due to the reduction of the energy consumption.

Finally, in Table 1, the principal results of the total analysis of the wing platform are summarized. As a conclusion, the analytic, numerical and experimental results were well-developed, since the results are close, giving the knowledge of the aerodynamic behavior of the UAV-SOLVENDUS in the established mission.

Table 1. Comparison of aerodynamic coefficients.

Aerodynamic parameters	Experimental	Numerical	Analytical
Maximum C_L	1.11	1.18	1.35
α for C_{Lmax}	15°	15°	10°
C_D at zero α	0.025	0.025	0.02
$\delta C_M/\delta\alpha$	−0.0072	−0.012	−0.099
Maximum (*L/D*)	22.23	24.88	34.34
α for $(L/D)_{max}$	4°	5°	4°

EMPENNAGE DESIGN

After meeting the aerodynamic and geometric characteristics of the wing platform, the empennage development was carried out in order to study the stability behavior of the UAV-SOLVENDUS. In this regard, 3 configurations were examined (conventional, v-tail, and twin-tail); the Twin-tail was selected and connected to the wing via booms. For the horizontal stabilizer, the initial surface area was calculated geometrically from the design conceptions. Therefore, the dimensions of the solar panels were considered, *i.e.* the chord of the horizontal stabilizer was limited by the size of the solar cells, regardless of the relative position that was assumed (long-wide). The dimensions of the solar cells are: 14.6 × 24.8 cm. Thus, the horizontal stabilizer surface (S_H) was 0.405 m^2. The airfoil chosen was a NACA 0012, determined by static-dynamic stability and control criteria. For the benefit of simplicity in manufacture, the vertical stabilizer uses a NACA 0009, and its surface area is 0.23 m^2.

According to the parameters shown in Fig. 9, a conceptual analysis of the longitudinal static stability was carried out, aiming to validate the configuration of the empennage design, where AC_w is the aerodynamic center of the wing, AC_t is the aerodynamic center of the tail and GC is the aircraft gravity center. The longitudinal stability contribution of the wing, the empennage, and the full configuration were calculated in order to maintain the moment coefficient derivative ($\partial Cm/\partial\alpha < 0$) for the UAV. Figure 10 shows the moment coefficient at the center of gravity *versus* the angle of attack of each mentioned surfaces. As a result, the trim angle for the UAV-SOLVENDUS is set at 4°. The static margin (ST) for this case was at 17% of the mean aerodynamic chord (MAC), and the neutral point (NP) was at 47% of the MAC. The 3 views of the final prototype of the UAV-SOLVENDUS with its principal dimensions are shown in Fig. 11.

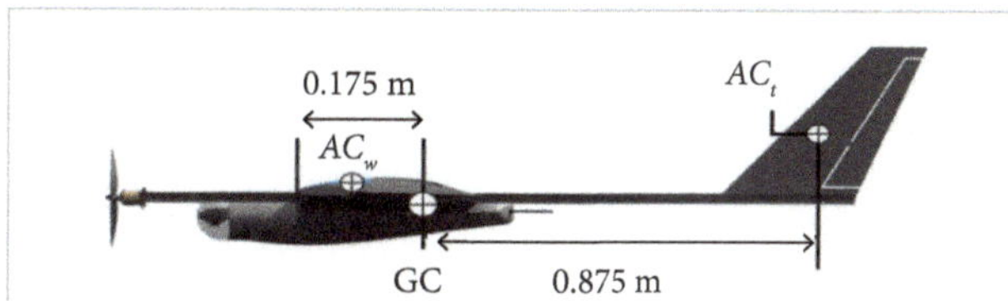

Figure 9. Static margin representation.

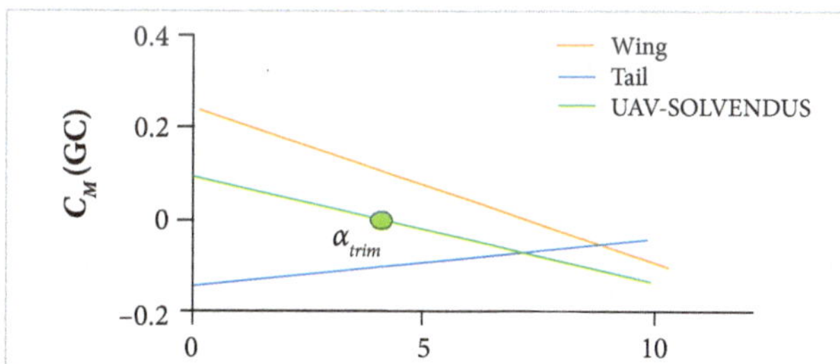

Figure 10. Stability contribution of the aerodynamic surfaces.

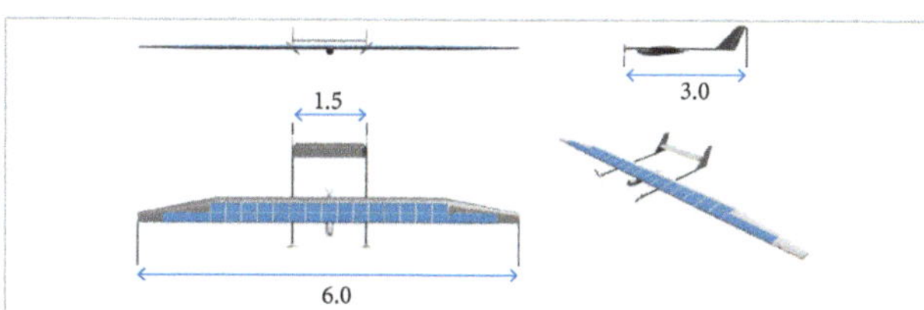

Figure 11. Dimensions of the final design of the UAV-SOLVENDUS in meters.

AIRCRAFT PERFORMANCE EVALUATION

Initially, the thrust required for the UAV-SOLVENDUS was estimated regarding the cruise velocity equal to 25 m·s^{-1} and the MTOW of 12 kg, determined in the conceptual design phase. Then:

$$T = W/(L/D) \tag{2}$$

For this case, the drag coefficient was analytically calculated, taking into account the form factor, the wet area, and the skin friction coefficient of the aerodynamic surfaces, in order to calculate the parasite drag and the characteristics of the wing configuration for the induced drag. As a result, the analytic drag polar equation for the UAV-SOLVENDUS was:

$$C_D = 0.032 + 0.0317 C_L^{\,2} \tag{3}$$

In this regard, the first estimation of the thrust required for this condition was $T_R = 6.75$ N. As a result, it was studied the thrust generated by different propellers for the engine Axi2820/12 (Axi Model Motors, 2015) in order to establish the differences or similarities in relation to the necessary force to move the aircraft in flight under the restrictions already mentioned. Thus, two propellers with 11 × 8″ were selected, providing a total thrust available $T = 77.91$ N and an efficiency $\eta_p = 80.2\%$. To obtain the capacity value of the electrical work generated by the engine Axi2820/12 (Axi Model Motors 2015) and the power required (P_R) for the aircraft, the following relations were established:

$$P = (I)^2 \times R \tag{4}$$

$$P_R = \sqrt{(2W^3\, C_D^{\,2})/(\rho S C_L^{3})} \tag{5}$$

where: P is the electrical power; ρ is the density; S is the wing surface; the values for electric current (I) vary between 15 to 25 A; the motor resistance (R) is set at 0.059 Ω.

The engine manufacturer (Axi Model Motors 2015) assumed these values as ideal. Therefore, these were selected as the initial parameters for calculating the engine performance. In Fig. 12, the thrust and power required by the UAV-SOLVENDUS are shown at different velocities. As can be observed, for the $(L/D)_{max}$, the thrust and power required presented the lower values available, and, for the cruise velocity, the thrust and power required for the UAV-SOLVENDUS were 15 N and 400 W, respectively.

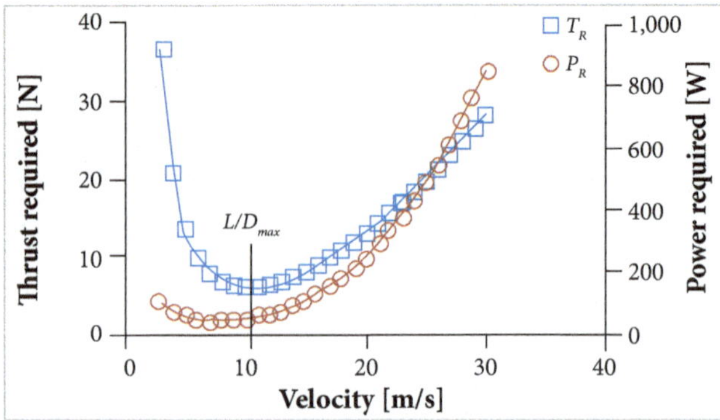

Figure 12. Thrust and power required by the UAV-SOLVENDUS.

PHOTOVOLTAIC SOLAR CELLS CONFIGURATION

The photovoltaic solar system has the purpose of supplying electrical energy to the internal systems of the aircraft. The motors are the components that present higher energy consumption. A motor Axi2820/12 was selected, which has a maximum capacity of 500 W and works with LiPo batteries of 3 and 4 cells. In this way, each motor operates with voltages ranging between 11.1 and 14.8 V, and the current varies between 45 and 30 A. The other elements that consume power are the avionics and servos, which operate at 5 and 6 V with currents of 4.5 A.

In order to determine the best arrangement according to the needs of the avionics and servos systems, it was necessary to carry out a characterization of each solar cell under real operating conditions. Figure 13 shows the behavior of solar radiation against time during the course of the tests. This reflected the drop in the level of solar radiation with dusk, confirming that the best time of the day for test flights is between 13:00 and 15:00 pm (Arai *et al.* 2014; Chang *et al.* 2014). Table 2 shows the environmental test conditions. In order to consider the variation of the solar radiation along the day, voltage and current were measured at different times of the day. In this

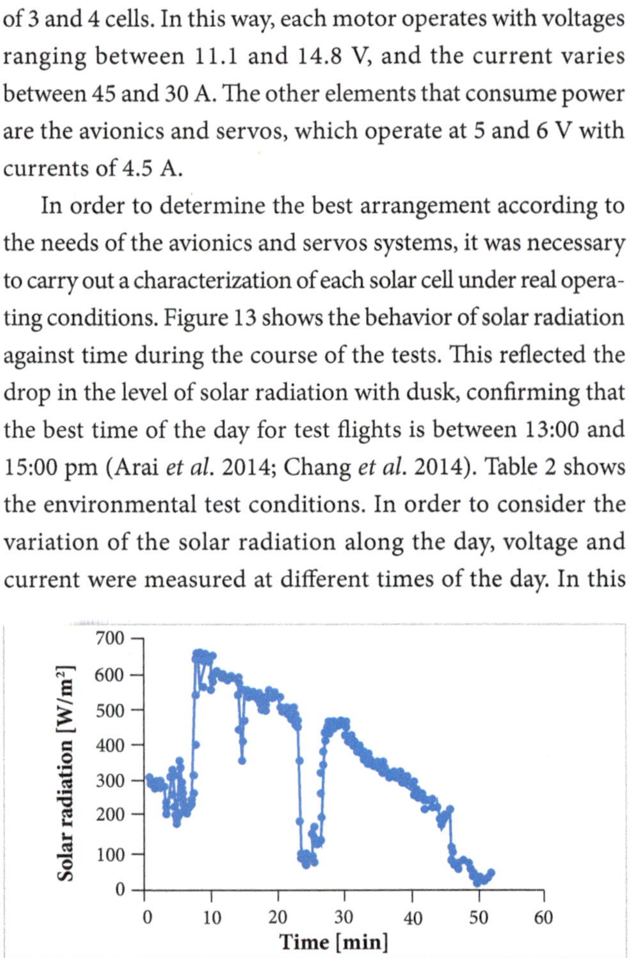

Figure 13. Radiation levels *versus* time.

manner, the voltage of the cells (V-cell) changed from 9 to 10 V (Fig. 14a), and the current (I-cell) changed with a rate of 0.1273 mA/(W·m^{-2}) (Fig. 14b).

Regarding avionics and servos requirements, a hybrid configuration was implemented, which has pairs of cells connected in series, in order to produce higher voltages (Fig. 15a), and all the pairs connected in parallel with the aim of increasing the current of the circuit (Fig. 15b). The cells were placed over the entire surface of the wing and the horizontal stabilizer, consisting of 45 pairs of cells that provide 4,500 mA with voltage ranging between 14 and 20 V. Figure 16 shows the scheme of the final configuration of the solar panel system.

On the other hand, in order to determine the efficiency of the cells, it was necessary to know their voltage-current curve characteristics. This curve is provided by the manufacturer PowerFilm RC7.2-75 PSA (PowerFilm Solar® 2006) and is shown in Fig. 17. The points *A* and *B* represent certain conditions given by the load, resistance and voltage. Bearing in mind that the temperature of the cells do not match the ambient temperature, it was necessary to estimate the variation of their temperature in relation to the environmental temperature. According to Mattos *et al.* (2013) and Skoplaki and Palyvos (2009), the efficiency of the cells at ambient temperature, between 19 and 20 °C, oscillates in a range from 4 to 7%. As there is variability in the electrical properties of

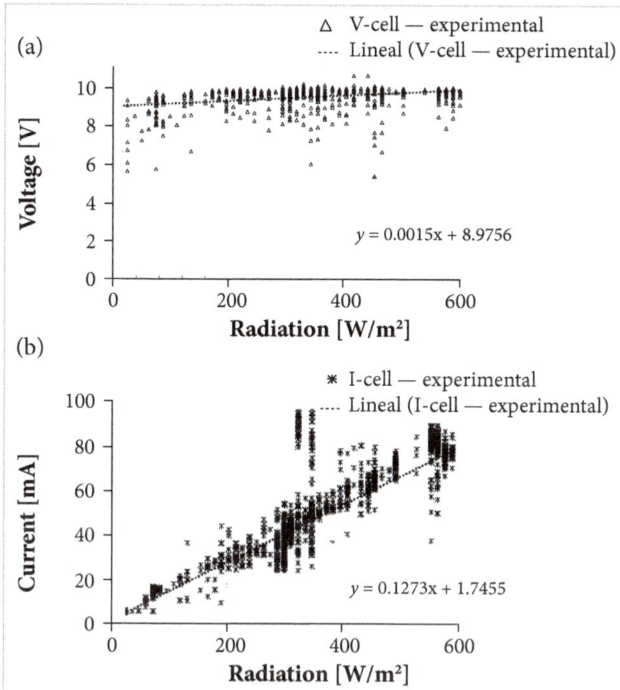

Figure 14. Cells voltages (a) e current (b) under operating conditions (temperature between 19 and 20 °C).

Table 2. Environmental test conditions of the photovoltaic solar panels.

Test	Environmental temperature (°C)	Time of the day (24 h)	Weather
1	19 – 20	14:05	Sunny
2	19 – 20	14:10	Sunny
3	19 – 20	14:15	Sunny
4	19 – 20	14:20	Sunny
5	19 – 20	14:25	Sunny
6	19 – 20	14:30	Sunny
7	19 – 20	14:35	Sunny
8	18 – 19	14:40 pm	Partly cloudy
9	18 – 19	14:45 pm	Partly cloudy
10	18 – 19	14:50 pm	Partly cloudy

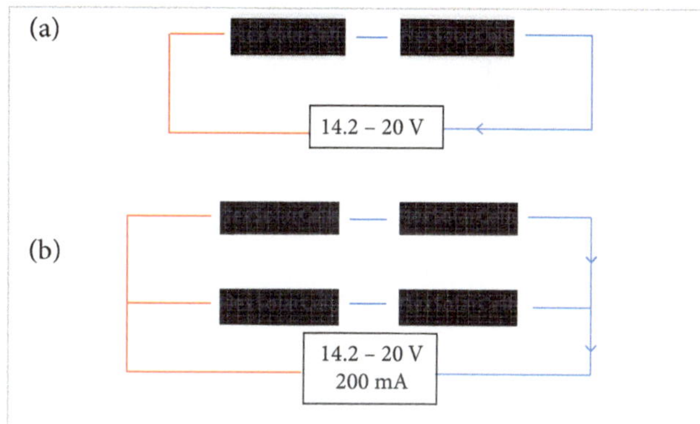

Figure 15. Cells connected (a) series; (b) hybrid array.

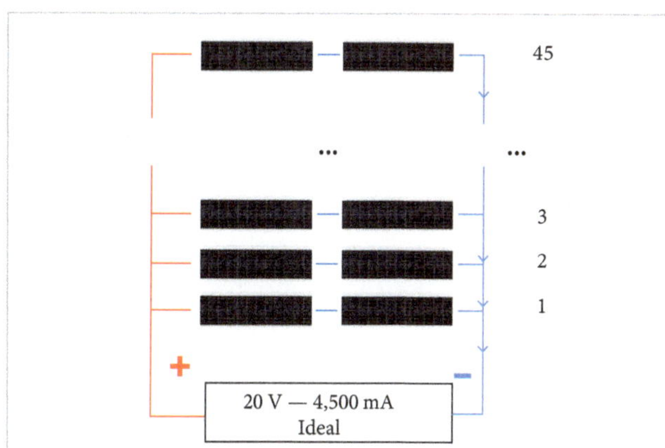

Figure 16. Scheme of the final configuration of the solar panels.

each cell, the selection of the cells arrangement was carried out taking into account that the cells connected in pairs have to provide similar efficiencies in the hybrid configuration. The pairs of cells for the hybrid configuration were made with those of similar efficiencies.

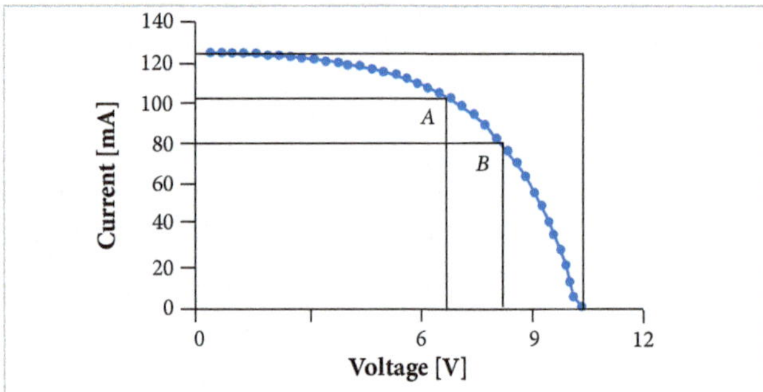

Figure 17. 100% Sun — standard test conditions, AM 1.5. Cell efficiency.

ELECTRONICS AND CONTROL SYSTEMS

The internal structure of the UAV requires electronic devices that allow identifying and establishing the attitude and position. For the attitude of the UAV-SOLVENDUS, environmental data are collected through a set of devices (Type COST, Commercial off-the-shelf), as inertial measurement unit (IMU), which perform the angular speeds (*P*, *Q*, and *R*) measurements in the movements of pitch, roll and yaw — with its respective Moment (M_a) in addition to the measurement of acceleration in the *x*, *y*, and *z* axes using accelerometers and gyroscopes (A_c) integrated with a micro-controller. The variables that are acquired from these sensors are subjected to an estimation process called Kalman filter, whereby it is possible to determine the attitude of the UAV.

In order to position the UAV and enable the navigation, a global position system (GPS) device was used. The altitude on the take-off and landing stages was determined by a sonar and an altimeter. High altitudes (above 2,000 m) were measured with a barometric altimeter. An additional sensor is the magnetometer or digital compass, which revealed the azimuth of the UAV. The scheme of these devices can be observed in Fig. 18. Considering that the power offered by the solar panels are only able to supply electrically the electronic control and aircraft navigation systems, the supply of the propulsion system must be performed by a set of auxiliary batteries (2S – 3S LiPo batteries), preloaded on the ground station. The electricity produced by the solar panels goes directly to the batteries that store and distribute energy to all the electronic components connected to this system. Batteries and panels are connected to a 12-V 2.2-A step-down voltage regulator D24V22F12 in order to keep a constant voltage of 12 V. The energy distribution of the electrical system is shown in Fig. 19.

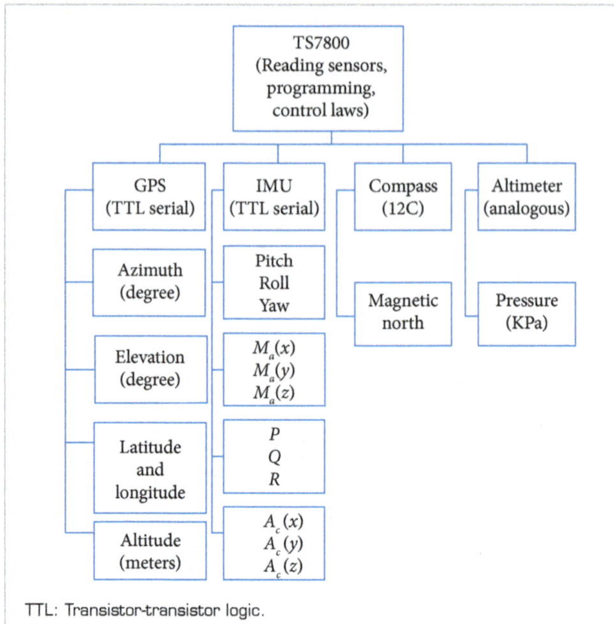

Figure 18. Scheme of the navigation system.

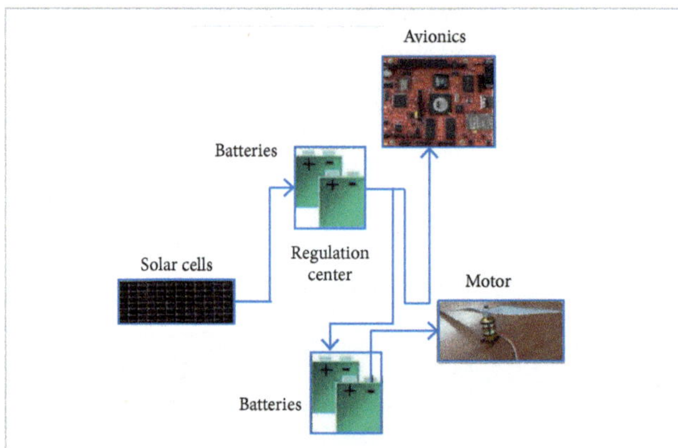

Figure 19. Energy distribution.

MANUFACTURE PROCESS

For structural design of the aircraft, it is important to consider the materials used for manufacturing. In this regard, for this project, the properties of the composite materials were studied, and it was decided to use carbon fiber. Depending on the component and its complexity, the methodology of using the materials and the assembly procedures can change; since a large part of the components were made of composite materials, manufacturing methods were defined, *e.g.* manual contact molding or vacuum packing.

Methodologies of vacuum packing were selected, whose advantages were:

- Allow the realization of parts in small and medium series, without restriction of shapes and sizes.
- Uniform distribution of the resin and reduction of the air inclusions, resulting in good mechanical properties.
- Good surface appearance of both sides.

A suitable process for the manufacture of the parts was carried out; it should be noted that the inputs must be placed in a correct and precise order, aiming to obtain a suitable finishing. Figure 20 shows the correct order of the inputs placed. Before to start the construction, the detail design of each part of the main surfaces was performed using CAD tools. With these designs, it was possible to adjust the distribution of the structural parts mainly in the wings and the empennage. Figure 21 shows the distribution of the ribs and stringers in the wing and the empennage. Jigs were used in order to control the location of the components during this process. Besides serving for correct alignment, jigs fulfil the function of temporary supports during assembly model.

After mounting the jigs, the ribs (which were formed by a core of light wood, 1.5 mm thick, and 2 layers of carbon fiber, 0.7 mm thick, forming a "sandwich", Fig. 22, with a thickness of 2.2 mm) were located on the jigs and joined to the beam. At the end of the process, 3 sections of 2 m each were built for the wing. The 3 wing sections were joined with screws for an easy assembly and disassembly. The fuselage manufacturing was made in sections. In the first instance, the fuselage CAD model was printed in a 3-D printer machine; this impression served to make the molds for manufacturing in composite material. Figure 23 shows the molds and the final manufacture of the fuselage. On the other hand, the UAV skin manufacture was performed in 6 sections, distributing inputs in the correct order as pointed in Fig. 20. As a result, each layer presented 2.20 × 0.40 m and weight of 173.4 g. Figure 24 shows the aircraft skin manufacture processes.

Each rib of the aircraft was check to ensure adherence of the skin with the wing structure and the empennage. During this process, the ribs and the main beam of the UAV were reinforced. The adhesion of the skin with the ribs of the wings was carried out with resin by the stringers designed; the union covered the leading edge and part of the main beam. Figure 25 shows the joint between the skin and ribs. Finally, tail booms were built with an "I" profile with carbon fiber and light wood cored, which were joined to the vertical stabilizers with 6 screws on each side. Figure 26 shows the final assembly of the UAV-SOLVENDUS ready to flight.

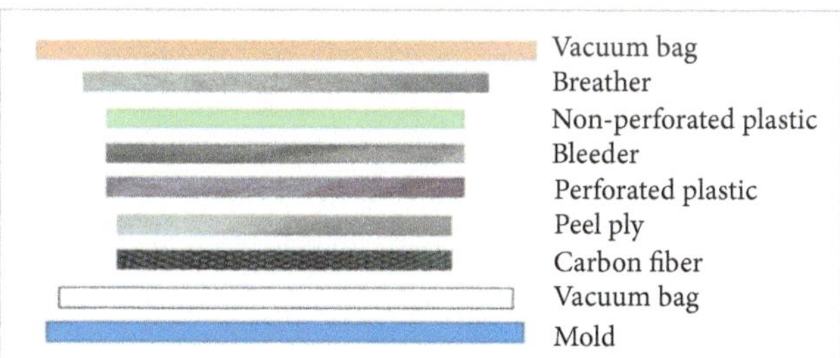

Figure 20. Materials and supplies ordered and used in vacuum.

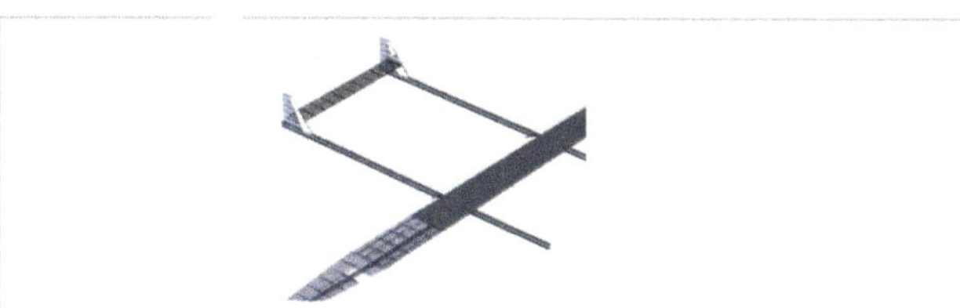

Figure 21. Distribution of ribs and stringers on the wing and empennage.

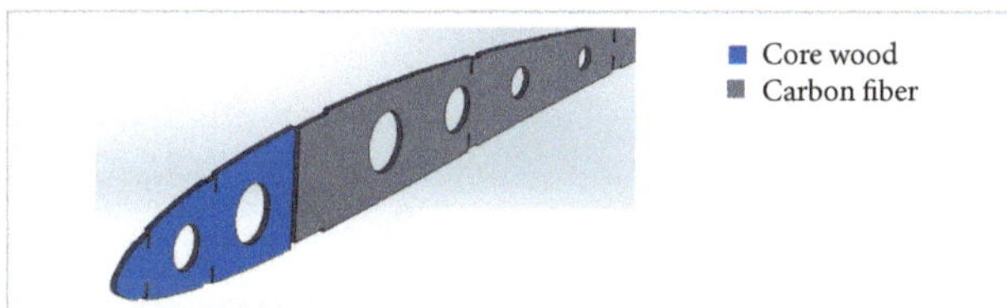

Figure 22. Wing rib detail.

Figure 23. Fuselage manufacture.

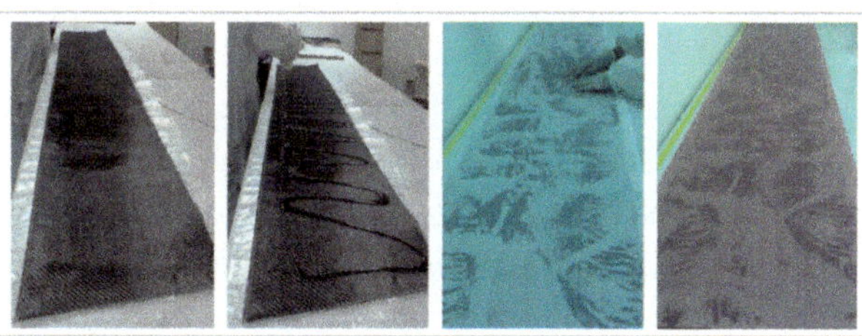

Figure 24. Skin manufacture.

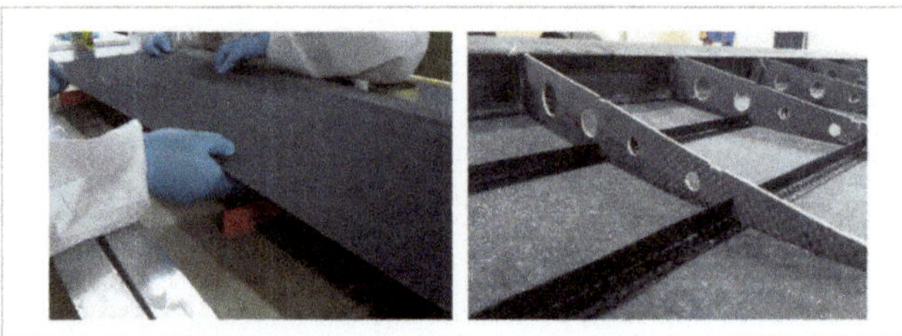

Figure 25. Joint between skin and ribs.

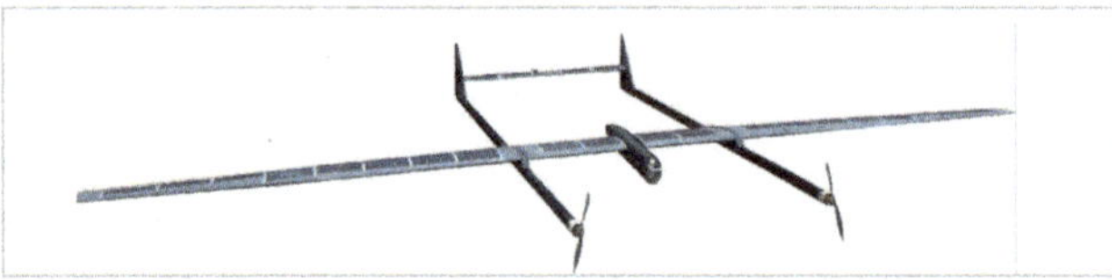

Figure 26. UAV-SOLVENDUS experimental prototype.

CONCLUSIONS

The first prototype of a light, solar-powered UAV for civilian surveillance missions was designed and successfully constructed, being one of the first projects that are registered in Colombia as viable due to the integration of independent complex systems. It has expanded the discussion of the development of unmanned systems designs with the aim of performing missions that could not be made with traditional aircraft. The results obtained with analytical, computational, and experimental methods have served to enrich the research exercise around the design of the UAV-SOLVENDUS. Therefore, it was possible to analyze and validate the main findings during the design process, which are summarized as follows:

- CFD tools were used in order to study and compare the experiments performed on the wing model. The aerodynamic efficiency (*L/D*) obtained in the wing design method reveals that it is maximum at 4°. Therefore, this angle of attack is recommended for better aerodynamic efficiency at cruise flight for the UAV.
- The combination of improvements proposed by the authors, as the increase in the wingspan and the wing surface, lead to a greater range and endurance of the aircraft. Therefore, the UAV-SOLVENDUS is a more economical and eco-friendly aircraft, compared with the other UAVs destined to perform similar missions.
- For this first prototype, an experimental study for the solar panel system characterization was executed, with the aim of analyzing the voltage, current, and power of UAV requirements according to different levels of radiation in real operating conditions. As a result, the most appropriate cells arrangement (hybrid) was chosen, in order to supply the power requirements for the avionics equipment, since the propulsion system must be energized with auxiliary batteries.
- The performance of the current solar power system is under evaluation. Until now, its behavior has been acceptable to supply the power necessary to the onboarding system, even though it has not been wholly satisfactory to supply energy to the propulsion system, and electrical batteries are still necessary.
- The manufacture process was developed based on ASTM D3039. The features offered by the carbon fiber were appropriate for the model; factors such as weight, strength, and behavior in environmental conditions were remarkably outstanding in the performed test.

Finally, through this investigation, it was possible to demonstrate the benefits of using renewable energies in the aeronautical engineering field, allowing to know new alternatives to the use of photovoltaic solar panels on aircraft models that seek to optimize the fuel consumption, keeping high values of endurance and range.

ACKNOWLEDGEMENTS

The authors express their gratitude to the Fundación Universitaria Los Libertadores, for the financial support, as well as to its Oficina de Relaciones Interinstitucionales (ORI), for the support to the Student Exchange Program. The Grupo de Investigación en Ciencias Aeroespaciales (GICA), for the cooperation and devotion on the development of this project. Prof.Fernando Martini Catalano, for the support in the preparation of the experiments at the Laboratório de Aerodinâmica, Dep. Engenharia Aeronáutica, Escola de Engenharia de São Carlos, Universidade de São Paulo.

AUTHOR'S CONTRIBUTION

Betancourth NJP, Villamarin JEP and Rios JJV conceived the idea and co-wrote the text. Cerón-Muñoz HD headed the current project and co-wrote the text. Bravo-Mosquera PD, aided on the calculations, prepared the figures and co-wrote the main text. All authors discussed the results and commented on the manuscript.

REFERENCES

Anderson Jr JD (1999) Aircraft performance and design. Boston: WCB/McGraw-Hill.

ANSYS® (2011) ANSYS FLUENT theory guide. Canonsburg: ANSYS, Inc.

Arai N, Harasani W, Hiraoka K (2014) Conceptual design of solar UAV for long endurance. St. Petersburg: International Council of Aeronautical Science.

Austin R (2010) Unmanned aircraft systems: UAVS design, development and deployment. Chichester: Wiley.

Axi Model Motors (2015) Axi2820/12 gold line, data sheet, technical specification, catalogue products; [accessed 2016 Feb 24]. https://www.modelmotors.cz/

Baoan L, Zhihua L, Xinjun L (2009) Research of UAV engine fault prediction based on particle filter. Proceedings of the 9th International Conference on Electronic Measurement and Instruments; Beijing, China.

Brandt SA, Gilliam FT (1995) Design analysis methodology for solar-powered aircraft. J Aircraft 32(4):703-709. doi: 10.2514/3.46780

Bravo-Mosquera P, Uribe A, Catalano F, Abdalla A (2015) Parametric conceptual design of a UAV for superficial volcano monitoring. Proceedings of the 23rd ABCM International Congress of Mechanical Engineering; Rio de Janeiro, Brazil.

Catalano F (2001) The new closed circuit wind tunnel of the aircraft laboratory of University of São Paulo. Proceedings of the 16th Brazilian Congress of Mechanical Engineering; Uberlândia, Brazil.

Cestino E (2006) Design of solar high altitude long endurance aircraft for multi payload and operations. Aero Sci Tech 10(6):541-550. doi: 10.1016/j.ast.2006.06.001

Chang M, Zhou Z, Wang R, Xu X (2014) A general design methodology for year-sound solar-powered stratospheric UAVs from low to middle latitudes. St. Petersburg: International Council of Aeronautical Science.

Fürrutter MK, Meyer J (2009) Small fuel cell powering an unmanned aerial vehicle. Proceedings of the AFRICON 2009; Naibori, Kenia.

Gao X, Hou Z, Guo Z, Chen X (2014) Design and performance test of a twin fuselage configuration solar-powered UAV. St. Petersburg: International Council of Aeronautical Science.

Herrera J, Silva N, Tellez C, Cerón H, Pedraza N (2014) Análisis aerodinámico computacional y experimental para el ala de un mini vehículo aéreo no tripulado (VANT). Proceedings of the 8th Congresso Nacional de Engenharia Mecânica; Uberlândia, Brazil.

Herwitz S, Dunagan S, Sullivan D, Higgins R, Johnson L, Zheng J, Aoyagi M (2003) Solar-powered UAV mission for agricultural decision support. Proceedings of the International Geoscience and Remote Sensing Symposium; Moffett Field, USA.

Kontogiannis G, Ekaterinaris J (2013) Design, performance evaluation and optimization of a UAV. Aero Sci Tech 29(1):339-350. doi: 10.1016/j.ast.2013.04.005

Lubkowski S, Jones B, Rojas E, Morris D (2010) Trade-off analysis of regenerative power source for long duration loitering Airship. Proceedings of the Systems and Information Engineering Design Symposium (SIEDS-IEEE); Charlottesville, USA.

MacCready PB, Lissaman PBS, Morgan WR, Burke JD (1983) Sun-powered aircraft designs. J Aircraft 20(6):487-493. doi: 10.2514/3.44898

Marta A, Gamboa P (2014) Long endurance electric UAV for civilian surveillance missions. St. Petersburg: International Council of Aeronautical Science.

Martínez P, Cerón H (2013) Análise computacional das características aerodinâmicas de um VANT em configuração box-wing. Proceedings of the 6th Congreso Internacional de Ingeniería Mecánica (CIMM); Barranquilla, Colombia.

Mattos BS, Secco NR, Salles EF (2013) Optimal design of a high-altitude solar-powered unmanned airplane. J Aerosp Technol Manag 5(3):349-361. doi: 10.5028/jatm.v5i3.223

Nickol L, Guynn M, Kohout L, Ozoroski T (2007) High altitude long endurance air vehicle analysis of alternatives and technology requirements development. Proceedings of the 45th Aerospace Sciences Meeting and Exhibit; Reno, USA.

Noth A (2008) Design of solar powered airplanes for continuous flight (PhD thesis). Zürich: ETH Zürich.

Panagiotou P, Tsavlidis I, Yakinthos K (2016) Conceptual design of a hybrid solar MALE UAV. Aero Sci Tech 53:207-219. doi: 10.1016.ast.2016.03.023

Parra J, Cerón H, Gómez A (2013) Pruebas aerodinámicas experimentales de aeronave con alta relación de aspecto a bajos Reynolds. Proceedings of the 6th Congreso Internacional de Ingeniería Mecánica (CIMM); Barranquilla, Colombia.

PowerFilm Solar® (2006) RC7.2-75 PSAF. Datasheet. Catalogue products; [accessed 2016 May 1]. http://www.powerfilmsolar.com/

Ramanathan V, Ramana MV, Roberts G, Kim D, Corrigan C, Chung C, Winker D (2007) Warming trends in Asia amplified by brown cloud solar absorption. Nature 448:575-578. doi: 10.1038/nature06019

Raymer DP (1992) Aircraft design: a conceptual approach. Washington: American Institute of Aeronautics and Astronautics.

Romeo G, Frulla G, Cestino E (2007) Design of a high-altitude long-endurance solar-powered unmanned air vehicle for multi-payload and operations. Proc IME G J Aero Eng 221(2):199-216. doi: 10.1243/09544100JAERO119

Roskam J (1997) Airplane aerodynamics and performance. Lawrence: DAR Corporation.

Skoplaki E, Palyvos JA (2009) On the temperature dependence of photovoltaic module electrical performance: a review of efficiency/power correlations. Sol Energ 83(5):614-624.

Song L, Evans JW (2000) Electrochemical-thermal model of lithium polymer batteries. J Electrochem Soc 147(6):2086-2095. doi: 10.1149/1.1393490

Sun K, Zhu M, Wang L, Liu H (2015) Analysis of the hybrid power system for high-altitude unmanned aircraft. Advances in Materials Science and Engineering 2015(2015):Article ID 380708 .

Tornado® (2001) A vortex lattice method implemented in MATLAB; [accessed 2016 May 4]. http://tornado.redhammer.se/

Wilkins G, Fourie D, Meyer J (2009) Critical design parameters for a low altitude long endurance solar powered UAV. Proceedings of the AFRICON 2009; Naibori, Kenia.

Youngblood JW, Talay TA, Pegg RJ (1984) Design of long-endurance unmanned airplanes incorporating solar and fuel cell propulsion. Proceedings of the 20th Joint Propulsion Conference; Cincinnati, USA.

Study of Aramid Fiber/Polychloroprene Recycling Process by Thermal Degradation

Igor Dabkiewicz[1], Ítalo Vinícius Soares e Silva[1], Jossano Saldanha Marcuzzo[1], Rita de Cassia Mendonça Sales Contini[1]

ABSTRACT: Aramid fiber is an important polymer applied as reinforcement in high-performance composites, which, due its exceptional properties, becomes an excellent impact absorption material. It has been broadly utilized in aeronautic industry and ballistic protection. In aircrafts, it is mainly used in secondary structures, such as fairings, floor panels, and bullet proof structures in helicopters, whereas, in ballistic protection industry, it is applied in automotive armor and bullet proof vest. Under environmental perspective, it is worrying the development and application of composites, which generate proportional discards of these materials, whether originated from manufacturing process, spare parts or end of life cycle. High-performance composite materials like those using aramid fiber are generally difficult to recycle due to their properties and the difficulty for the separation of the components, making their recycling economically unviable. From the characteristics of composite materials and environmental viewpoint, this paper presents a new aramid fiber recycling process. The main objective of this research was to study different recycling methods in aramid fibers/Neoprene® composites. To promote the Neoprene® degradation, it was used a pyrolysis oven with controlled atmosphere and CO_2 injection. For the degraded separation, it was designed a mechanical washing machine in which the most degraded separation occurred. To complete the materials separation, it was employed a manual cleaning process, and, at least to prove the efficacy of the process, it was applied a tensile test in the yarns.

KEYWORDS: Recycling, Aramid fiber, Composite, Polychloroprene, Pyrolysis.

INTRODUCTION

The usage of composite materials is increasing with the development of new technologies due to the necessity of materials that match the needs for strength and lightness (Chawla 2012). Among the composites, the polymeric ones are advanced engineering materials composed by reinforcement fibers like aramid, glass or carbon, combined with a matrix, generally made of thermo-rigid or thermo-plastic polymeric material (Cândido *et al.* 2006).

The aramid fiber is widely applied in functions that demand a high-performance material, in different sectors, like aeronautic and automotive sectors (Boukouvalas and Wiebeck 2007).

In the aeronautic sector, this is an important material present in components that require high strength-to-weight ratio, high strain-to-failure and high temperature resistance, being commonly utilized in fairings, engine containing rings, skins, floor panels, landing gear doors, radomes, and others that need structural efficiency and suitable dielectric properties (Baker *et al.* 2004).

Structures for ballistic protection have been developed over the years, and the materials and design of these structures have evolved enough. Aramid composites have been used worldwide in production of ballistic protection due to their high capacity to absorb energy mainly on armor components in aeronautic and automotive industries (Lopes *et al.* 2007). One of those composite armors is a combination of aramid fiber (Kevlar®) with a polychloroprene resin (Neoprene®). The aramid is responsible for the mechanical and thermal endurance, and the polychloroprene provides elasticity modulus, keeping the materials united, acting as a connection element

1.Centro Estadual Tecnológico Paula Souza – Faculdade de Tecnologia de São José dos Campos - Prof. Jessen Vidal – São José dos Campos/SP – Brazil.

Author for correspondence: Rita de Cássia Mendonça Sales Contini | Centro Estadual Tecnológico Paula Souza – Faculdade de Tecnologia de São José dos Campos - Prof. Jessen Vidal | Avenida Cesare Mansueto Giulio Lattes, 1.350 – Eugênio de Melo | CEP: 12.247-014 – São José dos Campos/SP – Brazil | Email: rita.sales@fatec.sp.gov.br

and protecting them against acids, ozone and other aggressive agents (Bittencourt 2011).

At the end of the life cycle, these components will be replaced by new ones, then the discarded part will be sent to disposal sites. This method of treating composites residues is becoming more expensive, and the pressure from society and environmental agencies is growing, being necessary a reassessment of the allocation of these materials (Naglis and D'Almeida 1998).

According to the Boeing Company (2007), it is expected that up to 7,200 commercial aircrafts will be retired from active service in the next 20 years and all should be available for recycling. However, in general, composites require complex recycling methods, often being sent to landfills (Cunliffe *et al.* 2003). On the other hand, aramid is a fiber with excellent mechanical properties that can be recycled and applied in another product, preserving natural resources and allowing economical gain.

The recycling of these materials is in constant development, and, nowadays, the more developed techniques are mechanical, chemical and thermal processes (Pickering 2006).

Pyrolysis is a thermal recycling process widely studied in fiber recovery. In this process, the thermal degradation of one or more constituent materials occurs, recovering the desired material for a new manufacture (Pickering 2006).

This research had the objective of showing that it is possible to carry out the separation of the components from a composite material made from aramid fiber and polychloroprene polymer without solvents usage, recycling in an ecologic way. Therefore, it was projected a washing machine capable of performing the separation of the polychloroprene degraded from the aramid fibers only with the help of simple cleaning materials.

MATERIALS AND METHODS

MATERIALS

The samples utilized in this research are strips from a composite applied in vehicle armor; the reinforcement fibers (Kevlar®) form bidirectional orientation cables, threaded in a perpendicular way one to another, arranged in 7 layers, and, between them, there are layers of the polymeric matrix (Neoprene®). Figure 1 shows, in a simplified form, the composition of the material with the respective layers. The samples were cut into strips with dimensions of 5.0 × 20.0 cm, weight of 2.8 kg, and then were taken to the oven where the pyrolysis process was performed.

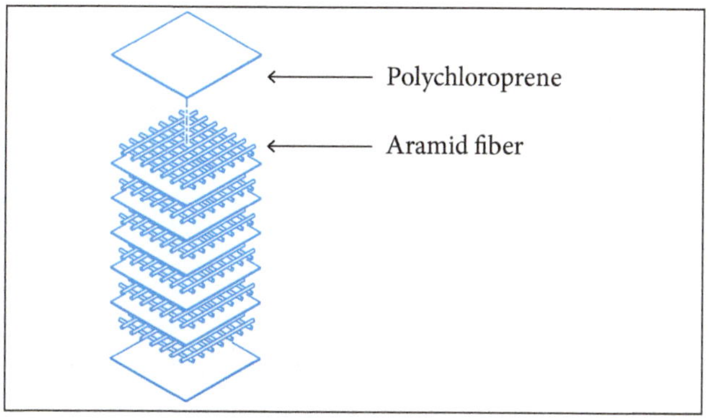

Figure 1. Material composition.

PYROLYSIS PROCESS

The pyrolysis was performed in 2 stages, in a controlled temperature oven, with injection of CO_2 in a volume of 300 mL·L^{-1}. In the first stage, the oven was programmed to a heating ramp of 300 °C, reaching this temperature and remaining at this value for 15 h (Fig. 2). At the end of this stage, the material was again weighed and analyzed.

In the second pyrolysis stage, the 7 layers of the material were separated as illustrated in Fig. 1, resulting in a larger contact surface. Then, the oven was heated at 300 °C for 1 h. After that, the temperature was increased at 350 °C for 15 min, remaining 5 h at this value (Fig. 3).

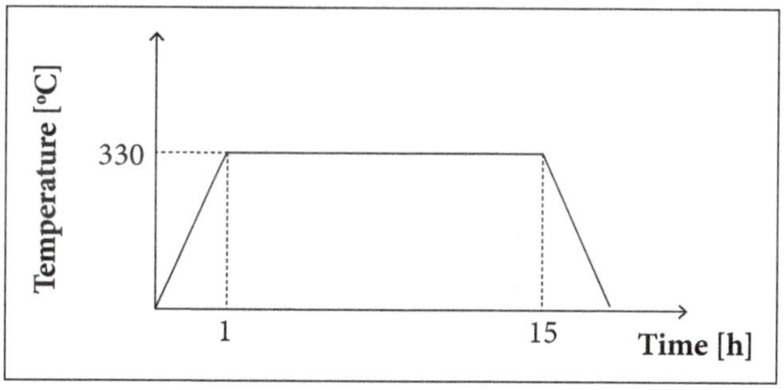

Figure 2. First pyrolysis stages.

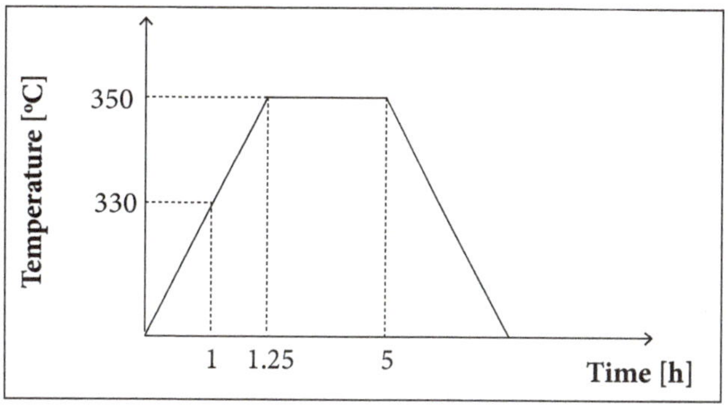

Figure 3. Second pyrolysis stage.

SAMPLES WASHING

After the pyrolysis process, the samples were subjected to the mechanical washing process for the degraded Neoprene® separation from the Kevlar®. In this process, 2 different techniques were employed: the mechanical washing and then the manual washing.

Mechanical Washing

This washing method was employed with the objective of performing the separation of degraded products from the pyrolysis. Then, for this purpose, it was built a mechanical action washer where metal spheres and rollers promote the required friction for the material separation. In this washing step, soap powder and water were utilized together with the spheres and rollers. In order to avoid water saturation, after 3 h of washing, the water and the soap were replaced, and the washing was performed for a further 3-h period.

Manual Washing

After the mechanical washing, the samples were rubbed with a sponge to remove possible traces of the degraded polychloroprene.

MECHANICAL TENSILE TEST

The mechanical tensile test was employed to determine the integrity of Kevlar® during each step of the recycling process; therefore, samples were collected from the starting material after pyrolysis, after the mechanical washing and after the manual washing.

The following aramid reference values were utilized: linear mass density of 1.27 cN/dtex, density of 1.44 g·cm^{-3}, and diameter of 12.0 μm. These values are comparable to the results obtained in the test (DuPont® 2011). The linear mass density of 1.27 cN/dtex was applied to determine the pre-tension, which set a pre-tensioning claw of 50.0 mg used in the Textechno Vibromat equipment in order to obtain the filament linear mass density.

Thus, with the linear mass density determined, the tensile test was performed using the Textechno Fafergrhaf equipment. For this test, ten samples of each step of the recycling process were collected: ten from the original material (before any recycling process), ten after pyrolysis and ten after washing.

The equipment was set to perform tests in filaments of 20 mm, with a speed of 50.0 m·min^{-1}, 10.0% of minimum and 30.0% of maximum elongation compared to the initial length of the filament and maximum strength of 45.0 cN.

RESULTS AND DISCUSSION

PYROLYSIS

After the first stage of pyrolysis, it was observed that the weight of the samples decreased 21.0% compared to the original material; in the second stage, after the delamination, the samples were weighed, and it was observed that they lost 0.20 kg, weighing at last 1.60 kg. At this point, it was observed that the polychloroprene degradation occurred more significantly, with a weight loss of 14.0% compared to the resulting material from the first stage (Table 1).

SAMPLE WASHING

In the mechanical washing process, tests were carried out with different washing times and different cleaning products to set the most efficient method. Thus, it was observed that the most efficient washing process refers to that in which the samples were treated during a total of 6 h. Thus, to avoid the water saturation, at the end of 3 h, the water and the soap were changed, and the washing was carried out for more 3 h. To execute the washing process, it was used public distribution water and Omo Multiaction® washing powder, which is a commercial cleaning product that reduces the surface tension of the water, allowing its deeper penetration into the material. At the same time, this washing powder is able to attract debris through its polarity, increasing the washing efficacy (Souza *et al.* 2010).

It was observed that the mechanical washing removed the most part of the degraded Neoprene® from the fibers (Fig. 4); however, it was necessary the use of manual washing for the complete removal of the degraded remains.

Table 1. Pyrolysis stages.

Pyrolysis	Mass loss (g)	Time (h)	Temperature (°C)	Flow rate of CO_2 (mL·L^{-1})	Results
Stage 1	680	16	330	300	The material became malleable
Stage 2	0.050	6	350	300	The polymeric matrix degraded and obtained a shattered appearance

After performing tests to determine the best manual washing method, the sponge was the one which presented the best efficiency, with which the fibers showed coloration near to their natural yellow (Fig. 5).

Figure 4. Material after mechanical washing.

Figure 5. Material after manual washing.

TENSILE TEST

Through the analysis of the tensile tests results, it was observed that the pyrolysis process employed in recycling degraded the fibers. The fiber became fragile, and values of tenacity and tensile strength became very close. The samples lost some of their properties, occurring a decrease of 64.0% in tensile strength, 64.0% in tenacity, 84.0% in elongation and 87.0% in the time required for rupture (Table 2).

It was observed that all plastic and almost the entire elastic region from the material were lost after pyrolysis, making the fiber a fragile material. One explanation for this degradation is the conditions and parameters adopted for the pyrolysis of the material and the properties of the fiber.

The aramid fiber begins to decompose at 427 °C, losing only humidity. In quick exposures, it would bear the maximum temperature employed in the process (350 °C); however, when the temperature exposure occurs over a long period, the aramid begins to decompose in a range between 147 and 177 °C (Cunliffe *et al.* 2003).

It can be observed that the exposure to elevated temperatures leads to the loss of the fibers' properties, making 22 h of exposure in the oven sufficient for their degradation.

CONCLUSIONS

The proposed material separation for this study was successfully done; however, the fibers suffered significant loss of mechanical properties, especially in their main features.

Regarding the toughness, a critical feature of the fiber for impact absorption, a decrease of 80.13% has occurred. The tensile strength decreased by 79.80%, also limiting the fiber application in structural efforts. The pyrolysis degraded the polychloroprene impregnated into the fibers satisfactorily, allowing the subsequent complete removal by the washing process; however, the aramid fibers were also partially degraded during the pyrolysis process, resulting in decreased mechanical strength.

The mechanical action washer did not completely removed the degraded polychloroprene as expected, requiring a subsequent removal from it with a manual method.

These conclusions allow making some considerations about this research. The loss of aramid fiber mechanical strength properties during the pyrolysis process has been caused by exposure to high temperatures for a long period. The mechanical washing proposed in this study is not able to perform the complete separation of the materials, being necessary to improve the washing method.

Table 2. Tensile test results.

Sample	Rupture time (s)	Linear mass density (dtex)	Tensile strength (cN)	Angular coefficient (°)	Tenacity (cN/dtex)	Stretching (%)
Starting material 1	4.59	1.00	21.49	10	21.49	8.52
Pyrolysed 1	0.58	1.00	7.73	10	7.73	1.48
Washed 1	0.63	1.00	12.35	10	12.35	1.73

REFERENCES

Baker AA, Dutton S, Kelly D, Kelly DW (2004) Composite materials for aircraft structures. 2nd ed. Reston: The American Institute of Aeronautics and Astronautics.

Bittencourt GA (2011) Efeito da radiação gama em blindagens balísticas compósitas de poliaramida (Master's thesis). Rio de Janeiro: Instituto Militar de Engenharia. In portuguese.

Boeing Company (2007) Aircraft & composite recycling. Environmental Technotes 12.

Boukouvalas NT, Wiebeck H (2007) Caracterização térmica de fios de poliaramida. Polímeros 17(4):284-288. doi: 10.1590/S0104-14282007000400006

Cândido G, Paiva JMF, Mayer S, Rezende MMC (2006) Avaliação da temperatura de transição vítrea de compósitos poliméricos reparados de uso aeronáutico. Polímeros 16(1):79-87. doi: 10.1590/S0104-14282006000100016

Chawla KK (2012) Composite materials: science and engineering. 3rd ed. Birmingham: Springer Science.

Cunliffe A, Jones N, Williams PT (2003) Recycling of fibre-reinforced polymeric waste by pyrolysis: thermo-gravimetric and bench-scale investigations. J Anal Appl Pyrol 70:315-338. doi: 10.1016/S0165-2370(02)00161-4

DuPont® (2011) Kevlar® Aramid Fiber Technical Guide. n. 32.

Lopes CMA, Gonçalves DP, Melo FCl (2007) Resistência balística de compósitos poliméricos laminados. Proceedings of the 9° Congresso Brasileiro de Polímeros. Campina Grande; Brazil.

Naglis MMM, D'Almeida JRM (1998) Avaliação dos modos de falha sob impacto de compósitos de matriz polimérica reforçados por fibras. Polímeros 8(1):54-60. doi: 10.1590/S0104-14281998000100008

Pickering SJ (2006) Recycling technologies for thermoset composite materials — current status. Compos Appl Sci Manuf 37(8):1206-1215. doi: 10.1016/j.compositesa.2005.05.030

Souza ACS, Cardoso CE, Fraguas Neto MR (2010) Determinação da concentração de fosfato em amostras de detergentes em pó utilizando detecção titulométrica. Revista Eletrônica TECCEN 3(4):19-28.

12

A Post-Accident Analysis of Civil Remotely-Piloted Aircraft System Accidents and Incidents

Graham Wild[1], Kellie Gavin[1], John Murray[1,2], Jose Silva[1], Glenn Baxter[1]

ABSTRACT: A sample of 152 accidents and incidents involving Remotely Piloted Aircraft Systems, more commonly referred to as "drones", have been analysed. The data was collected from a 10-year period, 2006 to 2015, conveniently sourced from a limited population owing to the scarcity of reports. Results indicate that safety occurrences involving Remotely Piloted Aircraft Systems (RPAS) have a significantly different distribution of contributing factors when sorted into distinct categories. This provides a thorough and up-to-date characterization of the safety deficiencies specific to RPAS. In turn, this contributes to the development of adequate safety management systems applicable to the RPAS sector. The majority of RPAS occurrences involved system component failures which were the result of equipment problems. Therefore, airworthiness instead of pilot licensing needs to be considered first when regulating the Remotely Piloted Aircraft System industry. "Human factors" and "loss of control in-flight" were found to be the second most common "contributing factor" and "occurrence category", respectively; Remotely Piloted Aircraft pilot licensing will help reduce the probability of these secondary occurrences. The most significant conclusion is that reporting systems must be implemented to address RPAS accidents and incidents specifically, such that more useful data is available, and further analysis is possible facilitating an improved understanding and greater awareness.

KEYWORDS: RPAS, UAS, UAV, Accidents and incidents, Aviation safety.

INTRODUCTION

There is a growing interest in aircraft that are controlled from a remote location without a pilot located in the aircraft itself. Referred to by many terms ("drones", UAV, UAS, etc.) the International Civil Aviation Organization (ICAO 2015) has recently adopted "Remotely Piloted Aircraft System" (RPAS) to identify these aircraft. The absence of a flight crew on-board the aircraft creates unique challenges in terms of the safety risks associated with the operation of RPAS. That is, the dimensions and applicability of the well-known Software, Hardware, Environment, Liveware, and Liveware (SHELL) model adopted by ICAO Circular 216-AN31 (ICAO 1989) will be significantly different from those associated with the operation of conventionally-piloted aircraft. In particular, the interrelationships between the elements constituting the framework of the SHELL model will be significantly different.

These unmanned aircraft, commonly referred to as "drones" in the defence sector, usually conjure up thoughts of Predator UCAVs firing Hellfire missiles in combat operations (Callam 2015). In the civil sector, the presence of RPAS is still a relatively new phenomenon (ICAO 2011; Skrzypietz 2012). From the 1990s, civil operation of RPAS was mostly seen in the Japanese agriculture industry (MarketLine 2014; Odido and Madara 2013). The ever-evolving nature of the aviation industry has supported a vast deviation of RPAS into civilian aviation. These areas include policing activities, wildlife and fisheries protection, environmental monitoring, surveillance, as well as search and rescue (Gupta *et al.* 2013). The aviation industry

1.RMIT University – School of Engineering – Aerospace Engineering and Aviation – Melbourne/Victoria – Australia. **2.**Edith Cowan University – School of Engineering – Aviation – Perth/Western Australia – Australia.

Author for correspondence: Graham Wild | RMIT University – School of Engineering – Aerospace Engineering and Aviation | GPO Box 2476 Melbourne, Victoria, 3001, Australia | Email: graham.wild@rmit.edu.au recognises the economic benefits of remotely piloted aircraft and acknowledges the many opportunities provided by their use in "dull, dirty, dangerous and demanding" tasks, that may otherwise impose high risk to a piloted aircraft (CASA date unknown, b).

The examination of past accident and incident cases can assist in the continuous improvement of safety, such that potential hazards, unsafe acts, and latent conditions are identified before they have disastrous effects (ICAO 2013). This vital data can lead to more informed decision-making by regulatory bodies around corrective actions moving forward and the allocation of resources (ICAO 2007). It also allows for the communication and dissemination of valuable safety information, which is key in fostering a positive safety culture in the industry (ICAO 2013). The definition of accident and incident adopted by ICAO will be used in the context of this paper. For simplicity, throughout this paper, the term "occurrence" will be used when referring to both accident and incident events. It is hoped that this research will assist in the reduction of accidents and incidents in the civil RPAS sector by analysing past occurrences and identifying common contributing factors. To this end, a sample of 152 civil RPAS accidents and incidents was analysed. The data set spanned a 10-year period, from 2006 to 2015. The data was sourced from multiple online databases and was then classified by type of occurrence, occurrence category, contributing factors, phase of flight, and time of flight. The primary research question posed in this paper is "what are the common factors in RPAS accidents and incidents in civil aviation between 2006 and 2015?".

LITERATURE REVIEW

Accidents and incidents are an unfortunate element of all sectors of the aviation industry, with the RPAS sector being no exception (Clothier and Walker 2015; ICAO 2013). The complexity of the systems and the many external influences on them mean that aiming for zero accidents is unrealistic. A more achievable approach to safety for operators and regulators is to focus on managing the potential hazards and risks associated with their operation to a level as low as reasonably possible (Clothier and Walker 2015; Xunguo *et al.* 2014). As supported by ICAO (2013) in their Safety Management Manual, it is clear that the collection of accident and incident data is a key step in the identification of potential hazards and risk areas.

The necessity of this research is exacerbated by the intense growth of the RPAS sector in recent years. Valavanis and Vachtsevanos (2015) attribute this growth to the steps taken by regulatory bodies such as the Federal Aviation Administration (FAA) and the European Commission to outline Civilian RPAS roadmaps. With the use of RPAS becoming more diverse and its development fast-tracked with lowering costs, there is a real need to remain proactive, ensuring the overwhelming benefits are not overshadowed by the potential risk to safety (AIA date unknown; Harrison 2013; Valavanis and Vachtsevanos 2015).

SAFETY TREND

With air travel commonly referred to as the safest form of travel, statistics published by Allianz (2014) show that the aviation sector's safety level has consistently increased over the decades with accident rates in recent years at their lowest. This positive safety trend has been attributed to advancements of technology and the process of continuous improvement adopted by the industry, which includes the study of accident and incident causation (Allianz 2014).

However, Allianz (2014) also highlights the very real risk that technological advancements such as RPAS may have on this safety position, with a report undertaken by the Joint Authorities for Rule-Making on Unmanned Systems (JARUS WG-6 2014) supporting this. RPAS operations are set to increase substantially in the future and focus on operations in non-segregated airspace, a requirement for the sector's future viability (AIA date unknown; European Commission 2014). As such, it is important that accidents and incidents involving these systems are mitigated before they eventuate — a sentiment supported by Clarke and Bennett Moses (2014).

Whilst it is not the aim of this research to determine if unmanned aircraft are, or are not, safer than manned aircraft; it is important to acknowledge the literature on both sides. AIA (date unknown) and Skrzypietz (2012) take the point-of-view that RPAS may in fact offer increased operational safety over conventional manned aircraft through "sense and avoid" technologies. Those studies suggest that the risks are lower given the removal of the human element in the cockpit. In contrast, Clarke and Bennett Moses (2014) suggest that the remoteness of the pilot in these systems may lead to a greater lack of situational awareness, and hence an increased safety risk. What this literature highlights is that more research needs to be undertaken on the impact of RPAS operations in civil environments.

THE IMPORTANCE OF IDENTIFYING COMMON FACTORS

There has been extensive research highlighting the potential of the emerging RPAS sector and the challenges it brings in the safety arena. The industry and regulatory bodies are working to address these many challenges (AIA date unknown). As previously discussed, the exploration of common factors in previous occurrences is a key input for helping to identify and influence relevant regulatory decisions and processes (ICAO 2013).

Previous research undertaken within different aviation environments has previously been conducted and highlights the effectiveness of this reactive method. Australia's Bureau of Air Safety Investigation (BASI 1996) completed a study into fatal accidents in the general aviation (GA) sector. Issues surrounding human factors were found to contribute to approximately 70 – 80% of accidents in this sector. A later report by Australia's Civil Aviation Safety Authority (CASA date unknown, a) into common factors in Australian GA accidents was able to reveal information including that:

- The high number of fatal accidents in the private flight category.
- Key factors in accidents of inadequate flight planning and aircraft handling.
- Of the flight planning management category, 17% could be attributed to unnecessary low-level flying.

The CASA's research was undertaken with a similar view to this present study, in that it will enable further detailed analysis to take place in the future (CASA date unknown, a). In 2010, further analysis took place addressing key CASA findings of GA accidents in the private flight category. Undertaken by the Australian Transport Safety Bureau (ATSB 2010), the report analysed this type of occurrence in more detail and was able to provide key safety information.

Research completed by Clothier and Walker (2015) used sample data from Tvaryanas *et al.* (2006). This study of military RPAS accidents and common failure categories identified common human factor elements as a cause for 60.2% of the 221 cases studied. This information allows risks to be identified with the aim of mitigating them before they eventuate (Clothier and Walker 2015; ICAO 2007, 2013). This study was significant as it represented the first post-accident analysis of a relatively large sample of RPAS accidents for the defence sector.

The relevance of post-accident, explorative research can also be witnessed in a recent report by Boyd (2015), who investigated the "causes and risk factors for fatal accidents in non-commercial twin engine piston general aviation aircraft". Having similar motives to the research herein, Boyd's report highlights the valuable information that can be attained through a post-accident review, and identified a potential deficiency in key training areas. Armed with this vital knowledge, regulatory bodies were able to make informed decisions, in Boyd's case about the multi-engine rating training syllabus. Safety bulletins were then disseminated and flight schools could review their training methods. Without post-accident analysis, vital statistics would not have been discovered, including that 70 – 80% of accidents are related to human elements (Clothier and Walker 2015), or that 53% of GA accidents during 1999 – 2000 correspond to private flight (CASA date unknown, a). Hence valuable safety information and regulations may not have been created or amended.

ACCIDENT AND INCIDENT REPORTING

The rules and regulations governing aviation activities are "as fundamental and rudimentary to the aviation industry as civil order is to modern society" (Bartsch 2015). Australia became one of the first countries to regulate the operation of RPAS in civil airspace with the introduction in 2002 of rules specifically for unmanned aerial activities, (CASA date unknown). Since then, significant progress has been made in the promulgation of rules for RPAS operations with the FAA in the USA putting into place policies in 2007 allowing the integration of RPAS into non-segregated airspace (FAA 2013). Work has continued to progress in Europe to assure harmonisation of regulations across the continents (European RPAS Steering Group 2013). ICAO is in the process of developing standards and recommended practices (SARPs) for RPAS operations and the RPAS Manual (RPASM) was published in 2015 to provide guidance for contracting States on RPAS integration into non-segregated airspace (Bartsch 2015).

Whilst all rules and regulations are important to guide this sector, the focus for this research is only on the importance of regulations pertaining to the reporting of RPAS accidents and incidents. Issues surrounding the existence of RPAS accident and incident reporting systems have been discussed recently by both Enomoto *et al.* (2013) and Clothier and Walker (2015). Both studies identified issues with the ability to collect valid accident and incident information due to inaccessibility, inconsistencies, and gaps in data, a limitation that is also noted in this current paper.

The opposite is apparent in the military RPAS sector where data is more publically accessible (Clothier and Walker 2015), therefore, studies of military occurrences have been more prominent. Enomoto *et al.* (2013) identified in their study a number of publicly available sources that contained or had the potential to capture civil RPAS occurrences, however these existed only in the US, Australia, United Kingdom, and Canada.

Voluntary reporting systems, such as the United States FAA Near Mid-Air Collision System (NMACS), exist as a means to encourage aviators to submit reports. Whilst the importance of these reporting systems are recognised, it has been suggested that not having controls or regulations for RPAS occurrence reporting facilitates the distortion of data (Goglia 2014). This then suggests that the subjectiveness of suspected RPAS occurrences can in fact degrade the ability to draw on these databases as a source of truth in implementing safety actions/recommendations. Instead of playing a role in the output of safety recommendations and regulations as is intended with these regulatory accident and incident databases, Goglia (2014) argues that they just allow for the collection of unsubstantiated claims of a "growing problem with small drones".

Regardless of whether the systems available are mandatory or voluntary, a key issue is the lack of consistency across ICAO member states and within the respective databases. Having appropriate regulations in place to address the type and quantity of data to be collected will no doubt assist as a first step. The approach taken by ICAO, the International Air Transport Association (IATA), and other organisations to harmonise their safety data reporting would also help in the comparison of safety data regarding accidents and incidents in RPAS.

METHODOLOGY

DESIGN

This study followed a mixed-method approach, specifically an exploratory design, in which the first phase saw the implementation of content analysis, a qualitative method. This was utilised in order to systematically select and examine information within relevant accident and incident case studies and then to identify common trends and themes (Leedy and Ormrod 2013). The cases were selected through convenience sampling. That is, no random sub-sampling was conducted and all of the cases found were included. Following this, the categorised data was then quantified and analysed to uncover common causal factors to determine if any statistical significance existed. That is, the second phase (the quantitative phase) of the mixed method was a causal-comparative study.

DATA COLLECTION

The 152 cases analysed and discussed in this paper were collected from a number of publically available accident investigation databases, safety reporting systems, and through a general website search. These included, but were not limited to:

- FAA Aviation Safety Information Analysis and Sharing System.
- NASA Aviation Safety Reporting System.
- Civil Aviation Authority.

The data collected focussed specifically on RPAS accidents and incidents whilst under civil operation only, between the years 2006 to 2015.

A number of reports were found (35 in total) through the data collection stage that identified airspace incursions or separation incidents involving RPAS. These reports were made to accident investigation bodies by numerous sources such as commercial airline pilots and air traffic controllers. Unfortunately these cases did not have conclusive evidence of RPAS involvement and the RPAS operator was not able to be identified. For this reason they have been excluded from the scope of this study. Instead, this research has only focused on the common "occurrence categories" and "contributing factors" that lead to RPAS incidents or accidents so that the frequency and type of these occurrences may be better understood and mitigated in future.

DATA CLASSIFICATION

Specific fields (variables) were identified and the cases were entered against these. These fields were selected and this method of collection was chosen after a review of previous research papers that performed similar activities in other aviation sectors. In particular, the annual Statistical Summary of Commercial Jet Airplane Accidents publication completed by Boeing was pivotal to inform the post accident analysis methodology. The ATSB Aviation Accident or Incident Notication form also provided relevant classifications for collection. The importance and relevance of selecting key categories in accident analysis was identifed by Boeing (2013), which suggested that the approach provides greater insight for the risk management and continuous improvement processes. As such, the ICAO Aviation Occurrence Categories were used to code the cases. A slight revision to these categories was made in this paper in order to simplify the collection. The System Component Failure/ Malfunction (SCF) categories under ICAO's standards are separated into 2 subcategories — SCF non powerplant events and SCF powerplant events. This study has simply combined the 2 and considers all SCF events together. The Occurrence Categories (OC) used in the coding were:

- System Component Failure (SCF).
- Loss of Control – Inflight (LOC-I).
- Navigation Error (NAV).
- Abnormal Runway Contact (ARC).
- Collision with obstacle(s) during takeoff and landing (CTOL).
- Air Traffic Management/Communication Navigation Survellence (ATM).
- Midair/Near Midair Collision (MAC).
- Controlled Flight Into Terrain (CFIT).
- Loss of Control – Ground (LOC-G).
- Turbulence (TURB).
- Unknown (UNK).

The second field identified was the Factors Contributing to Occurrence. The Contributing Factors (CF) were coded into specific elements in order to ensure a consistent approach. These elements were:

- Equipment Problems (EP).
- Human Factors (HF).
- Organisational Issues (OI).
- Environmental Issues (EI).
- Unknown (UNK).

These elements were adopted through review of other similar studies such as Pagán *et al.* (2006) and Boyd (2015). Additionally, research by Johnson and Holloway (2007) was drawn on to classify the high level contributing factors.

The phase of flight was also an important field identified. The Phases of Flight (PoF) used to code the cases investigated in this study included:

- Takeoff and climb out.
- Cruise or en-route.
- Descent, approach, and landing.
- Unknown.

It should be noted that other typical phases of flight exist, which have been omitted from this study. This is justifiable as there were no RPAS occurrences in these categories.

The final 2 fields coded were time of occurrence (ToO) and the occurrence type (OT). The ToO was coded as either night or day. The OT was coded as either an accident or an incident.

ANALYSIS

Once all 152 cases had been collated and classified, common trends were identified and frequencies of occurrences were noted (Leedy and Ormrod 2013). The first stage of this was to visually inspect data in charts to assist in the uncovering of significant information such as the primary CF and OC as well as the frequency of these cases. In order to determine statistical significance of this categorical data, Pearson's χ^2 test of independence was performed (Berman and Wang 2011). The test was employed when comparing data such as OC by type of occurrence and importantly when looking at the significance of OCs and contributing factors under different conditions such as PoF and ToO conditions (Boyd 2015; Pagán *et al.* 2006).

The quantitative data analysis involved Pearson's χ^2 tests for independence. The statistical hypotheses are given as:

$$H_0: p_{\mathrm{i},n} = p_{\mathrm{j},n}$$
$$H_A: p_{\mathrm{i},n} \neq p_{\mathrm{j},n} \tag{1}$$

where: subscripts *i* and *j* represent the 2 fields being compared (in this paper these are 2 of OC, CF, PoF, ToO, or OT); *p* is in reference to the proportions of the *n*-th category (there are 4 for OC, 4 for CF, 3 for PoF, 2 for ToO, and 2 for OT). H_0 is the null hypothesis and can therefore be expressed as: "the proportions of field *i* cases are equal for the proportions for field *j* cases". H_A is the alternative hypothesis, and, in contrast to this, "the proportions are not equal". The χ^2 is given by Berman and Wang (2011):

$$\chi^2 = \sum_{i=1}^{n} \sum_{j=1}^{m} \frac{\left(O_{i,j} - E_{i,j}\right)^2}{E_{i,j}} \tag{2}$$

where: there are *n* classifications for field *i*, and *m* classifications for field *j*.

The number of degrees of freedom, *df*, is given as $(n - 1)(m - 1)$ for each test. The critical value was then determined from the degrees of freedom using the χ^2 table, with a 95% confidence level. Finally, if χ^2 was less than the critical value H_0 was accepted, otherwise, H_0 was rejected.

RESULTS

CLASSIFICATIONS

In total, 152 cases involving RPAS accidents or incidents were collected that occurred in the 10-year period from 2006 to 2015. The breakdown of cases coded in terms of OC is shown in Fig. 1a. Across the cases, the most common OCs were found to be SCFs at 63% and LOC-I at 14%.

Figure 1b shows the breakdown of cases coded by the CF. EPs have the highest percentage, at 41%, while HFs are the second most common at 15%. These numbers, however, are skewed by the large number of UNK events (35%). Hence, if these are excluded there are 64% EPs, and 23% HFs.

The breakdown of cases coded by PoF is shown in Fig. 1c. Once again a relatively large number of reported cases did not include information about which PoF the occurrence happened. Excluding these UNK cases it is noted that 45% of cases occurred in cruise, 30% of cases occurred at takeoff, and 25% of cases occurred during landing.

Figure 1d shows the breakdown of cases coded by ToO. Note the majority of case reports did not indicate the time when the occurrence happened. The percentage of cases that occurred during the day was 70% while the percentage of cases that occurred at night was 30%, excluding the UNK cases.

Finally, the OT was quantified. As every occurrence could be classified as either an incident or an accident there were no UNK cases. The breakdown is shown in Fig. 1e. The incidents account for 74% of all cases while accidents account for 26%.

COMPARISON

Following the initial quantification of the coded data, the various fields were compared and contrasted in terms of their percentage distribution. Figure 2a shows the OCs percentage proportions for each of the other fields. Figure 1a shows that a large number of categories contain a relatively small number of cases; as such, these categories were grouped into an "other" category. This would not influence the nature of the proportional distribution. Visually the proportions for the OCs appear to be different for every field. Of note is that the majority of EPs correspond to SCFs. Next, the OIs are split between NAV and "other". Finally, for EIs there are a large number of "other" cases which includes environmental related occurrences such as TURB, WILD etc. For PoF relative to OT shows a large number of "other" for landing, which includes ARC, CTOL, etc. For ToO, there is a large number of NAV cases at night. For the OT, the inference is that a larger proportion of incidents involve SCFs, and a larger proportion of accidents involve NAV.

Next the CF was compared and contrasted to the remaining fields. For PoF, it can be seen in Fig. 2b that HFs are more common for landing, while EPs are more common for takeoff, and cruise cases show the most OIs, with no EIs.

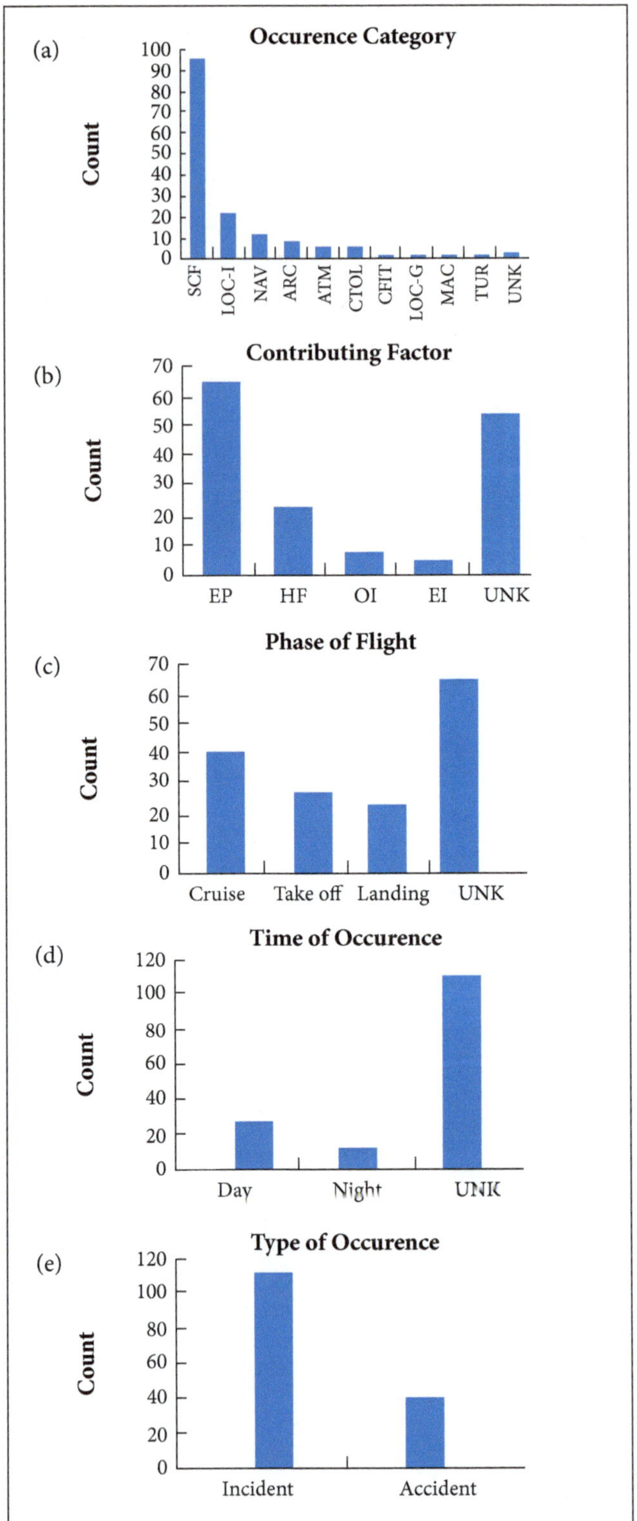

Figure 1. Pareto plot showing number of cases (2006-2015).

Figure 2c compares and contrasts PoF to ToO and OT. These show relatively similar proportions. This is important to consider as the data suggests that night time occurrences happen in cruise and not during takeoff or landing as one may hypothesise. When looking at the OT cases that are accidents they occur more frequently during takeoff and landing.

Finally, the comparison between ToO and OT is shown in Fig. 2d. The proportions are identical. It is worth noting that there is no specific reason why this would not be the case, and helps to suggest the random nature of the sample. It might be conceivable, however, to hypothesise that accidents at night would be more common.

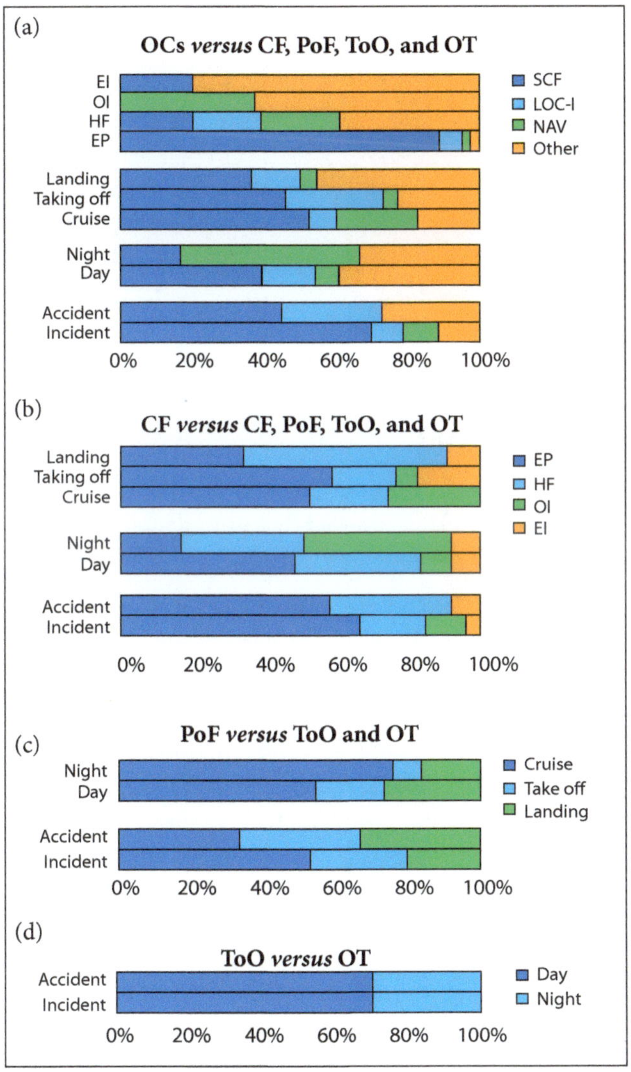

Figure 2. Percentage breakdown of cases, across the 10-year period (2006 – 2015). Note: UNK cases have been excluded.

ANALYSIS

To confirm the visual comparison from Fig. 2a to Fig. 2d, the Pearson χ^2 test for independence between each of the fields was undertaken. In total, there are 10 ways the 5 fields can be paired. These are all summarised in Table 1, in the top 2 rows. The table also shows the corresponding χ^2 test statistics, the degrees of freedom, the critical value, the p-value (the probability for the given test statistic with the stated degrees of freedom), and the conclusion. OCs and CF are discussed in detail below. At this point it is noted that the proportions of PoF between ToO is not statistically significant ($\chi^2 = 1.59$, $a = 0.05$, $df = 2$). That is, PoF cases are not more or less likely to occur because of the ToO. In fact, it may have been plausible to hypothesise that night operations would result in more occurrences during takeoff and landing, which is not supported by the findings. Similarly, the proportions for PoF between OT is not statistically significant ($\chi^2 = 4.44$, $a = 0.05$, $df = 2$). That is, PoF does not result in a greater or lesser proportion of accidents relative to incidents. There does appear to be a larger proportion of accidents during takeoff and landing, but the limited sample size means this conclusion is not statistically significant. Finally, of ToO and OT showed no statistical significance ($\chi^2 = 0.00$, $a = 0.05$, $df = 1$), as highlighted in the "Comparison" section.

Occurrence Category

A statistical significance was found between the OCs and the CFs ($\chi^2 = 75.7$, $a = 0.05$, $df = 9$), that is, at the 95% confidence level the null hypothesis is rejected and it is concluded that there is a difference in the distribution between CFs for different OCs. More directly, it is fair to state that CFs influence specific OCs in different ways. It can then be concluded that:

- EPs contribute the most to SCF.
- HFs are more significant for LOC-I cases.
- OIs are associated with "other" and NAV cases.
- EIs are more common for "other" (TURB and WILD etc.) cases.

A statistical significance was also found between the OC and PoF ($\chi^2 = 15.1$, $a = 0.05$, $df = 6$). So, at the 95% confidence level, the null hypothesis is rejected and it is concluded that there is a difference in the distribution between PoF for different OCs. That is, OC occur at different rates in the different PoF. Based on the proportions it can be concluded that:

- Cruise has a larger proportion of SCF.
- Takeoff has the largest portion of LOC-I cases.
- Landing has the most "other" (ARC, CTOL etc.) cases.

Table 1. Pearson's χ^2 test for independence comparing pairs of fields.

Field 1	Category				Factor			PoF		Time
Field 2	Factor	PoF	Time	Type	PoF	Time	Type	Time	Type	Type
c^2	75.7	15.1	10.8	18.8	19.02	6.98	5.5	1.59	4.44	0
α	0.05	0.05	0.05	0.05	0.05	0.05	0.05	0.05	0.05	0.05
df	9	6	3	3	6	3	3	2	2	1
Critical	16.92	12.59	7.81	7.81	12.59	7.81	7.81	5.99	5.99	3.84
p-value	1E-12	0.020	0.013	0.0003	0.004	0.07	0.139	0.45	0.11	1
Conclusion	Reject	Reject	Reject	Reject	Reject	Accept	Accept	Accept	Accept	Accept

A statistical significance was also found between the OCs and ToO ($\chi^2 = 10.8$, $a = 0.05$, $df = 3$). So, at the 95% confidence level, the null hypothesis is rejected and it is concluded that there is a difference in the distribution of the ToO between the types of OC. That is, OCs have different rates during the day and night. Based on the proportions it can be concluded that:

- Day-time cases have a larger proportion of SCF.
- Night-time cases have a significantly larger proportion of NAV cases.

A statistical significance was also found between the OC and OT ($\chi^2 = 18.8$, $a = 0.05$, $df = 3$). So, at the 95% confidence level, the null hypothesis is rejected and it is concluded that there is a difference in the distribution between OT for the different OCs. That is, the end result (an incident or an accident) varies based on the OCs. It can then be concluded that:

- A larger proportion of incidents are associated with SCFs.
- A relatively larger proportion of accidents result from LOC-I.

Contributing Factor

A statistical significance was found between the CF and PoF ($\chi^2 = 19.0$, $a = 0.05$, $df = 6$). So, at the 95% confidence level, the null hypothesis is rejected and it is concluded that there is a difference in the distribution between PoF across the CFs. That is, CFs are involved at different rates in the different PoF. Based on the proportions it can be concluded that:

- Cruise has the largest proportion of OIs and no recorded EIs.
- Takeoff has the largest proportion of EPs and the smallest portion HFs.
- Landing has the largest proportion of HFs contributions and no OIs.

A breakdown of CFs across OT showed no statistical significance ($\chi^2 = 5.50$, $a = 0.05$, $df = 3$). Hence, it can be assumed that the CFs are similar for cases if they are coded as accidents or incidents. Similarly the division of CFs by ToO showed no statistical significance ($\chi^2 = 6.98$, $a = 0.05$, $df = 3$). However, this results in a p-value of 0.07, which is more than 90% significant. The most likely type of error associated with a χ^2 test is a Type II Error (accepting H_0 when it should be rejected). As such, the division of CFs by ToO should be considered borderline. Therefore it is worth noting the potential that OIs contribute to the majority of night-time cases.

DISCUSSION

FINDINGS

The data collected revealed that 61% of all occurrences were attributed to SCFs, with a significant gap between the next closest OC factor, LOC-I. A report by the ATSB on the accidents and incidents in an Australian aviation context over the period 2003 to 2012 showed markedly different results than those found in the civil RPAS sector. The ATSB report revealed that for the Australian GA and Air Transport sectors, NAV and wildlife (WILD) occurrences dominated (ATSB 2013). Seemingly the RPAS sector most resembles that of the Recreational Aviation (RA) sector in terms of occurrence categories. One reason behind this similarity could be the less restrictive certification standards required in these 2 sectors in terms of airworthiness and design (Brandon 2014; Johnson 2010).

Whilst it is recognised that a zero occurrence rate is not realistic and that there are no preferable OCs, the high frequency of SCF occurrences certainly highlights an area that deserves further detailed research. It is disconcerting to find that when disregarding the UNK category, 93% of SCF cases were contributed to by EPs. These included loss of data link, software malfunctions and flight control issues to name a few. Examples of these issues identified in cases collected include:

- Case Reference No 34a "Engine mount failure resulting in propeller damage on landing".
- Case Reference No 34g "Iridium C2 modem failure. (One of 4 modems)".
- Case Reference No 34u "Anomalous Embedded GPS/ Inertial Navigation System (EGI) GPS degradation".

The types of accidents and incidents that occur in RPAS operations have been influenced by the greater reliance on and complexity of (Gupta *et al.* 2013; Hobbs and Herwitz 2006):

- Communication links.
- Navigation hardware.
- Software.

The same can also be said for the CFs identified in these accidents and incidents. This study has revealed that although HFs have been widely attributed to over 2/3 of aviation accidents and incidents (BASI 1996; Clothier and Walker 2015; Skrzypietz 2012), the same cannot be said regarding the civil RPAS sector. Instead, equipment failures (41%) have appeared to be more of a primary instigator in occurrences. This is clearly in contrast to other sectors. Interestingly, despite clear differences between the civil and military RPAS sectors, failure of aircraft components (66%) also had more of a significance over that of human error (34%) within the military as well (Williams 2004).

It appears suggestions by the AIA (date unknown) and also Skrzypietz (2012), that the removal of the on-board pilot in these systems should result in a reduced risk, maybe supported by the results in this study. It is unknown whether EPs based SCFs and LOC-I occurrences are due to design flaws or if system maintenance is to blame, which could be HF and OI induced. Hobbs and Herwitz (2006) suggest that the level of knowledge and experience of RPAS maintenance personnel may not be at the level required due to the complexity and diversity of the systems and the infancy of the sector, potentially contributing to the problem.

Reviewing cases in more detail revealed a statistical significance between the PoF and CFs, with EPs more prevalent in cruise whilst the landing phase saw HFs dominate. Further research needs to be undertaken in order to determine the main types of EPs experienced in these cases; however, initial review indicates that just under a third involved "lost link" issues. The prevalence of these types of failures may be rationalised by types of communication links utilised in different phases (Kaliardos and Lyall 2015). During critical phases such as takeoff and landing, line of sight links resulting in less latency and reduced impact of degradation are used; whereas phases such as cruise that require less manoeuvrability often rely on satellite based control links which are more susceptible to degradation given the increased latency (Kaliardos and Lyall 2015).

It was not surprising to discover that HFs played the primary role in cases during the landing phase. Cited as one of the most critical flight phases, both unmanned and manned aircraft share this unfortunate quality (Huh and Shim 2010). Evidence of this is can be seen in the military RPAS sector with Williams (2004) reporting pilot landing errors as the clear leader in HF issues. Similar findings are also apparent in the air transport sector, with Boeing (2013) reporting that 47% of fatal accidents occur in the final approach and landing phases. A report by Huh and Shim (2010) attributes HF occurrences in RPAS operations to the complexities in situational awareness brought about by the remoteness of the pilot. Without data on the pilot in commands flight experience and time on type, it is hard to determine whether workload or insufficient training contributed to these events.

LIMITATIONS

There are 2 limitations of the results presented, the number of cases, and the limited information about the cases. The number of cases, spread over a 10-year period, prohibits an analysis based on changes over time for different classifications. That is, if we consider PoF (3 groups) and year of occurrence (10 groups), breaking the data down into both (30 groups) will result in some groups having "0" entries, which precludes the ability to conduct a χ^2 test. With a larger sample size, such that no group had "0", statistical analysis could be undertaken. The same is true if we consider any combination of classifications (PoF, ToF, OT, and OC).

The more pressing limitation is the lack of data in the case reports. That is, other useful information such as make and model of aircraft (even fixed wing/rotor wing type), operating/ takeoff mass and physical dimensions, or operation category are typically not included in publically available databases for civil RPAS occurrences. As such, the data is presented as holistic coverage of the civil RPAS sector, collecting as many reports as possible at this time.

RECOMMENDATIONS

Based on the aforementioned limitations, the first recommendation to be made is that reporting of occurrences, particularly those concerning accidents and serious incidents, should be enforced by legislation applicable to the RPAS sector. Even though most civil aviation authorities have been revising their regulatory framework to incorporate this need, there is still a legislative gap to be addressed which precludes the effective reporting of all categories of RPAS, particularly small UAVs which regardless of their limited dimensions might still pose a serious risk to the operation of other aircraft and people on the ground. Concurrently, regulators should also invest in safety promotion actions tailored to the RPAS sector as a mean to foster operators to pro-actively report safety occurrences involving unmanned aircraft via the existing voluntary reporting systems irrespective of their perceived severity.

As the current study was based on data sourced from publicly available databases fed by voluntary reporting systems, some of the data used herein was lacking details on the aircraft size and maximum takeoff weight, which prevented analysing the results taking into consideration the operation categories typically used in similar studies for manned aircraft, such as (Evans 2015). As such, if governments and the RPAS sector implement and adopt reporting systems for accidents and incidents, a more in depth analysis of the RPAS sector will be possible. This will help to improve safety in the RPAS industry, by developing a greater understanding and awareness of the specific nature of accidents and incidents in different RPAS operation categories, which is currently beyond the scope of available data.

CONCLUSION

With a growth of RPAS operation forecast, and the alleged incident involving a "drone" and a British Airways A320 at Heathrow Airport (Stevenson 2016), it is imperative to explore RPAS accidents and incidents. Furthermore, common factors were identified and studies such as this need to be continually undertaken in order to ensure the ongoing safety of the community and the sustainability of this thriving sector. Globally, there has been a great deal of resources allocated by regulatory bodies to manage the exponential growth of the civil RPAS sector. However, in the area of accident and incident investigation and regulations in particular, it is possible to suggest that greater focus on data collection in the early stages of growth would have assisted greatly in developing a more targeted approach. Numerous studies have also been completed independently, identifying potential hazards and risks of these systems. In order to complement these proactive studies it is important that the reviews of past accidents and incidents are not forgotten as a vital source of data as they can provide an abundance of information used to validate these studies and identify lessons to be learnt.

This study aimed to identify common factors in RPAS accidents and incidents in civil aviation in order to assist in the process of mitigating these occurrences. This was done through posing the primary question "what are the common factors in RPAS accidents and incidents in civil aviation between 2006 and 2015?".

The analysis uncovered that the majority of occurrences were found to have involved SCFs with EPs dominating as primary CF. This led to the recognition that civil operated RPAS have distinct differences between other sectors of the industry such as GA and air transport. The result of this is that lessons learnt through post-accident and incident analysis in other sectors are less able to be transferred to the RPAS sector. Instead it was found that similarities were seen between the military and civil RPAS sectors, and hence RPAS is unique in the aviation industry. Currently, the industry trend to license RPAS operators will, in effect, focus on HFs issues. Specifically, the recommendation from this research is that regulators need to focus primarily on airworthiness requirements, which are still yet to be formalised for civil RPAS (Clothier *et al.* 2015).

AUTHOR'S CONTRIBUTION

Wild G completed the abstract, data presentation, data analysis, and results, as well as finalised the methodology. Gavin K completed the data collection and drafted the methodology. All authors contributed to the introduction, literature review, discussion, and conclusion.

REFERENCES

AIA (date unknown) Unmanned aircraft systems: perceptions & potential. Aerospace Industries Association; [accessed 2015 Mar 22]. http://www.aia-aerospace.org/assets/AIA_UAS_Report_small.pdf

Allianz (2014) Global Aviation Safety Study; [accessed 2015 Mar 20]. http://www.allianz.com.au/aalaus/aalaus.nsf/docs/F5E0B53FEC5B210DCA257CA10012C7EB/$FILE/Allianz%20Global%20Corporate%20&%20Specialty%20Global%20Aviation%20Safety%20Study.pdf

ATSB (2010) Improving the odds: trends in fatal and non-fatal accidents in private flying operations. Canberra: Australian Transport Safety Bureau, ACT.

ATBS (2013) Aviation occurence statistics: 2003 to 2012. Australian Transport Safety Bureau; [accessed 2015 Apr 12]. https://www.atsb.gov.au/media/4355945/ar-2013-067_final.pdf

Bartsch (2015) Unmanned and uncontrolled: the commingling theory and the legality of unmanned aircraft system operations. J Aeronaut Aerospace Eng 4(1):1-27. doi: 10.4172/2168-9792.1000140

Berman E, Wang X (2011) Essential statistics for public managers and policy analysts. New York: SAGE Publications.

Boeing (2013) Statistical summary of commercial jet airplane accidents: worldwide operations — 1959 – 2013; [accessed 2015 Apr 16]. http://www.boeing.com/news/techissues/pdf/statsum.pdf

Boyd DD (2015) Causes and risk factors for fatal accidents in non-commercial twin engine piston general aviation aircraft. Accid Anal Prev 77:113-119. doi: 10.1016/j.aap.2015.01.021

Brandon J (2014) Overview of the legislative framework enabling sport and recreational aviation; [accessed 2015 Jun 13]. http://www.recreationalflying.com/tutorials/regulations/regulations.html#certs

BASI (1996) Human factors in fatal aircraft accidents. Department of Transport and Regional Development [accessed 2015 Apr 12]. https://www.atsb.gov.au/media/28363/sir199604_001.pdf

Callam A (2015) Drone Wars: armed unmanned aerial vehicles; [accessed 2016 Sept 16]. http://www.iar-gwu.org/node/144

CASA (date unknown, a) A Preliminary analysis of fatal general aviation accidents in Australia: 1991 to 2000; [accessed 2015 Apr 16]. http://casa.gov.au/wcmswr/_assets/main/media/download/accidentreport.pdf

CASA (date unknown, b) RPA background; [accessed 2015 Mar 20]. http://www.casa.gov.au/scripts/nc.dll?WCMS:STANDARD::pc=PC_101984

Clarke R, Bennett Moses L (2014) The regulation of civilian drones' impacts on public safety. Computer Law & Security Review 30(3):263-285. doi: 10.1016/j.clsr.2014.03.007

Clothier RA, Walker RA (2015) Safety risk management of unmanned aircraft systems. In: Valavanis KP, Vachtsevanos GJ, editors. Handbook of Unmanned Aerial Vehicles. Dordrecht: Springer Netherlands. p. 2229-2275.

Clothier RA, Williams BP, Coyne J, Wade M, Washington A (2015) Challenges to the development of an airworthiness regulatory framework for unmanned aircraft systems. Proceedings of the 16th Australian International Aerospace Congress; Melbourne, Australia.

Enomoto F, Bushnell D, Denney E, Pai G, Schumann J (2013) Preliminary recommendations for the collection, storage, and analysis of UAS safety data; [accessed 2015 Mar 9]. http://ntrs.nasa.gov/archive/nasa/casi.ntrs.nasa.gov/20140011531.pdf

European Commission (2014) Communication from the Commission to the European Parliament and the Council; [accessed 2015 Mar 20]. http://eur-lex.europa.eu/legal-content/EN/TXT/PDF/?uri=CELEX:52014DC0207&from=EN

ERPASG (2013) Roadmap for the integration of civil Remotely-Piloted Aircraft Systems into the European Aviation System: Annex 1. European Commission; [accessed 2015 Mar 19]. http://ec.europa.eu/enterprise/sectors/aerospace/files/rpas-roadmap-annex-1_en.pdf

Evans JK (2015) Differences in characteristics of aviation accidents during 1993-2012 based on aircraft type. NASA/CR 2015-218999. Hampton: NASA Langley Research Center.

FAA (2013) Integration of Civil Unmanned Aircraft Systems (UAS) in the National Airspace System (NAS) Roadmap; [accessed 2015 May 30]. https://www.faa.gov/uas/media/UAS_Roadmap_2013.pdf

Goglia J (2014) FAA list of reported drone incidents indicates a problem. But it's not what you think. Forbes; [accessed 2015 May 13]. http://www.forbes.com/sites/johngoglia/2014/11/27/faa-list-of-reported-drone-incidents-indicates-a-problem-but-its-not-what-you-think/

Gupta SG, Ghonge MM, Jawandhiya PM (2013) Review of Unmanned Aircraft System (UAS). International Journal of Advanced Research in Computer Engineering & Technology 2(4):1646–1658; [accessed 2015 Mar 29]. http://www.uxvuniversity.com/wp-content/uploads/2014/04/Review-of-Unmanned-Aircraft-System-UAS.pdf

Harrison GJ (2013) Unmanned Aircraft Systems (UAS): manufacturing trends. Congressional Research Service, Library of Congress; [accessed 2015 Mar 22]. http://www.fas.org/sgp/crs/natsec/R42938.pdf

Hobbs A, Herwitz SR (2006) Human challenges in the maintenance of Unmanned Aircraft Systems. FAA and NASA Report; [accessed 2015 Jun 13]. http://human-factors.arc.nasa.gov/publications/UAV_interimreport_Hobbs_Herwitz.pdf

Huh S, Shim D (2010) A vision-based automatic landing method for fixed-wing UAVs. J Intell Robotic Syst 57(1-4):217-231. doi: 10.1007/s10846-009-9382-2

ICAO (1989) ICAO Cir 216 — Human Factors Concepts; [accessed 2015 Mar 27]. http://www.icao.int/Meetings/UAS/Documents/Circular%20328_en.pdf

ICAO (2007) Addressing Unmanned Aircraft System (UAS) accident investigation and prevention by ICAO Member States; [accessed 2015 Mar 9]. http://www.icao.int/Meetings/AMC/MA/Assembly%2036th%20Session/wp217_en.pdf

ICAO (2011) ICAO Cir 328 — Unmanned Aircraft Systems (UAS); [accessed 2015 Mar 27]. http://www.icao.int/Meetings/UAS/Documents/Circular%20328_en.pdf

ICAO (2013) Safety Management Manual (Doc 9859); [accessed 2015 Mar 19]. http://www.icao.int/safety/SafetyManagement/Documents/Doc.9859.3rd%20Edition.alltext.en.pdf

ICAO (2015) Manual on Remotely Piloted Aircraft Systems. vol. 10019. Montreal: International Civil Aviation Organization.

JARUS WG-6 (2014) Working Group 6: UAS Systems Safety Analysis (1309). European Aviation Safety Agency; [2015 Mar 22]. https://www.eurocockpit.be/sites/default/files/amc-rpas_1309_scoping-paper_issue-1.pdf

Johnson C (2010) Computational concerns in the integration of unmanned airborne systems into controlled airspace. Computer Safety, Reliability, and Security 6351:142-154. doi: 10.1007/978-3-642-15651-9_11

Johnson C, Holloway C (2007) A longitudinal analysis of the causal factors in major aviation accidents in the USA from 1976 to 2006; [accessed 2015 May 20]. http://www.dcs.gla.ac.uk/~johnson/papers/Johnson_Holloway_Longitudinal_Study.pdf

Kaliardos B, Lyall B (2015) Human factors of Unmanned Aircraft System Integration in the National Airspace System. In: Valavanis KP, Vachtsevanos GJ, editors. Handbook of Unmanned Aerial Vehicles. Dordrecht: Springer Netherlands. p. 2135-2158.

Leedy P, Ormrod JE (2013) Practical research: planning and design. 10th edition. Boston: Pearson Education.

MarketLine (2014) Unmanned Aerial Vehicles — the economic case for drones; [accessed 2015 Mar 26]. http://search.ebscohost.com/login.aspx?direct=true&db=bth&AN=97187491&site=ehost-live&scope=site

Odido D, Madara D (2013) Emerging technologies: use of Unmanned Aerial Systems in the realisation of Vision 2030 Goals in the Counties. Int J Appl Sci Tech 3(8):107-127.

Pagán BJ, de Voogt AJ, van Doorn RR (2006) Ultralight aviation accident factors and latent failures: a 66-case study. Aviation, Space, and Environmental Medicine 77(9):950-952.

Skrzypietz T (2012) Unmanned Aircraft Systems for Civilian Missions Brandenburg Institute for Society and Security; [accessed 2015 Jun 4]. http://www.bigs-potsdam.org/images/Policy%20Paper/PolicyPaper-No.1_Civil-Use-of-UAS_Bildschirmversion%20interaktiv.pdf

Stevenson B (2016) A320 collision heightens UAV safety concerns. Flight Global; [accessed 2016 Apr 21]. https://www.flightglobal.com/news/articles/a320-collision-heightens-uav-safety-concerns-424419/

Tvaryanas AP, Thompson WT, Constable SH (2006) Human factors in remotely piloted aircraft operations: HFACS analysis of 221 mishaps over 10 years. Aviation, Space, and Environmental Medicine 77(7):724-732.

Valavanis KP, Vachtsevanos GJ, editors (2015) Handbook of Unmanned Aerial Vehicles. Dordrecht: Springer Netherlands; Chapter 6, Military and Civilian Unmanned Aircraft. p. 93-103.

Williams K (2004) A summary of Unmanned Aircraft accident/incident data: human factors implications. DOT/FAA/AM-04/24. Washington: Federal Aviation Administration.

Xunguo L, Fulton NL, Horn MET (2014) Quantification of high level safety criteria for civil Unmanned Aircraft Systems. Proceedings of the IEEE Aerospace Conference; Montana, USA.

13

Application of Multidisciplinary Design Optimization on Advanced Configuration Aircraft

Yalin Pan[1], Jun Huang[1], Feng Li[2], Chuxiong Yan[1]

ABSTRACT: An optimization strategy is constructed to solve the aerodynamic and structural optimization problems in the conceptual design of double-swept flying wing aircraft. Aircraft preliminary aerodynamic and structural design optimization is typically based on the application of a deterministic approach of optimizing aerodynamic performance and structural weight. In aerodynamic optimization, the objective is to minimize induced drag coefficient, and the structural optimization aims to find the minimization of the structural weight. In order to deal with the multiple objective optimization problems, an optimization strategy based on collaborative optimization is adopted. Based on the optimization strategy, the optimization process is divided into system level optimization and subsystem level optimization. The system level optimization aims to obtain the optimized design which meets the constraints of all disciplines. In subsystem optimization, the optimization process for different disciplines can be executed simultaneously to search for the consistent schemes. A double-swept configuration of flying wing aircraft is optimized through the suggested optimization strategy, and the optimization results demonstrate the effectiveness of the method.

KEYWORDS: Aerodynamic performance, Structural optimization, Multiple objective optimization, Collaborative optimization, Flying wing.

INTRODUCTION

Flying wing configuration has been considered as an ideal configuration of the future unmanned aerial vehicles (UAV) due to its potential benefits over conventional configurations in stealth capability, aerodynamic performance, and structural efficiency. Several next-generation UAVs are of flying wing designs, such as the X-45; X-47B; nEURO; etc. (Song *et al.* 2014; Li *et al.* 2014; Bolsunovsky *et al.* 2001).

Compared with the conventional configuration, flying wing aircraft has become the research hotspot of advanced aircraft in recent years (Zhou and Liu 2015), and the number of flying wing aircrafts which have been developed successfully is far less than the number of aircrafts with the conventional configuration. The lack of statistics and practical experience about flying wing configuration posed great difficulties in aircraft conceptual design. It has been proved that multidisciplinary design optimization (MDO) is an effective technique to deal with these problems. Now, it has been widely used in the conceptual design of traditional layout aircraft (Piperni *et al.* 2007; Lee *et al.* 2007; Lambe and Martins 2016).

Aircraft optimization problems involve multiple objectives and should be treated as multi-objective optimization (MOO) problems. As a classic example, aircraft preliminary aerodynamic and structural design optimization is typically based on the application of a deterministic approach of maximizing lift-to-drag ratio under cruising or other flight conditions and minimizing structural weight due to applied air load in an optimization process (Gou and Song 2006; Gao *et al.* 2003; Molinari *et al.* 2014). Thus, aerodynamic and structural

1.Beihang University – School of Aeronautic Science and Engineering – Beijing – China. **2.**China Academy of Aerospace Aerodynamics – Beijing – China.
Author for correspondence: Yalin Pan | Beihang University – School of Aeronautic Science and Engineering | Beijing 100191 – China | Email: pyalin@buaa.edu.cn

optimization problem is a 2-objective optimization problem in aircraft design. In aerodynamic design, in order to minimize the induced drag coefficient, the shape of the lift coefficient distribution along the spanwise direction is expected to be close to an ellipse. However, under this kind of load distribution, the structural weight is not always the lightest during the process of structural design. Thus, in terms of load distribution, the aerodynamic performance and the structural weight are usually conflicting. In MOO, the optimization objectives belong to different disciplines always in conflict with each other. It is impossible to reach the optimal results for all the optimization goals at the same time. The optimization results ought to be a set of optimal solutions rather than an optimal solution. This set is known as the Pareto optimal set, and its corresponding tradeoff in objective space is known as the Pareto optimal frontier, which is made up of the Pareto optimal points (Sanghvi *et al.* 2014; Huang et al. 2007; Hu and Yu 2009). Designers could select an optimized scheme which satisfies the requirements of all disciplines from the Pareto optimal set. In this article, an optimization strategy based on collaborative optimization is proposed to deal with the MOO problem in flying wing aircraft conceptual design. The UAV configuration used as a basis for the MOO is shown in Fig. 1. The take-off weight of the aircraft is 20,000 kg and it is used for intelligence, surveillance, and reconnaissance missions as a high-altitude long-range UAV.

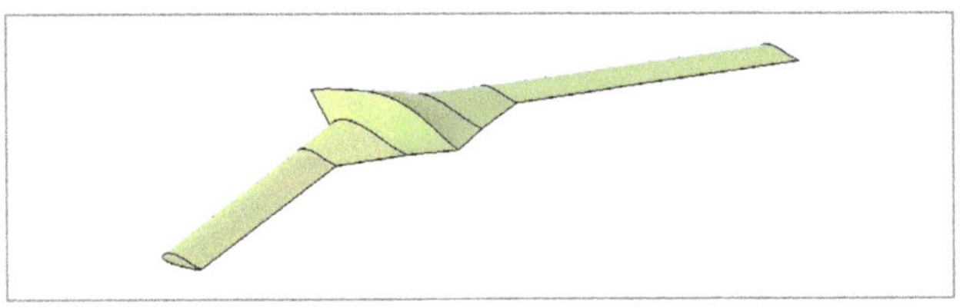

Figure 1. CAD model of flying wing aircraft.

STATEMENT OF THE OPTIMIZATION PROBLEM

The platform of the UAV is shown in Fig. 2. The parameters as shown in the figure are the outline ones and they are used to determine the planform of the UAV. The chord lengths at different locations along spanwise direction are described by b_r, b_2, and b_t. The wing span for different sections are described by l_1, l_2, and l_3. In order to describe the sectional shape of the UAV, section 1, section 2, section 3, and section 4 (as shown in Fig. 2) are defined as the master sections, whose profiles are displayed in Fig. 3 and Fig. 4; these sections are used to fit the contour surface of the aircraft. The lift and the induced drag of the UAV are mainly concerned with the airfoil camber. The outline parameters are fixed in the original design. It is expected that the minimum of the induced drag coefficient of the UAV can be found by adjusting the mean lines of these sections.

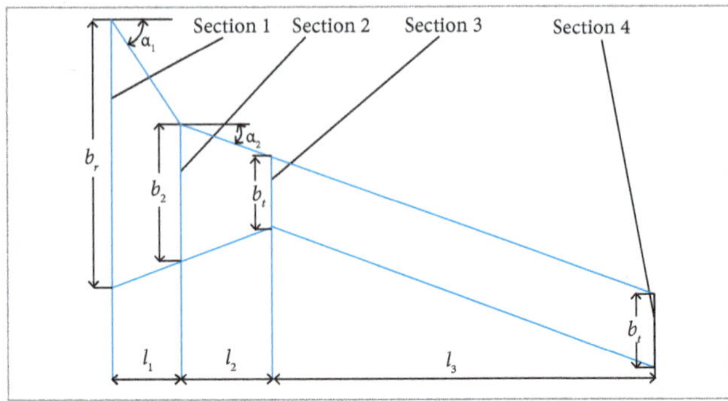

Figure 2. Half of flying wing configuration platform.

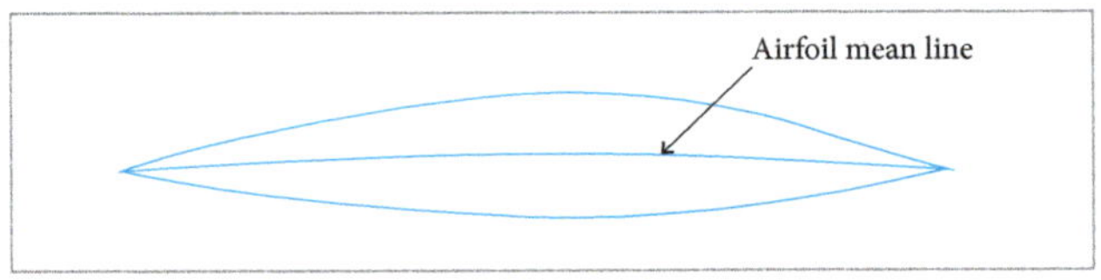

Figure 3. Profile of section 1.

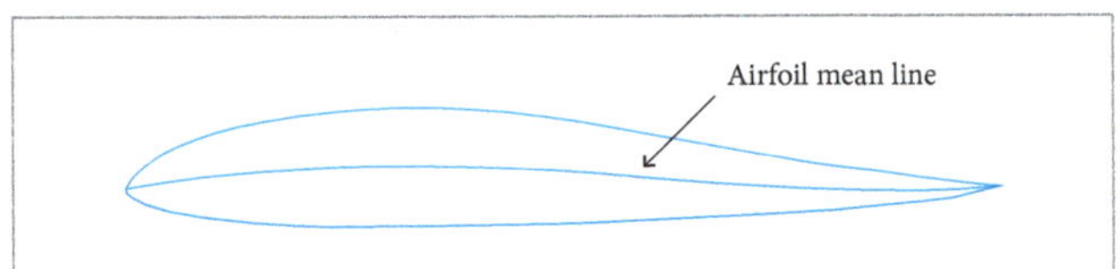

Figure 4. Profile of the other master sections.

Constrained by the material allowable stress and structural deformation, it is required to find the minimum of structural weight by adjusting the structural dimensions. The structure of the UAV contains outer skins and inner structural layout, as shown in Fig. 5. The inner structural layout of the UAV is displayed clearly in Fig. 6. For inner structure, the fuselage (Part 1) is composed by longitudinal beams and reinforced frames, the wing closed to the fuselage (Part 2) includes wing spars and reinforced ribs, and the outboard wing (Part 3) includes wing spars and wing ribs. The structural dimensions which are used as design variables include the thickness of skin and rib, the area of spar cap and rib cap, etc.

As already mentioned, the optimization problem of this UAV conceptual design can be formulated as follows:

- Objectives: (1) Minimized induced drag coefficient; (2) Minimized structural weight.

- Design variables: (1) Parameters which are used to describe the airfoil mean lines of the UAV; (2) Parameters which are used to define the structure of the UAV.
- Constraints: (1) Aerodynamic requirements; (2) Structural requirements.

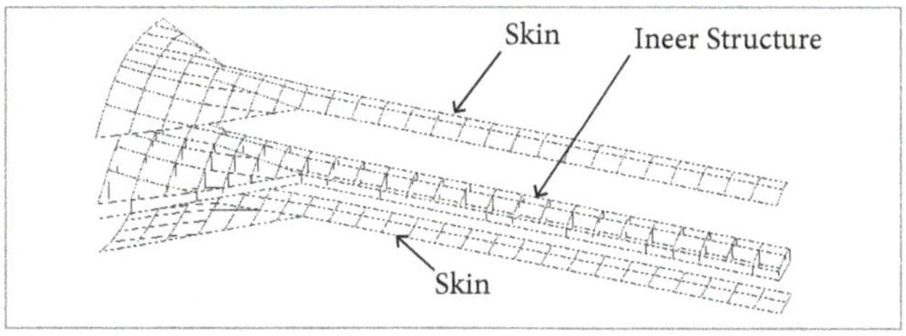

Figure 5. Structural layout of the UAV.

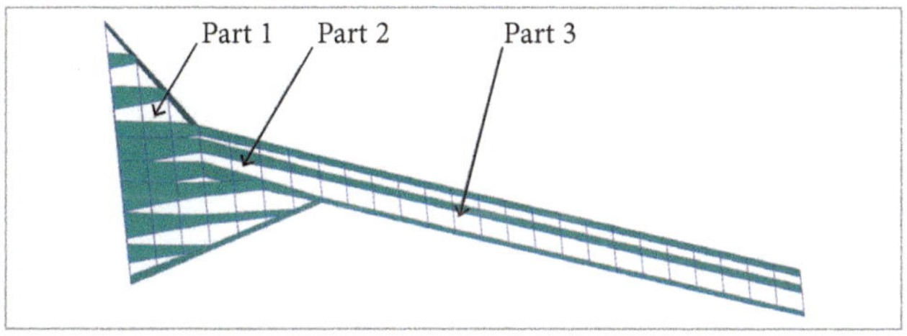

Figure 6. Inner structure of the UAV.

OPTIMIZATION STRATEGY

Aerodynamic and structural optimization problem is a typical MOO problem in aircraft design, and the 2 disciplines are always in conflict. The improvement of aerodynamic performance often brings an increase in structural weight, and the decrease in it usually causes the expense of aerodynamic performance (Chu 2011; Ma *et al.* 2009). In MOO, there often exists a set of optimal solutions, and none can be said to be better than any other without any further information. In order to coordinate the relationship between aerodynamic and structure of the UAV, an optimization strategy based on collaborative optimization is used to deal with the MOO problem. The optimization process can be divided into 2 levels which include the system level optimization and the subsystem level optimization. The system level optimization transfers the global design variables to the subsystem level optimization. In the subsystem level optimization, these global design variables are defined as target values. The state variables of each discipline are optimized to approximate these target values through optimization algorithm. Thus, the optimized designs for different disciplines are consistent.

The system level optimization aims to search for the design scheme with minimum induced drag coefficient and structural weight of the UAV. The constraint function of system optimization is used to make the schemes of different subsystem optimizations consistent. In this article, only the state variables of aerodynamic discipline which are used to describe the load distribution on the aircraft surfaces are defined as the global design variables (*i.e.* the design variables in system level optimization). The design parameters which only have impact to 1 discipline are defined as local design variables (*i.e.* the design variables in subsystem level optimization). Since the airfoil mean lines of the master sections have influence mainly on aerodynamic performance but have little effect on structure weight, the parameters which are used to describe the airfoil mean lines are local design variables. The structural dimensions have influence mainly on structure weight and impact on aerodynamic performance slightly, so these parameters are also local design variables. The load distribution can be described by lift coefficient curve and it is generated by fitting the lift coefficient of the sections along its spanwise direction. The lift coefficient curve is described by a cubic polynomial function. The fitting function is shown as follows:

$$Cl\,(\eta) = a_3{\cdot}\eta^3 + a_2{\cdot}\eta^2 + a_1{\cdot}\eta^1 + a_0 \quad (1)$$

where: $\eta = y/b$; y represents the coordinates of the UAV spanwise direction; b is half of the span; the shape of the load distribution curve is determined by the function coefficients (*i.e.* a_0, a_1, a_2, and a_3). The global design variables can be represented by a_0, a_1, a_2, and a_3.

The subsystem level optimization is integrated with the system level optimization, and the goal of aerodynamic optimization is to achieve the minimum of difference between the state variables and global design variables under the constraints of aerodynamic characteristics by changing the camber curve shapes of master sections airfoils. Constrained by the material allowable stress and structural deformation, the structure optimization aims to find the minimum of structural weight by adjusting the structural dimensions. The frame of the method for aerodynamic and structural multiple-objective design optimization is depicted in Fig. 7.

In the optimization strategy, the non-dominated sorting genetic algorithm (NSGA-ii) is adopted for system optimization. The formulation of system level optimization is stated as follows:

- Objectives: minimized induced drag coefficient (C_{Di}) and structural weight (W).
- Design variables: parameters which are used to define the lift coefficient distribution along spanwise direction, such as a_0; a_1; a_2 and a_3.

- Constraints: the scheme optimized by the subsystem optimization is consistent with the design offered by the system ($J_1 = 0$).

The sequential quadratic programming algorithm is used for aerodynamic optimization. The formulation of aerodynamic optimization problem is as follows:

- Given conditions: cruise Mach number is 0.8, and cruise altitude is 18 km.
- Objective: to minimize the difference between the state variables and the global design variables, $J_1 = \sum_{i=0}^{3} (a_i - a_i^0)^2$.
- Design variables: the parameters for describing the airfoil mean lines of master sections.
- Constraints: to design lift coefficient ($C_L = 0.362$).

The sequential quadratic programming algorithm is adopted for structural optimization. The formulation of the structural optimization problem is as follows:

- Objective: minimized structural weight (W).
- Design variables: (1) the areas of spar caps, ribs, and reinforced frames; (2) the thicknesses of the webs of spars, ribs, and reinforced frames; (3) the thicknesses of wing skins; (4) the stiffener areas of the webs. Eighty-six dimensions in total are used as design variables.
- Constraints: (1) the axial stress of the rods ≤ 450 MPa; (2) the shear stress of the plates ≤ 250 MPa; (3) the displacement of wing tip ≤ 5% of the semi span of the wing.

The final result can be obtained by iterating the system level and the subsystem level optimizations until converging to the optimum values. The subsystem optimizations are executed as the global variables are updated.

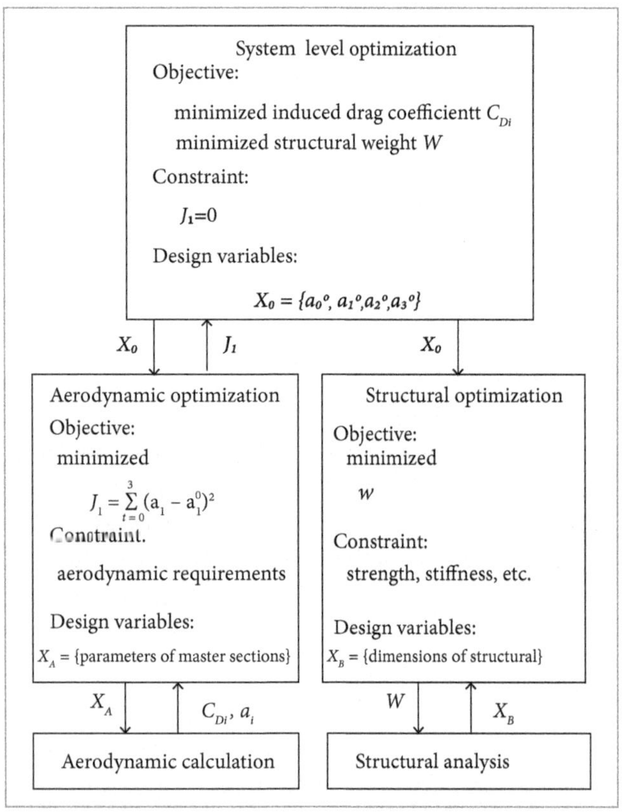

Figure 7. Optimization strategy for multi-objective optimization in aircraft conceptual design.

AIRCRAFT MODEL AND ANALYSIS METHODS

Executing optimization process automatically is necessary for solving complex optimization problems. To implement the procedure of aerodynamic and structural optimization for the UAV conceptual design, some important technologies, like parametric geometry description and automatic execution of aerodynamic computing and structure calculation, are essential. These approaches which will be used in the optimization are explained in the following subsections.

GENERATING PARAMETRIC MODEL

As for all optimization tasks, the complexity of the problem is directly coupled to the parameterization of the geometry. Of highest relevance is the number of parameters that are required to ensure valid modeling. The most important characteristic of the CAD model is to be highly flexible in order to represent a variety of designs as large as possible. Secondly the model must be robust and reliable, since there will not be a specialist manually entering new parameters and supervising the update process (Amadori *et al.* 2008; Sripawadkul *et al.* 2010; Wang *et al.* 2013). Flying wing configuration has the characteristics of simple shape and blended wing body, and the UAV can be seen as a special wing which is connected together by 3 segments. The important content of the CAD model parameterization is to describe parametric airfoils shape of the UAV.

In the article, the airfoil profile is described by the superposition of camber distribution and thickness distribution. The function which is used to explain the camber distribution is shown as follows:

$$f(x) = c_1 \cdot \sin(\pi \cdot x) + c_2 \cdot \sin(\pi \cdot x^{1.5}) + c_3 \cdot \sin(\pi \cdot x^2) + c_4 \cdot \sin(\pi \cdot x^{2.5}) \quad (2)$$

The coefficients of the function are derived from four characteristic parameters of the airfoil mean line, such as relative camber (C), relative camber location (XC), the angle between leading edge of camber line and chord line (α_{LE}), and the angle between trailing edge of camber line and chord line (α_{TE}). The geometric significance of these parameters is shown in Fig. 8.

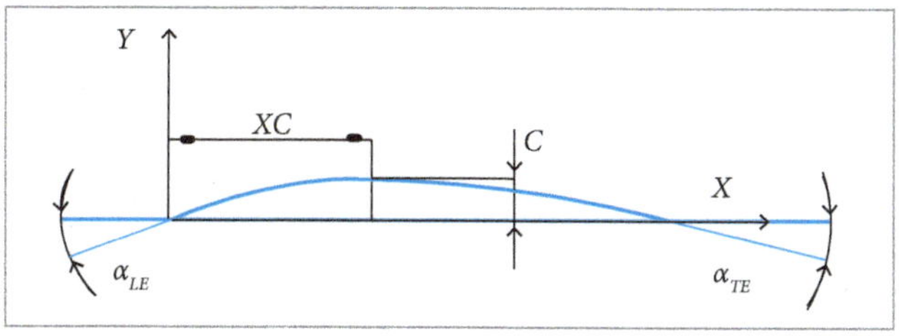

Figure 8. The meaning of camber parameters.

AERODYNAMIC COMPUTING

In order to get the induced drag coefficient and lift coefficient distribution along spanwise direction corresponding to design lift coefficient of the UAV, a panel code (Panair) is adopted. Panel codes are numerical schemes for solving the Prandtl-Glauert equation for linear, inviscid, irrotational flow about aircraft flying at subsonic or supersonic speeds (Lehmkuehler *et al.* 2012). Compared to CFD codes, Panair has advantages in terms of speed and ease of meshing. The surface mesh information of the UAV required for aerodynamic analysis is shown in Fig. 9. Pointwise software is used to divide quad surface mesh of the UAV model. An auxiliary numerical code is written to transform the surface mesh into the file which will be transmitted to Panair.

Aerodynamic performance is calculated through the following steps: (1) generating the mesh file of the UAV in Pointwise based on CAD model; (2) generating the file which will be transferred to Panair; (3) executing Panair program to calculate the lift coefficient of the UAV at different attack angles; (4) calculating the attack angle corresponding to design lift coefficient; (5) computing the aerodynamic performance of the UAV at design point; (6) deleting the files which are generated in the procedure.

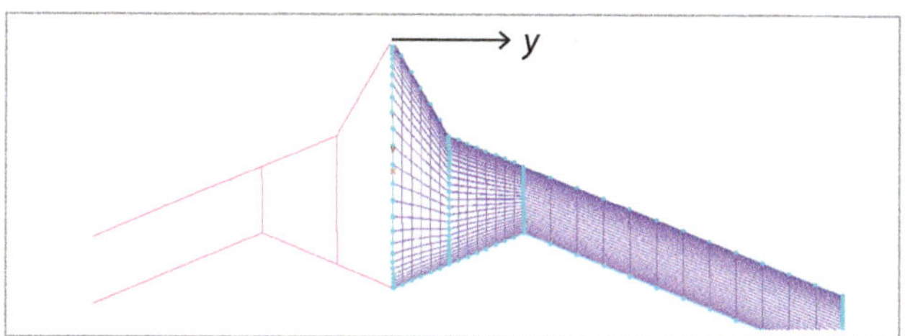

Figure 9. Aerodynamic model of the UAV.

Through aerodynamic performance calculation, the lift drag ratio and the lift coefficients of the sections along the UAV spanwise direction corresponding to design lift coefficient can be generated and outputted.

STRUCTURAL ANALYSIS

The structural model is more complex than the aerodynamic one due to the structural layout in the inertial model of the UAV. The features of the structural layout are defined by the number and positions of the spars, as well as the number and orientation of the ribs. These parameters are used for splitting the geometric model surface of the UAV in CATIA. Then the split model is transmitted to MSC.PATRAN to generate the structural layout model and the finite element model (FEM), which is shown in Fig. 10. In the FEM of the structure, the skins and webs of spars, ribs, and reinforced frames are modeled by plates, and the spars and the stiffeners for webs are modeled by rods. The dimension parameters of the FEM include thickness of skin and rib, area of spar caps and rib caps, etc.

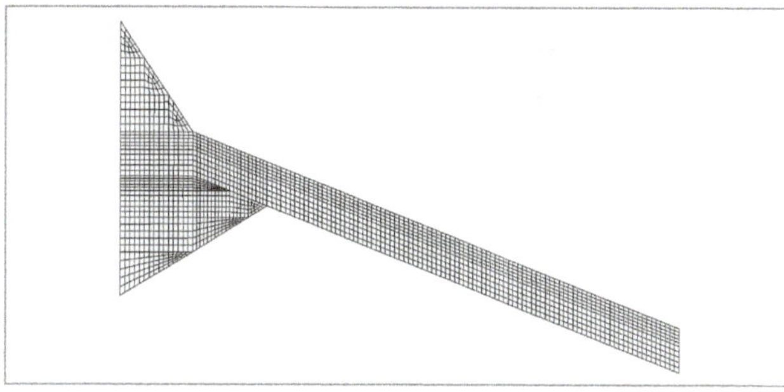

Figure 10. Finite element model for the UAV structure.

By using the FEM of the UAV structure, the optimization process can be used. The structural optimization is carried out by running MSC.NASTRAN software. It aims to search for the values of structural design variables which minimize structural weight in the condition when the material allowable stress, structural deformation, and geometry dimensions are satisfied. The structural weight optimized is necessary for overall performance calculation.

OPTIMIZATION PROCESS

In this article, iSIGHT is adopted to integrate all the software and programs we used (Koch *et al.* 2002). In order to improve the optimization efficiency, a surrogate model is constructed before MDO. Since aerodynamic problems are often highly non-linear, the surrogate model chosen in this article is the Radial Basis Function Neural Network (RBFNN). The input data of the surrogate model including the camber parameters of master sections and the output parameters are a_0; a_1; a_2; a_3 and C_{Di}. In the surrogate model, 370 sample points are selected randomly in design space to construct the surrogate model. Other 185 sample points are selected as the error analysis points of the surrogate model. In order to improve the credibility of the surrogate model, 370 new sample points are selected to reconstruct the surrogate model. The error analysis results of the final surrogate model for the output parameters are shown in Table 1. Six new sample points which are selected from the design space are calculated by Panair and the surrogated model, respectively, and the results are shown in Table 2. Once the surrogate model is built with given exact function data, it can efficiently be in the exploration of the design space and the MDO by replacing the original code.

Figure 11 shows the detailed procedure for dealing with the MOO problem of the UAV. The optimization process is carried out step by step following the flowchart. Each system level optimization iterates, during which the global variables can be updated and then the subsystem level optimization is completed once.

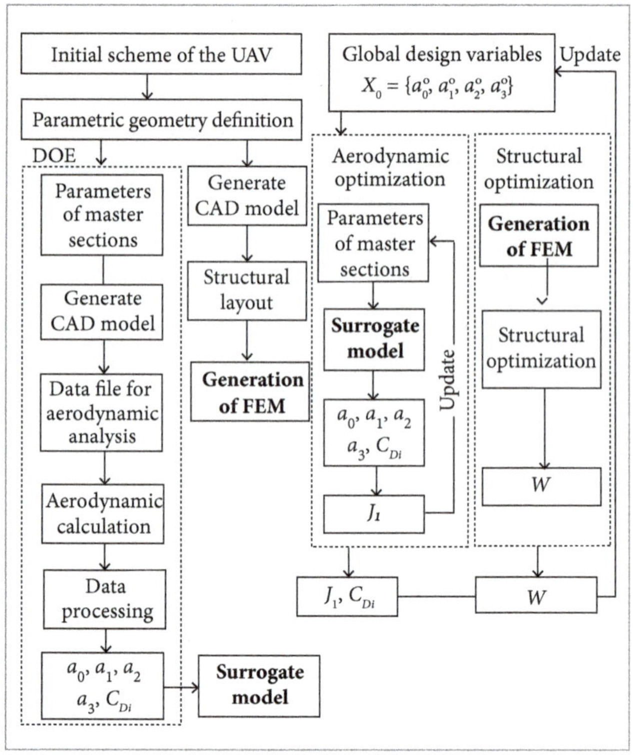

Figure 11. Flowchart of aerodynamic and structural multidisciplinary design optimization.

Table 1. Approximation error analysis.

Variables	Average	Maximum	Root Mean Square	R-squared
a_0	0.00508	0.02527	0.00677	0.99909
a_1	0.00456	0.02435	0.006	0.99912
a_2	0.00423	0.02318	0.00551	0.99911
a_3	0.00423	0.02274	0.00545	0.99909
C_{Di}	0.0134	0.04002	0.01572	0.9927

Table 2. Comparison of calculation results by Panair and RBFNN model.

Responses		1	2	3	4	5	6
a_0	Panair	2.028481	2.135833	2.099838	2.129787	2.089403	2.056113
	RBFNN	2.029884	2.135446	2.097253	2.130500	2.092359	2.056392
a_1	Panair	−3.284242	−3.56747	−3.718978	−3.458064	−3.533423	−3.257147
	RBFNN	−3.287965	−3.56777	−3.70466	−3.461285	−3.549795	−3.262633
a_2	Panair	2.760191	2.858455	3.431695	2.571950	3.064149	2.578853
	RBFNN	2.760020	2.860046	3.409363	2.578452	3.089931	2.589869
a_3	Panair	−1.422091	−1.35417	−1.733668	−1.181949	−1.54088	−1.298712
	RBFNN	−1.419876	−1.35518	−1.72302	−1.185902	−1.553252	−1.304347
C_{Di}	Panair	0.003510	0.003480	0.003470	0.003410	0.003470	0.003480
	RBFNN	0.003505	0.003478	0.003472	0.003405	0.003473	0.003477

OPTIMIZATION RESULTS

Based on the given optimization strategy, the optimization process is executed automatically. Figure 12 provides the Pareto fronts which are the solutions of minimizing both the induced drag coefficient and structural weight. An optimal point can be selected from the Pareto optimal solution set according to the design requirements.

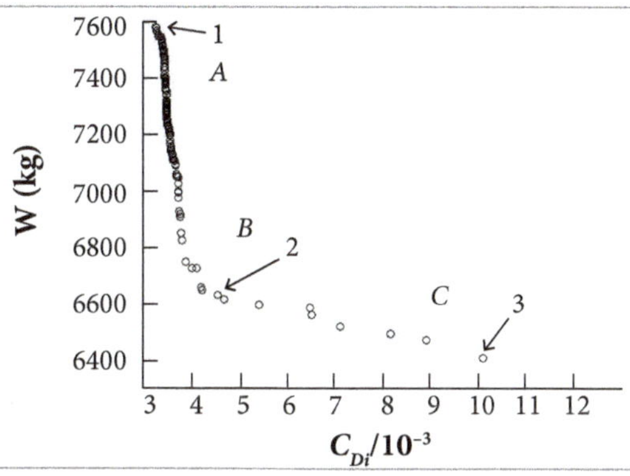

Figure 12. Pareto optimal front obtained from system level optimization.

The Pareto optimal front can be divided into 3 typical subsets, including subset A, subset B, and subset C. These subsets correspond to 3 typical subsets of the Pareto optimal solution set, respectively. The subset A is a set of optimization solutions with smaller induced drag coefficient. The subset C represents the optimization solutions which have lighter structural weight. The subset B lying between subsets A and C is the trade-off between aerodynamic performance and structural weight, so it has both relatively smaller induced drag coefficient and lighter structural weight. The choice for a solution from each subset as the typical sample and the aerodynamic performance and structural weight of selected samples are listed in Table 3, which shows that the aerodynamic performance of design 1 is better than the 2 other solutions; design 3 is the best one with respect to structural weight; and design 2 is a solution which is a compromise of the induced drag coefficient and structural weight.

Table 3. Typical solutions in Pareto optimal front.

Number of the sample	C_{Di}	*W* (kg)
1	0.00323511	7,582
2	0.00454684	6,632
3	0.01012	6,409

CONCLUSION

In this paper, an investigation has been made to study aerodynamic and structural optimization of flying wing aircraft. A multi-objective optimization strategy based on collaborative optimization strategy is proposed. During the optimization process, the parallel computing in subsystem optimization is used, which improves the efficiency of optimization. The optimization result is a Pareto optimal set, which provides the designer with more options. A flying wing configuration of aircraft is optimized by the strategy, and the optimization results demonstrate that the present method can efficiently find the Pareto optimal set.

AUTHORS' CONTRIBUTIONS

Pan Y and Huang J conceived the idea and co-wrote the main text; Li F and Yan C constructed the numerical models for multidisciplinary analysis; Yan C processed the calculation data. All authors discussed the results and commented on the manuscript. This paper was carried out through the collaboration of all authors.

REFERENCES

Amadori K, Jouannet C, Krus P (2008) Aircraft conceptual design optimization. Proceedings of the 28th Congress of the International Council of the Aeronautical Sciences; Anchorage, USA.

Bolsunovsky AL, Buzoverya NP, Gurevich BI, Denisov VE, Dunaevsky AI, Shkadov LM, Sonin OV, Udzhuhu AJ, Zhurihin JP (2001) Flying wing — problems and decisions. Aircraft Des 4(4):193-219. doi: 10.1016/S1369-8869(01)00005-2

Chu L (2011) Integration design of aerodynamic, structure and stealthy performance for joined-wing. Aeronautical Science and Technology 2:57-60.

Gao ZH, Xia L, Li T, Zhao X (2003) Investigation into collaborative optimization design techniques of aircraft aerodynamics and stealth performances. Aircraft Des 3:1-5. doi: 10.3969/j.issn.1673-4599.2003.03.001

Gou ZQ, Song BF (2006) The research of structural optimization with the constraints of aerodynamics and radar cross section (RCS) of aircraft. Systems Engineering - Theory & Practice 3:135-140.

Hu TY, Yu XQ (2009) Aerodynamic/stealthy/structural multidisciplinary design optimization of unmanned combat air vehicle. Chin J Aeronaut 22(4):380-386. doi: 10.1016/S1000-9361(08)60114-4

Huang CH, Galuski J, Bloebaum CL (2007) Multi-objective Pareto concurrent subspace optimization for multidisciplinary design. AIAA J 45(8):1894-1906. doi: 10.2514/1.19972

Koch PN, Evans JP, Powell D (2002) Interdigitation for effective design space exploration using iSIGHT. Structural and Multidisciplinary Optimization 23(2):111-126. doi: 10.1007/s00158-002-0171-9

Lambe AB, Martins JR (2016). Matrix-free aerostructural optimization of aircraft wings. Structural and Multidisciplinary Optimization 53(3):589-603. doi: 10.1007/s00158-015-1349-2

Lee D, Gonzalez LF, Srinivas K, Auld D, Periaux J (2007) Multi-objective/multidisciplinary design optimisation of blended wing body UAV via advanced evolutionary algorithms. Proceedings of the 45th AIAA Aerospace Sciences Meeting; Reno, USA.

Lehmkuehler K, Wong K, Verstraete D (2012) Design and test of a UAV blended wing body configuration. Proceedings of the 28th Congress of the International Council of the Aeronautical Sciences; Brisbane, Australia.

Li YH, Huang Y, Su JC (2014) Aerodynamic analysis and design numerical simulation of the vortex flow over a common research model with flying-wing configuration. Appl Mech Mater 540:138-142. doi: 10.4028/www.scientific.net/AMM.540.138

Ma TL, Ma DL, Zhang S (2009) Method for aerodynamic and structure integrative design. Journal of Beijing University of Aeronautics and Astronautics 35(11): 1366-1383. doi: 10.13700/j.bh.1001-5965.2009.11.011

Molinari G, Arrieta AF, Ermanni P (2014) Aero-structural optimization of three-dimensional adaptive wings with embedded smart actuators. AIAA J 52(9):1940-1951. doi: 10.2514/1.J052715

Piperni P, Abdo M, Kafyeke F, Isikveren AT (2007) Preliminary aerostructural optimization of a large business jet. J Aircraft 44(5):1422-1438. doi: 10.2514/1.26989

Sanghvi RC, Vashi AS, Patolia HP, Jivani RG (2014) Multi-objective optimization of two-stage helical gear train using NSGA-II. Journal of Optimization 2014:1-8. doi: 10.1155/2014/670297

Song L, Yang H, Zhang Y, Zhang H, Huang J (2014) Dihedral influence on lateral-directional dynamic stability on large aspect ratio tailless flying wing aircraft. Chin J Aeronaut 27(5):1149-1155. doi: 10.1016/j.cja.2014.08.003

Sripawadkul V, Padulo M, Guenov M (2010) A comparison of airfoil shape parameterization techniques for early design optimization. Proceedings of the 13th AIAA/ISSMO Multidisciplinary Analysis and Optimization Conference; Fort Worth, USA.

Wang WJ, Huang J, Song L (2013) A geometric representation of airfoil based on applied parameters. Aircraft Des 33(3):1-4.

Zhou HX, Liu B (2015) Characteristics analysis and optimization of flying-wing vehicle structure. Adv Mater Res 1077:177-184. doi: 10.4028/www.scientific.net/AMR.1077.177

14

Orthotropic Elastic Properties Assessment of Sandwich Laminates

Paulo Pedro Kenedi[1,2], Lucas Lisboa Vignoli[3], Brenno Tavares Duarte[2], Fernando Cesar de Abreu Matos[1], Humberto Oberosler Terço Dias[1]

ABSTRACT: Sandwich structures are nowadays widely used due its excellent characteristics of matching light weight with high bend stiffness. However the complex nature of its composition turns the estimation of its mechanical elastic properties challenging. The objective of the present article is to propose an analytical/experimental model to estimate the orthotropic elastic properties of this composite material. To accomplish this objective a theory of anisotropic plates is used together to an extensometric experimental approach. A sandwich laminate specimen, instrumented with strain gages, was submitted through pure bending loading by a mechanical testing machine on a 4-point bending apparatus. The result of this research was the determination of 4 orthotropic elastic properties of the sandwich laminate specimen, namely: the longitudinal and transversal Young moduli, the shear modulus, and the Poisson's ratio.

KEYWORDS: Sandwich laminates, Orthotropic properties, Analytical model.

INTRODUCTION

Composites materials, as sandwich laminates, have been widely used for structural purposes. These structures are anisotropic, combining external fibers (with high strength) and an internal core (with low stiffness). This combination of materials contributes to turn its stress analysis not as straightforward.

Much effort has been done to use sandwich laminates as an engineering material. Authors as Bledzki *et al.* (1999) identified elastic properties of unidirectional glass/epoxy laminates through the measuring of its natural frequencies in experimental vibration experiments. Lamboul *et al.* (2013) investigated the potential of ultrasonic Lamb waves to estimate the initial elastic modulus of glass fiber woven composite materials. The elastic modulus of the undamaged materials was estimated using time of flights (TOF) of extensional waves (also known as Lamb S0 mode) generated by a piezoelectric material. Rikards *et al.* (2001) and Rikards *et al.* (1999) used mixed numerical/experimental techniques to estimate mechanical properties of composites laminates. In the experimental part it was used real-time holography, and in the numerical part it was used the finite element method. Lecompte *et al.* (2007) have also used a mixed numerical-experimental method to identify 4 in-plane orthotropic engineering constants of composite plate materials. It was used a biaxial tensile test, performed on a cruciform test specimen. A heterogeneous displacement field was observed by a CCD camera and measured by a digital image correlation (DIC) technique. The measured displacement field and the subsequently computed strain field were compared with a finite element simulation of the same experiment. An optimization technique was then used to minimize the errors

1.Centro Federal de Educação Tecnológica Celso Suckow da Fonseca – Departamento de Engenharia Mecânica – Laboratório de Compósitos e Adesivos – Rio de Janeiro/RJ – Brazil. **2.**Centro Federal de Educação Tecnológica Celso Suckow da Fonseca – Programa de Pós-Graduação em Engenharia Mecânica e Tecnologia de Materiais – Laboratório de Compósitos e Adesivos – Rio de Janeiro/RJ – Brazil. **3.**Universidade Federal do Rio de Janeiro – Centro de Mecânica Não-Linear – Departamento de Engenharia Mecânica – Rio de Janeiro/RJ, Brazil.

Author for correspondence: Paulo Pedro Kenedi | Centro Federal de Educação Tecnológica Celso Suckow da Fonseca – Departamento de Engenharia Mecânica – Laboratório de Compósitos e Adesivos | Avenida Maracanã, 229 – Maracanã | CEP: 20.271-110 – Rio de Janeiro/RJ – Brazil | Email: paulo.kenedi@cefet-rj.br

between experimental and numerical approaches resulting in the estimation of 4 independent apparent engineering constants: E_x, E_y, G_{xy}, and ν_{xy}.

Other authors as Tamilarasan *et al.* (2015) evaluated mechanical properties of sandwich laminates made up of carbon fiber with epoxy resin, used as a matrix, and aluminum (AA6061-T6), used as sandwich plate, through the realization of multiple tests as: uniaxial tensile, flexural, and impact. The tests results were used by the authors to describe, in a macroscopic perspective, the different material behavior associated to different levels of strengths (axial, flexural, and impact), including the failure mechanisms.

The research of Hu and Karki (2015) involved a 2-step modeling approach to estimate the mechanical properties of 3-D carbon fiber-reinforced epoxy composites. In step 1, the microstructures of the composites were represented by 3-D microscale cylindrical, square, or hexagonal prismatic representative volume element (RVE) and were subject of multiple tests as: uniaxial tensile, lateral expansion, and transverse shear. In step 2, the overall composites were represented by a 3-D macroscale unit cell that was discretized into elements. Each element was assigned with appropriate properties, taking into account the carbon fiber orientation and volume fraction. The averaged mechanical properties of the RVEs from step 1 were input into each element accordingly. The effective orthotropic mechanical properties of the composites were predicted by uniaxial tensile tests and shear tests on the unit cell.

To improve the understanding of mechanical behavior of sandwich laminates, this article proposes a combined model, using analytical and experimental approaches, to estimate 4 orthotropic mechanical elastic properties of sandwich laminates. The analytical part uses a theory of anisotropic plates and the experimental one, strain gages to obtain the strains generated on a sandwich laminate specimen, submitted to pure bending on a 4-point bending test. Note that although, in technical literature, it is more common the utilization of unidirectional composite materials, in this article a woven (fabric) sandwich laminate was used to obtain the longitudinal and the transversal Young moduli, the shear modulus, and the Poisson's ratio mechanical properties. In the next section the analytical models are described.

ANALYTICAL MODELS

Some initial information, as the 4-point bending apparatus and the specimen principal dimensions are provided before the analytical models description are given. Figure 1 shows a schematic representation of the sandwich laminate specimen, instrumented with strain gages, submitted to pure bending in its central region (between upper roller anvils), on a 4-point bending apparatus.

Note in Fig. 1a that the dark blue color represents the external polymeric matrix, with high strength carbon fibers (0°/90°)/epoxy (with 0.6 mm of thickness each) and the light blue color represents the low stiffness PVC foam (with 3.8 mm of thickness). The distance between the upper roller anvil and the immediate adjacent lower roller anvil is $l = 50$ mm and the distance between the lower roller anvils is $L = 170$ mm.

Figure 2 shows the sandwich laminate specimen schematic drawing, with specimen nominal dimensions and the used coordinate systems. θ is an angle, in counterclockwise direction, between local and global systems, beginning with $\theta = 0°$ on global axis direction 11.

Two analytical models are provided to estimate orthotropic elastic constants of composite sandwich specimens, both considering Euler-Bernoulli hypothesis. The first model is a simplified approach which disregards the core material stiffness and the second one is a complete analytical solution that considers

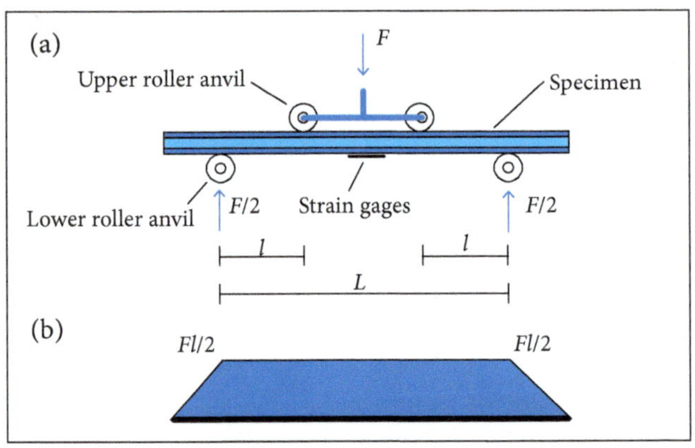

Figure 1. (a) Schematic representation of the 4-point bending apparatus with specimen (lateral view); (b) Correspondent bending moment diagram.

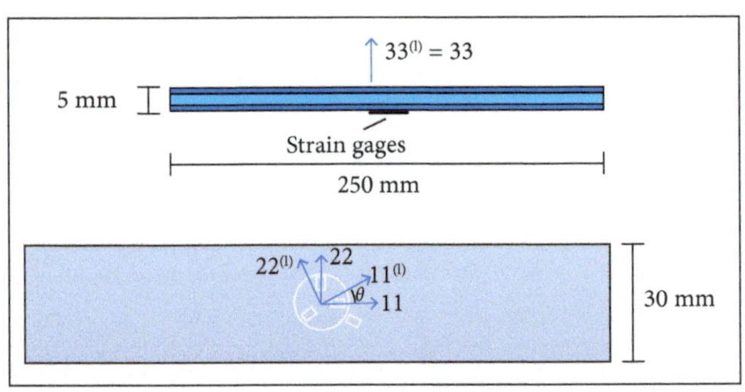

Figure 2. Composite specimen schematic drawing at a third angle projection (unscaled).

the stiffness of every layer, including the different elastic moduli, in tension and in compression, of the core material. Both models depend on the experimental strain acquisition. Regardless of the layers anisotropy, it is assumed that the sections remain plane and perpendicular to the neutral axis, being enough to describe the normal strain variation along the cross section. A constitutive relation and the equilibrium requirements are additionally provided to implement the stress calculation, as in Lekhnitskii (1981).

SIMPLIFIED ANALYTICAL MODEL

The simplified model is divided in stress and strain parts. First, the stress is estimated by application of a well-known mechanics of solids equation and the proposal of a constitutive relationship. Then, with the input of experimental strains, 4 elastic orthotropic constants are estimated.

At the central region of a 4-point bending setup, between upper roller anvils, the shear force vanishes and it is considered that this region is only submitted to pure bending loading. Hence, the unique non-null global stress component on the bottom of the beam in this region is, as in Crandall *et al.* (1978),

$$\sigma_{11} = \frac{M_{22}}{I_{22}} x_3 \tag{1}$$

where: x_3 is the vertical distance between the neutral axis and a given point of interest; M_{22} is the pure moment around 22 axis (see Fig. 2); I_{22} is the cross section second moment of area relative to 22 axis:

$$\begin{aligned} M_{22} &= Fl/2 \\ I_{22} &= \frac{b}{12}\left[\left(t_c + 2t_l\right)^3 - \left(t_c\right)^3\right] \\ x_3 &= t_l + t_c/2 \end{aligned} \tag{2}$$

where: F is the applied load; l is the distance between the upper roller anvil and the immediate adjacent lower roller anvil (see Fig. 1a); t_1 is the woven lamina thickness; t_c is the core thickness; b is the specimen width.

Note that for this simplified analytical model it is supposed that all loads are only resisted by the woven laminas, since the core material has a much lower stiffness (Kassapoglou 2010). This simplified approach turns the σ_{11} estimation straightforward.

The normal local strains, at a general given angle θ, in relation to the global axes, shown in Fig. 2, can be estimated as in Lekhnitskii (1981):

$$\varepsilon_{11}^{(l)} = C_{11}^{(l)}\sigma_{11}^{(l)} + C_{12}^{(l)}\sigma_{22}^{(l)} + C_{16}^{(l)}\sigma_{12}^{(l)} \tag{3}$$

where the local stress components $\sigma_{11}^{(l)}$, $\sigma_{22}^{(l)}$, and $\sigma_{12}^{(l)}$ can be obtained through the use of Mohr's circle as in Crandall *et al.* (1978):

$$\begin{aligned} \sigma_{11}^{(l)} &= \frac{\sigma_{11}}{2}\left(1 + \cos 2\theta\right) \\ \sigma_{22}^{(l)} &= \frac{\sigma_{11}}{2}\left(1 - \cos 2\theta\right) \\ \sigma_{12}^{(l)} &= -\frac{\sigma_{11}}{2}\sin 2\theta \end{aligned} \tag{4}$$

and the local compliance components can be estimated as in Staab (2015):

$$C_{11}^{(l)} = C_{11}\cos^4\theta + \left(2C_{12} + C_{66}\right)\sin^2\theta\cos^2\theta + C_{22}\sin^4\theta \tag{5}$$

$$C_{12}^{(l)} = C_{12} + \left(C_{11} + C_{22} - 2C_{12} - C_{66}\right)\sin^2\theta\cos^2\theta \tag{6}$$

$$\begin{aligned} C_{16}^{(l)} = {} & \left(2C_{11} - 2C_{12} - C_{66}\right)\sin\theta\cos^3\theta - \\ & -\left(2C_{22} - 2C_{12} - C_{66}\right)\sin^3\theta\cos\theta \end{aligned} \tag{7}$$

Considering that the material coordinate system is coincident with the global coordinate axes, the reduced global compliance matrix C, also considering just the in-plane components and its orthotropic symmetry for a woven lamina, can be written as in Staab (2015):

$$C = \begin{bmatrix} C_{11} & C_{12} & C_{16} \\ C_{12} & C_{22} & C_{26} \\ C_{16} & C_{26} & C_{66} \end{bmatrix} = \begin{bmatrix} 1/E_1 & -\nu_{12}/E_1 & 0 \\ -\nu_{12}/E_1 & 1/E_2 & 0 \\ 0 & 0 & 1/G_{12} \end{bmatrix} \tag{8}$$

where: E_1 and E_2 are the lamina elastic moduli for, respectively, 11 and 22 axis global directions; G_{12} is the lamina shear modulus; ν_{12} is the lamina Poisson's ratio.

Using the compliance matrix components, shown in Eq. 8, Eqs. 5 – 7 can be rewritten for plane stress case, as in Staab (2015):

$$C_{11}^{(l)} = \frac{1}{E_1}\cos^4\theta + \left(-2\frac{\nu_{12}}{E_1} + \frac{1}{G_{12}}\right)\sin^2\theta\cos^2\theta + \frac{1}{E_2}\sin^4\theta \tag{9}$$

$$C_{12}^{(l)} = -\frac{\nu_{12}}{E_1} + \left(\frac{1 + 2\nu_{12}}{E_1} + \frac{1}{E_2} - \frac{1}{G_{12}}\right)\sin^2\theta\cos^2\theta \tag{10}$$

$$C_{16}^{(l)} = 2\left(\frac{1+\nu_{12}}{E_1} - \frac{1}{2G_{12}}\right)\sin\theta\cos^3\theta - \tag{11}$$
$$- 2\left(\frac{1}{E_2} + \frac{\nu_{12}}{E_1} - \frac{1}{2G_{12}}\right)\sin^3\theta\cos\theta$$

Notice that Eq. 3 can be rewritten in a matricial form, with 4 lines, one per strain gage of angle θ_n.

$$\begin{bmatrix} \varepsilon_{11}^{(l)}\big|_{\theta_1} \\ \varepsilon_{11}^{(l)}\big|_{\theta_2} \\ \varepsilon_{11}^{(l)}\big|_{\theta_3} \\ \varepsilon_{11}^{(l)}\big|_{\theta_4} \end{bmatrix} = \begin{bmatrix} f_1(\theta_1) & f_2(\theta_1) & f_3(\theta_1) & f_4(\theta_1) \\ f_1(\theta_2) & f_2(\theta_2) & f_3(\theta_2) & f_4(\theta_2) \\ f_1(\theta_3) & f_2(\theta_3) & f_3(\theta_3) & f_4(\theta_3) \\ f_1(\theta_4) & f_2(\theta_4) & f_3(\theta_4) & f_4(\theta_4) \end{bmatrix} \begin{bmatrix} 1/E_1 \\ \nu_{12}/E_1 \\ 1/E_2 \\ 1/G_{12} \end{bmatrix} \frac{\sigma_{11}}{2} \tag{12}$$

where:

$$f_1(\theta) = 2\cos^2\theta\left(2\cos^2\theta - 1\right)^2$$
$$f_2(\theta) = 2\cos^3 2\theta - \cos 2\theta - 1 \tag{13}$$

$$f_3(\theta) = -8\sin^4\theta\left(\sin^2\theta - 1\right)$$
$$f_4(\theta) = \cos 2\theta \sin^2 2\theta \tag{14}$$

The final equation can be obtained rearranging Eq. 12, substituting the result of Eq. 1 and the strain-gages angles (from θ_1 to θ_4), and its respective experimental values (from $\varepsilon_{11}^{(l)}|_{\theta_1}$ to $\varepsilon_{11}^{(l)}|_{\theta_4}$).

$$\begin{bmatrix} 1/E_1 \\ \nu_{12}/E_1 \\ 1/E_2 \\ 1/G_{12} \end{bmatrix} = \begin{bmatrix} f_1(\theta_1) & f_2(\theta_1) & f_3(\theta_1) & f_4(\theta_1) \\ f_1(\theta_2) & f_2(\theta_2) & f_3(\theta_2) & f_4(\theta_2) \\ f_1(\theta_3) & f_2(\theta_3) & f_3(\theta_3) & f_4(\theta_3) \\ f_1(\theta_4) & f_2(\theta_4) & f_3(\theta_4) & f_4(\theta_4) \end{bmatrix}^{-1} \begin{bmatrix} \varepsilon_{11}^{(l)}\big|_{\theta_1} \\ \varepsilon_{11}^{(l)}\big|_{\theta_2} \\ \varepsilon_{11}^{(l)}\big|_{\theta_3} \\ \varepsilon_{11}^{(l)}\big|_{\theta_4} \end{bmatrix} \frac{2}{\sigma_{11}} \tag{15}$$

With Eq. 15 it is possible to estimate 4 elastic orthotropic constants (E_1, E_2, ν_{12}, and G_{12}). A flowchart is shown in Fig. 3 as a form to review the steps necessary to apply the simplified analytical model.

The flowchart of Fig. 3 shows the straightforward utilization of the simplified analytical model. In fact to estimate the 4 orthotropic elastic constants E_1^{SAM}, E_2^{SAM}, ν_{12}^{SAM}, G_{12}^{SAM} the angles of the strain gages have to be entered in relation to the specimen longitudinal axis θ_1, θ_2, θ_3, and θ_4; the correspondent experimental strain values $\varepsilon_{11}^{(l)}|_{\theta_1}$, $\varepsilon_{11}^{(l)}|_{\theta_2}$, $\varepsilon_{11}^{(l)}|_{\theta_3}$, and $\varepsilon_{11}^{(l)}|_{\theta_4}$; and the estimative of longitudinal bottom specimen stress σ_{11} into the linear system of Eq. 15 to obtain the estimative E_1^{SAM}, E_2^{SAM}, ν_{12}^{SAM}, G_{12}^{SAM}. The index "SAM" is the abbreviation of simplified analytical model. The complete analytical model is provided in the next subsection.

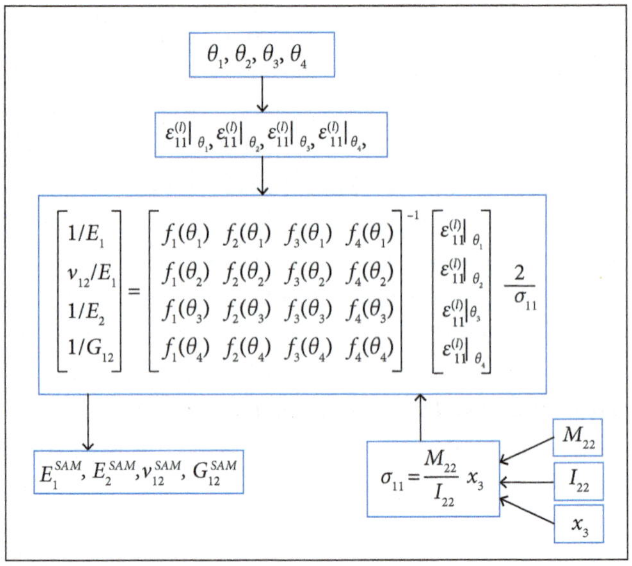

Figure 3. Simplified analytical model flowchart.

COMPLETE ANALYTICAL MODEL

In this section the influence of the core on the estimative of the elastic orthotropic constants is carefully studied. To accomplish this objective a complete analytical model is presented considering the core asymmetric tension/compression behavior, as in Kenedi *et al* (2016). The procedure presented next is based on solid mechanics theory and further discussions can be found in Crandall *et al.* (1978) as well as in Vasiliev and Morozov (2001).

In Fig. 4 $E_c^{(t)}$ and $E_c^{(c)}$ are, respectively, the core elastic modulus in tension and in compression. As the core tension properties are different from the compression ones, the stress distribution is no more symmetric, turning the neutral axis position $\overline{x}_3$ (see Fig. 4) an additional parameter to be taken

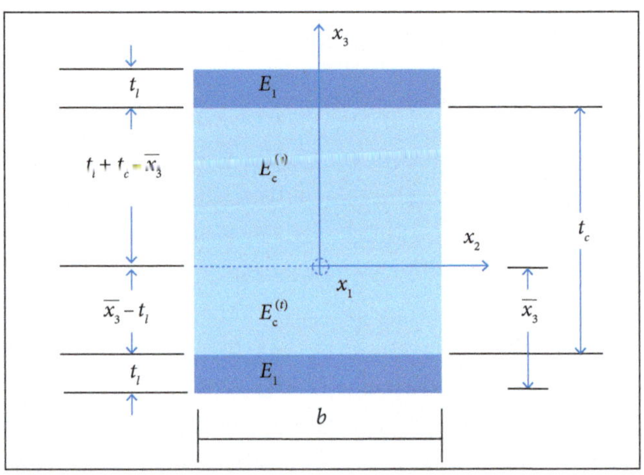

Figure 4. Sandwich laminates specimen transversal cross-section dimensions.

into account by imposing the equilibrium conditions along the beam cross section. In fact $E_c^{(t)}$ and $E_c^{(c)}$ must be used as 2 limiting conditions with minimum and maximum properties, called, respectively, in this article, as lower and upper core bounds, which are available in Table 1. These lower and upper properties cover the range of elastic moduli available in technical literature for the core material, for instance, PVC foam. Note that to use both models there is no need for having any estimative of the lamina elastic moduli E_1, which can be considered an advantage due to the great variability of this property in function of parameters like: fraction volume, epoxy type, fabric patterns and fabrication process.

Table 1. Range of elastic modulus core material.

	Lower bound core (lo)	Upper bound core (up)
$E_c^{(t)}$ (MPa)	49	320
$E_c^{(c)}$ (MPa)	40	400

Source: Diab Group and Divinycell (2016).

Taking the asymmetric characteristic into account, considering that the section remains plane and perpendicular to the neutral axis and using the linear elastic constitutive relation, the force equilibrium along the cross section can be written as (Crandall *et al.* 1978):

$$\int \sigma_{11} dA = \int E\,\varepsilon_{11} dA = \frac{1}{\rho}\int E x_3 dA = 0 \tag{16}$$

where: ρ is the radius of curvature, which is supposed to be constant on the cross section.

Equation 16 is the definition of the centroid in non-homogeneous materials, which can be rewritten as:

$$\begin{aligned}&\left(E_c^{(t)} - E_c^{(c)}\right)\overline{x_3}^2 + 2\left[\left(2E_1 - E_c^{(t)} + E_c^{(c)}\right)t_l + E_c^{(c)} t_c\right]\overline{x_3} -\\ &-\left[\left(4E_1 - E_c^{(t)} + E_c^{(c)}\right)t_l^2 + 2\left(E_1 + E_c^{(c)}\right)t_c t_l + E_c^{(c)} t_c^2\right] = 0\end{aligned} \tag{17}$$

Note that Eq. 17 is a second-order polynomial function but it is not completely defined once E_1 is still unknown. $E_c^{(t)}$ and $E_c^{(c)}$ are first substituted by lower bounds and then by upper ones to verify the solution range. Since the neutral axis expression is obtained, the moment equilibrium on the cross section, following the same assumptions used for the force equilibrium, presented previously, is defined as Crandall *et al.* (1978):

$$\int x_3 \sigma_{11} dA = \frac{1}{\rho}\int E x_3^2 dA = M_{22}, \tag{18}$$

defining equivalent flexural rigidity as:

$$EI_{eq}\left(E_c^{(c)}, E_c^{(t)}, E_1\right) = \int E x_3^2 dA \tag{19}$$

This quantity can be computed solving the integral or alternatively using the parallel axis theorem as

$$\begin{aligned}EI_{eq}\left(E_c^{(c)}, E_c^{(t)}, E_1\right) = E_1 I_{22}^{(l,t)} + E_c^{(t)} I_{22}^{(f,t)} +\\ + E_c^{(c)} I_{22}^{(f,c)} + E_1 I_{22}^{(l,c)}\end{aligned} \tag{20}$$

where:

$$\begin{aligned}I_{22}^{(l,t)} &= \frac{bt_l^3}{12} + (bt_l)\left(\overline{x_3} - \frac{t_l}{2}\right)^2\\ I_{22}^{(l,c)} &= \frac{bt_l^3}{12} + (bt_l)\left(\frac{3t_l}{2} + t_c - \overline{x_3}\right)^2\end{aligned} \tag{21}$$

$$\begin{aligned}I_{22}^{(f,t)} &= \frac{b}{12}\left(\overline{x_3} - t_l\right)^3 + \left[b\left(\overline{x_3} - t_l\right)\right]\left(\frac{\overline{x_3} - t_l}{2}\right)^2 \quad I_{22}^{(f,c)} =\\ &= \frac{b}{12}\left(t_l + t_c - \overline{x_3}\right)^3 + \left[b\left(t_l + t_c - \overline{x_3}\right)\right]\left(\frac{t_l + t_c - \overline{x_3}}{2}\right)^2\end{aligned} \tag{22}$$

The upper indexes of the second moment of area are related to: the first index for the material l represents the lamina and f represents the foam (at core); the second index for the stress state t represents tension and c represents compression. Note that to maintain compact the notation in Eqs. 21 and 22, $\overline{x_3}$ ($E_c^{(c)}$, $E_c^{(t)}$, E_1) and $I_{22}^{(,)}$ ($E_c^{(c)}$, $E_c^{(t)}$, E_1) are written, respectively, as $\overline{x_3}$ and $I_{22}^{(,)}$.

Thus, the radius of curvature expression is obtained from direct substitution in Eq. 18. Hence, the normal stress on the laminas of the sandwich laminate is calculated as

$$\sigma_{11}\left(E_c^{(c)}, E_c^{(t)}, E_1\right) = E_1 \frac{M_{22}}{EI_{eq}\left(E_c^{(c)}, E_c^{(t)}, E_1\right)}\overline{x_3}\left(E_c^{(c)}, E_c^{(t)}, E_1\right) \tag{23}$$

In Fig. 5 a flowchart of the complete analytical model is presented and shows a little more laborious implementation than the flowchart of Fig. 3. Note that this flowchart needs the results of the former one (Fig. 3), E_1^{SAM}, E_2^{SAM}, ν_{12}^{SAM}, G_{12}^{SAM}, as an initial trial to solve a linear system through the use of the "find" command of Mathcad software to calculate E_1 and, consequently, E_2, ν_{12}, and G_{12}. In fact, the complete analytical model has to be applied twice: first the lower bond core properties are used (with an index $_{lo}$), shown in Table 1, for the $E_c^{(t)}$ and $E_c^{(c)}$ properties to estimate: $\overline{x}_3$ and EI_{eq} $(E^{(c)}_{c\_lo}, E^{(t)}_{c\_lo}, E_1)$, which are substituted in $\sigma_{11}(E^{(c)}_{c\_lo}, E^{(t)}_{c\_lo}, E_1)$. Inputting the angles of the strain gages θ_1, θ_2, θ_3, and θ_4 and the correspondent experimental strain values $\varepsilon^{(l)}_{11}|_{\theta_1}, \varepsilon^{(l)}_{11}|_{\theta_2}, \varepsilon^{(l)}_{11}|_{\theta_3}, \varepsilon^{(l)}_{11}|_{\theta_4}$ into Eq. 15 and activating the "find" command of Mathcad software, the $E^{CAM}_{1\_lo}$, $E^{CAM}_{2\_lo}$, $\nu^{CAM}_{12\_lo}$, $G^{CAM}_{12\_lo}$ results are obtained. Note that the index "CAM" is the abbreviation of complete analytical model. At the second time the upper bond core properties are used (with an index "up"), as shown in Table 1, for the $E_c^{(t)}$ and $E_c^{(c)}$ properties to estimate: $\overline{x}_3$ and EI_{eq} $(E^{(c)}_{c\_up}, E^{(t)}_{c\_up}, E_1)$, which are substituted in $\sigma_{11}(E^{(c)}_{c\_up}, E^{(t)}_{c\_up}, E_1)$; the same sequence that is done in the first time is repeated, resulting in $E^{CAM}_{1\_up}$, $E^{CAM}_{2\_up}$, $\nu^{CAM}_{12\_up}$, $G^{CAM}_{12\_up}$.

In the next section the experimental approach necessary to obtain 4 strain measurements in 4 distinct angles is explained.

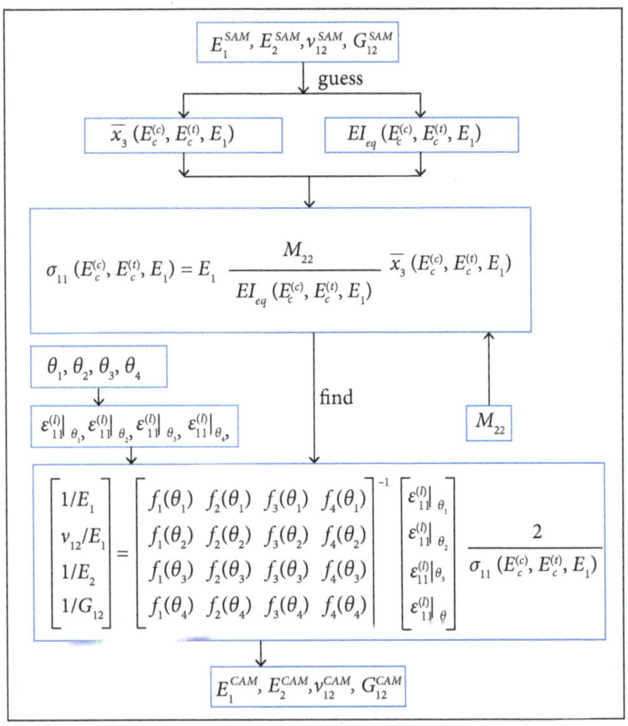

Figure 5. Complete analytical model flowchart.

EXPERIMENTAL APPROACH

In this section it is made a description of the equipment utilized in the experimental approach. All the sensors, transducers, data acquisition equipment, and materials testing machine used in the experiment to obtain the strain signals are described in details. In Fig. 6 it is shown the 4-point bending test apparatus, used in an INSTRON material testing machine. It was used a 5-kN load cell and the LVDT of the test machine. The test was conducted monotonically under load control as stated in ASTM C393-00.

Note, in Fig. 6, the instrumented specimen is lying on a 4-point bending apparatus. The strain gages are bonded at the bottom side of the specimen, as can be seen in Fig. 6b. It is possible to see a detailed view of the strain gages specimen instrumentation in Fig. 7a, where 3 of these strain gages are provided by an extensometric rosette and the fourth strain gage is provided by a uniaxial one. In Fig. 7b the denomination and the angle of each strain gage are presented.

A Spider 8 data acquisition system was used together with a HBM Catman Software to set up the experimental strain data acquisition. It was used a Kyowa extensometric rosette KFG-3-120-D28-11, as well as a Kyowa uniaxial strain gage

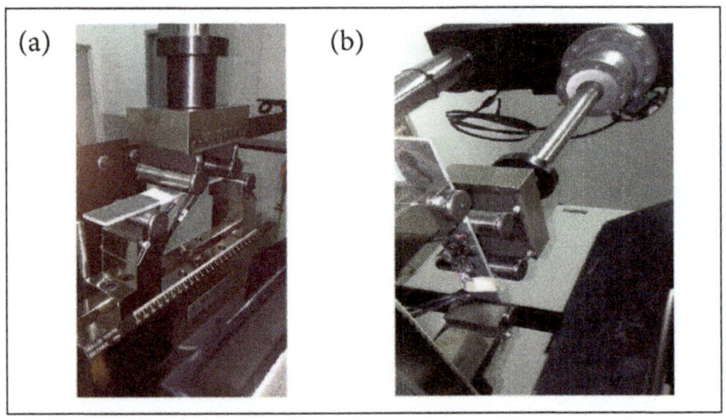

Figure 6. 4-point bending apparatus with instrumented specimen. (a) General view; (b) Specimen underside view.

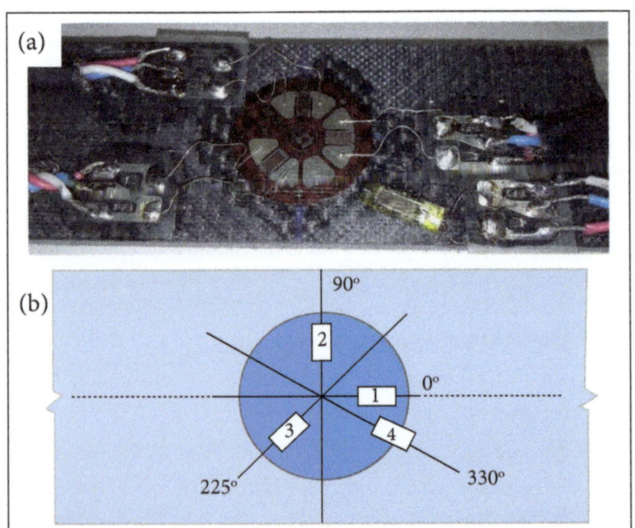

Figure 7. (a) Strain gages details; (b) Strain gage positions.

KFG-5-120-C1-11, bonded, at a sanding-prepared surface of the bottom lamina of the sandwich laminate specimen, with the utilization of a proper cyanoacrylate glue. Note in the detailed view (Fig. 7a), the utilized cabling. More information about the utilization of strain gages to measure strains on composites materials can be accessed at the fourth chapter of Staab (2015).

The objective of this test was to experimentally acquire the strain signals at 4 different angles, 0° (longitudinal and aligned to global axis 11), 90° (transversal), 225° and 330° (both inclined in relation to specimen longitudinal axis). The pure bending load was selected in order to make the strain gages positioning less critical, principally on the longitudinal direction.

Note that for analytical models, with the objective of using a more compact notation the strain gages were numbered from 1 to 4. The index equivalence between analytical and experimental notation is: 1 = 0°, 2 = 90°, 3 = 225°, and 4 = 330°, as also can be seen in Fig. 7b.

RESULTS AND DISCUSSION

In this section the experimental strain signals outputs of the 4 strain gages are provided. Figure 8 shows the strain signals outputs of the strain gages bonded at the central part of the bottom side of the specimen submitted to a pure bending on a 4-point bending test.

The data acquisition rate used in this test was selected on 50 Hz for all the 4 channels, or, in other words, 50 samples per second for each channel. Many of the acquired points of Fig. 8 were suppressed, without any loss of relevant information, to turn the experimental results clearer. It was imposed

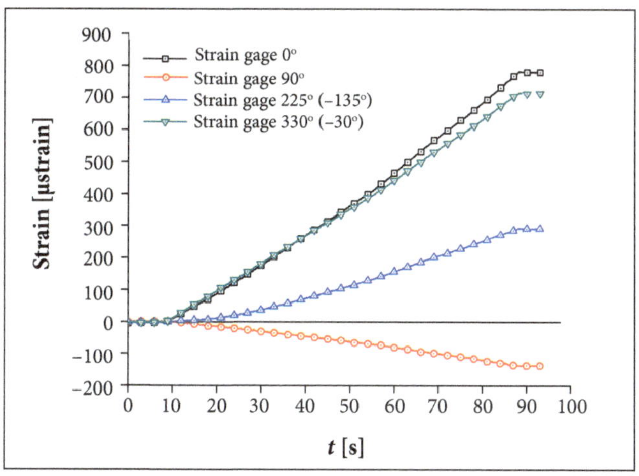

Figure 8. Strains experimental results.

a displacement rate of 1 mm/min until the load reached the force plateau of 40 N and a correspondent transversal displacement of 1.3 mm.

Note that the monotonic increasing transversal force was arbitrarily stopped at a plateau of 40 N and transversal displacement of 1.3 mm because the specimen was perceptively beginning to assume a curved configuration, which could conflict with the applicability of the mechanics of solids theory, used for instance in Eqs. 1 and 2. It would not be expected to have any perceptible geometric differences between loaded and unloaded configurations. Also it was used the high-quality load cell, already mounted on the test machine, which provided a load signal with a very low noise level, even for such low loads.

As expected the major strains occurred at 0° strain gage, aligned to the longitudinal direction (parallel to 11 global axis); at 225° and 330° the strains were at intermediary levels and, at 90° strain gage, which is the transversal one (parallel to 22 global axis), under the influence of Poisson's rate effect, it resulted in negative strains.

For this study the following strains results were used for both analytical models: $\varepsilon_{11}^{(l)}|_{\theta_1} = 780\ \mu\varepsilon$, $\varepsilon_{11}^{(l)}|_{\theta_2} = -136\ \mu\varepsilon$, $\varepsilon_{11}^{(l)}|_{\theta_3} = 292\ \mu\varepsilon$ and $\varepsilon_{11}^{(l)}|_{\theta_4} = 715\ \mu\varepsilon$.

Table 1 shows the elastic modulus range of the core material (PVC foam) obtained in technical literature (Diab Group and Divinycell 2016).

Note that as previously explained in the section "Complete Analytical Model", there is a need for the lower bond properties (with an index "lo") $E_c^{(t)}$ and $E_c^{(c)}$ and the upper bond properties (with an index "up") $E_c^{(t)}$ and $E_c^{(c)}$. These properties were provided in Table 1: $E_{c\_lo}^{(t)} = 49$ MPa, $E_{c\_lo}^{(c)} = 40$ MPa, $E_{c\_up}^{(t)} = 320$ MPa, and $E_{c\_up}^{(c)} = 400$ MPa. As already commented to use both models there is no need for having any estimative of the lamina elastic moduli E_1, which is an advantage since the great variability of this property is a function of parameters as: fraction volume, epoxy type, fabric patterns, and fabrication process.

In Table 2 there are the results of application of the models presented in section "analytical models" to estimate the orthotropic elastic properties of the sandwich laminates specimens: E_1, E_2, G_{12}, and ν_{12}.

Analyzing the proposed model results written in Table 2, it is clear that the simplified analytical model produced good results if compared to the range of results of the complete analytical model (between lower and upper bounds). Nevertheless the complete analytical model could give a better contribution to the estimation of elastic mechanical properties of sandwich laminates which have a stiffer core.

Table 2. Orthotropic elastic properties of the sandwich laminates specimens.

Model	Mechanical properties			
	E_1 (GPa)	E_2 (GPa)	ν_{12}	G_{12} (GPa)
Simplified (SAM)	18.3	19.8	0.17	5.15
Complete (CAM) (lower bound)	18.3	19.8	0.17	5.16
Complete (CAM) (upper bound)	18.6	20.2	0.17	5.25

CONCLUSIONS

Two combined analytical/experimental models were proposed to estimate 4 orthotropic mechanical elastic properties of sandwich laminates. The analytical part uses a theory of anisotropic plates and the experimental one, strain gages to obtain the strains generated on a sandwich laminate specimen submitted to a pure bending load on a 4-point bending test. As a result, the 4 orthotropic mechanical elastic properties of sandwich laminates were estimated: the longitudinal and the transversal Young moduli, the shear modulus, and the Poisson's ratio. In short, it was proposed a relatively inexpensive and straightforward form of estimating 4 orthotropic mechanical elastic properties of sandwich laminates, which adds a little contribution to the development of this challenging area of composite materials.

AUTHOR'S CONTRIBUTION

Kenedi PP and Vignoli LL cowrote the main text. Duarte BT, Matos FCA e Dias HOT supportedthe experimental parts. All authors discussed the results and commented on the manuscript.

REFERENCES

Bledzki AK, Kessler A, Rikards R, Chate A (1999) Determination of elastic constants of glass/epoxy unidirectional laminates by the vibration testing of plates. Compos Sci Technol 59(13):2015-2024. doi: 10.1016/S0266-3538(99)00059-7

Crandall SH, Dahl NC, Ladner TJ (1978) An introduction to mechanics of solids. 2nd Edition [with SI units]. New York: McGraw-Hill International Editions.

Diab Group, Divinycell H (2016) Technical Data No: H Feb 2016 rev16 SI; [accessed 2016 Aug 30]. https://www.diabgroup.com

Hu Z, Karki R (2015) Prediction of mechanical properties of three-dimensional fabric composites reinforced by transversally isotropic carbon fibers. J Compos Mater 49(12):1513-1524. doi: 10.1177/0021998314535960

Kassapoglou C (2010) Design and analysis of composite structures with applications to aerospace structures. Hoboken: Wiley.

Kenedi PP, Vignoli LL, Duarte DT, Matos FCA, Dias HOT (2016) Bending response of sandwich laminates. Proceedings of the BCCM-3 – Brazilian Conference on Composite Materials; Gramado, Brazil.

Lamboul B, Bai G, Roche JM, Grail G (2013) Estimation of elastic properties and damage monitoring in 2D woven composites using Lamb waves. Proceedings of the 13th Int. Symposium on Nondestructive Characterization of Materials (NDCM-XIII); Le Mans, France.

Lecompte D, Smits A, Sol H, Vantomme J, Hemelrijck (2007) Mixed numerical-experimental technique for orthotropic parameter identification using biaxial tensile tests on cruciform specimens. Int J Solids Struct 44(5):1643-1656. doi: 10.1016/j.ijsolstr.2006.06.050

Lekhnitskii SG (1981) Theory of elasticity of an anisotropic body. Moscow: Mir Publishers.

Rikards R, Chate A, Gailis G (2001) Identification of elastic properties of laminates based on experiment design. Int J Solids Struct 38 (30-31):5097-5155. doi: 10.1016/S0020-7683(00)00349-8

Rikards R, Chate A, Kessler A, Steinchen W, Bledzki AK (1999) Method for identification of elastic properties of laminates based on experiment design. Composites Part B 30(3):279-289. doi: 10.1016/S1359-8368(98)00059-6

Staab GH (2015) Laminar composites. 2nd Edition. Oxford: Elsevier.

Tamilarasan U, Karunamoorthy L, Palanikumar K (2015) Mechanical properties evaluation of the carbon fibre reinforced aluminium sandwich composites. Mater Res 18(5):1029-1037. doi: 10.1590/1516-1439.017215

Vasiliev VV, Morozov EV (2001) Mechanics and analysis of composite materials. New York: Elsevier.

Evaluation of Filler Effects on the Dynamic Mechanical Behavior of HTPB-Elastomer Used as Binder in Exemplary Composite Formulations

Mauricio Ferrapontoff Lemos[1], Günter Mussbach[2], Manfred August Bohn[3]

ABSTRACT: This study aims to evaluate the dynamic mechanical properties of a binder, used in composite propellants, with increasing amount of different solid fillers. Dynamic Mechanical Analysis (DMA) is performed to determine the region of glass-rubber transition temperature (T_g). Further parameters were obtained from the description of the loss factor curves (tanδ) with models based on exponentially modified Gauss (EMG) distribution function. Polyurethane binders employing the polyol HTPB and diisocyanate were filled with 20, 40, and 60 mass-% of ammonium perchlorate (AP), aluminum (Al) or cyclotrimethylene trinitramine (RDX) particles, using fine and coarse mean sizes. A special turning device manufactured by Fraunhofer ICT was installed inside the curing oven in order to avoid sedimentation during curing. The cured composites were characterized by DMA in torsion mode from -100 °C to 70 °C. The good homogeneous distribution of fillers in the samples was evaluated by computational X-ray micro-tomography. The data on tanδ showed a complex structure describable at first with 2 overlapping peaks. The peak at lower temperatures is due to mobility of unrestricted binder parts (T_g^{unr}) and the second broader peak at high temperature is due to mobility of restricted binder parts (T_g^{res}). Results of tanδ, calculated areas of EMG curves and apparent activation energies of the distinct relaxation processes indicate that Al particles have higher interaction with HTPB-based polyurethane than AP and RDX. Probably Al is bonded to the binder network via isocyanate coupling with OH groups present in the surface. Ammonium perchlorate and cyclotrimethylene trinitramine particle sizes and its morphology influenced the viscoelastic properties. AP and RDX cause more changes in intensity of first peak of loss factor than Al. Increasing amount of these both particles enhances storage (G') and loss (G'') shear moduli. As a whole tanδ intensity is lowered in the main peak and Al showed an intensity increase in the second apparent peak.

KEYWORDS: DMA, Ammonium perchlorate, Aluminum, Cyclotrimethylene trinitramine, HTPB loss factor modelling.

INTRODUCTION

A composite solid propellant (CSP) is an energetic material characterized by the release of high amount of energy during burning, generating hot gases and low molar mass molecules (Davenas 1993; Kubota 2007; Cerri *et al.* 2009). The CSP consists of a polymer matrix (binder) containing a high amount of solid particles. Polybutadiene-based polyols such as hydroxyl-terminated polybutadiene (HTPB) are today the main pre-polymers used for CSP, which are cross-linked with polyisocyanates to give a polyurethane elastomer (Cerri *et al.* 2009; Hufferd and Fitzgerald 1976). The main solid particles used as fillers are oxidizers, such as nitrates and perchlorates, which supply oxygen for the combustion reactions. Another important filler is the aluminum powder, which acts as fuel during the combustion by its strong exothermic oxidation reaction and cyclotrimethylene trinitramine (RDX) particle that can be used in some specific HTPB-based formulations as well as in plastic bonded explosives (PBX) (Davenas 1993; Kubota 2007; Cerri *et al.* 2009; Hufferd and Fitzgerald 1976; Lemos and Bohn 2016). The binder main functions are to give dimensional stability and viscoelastic properties, assuring structural integrity to CSP and minimizing or even avoiding combustion instabilities (Davenas 1993; Kubota 2007; Hufferd and Fitzgerald 1976).

The mechanical properties of composites with elastomeric binders depend strongly on the surrounding temperatures and on the deformation frequency of the applied mechanical load. In terms of low temperature properties, the elastomer

1.Instituto de Pesquisas da Marinha – Grupo de Tecnologia de Materiais – Rio de Janeiro/RJ – Brazil. 2.Bayern-Chemie GmbH – Aschau am Inn – Germany. 3.Fraunhofer Institute for Chemical Technology – Pfinztal – Germany.

Author for correspondence: Mauricio Ferrapontoff Lemos | Instituto de Pesquisas da Marinha – Grupo de Tecnologia de Materiais | Rua Ipiru, 2 – Cacuia | CEP: 21.931-095 – Rio de Janeiro/RJ – Brazil | Email: mauricio.lemos@marinha.mil.br must be able to purvey a very low glass-rubber transition temperature (T_g) for safe operation (Davenas 1993; Kubota 2007; Cerri *et al.* 2009; Hufferd and Fitzgerald 1976; Lemos and Bohn 2016; Fluke 1969; Bohn 2016; Bohn *et al.* 2014; Cerri and Bohn 2010; Cerri *et al.* 2013). The loss factor (tanδ) as function of temperature, obtained in DMA, defines the region of glass-rubber transition. The glass-rubber transition temperature T_g is defined as temperature at the maximum in loss factor curve. However, the tanδ curve of CSP consists of several sub-transitions referring to different fractions of the elastomer binder with different mobility (Cerri *et al.* 2009; Bohn 2016; Bohn *et al.* 2014; Cerri and Bohn 2010; Cerri *et al.* 2013; Azoug *et al.* 2015). A modeling of the loss factor curve with suitable distribution functions is needed to separate these effects in tanδ and to characterize them quantitatively in terms of the maximum peak temperatures and by other parameters as areas (intensities), peak widths, and peak skewness (Bohn 2016; Bohn *et al.* 2014; Cerri and Bohn 2010; Cerri *et al.* 2013; Azoug *et al.* 2015).

A CSP grain is subjected to several dynamic mechanical load stresses during transport, ignition and even during storage in a broad temperature range. In several investigations the viscoelastic response is reported under dynamic loads at frequencies from 0.1 up to 110 Hz from −40 °C to 50 °C (Hufferd and Fitzgerald 1976). Deformation frequency of applied stress or strain imparts changes in mechanical response of a viscoelastic binder, especially in the T_g region (Cerri *et al.* 2009; Hufferd and Fitzgerald 1976; Bohn *et al.* 2014; Cerri *et al.* 2013; Ornaghi Jr *et al.* 2010). Previous study showed an undesirable δ intensity increase and a shift of the peak to higher temperatures with increasing deformation rate in CSP, caused by molecular frictions, thus indicating the possibility of high energy dissipation and therefore a heating up of the material (Bohn 2016; Bohn *et al.* 2014; Cerri *et al.* 2013). These effects have severe implications for these materials already at moderate low temperatures. It is important to evaluate to which extend each type of solid particle in a CSP formulation affects these relaxations and dissipative effects, since each particle has different interaction possibilities with the binder. This paper aims at a first step of investigation to find out how the individual fillers interact with the HTPB-based matrix. Experimental investigation is performed in order to evaluate the influence of the type of fillers on the shape and intensity of loss factor and on its shift with deformation frequency.

METHODOLOGY

SAMPLE PREPARATION

The composite samples consisted of a polyurethane binder filled with 20, 40 or 60 mass-% of different filler types. The polyurethane was obtained by the reaction of the polyol Hydroxyl-Terminated Polybutadiene type HTPB R45 HTLO (Sartomer, Polybd, Oakland, USA) with the curing agent Isophorone Diisocyanate – IPDI (Evonik, Marl, Germany) in a ratio of NCO to OH groups (R_{eq}) adjusted to 0.85. Plasticizer Dioctyl Adipate – DOA (BASF, Ludwigshafen, Germany) was added in 5 mass-% relative to total binder content. As fillers were used the ammonium perchlorate (AP), with particle size d(0.5) equal to 42.83 μm (fine) and 202.28 μm (coarse); RDX from Dyno/Norway with 60.86 μm (fine) and 233.26 μm (coarse); and Al (Toyal, USA) with d(0.5) of 20.69 μm (type X81, fine) and 130.25 μm (coarse). The composite samples were manufactured at Fraunhofer ICT in a planetary centrifugal vacuum mixer (Thinky Corporation, Tokyo, Japan). Operational conditions were 1,600 rpm rotation speed and 3 kPa vacuum during few minutes of mixing time. The temperature in the mix was kept below 40 °C. The formulations were produced in small batches and poured into a cylindrical glass recipient, sealed and cured in air at 60 °C for 96 h. The binder was protected by the addition of the antioxidant Vulkanox^{TM} BKF.

Just after mixing, the particles were homogeneously distributed all over the sample volume. However, the particles tended to sediment in the liquid binder mix in early stages of the curing process, when the viscosity was not yet high. This problem was even more critical for samples with lower amounts of solid loading, such as 20 and 40 mass-% of fillers. Thus, in order to keep the fillers homogeneously distributed during all stages of curing process, a special turning device was manufactured at Fraunhofer ICT and installed inside the curing oven (Fig. 1).

The turning machine was optimized to work safely with explosive materials. It provides a rectangular frame for adapting samples within a standard laboratory circulation oven UFE500 (Memmert, Germany). The frame was rotated not permanently to avoid distribution gradients of the filler particles due to inertia forces. It was rotated stepwise by 180 degrees, controlled by encapsulated, low-voltage magnetic switches. A belt drive in combination with a screw wheel transmission was powered by a brushless AC-motor. An additional transmission was necessary for suppressing rattle-effects. The switches and the motor were connected to a programmable electronic circuit, which allows controlling the rotation speed and the holding time. The circuit was equipped with a counter to monitor the number of rotation cycles, useful for instance in case of failures in power network. The turning time is typically about 4 s and the holding time was set to 60 s. After curing, the glasses were broken and the DMA samples cut from the middle of the cured composite.

(a)
(b)
(c)

Figure 1. Design of the turning machine, schematic sketches (a). Installation and operation in curing oven shown in photographs (b and c) at Fraunhofer ICT. The device was used successfully for the curing of elastomers with low degree of filling of solid particles to avoid sedimentation.

SAMPLE CHARACTERIZATION

A microcomputer tomography Micro-CT in-vivo Skyscan 1076 (Bruker, Germany) was used to evaluate the distribution of the fillers in the samples taken for DMA measurements.

Dynamic mechanical analysis measurements in torsion mode were carried out with a DMA ARES™ (Advanced Rheometric Expansion System) manufactured by the former Rheometric Group of the company Rheometric Scientific, Inc. (this group belongs now to Waters, Inc., BU TA Instruments, New Castle, NJ, USA). The following parameters were determined: storage shear modulus G', loss shear modulus G", loss factor $\tan\delta = G''/G'$ as well as the phase angle δ and the torque. Measurements were performed in torsion mode from −100 °C to +70 °C, with heating rate of 1 °C/min and soak time of 28 s. At −100 °C a pre-strain sweep test was performed to determine the value of strain control, in order to stay in the linear viscoelastic region during the measurements. At each temperature step, the specimens were measured at 4 sinusoidal deformation frequencies (0.1; 1; 10 and 30 Hz). The geometrical dimensions of the rectangular samples were about 40 mm length, 10 mm width and 5 mm thickness. In the cases where the material was too soft (as with pure binder and binder with 20 and 40 mass-% of filler), the signal scattered and a second strain sweep was necessary at −40 °C to determine a higher second strain control to be used from about −40 °C up to +70 °C.

Scanning electron microscopy (SEM) images were obtained by Supra 55VP (Zeiss, Germany).

RESULTS AND DISCUSSIONS

In Fig. 2, a sample with 20 mass-% of AP was used to show how the particles are homogeneously distributed in the elastomers after curing using the turning machine. The Micro-CT X-ray scan, shown in Fig. 2b, demonstrates the homogeneity of the distribution of the AP particle.

(a) (b)

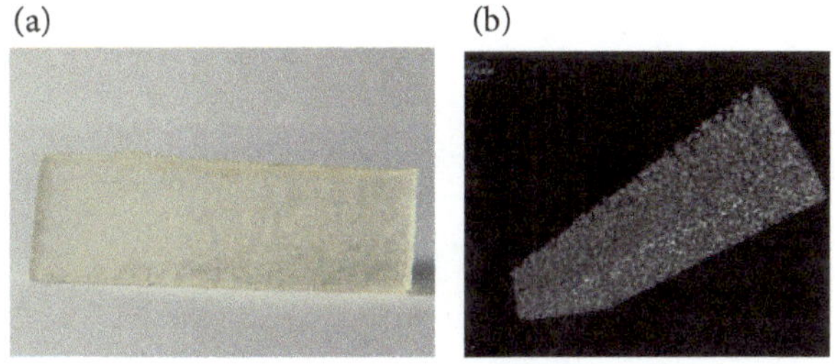

Figure 2. Images from DMA sample with 20 mass-% of AP. (a) Normal optical photograph and (b) 3-D image taken from a Micro-CT X-ray scan, showing AP particles in gray.

At first and to model the loss factor curves, a suitable base line correction (BLC) was applied to separate the part standing for only dissipation of applied deformation energy (Bohn 2016; Bohn *et al.* 2014; Cerri and Bohn 2010) from the part containing information on the molecular rearrangement processes in going from energy elastic to entropy elastic state. Further, the modeling allows then to separate the binder contributions to the loss factor curve in terms of differences in mobility of binder molecules and to quantify the influence of fillers on the loss factor. In Fig. 3, the base line corrected loss factor curves at all deformation frequencies are shown for the samples with increasing amount of AP, RDX and Al. Each sample has 2 apparent peaks, with tailing towards higher temperatures. Table 1 shows the values of T_g for each tanδ peak for each frequency. The values of the differences between the T_g of composite and of pure binder are taken as reference and are also presented. Samples with 60 mass-% of RDX had no defined apparent second peak in loss factor curves at deformation frequency higher than 0.1 Hz. This is a general feature also shown with HMX as filler (Bohn *et al.* 2014).

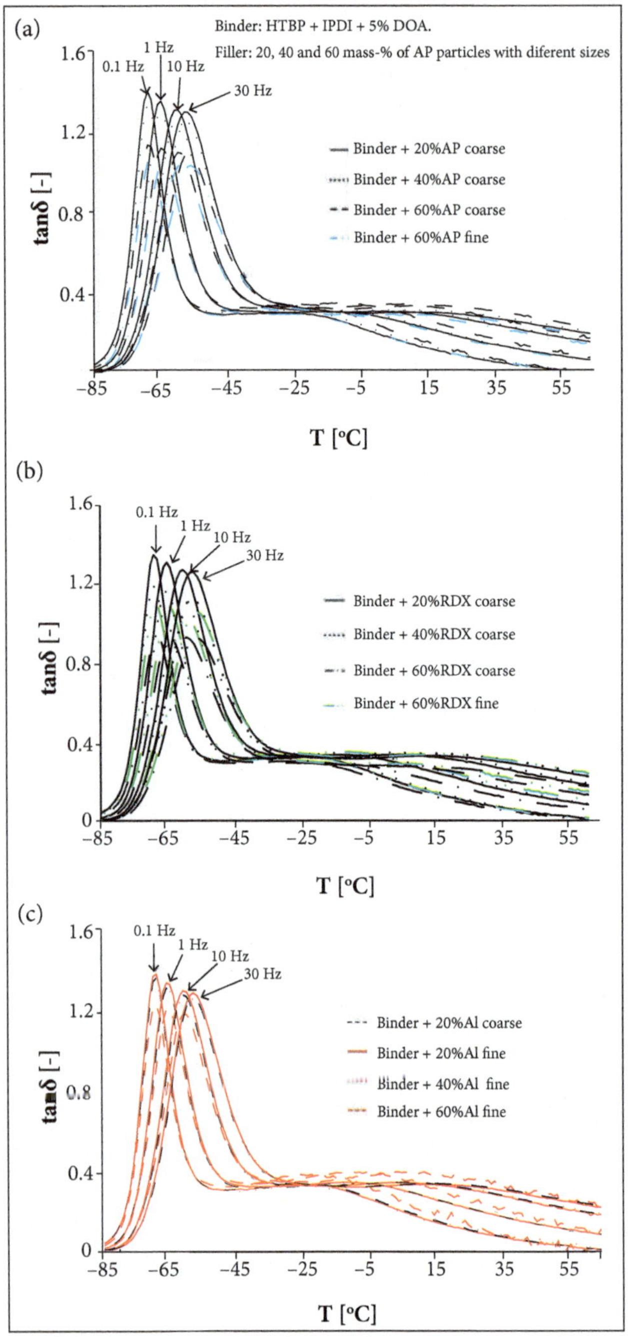

Figure 3. Deformation frequency dependence of the loss factor curve of composites samples with 20, 40, and 60 mass-% of AP (a), RDX (b) and Al (c) fine and coarse particles.

The first peak at lower temperatures represents the "mobility unrestricted glass-rubber transition" with the temperature T_g^{unr}, and it is attributed to the main transition of the soft segment regions of the HTPB chains in the temperatures range of −80 to −60 °C. The second broader peak originates from the transition regions of the hard segments of the polyurethane network (cross-link sites) and in part of the binder molecules (soft and/or hard segments), which have restricted mobility, for example, by filler-binder interaction. Therefore, this peak is called "restricted" and the associated maximum temperature is named T_g^{res} (Cerri *et al.* 2009; Bohn 2016; Cerri *et al.* 2013; Azoug *et al.* 2015). Each transition region corresponds to molecular rearrangements of fractions of the binder with different mobility. T_g^{unr} increased slightly at higher temperatures by increasing the amount of AP and RDX particles but this decreased more considerably the tanδ intensity of the first peak especially at higher frequencies of applied strain. In Fig. 4, the curves are shown comparing loss factor curves at frequencies of 0.1 and 30 Hz. The curves of the samples with 60 mass-% of AP, RDX, and Al (fine) are indicated. Despite having less effect on the first peak, the Al addition imparts more intensity to tanδ of the second peak, and also shifted it to higher temperatures (see T_g^{res} values of the lines of 60 mass-%-filled samples in Table 1). A distinguishable

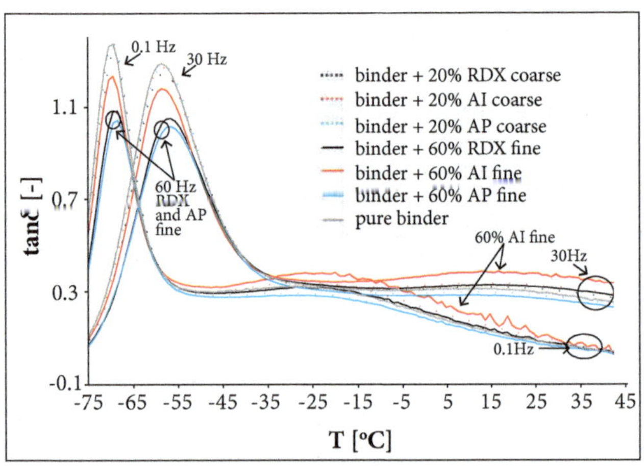

Figure 4. Loss factor curves of pure binder and of composites samples with 20 and 60 mass-% of AP, RDX, and Al coarse and fine particles. DMA in torsion mode at frequencies of 0.1 and 30 Hz.

Table 1. Values of T_g^{unr} and T_g^{res} for the first and second peaks, respectively, of tanδ curve, at different deformation frequencies.

Sample content mass (%)	Values of T_g^{unr} (°C) for each test frequency				Difference to reference (°C)				Ea_f^* [kJ/mol]	ln (f_0) [Hz]
	0.1 Hz	1 Hz	10 Hz	30 Hz	0.1 Hz	1 Hz	10 Hz	30 Hz		
Binder alone	−69.42	−65.58	−60.72	−57.95	Reference				180.7	45.43
20%AP coarse	−68.86	−65.24	−60.28	−57.51	0.56	0.34	0.44	0.44	182.6	45.79
40%AP coarse	−68.78	−65.04	−60.11	−57.27	0.64	0.54	0.61	0.68	180.9	45.34
60%AP coarse	−68.63	−64.76	−59.62	−56.74	0.79	0.82	1.10	1.21	175.4	43.91
60%AP fine	−68.29	−64.38	−59.06	−56.16	1.13	1.20	1.66	1.79	172.3	43.04
20%RDX coarse	−68.95	−65.02	−60.09	−57.37	0.47	0.56	0.63	0.58	179.8	45.07
40%RDX coarse	−68.97	−65.01	−59.91	−57.22	0.45	0.57	0.81	0.73	176.7	44.30
60%RDX coarse	−69.58	−64.28	−59.02	−55.94	−0.16	1.30	1.70	2.01	154.9	38.78
60%RDX fine	−69.44	−64.56	−59.19	−56.19	−0.02	1.02	1.53	1.76	158.6	39.71
20%Al coarse	−68.90	−65.65	−60.76	−57.91	0.52	−0.07	−0.04	0.04	187.3	47.05
20%Al fine	−69.54	−65.71	−60.81	−57.95	−0.12	−0.13	−0.09	0	179.1	45.00
40%Al fine	−69.23	−65.69	−60.71	−57.77	0.19	−0.11	0.01	0.18	180.6	45.34
60%Al fine	−69.01	−65.56	−60.60	−57.80	0.41	0.02	0.12	0.15	184.0	46.18
Sample content mass (%)	Values of T_g^{res} (°C) for each test frequency				Difference to reference (°C)				Ea_f^* [kJ/mol]	ln (f_0) [Hz]
	0.1 Hz	1 Hz	10 Hz	30 Hz	0.1 Hz	1 Hz	10 Hz	30 Hz		
Binder alone	−28.04	−13.65	2.52	10.87	Reference				84.7	17.06
20%AP coarse	−29.42	−15.63	2.00	11.03	−1.38	−1.98	−0.52	0.16	80.7	16.33
40%AP coarse	−28.33	−14.14	2.94	12.46	−0.29	−0.49	0.42	1.59	81.3	16.37
60%AP coarse	−24.92	−11.96	4.22	14.98	3.12	1.69	1.70	4.11	85.4	17.03
60%AP fine	−29.25	−16.82	0.23	8.49	−1.21	−3.17	−2.29	−2.38	85.2	17.31
20%RDX coarse	−25.36	−12.23	4.57	14.24	2.68	1.42	2.05	3.37	85.1	17.00
40%RDX coarse	−24.34	−10.94	6.36	15.75	3.70	2.71	3.84	4.88	84.6	16.82
60%RDX coarse	−25.38	Not well defined			2.66	Not well defined				
60%RDX fine	−24.05	−11.76	4.96	15.47	3.99	1.89	2.44	4.60	86.1	17.14
20%Al coarse	−26.01	−11.7	6.13	15.09	2.03	1.95	3.61	4.22	81.7	16.29
20%Al fine	−25.88	−12.44	3.76	13.54	2.16	1.21	1.24	2.67	85.5	17.10
40%Al fine	−25.91	−11.74	4.35	11.32	2.13	1.91	1.83	0.45	88.7	17.74
60%Al fine	−23.25	−10.33	7.38	15.8	4.79	3.32	4.86	4.93	86.5	17.14

change in the first peak with the addition of AP and RDX in comparison to Al might be related with volumetric effects. The densities of AP and RDX are 1.95 g/cm^3 and 1.82 g/cm^3, respectively, whereas Al has 2.7 g/cm^3. So the use of an equal mass results in a higher occupied volume by the particles in the composites with AP and RDX, imparting more geometrical hindrance to chain mobility than by the Al. On the other hand, Al filled samples present a higher intensity on the second peak, which might be explained by a higher filler-binder interactions.

The molecular rearrangements during glass-rubber transition enhance the thermo-mechanical energy loss inside the polymer. Therefore, loss shear modulus increases. During glass-rubber transition, when the molecules are stimulated to pass a threshold of energy, long-range chain mobility takes place along the segments and the loss factor is at its maximum. By increasing the deformation frequency in DMA, the chains have less time to move and rearrange in order to accommodate the mechanical energy. So when heating up the samples from very low temperatures, the energy elastic state is retained longer, up to higher temperatures. Often this effect is also named strain rate hardening.

The deformation frequency dependence of the thermo-mechanically activated glass-rubber transition can be expressed by Arrhenius parameterization. With the use of an Arrhenius type equation (Eq. 1), an apparent activation energy Ea_f^* of the distinct relaxation process can be estimated.

$$f = f_0 \cdot \exp\left(-\frac{Ea_f}{R \cdot T_g(f)}\right) \quad (1)$$

where: f is the applied deformation frequency [Hz]; f_0 is the apparent pre-exponential factor [Hz]; Ea_f^* is the apparent activation energy [kJ/mol]; R is the general gas constant [8.31441 J/K·mol]; T_g is the glass-rubber transition temperature [K] as function of deformation frequency f.

In Table 1 the values of Ea_f^* and f_0 are listed. In Fig. 3 the shift of the loss factor curve to higher temperatures can be seen with increasing deformation frequency for the samples with increasing amount of AP, RDX, and Al.

The Ea_f^* values of the distinct relaxation process change slightly with filler addition. With AP addition the Ea_f^* of first peak (T_g^{unr}) increases initially (20 mass-%), but then decreases with increasing amount of filler. RDX addition, however, decreases continuously the Ea_f^* of T_g^{unr}. An opposite trend is found with the values of activation energies of the T_g^{res}. The addition of Al particles imparted overall slightly higher values of Ea_f^{**} among all, which is an indication of higher interaction with the polymer matrix.

In Fig. 5, the elastic (storage) and loss shear modulus (G' and G") are presented for all the samples tested at deformation frequency of 10 Hz. Before the transition, G' and G" are very similar for the composites. Only G' of samples with 60 mass-% is significantly higher. Increasing amount of AP and RDX particles enhances G' and G", although with AP this happens mainly after T_g^{unr}. Al produces this effect more pronounced at lower temperatures. After the first transition the differences increase, and it is notable that the increase in solid filler content in the composites increases both G' and G". Storage modulus increase is related with a reinforcement of the material, whereas loss modulus increase with increasing solid fillers in composites is known to be caused by an enhanced internal friction in dynamic mechanical loading experiments (Bohn *et al.* 2014; Ornaghi Jr *et al.* 2010). Regarding particle sizes, the use of fine AP and fine RDX incremented G' more than G".

The quantification of the effects (interaction and amount) of each filler on the loss factor curve is possible with the use of a model based on exponentially modified Gauss (EMG) distribution functions. The EMG is a convolution between Gauss distribution and exponential decay function. Previous studies have described the model in details (Bohn 2016; Bohn *et al.* 2014; Cerri and Bohn 2010; Azoug *et al.* 2015). The application

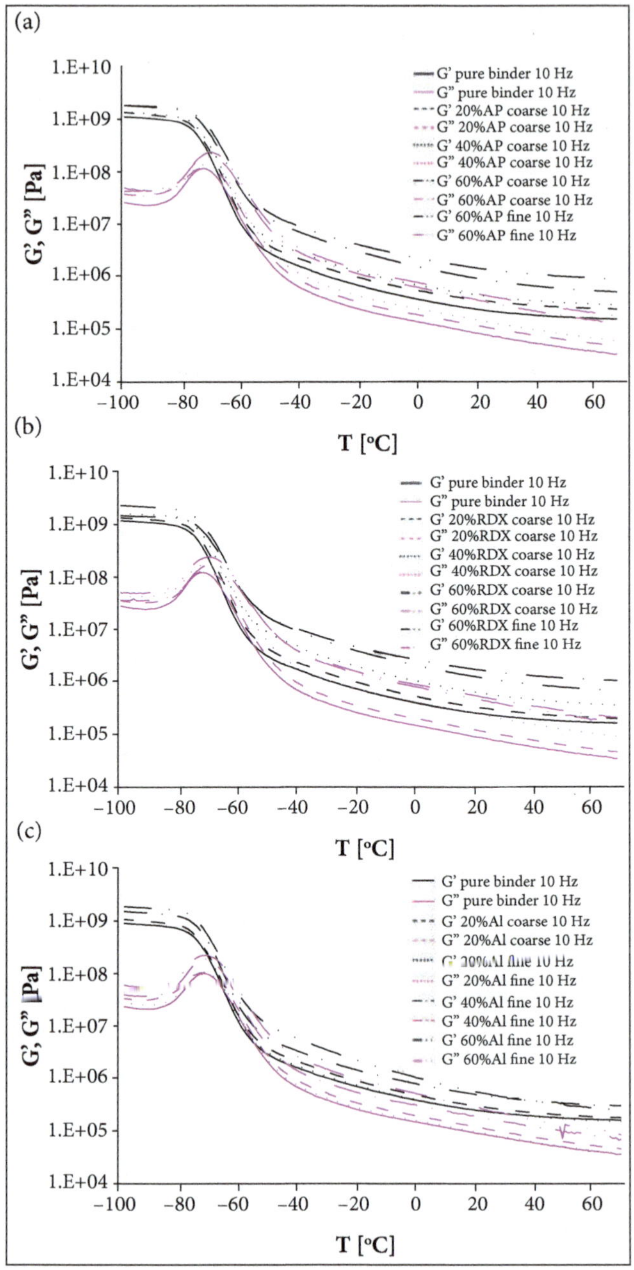

Figure 5. G' and G" curves of all samples measured at 10 Hz deformation frequency of applied strain. Samples consisting of binder with increasing amount (20, 40, and 60 mass-%) of (a) AP, (b) RDX, and (c) Al.

of EMGs provides parameters of the sub-transition regions of the loss factor curve: the peak areas (A_i); the half peak width at half height of the Gaussian part of corresponding EMG_i (w_i); the temperature at peak maximum of Gaussian part of the EMG_i (Tc_i) and the relaxation parameter τ_i in the exponential part of the EMG_i (herein named To). Figure 6 shows 3 EMG functions to describe each of the tanδ data of the pure binder (a) and of the composite filled with 60 mass-% of Al particles (b). This was done for all samples at all deformation frequencies. Due to limited space only the parameters obtained for the samples filled with 60 mass-% particles are presented in Table 2.

Two parameters are considered to be more important for composites and are discussed in more detail. One is the parameter A_i, which is the area of the loss factor curve. When the intensity of the glass-rubber transition expressed by A is small, the reasons can be: (1) more hindrance in mobility for the pre-polymer or (2) a more elastic behavior of the polymer means a more dominant elastic (storage) shear modulus. The other parameter of importance is the peak temperature Tc_i.

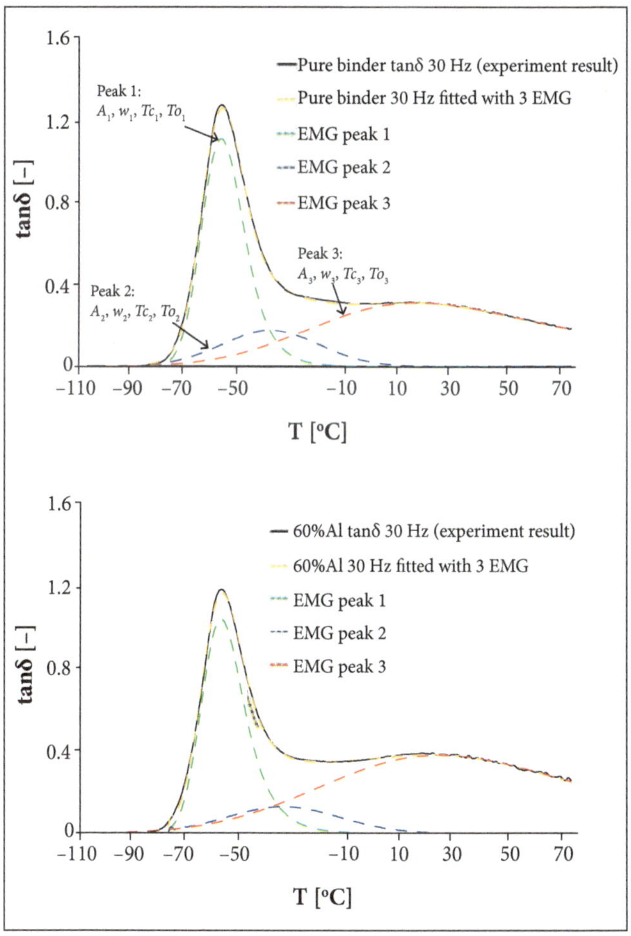

Figure 6. Description of loss factor with 3 EMG functions of the pure binder (a) and of the composite with 60 mass-% of Al (b). This resolves 3 sub-peaks or 3 binder fractions with different mobility.

Regarding the sum of the peak areas (ΣA_i), filling the binder with AP coarse particles had nearly no effect. Fine AP however had caused an overall depreciation of A_i parameters. This means that the polymer gained more in the elastic than in the loss shear modulus — one can say the sample is more elastic and less viscous. RDX has a reversed effect: the filling with coarse particle size had decreased the sum of A_i parameters. SEM analysis of AP fine and RDX coarse particles (Fig. 7b and 7c) showed that both are less round (means have more irregular geometry and edges), whereas coarse AP and fine RDX (Fig. 7a and 7d) had an apparent geometry closer to a sphere or are rounded particles. This is an indicative that the sum of A_i parameters, meaning the overall area of the loss factor curve, seems more affected by the AP and RDX particle shapes than by their size.

A_3 is the area of the peak attributed to the transition of parts of the binder with more restriction to chain mobility, i.e. in filler-binder interfaces for example. It is evident from the results that the addition of Al particles had a stronger effect on increasing the value of this parameter than the other particles, and this effect is even more pronounced when deformation frequency is increased.

On the surface of aluminum particles there is the possibility of forming Al hydroxide due to contact of the metallic Al with air (oxygen) and humidity. It was already discussed in the literature that the OH on Al may capture one end of isocyanate molecules (Cerri *et al.* 2009; Cerri *et al.* 2013), leaving less isocyanate groups to react with pre-polymer and more not bonded polyol chain ends are available. This may change the R_{eq} in the binder near the Al particles. According to literature, the second apparent loss factor peak in HTPB-based solid propellants, consisting mainly by the EMG-3 (peak 3), can be caused by 2 effects:

1. From the flow of free polymer chains in the polymer network with a reptation mechanism. If more polymer or polyol chains are available to move near the particle due to isocyanate "capture" near the Al surface, one could expect an increase in A_3 value (Azoug *et al.* 2015).
2. From a more intense intermolecular interaction between Al and HTPB than between the polyol and AP. Due to stronger interactions, the Al exert more hindrance on the mobility of the HTPB chains in the polymer shell around the particles (Nardai and Bohn 2015). As a first consequence the glass-rubber transition temperature of

Table 2. Values of EMG parameters of samples with 60 mass-% coarse and fine particles in comparison with the pure binder.

	Pure binder				60 mass-% AP coarse				60 mass-% AP fine			
f (Hz)	0.1	1	10	30	0.1	1	10	30	0.1	1	10	30
A_1	13.3	16.7	20.3	22.5	10.8	13.5	16.2	18.0	11.0	13.9	17.0	22.5
A_2	2.6	5.6	7.2	7.8	2.8	4.8	8.2	8.3	3.7	7.9	7.6	6.7
A_3	24.8	29.0	39.0	42.3	25.2	30.9	38.7	45.8	19.9	21.3	31.3	34.1
$\sum A_i$	40.8	51.3	66.5	72.7	38.8	49.3	63.2	72.2	34.5	43.1	56.0	63.3
w_1	4.2	5.2	5.6	6.3	4.3	5.2	5.6	6.3	4.5	5.5	6.1	6.7
w_2	2.0	2.7	16.4	17.5	2.1	2.8	17.0	17.5	2.3	3.1	16.5	6.6
w_3	25.3	26.3	26.7	28.5	26.8	27.7	25.5	28.1	24.7	25.6	23.9	30.6
Tc_1	−70	−66	−65	−63	−69	−66	−64	−61	−69	−65	−63	−65
Tc_2	−63	−58	−46	−43	−62	−57	−46	−42	−61	−56	−43	−57
Tc_3	−43	−33	−23	−20	−38	−32	−20	−17	−39	−28	−19	−12
To_1	0.9	1.0	5.6	6.4	0.8	1.0	5.5	6.1	0.9	1.0	5.8	23.8
To_2	9.6	17.1	2.7	3.0	11.4	16.6	3.2	3.0	15.4	26.8	3.3	1.0
To_3	22.3	37.6	60.1	70.8	17.6	35.1	59.3	75.1	20.9	34.5	57.9	66.6
	60 mass-% RDX coarse				60 mass-% RDX fine				60 mass-% Al fine			
f (Hz)	0.1	1	10	30	0.1	1	10	30	0.1	1	10	30
A_1	9.7	12.1	16.2	19.2	10.8	13.7	17.4	19.4	13.4	14.9	17.7	21.2
A_2	3.3	6.3	7.3	6.1	3.0	5.4	10.1	13.4	2.6	5.2	7.5	6.5
A_3	21.3	22.9	28.6	33.5	23.9	32.5	35.6	37.4	25.4	35.6	48.2	49.8
$\sum A_i$	34.3	41.2	52.2	58.8	37.7	51.6	63.1	70.2	41.4	55.7	73.4	77.4
w_1	4.5	5.5	6.0	6.7	4.3	5.4	6.8	7.6	4.2	5.2	5.1	6.0
w_2	2.2	3.1	17.4	17.3	2.2	3.1	4.1	4.8	1.7	2.9	21.1	20.0
w_3	25.9	27.0	23.8	27.4	26.2	26.7	31.6	34.6	25.8	27.0	30.1	34.2
Tc_1	−69	−65	−64	−62	−69	−65	−61	−58	−70	−66	−65	−63
Tc_2	−61	−57	−41	−34	−62	−57	−50	−47	−63	−58	−49	−39
Tc_3	−40	−31	−18	−14	−39	−33	−16	−5	−41	−30	−20	−11
To_1	0.8	1.0	6.8	8.2	0.8	1.0	1.4	1.5	0.8	1.0	6.6	7.5
To_2	15.5	22.4	3.3	3.2	11.6	17.9	29.3	35.5	9.9	16.0	3.0	3.1
To_3	19.7	32.8	54.4	65.3	19.3	43.3	54.6	60.1	21.0	35.8	59.3	56.6

such binder fractions is shifted to higher temperature values compared to not mobility restricted HTPB chains (see Tc_3 parameters). Secondly, because of the increased binder part in this fraction, it creates a relative intense glass-rubber transition, since the mobility restriction imparted by Al is not high enough to avoid the transition to take place, meaning that no rigidity for the polymer chains is caused.

On the other hand, this effect of reducing mobility can be also the case, if one has increasing cross-linking between the

Figure 7. SEM images (magnification of 50X) of the particles loaded in the CSP in this study. (a) Coarse AP, rounded; (b) Fine AP, non-rounded (AP) particles; (c) Coarse RDX, non-rounded; (d) Fine RDX, rounded; (e) Coarse Al, rounded; (f) Fine Al, rounded. Scale of 100 μm.

polymer chains in the neighborhood of particles, which form then a more rigid amorphous region and the transition intensity is reduced, but the storage modulus increases. Another effect is the stronger chemical bonding of Al particles to the network, when isocyanate attaches to the Al surface via its OH groups. This enlarges the polymer shell around Al particles and the intensity of A_3 is increased.

Regarding the Tc parameters, Tc_1 is nearly not affected, independent of particle size or type. Tc_3 is increased, and the increase is more pronounced at higher frequencies, especially with fine particle sizes. This is an indication that an increase in solid filler tend to enhance chain mobility restrictions, affecting the T_g^{res}, corroborating the results shown in Table 1.

Results of tanδ intensity as calculated areas of EMGs fitting curves and results of apparent activation energies of the distinct relaxation process during T_g^{res} with increasing frequency indicates that Al particles have higher interaction possibilities with HTPB-based polyurethane binder than AP and RDX, when no bonding agents are used. Probably Al is bonded to the binder network via isocyanate coupling, because on Al surfaces OH groups are always present. AP and RDX particle sizes and their shape or their particle morphology influence the viscoelastic properties. Increasing amount of AP and RDX particles enhances storage (G') and loss (G") shear moduli, but AP causes this effect mainly during and after T_g^{unr}. Al produces this effect more pronounced at lower temperatures in the range of the second loss factor peak.

CONCLUSIONS

AP and RDX impart more changes in tanδ intensity of the first loss factor peak than Al particles. The main glass-rubber transition temperature (T_g^{unr}) is nearly not affected by the addition of particles.

AUTHOR'S CONTRIBUTION

Lemos MF and Mussbach G performed the experiments; Lemos MF and Bohn MA conceived the ideas and co-wrote the main text. All authors discussed the results and commented on the manuscript.

REFERENCES

Azoug A, Nevière R, Constantinescu A (2015) Molecular origin of the influence of the temperature on the loss factor of solid propellant. Propellants Explos Pyrotech 40(4):469-478. doi: 10.1002/prep.201400060

Bohn MA (2016) Influences on the loss factor of elastomer binder for composite materials and its modelling. Proceedings of the 3rd Brazilian Conference on Composite Materials; Gramado, Brazil.

Bohn MA, Mußbach G, Cerri S (2014) Modeling of loss factors of elastomer binders of high explosive charges and composite rocket propellants to separate binder fractions with different molecular mobility used to follow aging. Proceedings of 10th International Symposium on Special Topics in Chemical Propulsion & Energetic Materials; Poitiers, France.

Cerri S, Bohn MA (2010) Separation of molecular mobility ranges in loss factor curves by modeling with exponentially modified gauss distributions. Proceedings of the 41st International Annual Conference of ICT; Karlsruhe, Germany.

Cerri S, Bohn MA, Menke K, Galfetti L (2009) Ageing behavior of HTPB based rocket propellant formulations. Cent Eur J Energ Mater 6(2):149-165.

Cerri S, Bohn MA, Menke K, Galfetti L (2013) Ageing of HTPB/AL/AP rocket propellant formulations investigated by DMA measurements. Propellants Explos Pyrotech 38(2):190-198. doi: 10.1002/prep.201200186

Davenas A (1993) Solid rocket propulsion technology. Oxford; New York: Pergamon Press.

Fluke GA (1969) Composite solid propellant processing techniques. In: Boyars C, Klager K, editors. Propellants manufacture, hazards, and testing. Vol. 88. Washington: American Chemical Society. p. 165-187.

Hufferd WL, Fitzgerald JE (1976) Development of a solid propellant viscoelastic dynamic model. Atlanta: J. E. Fitzgerald & Associates.

Kubota N (2007) Propellants and explosives. Weiheim: Wiley-VCH Verlag GmbH & Co.

Lemos MF, Bohn MA (2016) Evaluation of the effect of plasticizers on the DMA loss factor, the thermal and mechanical properties of Desmophen® 2200-based elastomers used for composite propellants. Proceedings of the 18th International Seminar on New Trends in Research of Energetic Materials; Pardubice, Czech Republic.

Nardai MM, Bohn MA (2015) Cohesion properties in PBX and composite propellants – computational results and experimental aspects. Proceedings of the 46th International Annual Conference of ICT; Karlsruhe, Germany.

Ornaghi Jr HL, Bolner AS, Fiorio R, Zattera AJ, Amico SC (2010) Mechanical and dynamic mechanical analysis of hybrid composites molded by resin transfer molding. J Appl Polymer Sci 118(2):887-896. doi: 10.1002/app.32388

16

Aerodynamics of Harmonically Oscillating Aerofoil at Low Reynolds Number

Aiman Hakim Abdul Rahman[1], Nik Ahmad Ridhwan Nik Mohd[1], Tholudin Mat Lazim[1], Shuhaimi Mansor[1]

ABSTRACT: Two-dimensional flows over harmonically oscillating symmetrical aerofoil at reduced frequency of 0.1 were investigated for a Reynolds number of 135,000, with focus on the unsteady aerodynamic forces, pressure and vortex dynamics at post-stall angles of attack. Numerical simulations using ANSYS® FLUENT CFD solver, validated by wind tunnel experiment, were performed to study the method of sliding mesh employed to control the wing oscillation. The transport of flow was solved using incompressible, unsteady Reynolds-Averaged Navier-Stokes equations. The 2-equation k-ε realizable turbulence model was used as turbulence closure. At large angle of attack, complex flows structure developed on the upper surface of the aerofoil induced vortex shedding from the activity of separated flows and interaction of the leading edge vortex with the trailing edge one. This interaction at some stage promotes the generation of lift force and delays the static stall. In this investigation, it was found that the sliding mesh method combined with the k-ε realizable turbulence model provides better aerodynamic loads predictions compared to the methods reported in literature.

KEYWORDS: Aerofoil, Unsteady, Low Reynolds number, Harmonic oscillation, Post-stall, Computational fluid dynamics.

INTRODUCTION

Among the earliest documented phenomena of dynamic stall, there is the report of Kramer (1932) summarizing his experimental investigation on the effect of vertical gust on a wing stress. The study is also the first to recognize the dynamic features associated with the rapid variations of incidence experienced by an airfoil (Wernert *et al.* 1996).

Nowadays, dynamic stall is commonly recognized as the breakdown of the flow field around an oscillating slender lifting surface operating beyond its critical angle of attack (McCroskey 1981). Over a past few decades, the dynamic stall event has become one of the major problems in aeronautics and astronautics. Dynamic stall is a condition where the evolution of the flow field is rather different from that obtained for static stall and contains complex unsteady flow phenomenon that produces a detrimental effect in many aerodynamic systems. Maneuvering aircraft, helicopter rotor blades, wind turbines, flapping wing of insect, compressor and turbine blades are some examples of engineering application involving unsteady flow and unsteady load during operation. Unsteady flow is known to produce a negative effect on aerodynamic load, induced vibration, flutter, buffeting (Kim and Chang 2009) and is also known as one of the limiting factors of the aircraft flight speed and operation of mechanical system (Lee and Gerontakos 2004). Recently, oscillating wing has gained a particular interest in the field of biomimetic studies aiming at harvesting propulsive force from circulation flow to a small scaled vehicle.

The phenomenon of dynamic stall can be categorized into pre- and post-stall depending on the viscous-inviscid interaction (McCroskey 1981; Kim and Chang 2009; Lee and Gerontakos 2004; McCroskey *et al.* 1976). Beyond the stall,

1. Universiti Teknologi Malaysia – Faculty of Mechanical Engineering – Aeronautical Laboratory – Skudai/Johor – Malaysia.

Author for correspondence: Nik Ahmad Ridhwan Nik Mohd | Universiti Teknologi Malaysia – Faculty of Mechanical Engineering – Aeronautical Laboratory | 81310 Skudai/Johor – Malaysia | Email: ridhwan@utm.my the separation point moves toward the leading edge forming a complicated flow structure such as leading edge vortex (LEV), free shear layer, and viscous-inviscid interaction that plays an important role in the characteristics of dynamic stall (Kim and Chang 2009). In helicopter industry, the development of LEV associated with the increase in the pitching moment of rotor blades is undesirable. However, the delay in the LEV detachment from the aerofoil surface provides an increase in dynamic lift to the aircraft (Lee and Su 2011).

Accurate prediction of unsteady flow and dynamic stall is still one of challenging tasks in experimental and computational fluid dynamics. Numerous experimental papers (McCroskey 1981; Kim and Chang 2009; Lee and Gerontakos 2004; McCroskey *et al.* 1976; Panda and Zaman 1994; Coton and Galbraith 1999) and numerical investigations (McCroskey and Pucci 1982; Ericsson and Reding 1970; Gaitonde and Fiddes 1993; Dumlupinar and Murthy 2011; Barakos and Drikakis 1999) have been carried out and show that the unsteady flow can be separating or reattaching over a large portion of the upper surface of an aerofoil, features which reverse flow circulation, formation and shedding of leading and trailing vortex system that induces non-linear fluctuating pressure field and produce transient variation in forces and moments that are different to its steady-state conditions (Lee and Gerontakos 2004). The flow field of an aerofoil periodically pitched about a fixed-axis is primarily influenced by the amplitude α_o, the mean angle α_m and the frequency of oscillation f. The oscillation frequency is known as the most influential parameter and is often presented in terms of reduced frequency — the ratio of 2 time scales: one imposed by the pitching motion and the other by the free stream velocity and the aerofoil chord (Panda and Zaman 1994).

Wang *et al.* (2010), in their study, employed the dynamic mesh approach for the simulation of harmonically pitching aerofoil following Wernert *et al.* (1996) for moderate Reynolds number and Lee and Gerontakos (2004) for low Reynolds number cases. The used dynamic mesh technique consists of 2 sub-domains that communicate through a pair of circular common interfaces. The commercial ANSYS® (2013) FLUENT pressure-based solver that solves full unsteady Reynolds-Averaged Navier-Stokes governing equation for incompressible flow problem was used. The SIMPLE algorithm was used for the pressure-velocity coupling and the 2nd-order upwind scheme was utilized for the velocity and the turbulence quantities. To accelerate the convergence of the solution, the algebraic multigrid scheme (AMG) with a V-cycle type for the pressure and a flexible type for the momentum was applied. The turbulence in the flow was calculated using the standard k-ω and shear stress transport (SST) k-ω turbulence models with a modified turbulent viscosity damping.

In Gharali and Johnson (2013), the instantaneous aerodynamic load of a NACA 0012 aerofoil in pure pitching oscillation in a steady freestream was numerically simulated and validated against the experimental research of Lee and Gerontakos (2004), in which the k-ω SST turbulence model with low Reynolds correction similar to that of Wang *et al.* (2010) was used. The C-grid topology was employed with a total of 20,000 cells and 500 nodes wrapped around the aerofoil. The stability issue that is very sensitive to the value of turbulence intensity (Wang *et al.* 2010) and mesh skewness (Gharali and Johnson 2013) is also highlighted.

In this paper, the challenges and issues of modelling and understanding the unsteady aerodynamics of an aerofoil at high incidence angles and highly unsteady flow using CFD are investigated and presented. The test case is a 2-D aerofoil undergoing fast harmonic oscillations. The solution of the CFD using sliding mesh is then compared against dynamic mesh method and available experimental data.

EXPERIMENTAL METHODS

Wind tunnel or water tunnel becomes one of standard testing procedures in aerodynamic load measurements. To date, various experimental studies on oscillating aerofoil were reported from simple to complicated experimental setup. In this investigation, experimental data obtained from the McGill University's low speed wind tunnel are referred for validation (Gerontakos 2004; Lee and Gerontakos 2004).

In this experiment, dynamic stall and flow characteristics over an oscillating symmetrical NACA 0012 aerofoil at the Reynolds number of 135,000 were investigated. The NACA 0012 aerofoil was made of solid aluminum, with a chord length c of 15 cm and a span of 37.5 m. The origin of the coordinate was located at the leading edge of the aerofoil, which was equipped with 2 endplates with a sharp leading edge to eliminate the 3-D effect at the tip of the wing. The aerofoil pitching axis is located at the ¼ chord. In the test section, the measured turbulent intensity was 0.08% at a free stream velocity u_∞ = 35 m/s.

A miniature hot wire probe (DISA P11) with a Dantec 56C17 constant-temperature anemometer (CTA) was used for wake measurement. The probe was located at 1-chord downstream of the trailing edge of the aerofoil. For surface pressure measurement, 61 pressure taps (0.35 mm in diameter), in conjunction with 7 fast-response miniature pressure transducers (type YQCH-250-1) staggered 1.5 mm apart in the streamwise direction, were used.

NUMERICAL SIMULATION

In a numerical approach, a similar aerofoil section, as in the wind tunnel experiment of Lee and Gerontakos (2004), was used for validation studies. For unsteady, oscillating aerofoil problem, hybrid unstructured mesh was employed with 2 blocks consisting of stationary and rotating meshes. The Reynolds number used in this investigation was 135,000.

For consistency, the inlet turbulence intensity in CFD was set at 0.08%, similarly to the experiment of Lee and Gerontakos (2004). However, as reported by Wang *et al.* (2010), the value of the inlet turbulent intensity has some influence on the stability of the solution. Around the aerofoil, the no-slip boundary condition was employed and wrapped by a boundary layer mesh, which contains 25 layers with the first cell height set at $y^+ \leq 1$. To ensure that the viscous effect is properly captured, an improved method for calculating turbulent boundary layer based on the reference temperature method for turbulence flow was used, resulting in the 1st cell height of less than 5.8×10^{-5} m (Meador and Smart 2005).

In this paper, the ANSYS® (2013) FLUENT CFD that solves unsteady Reynolds-Averaged Navier-Stokes equations for incompressible flow problem on unstructured, cell-centered control volume was used. The main purposes of this CFD are the assessment of the sliding mesh method used and blind test for wind tunnel experiment verification. The used CFD solver takes momentum and pressure as the primary variable and solves pressure correction and momentum sequentially. This type of solver is also known as pressure-based. For time-dependent flow problem, the pressure implicit with splitting of operator (PISO) scheme for pressure-velocity coupling was used similarly to the study of Gharali and Johnson (2013). For the selection of turbulence models, similar studies were found in literature using the 2-equation k-ω turbulent model with the transport of shear stress (Gharali and Johnson 2013; Hutchinson *et al.* 2010; Wang *et al.* 2010). In Gharali and Johnson (2013), the k-ω turbulence with SST model was used with additional enhancement in the calculation of the turbulent viscosity damping for low Reynolds number problem.

On the other hand, in this paper, the realizable k-ε turbulent model (Shih *et al.* 1995) with enhanced wall treatment and mesh with more than 27,000 points was used and found adequate to resolve the viscous-inviscid interaction as well as able to capture some detail wake structure developed during upstroke and downstroke phases for an aerofoil in post-stall condition. The selection of the realizable k-ε turbulent model was made due to the capability of this turbulent model to accurately predict flows involving rotation, boundary layers under strong adverse pressure gradient, separation and recirculation which often occur around aerofoil undergoing fast dynamic oscillation and deep stall.

The schematic of sliding mesh used in the present study is shown in Fig. 1. The computational domain is divided into 2 blocks. The first contains stationary mesh and the second, a mesh that rotates at the same frequency as the aerofoil.

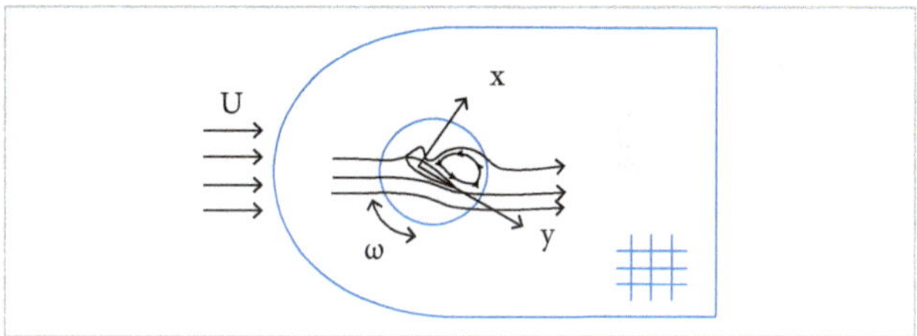

Figure 1. Schematics of the numerical setup.

The harmonic pitching motion of aerofoil is controlled by sliding mesh method that moves rigidly in a dynamic mesh zone (ANSYS® 2013). For the 2-D problem, the cell zones are connected each other through non-conformal interfaces. The transport equation for a general scalar quantity (ϕ) in moving boundary is solved according to the following equation:

$$\frac{\partial}{\partial t}\int_V \rho\phi dV + \int_{\partial V} \rho\phi(\vec{u} - \vec{u_g}) \cdot d\vec{A} = = \int_{\partial V} \Gamma\nabla\phi \cdot d\vec{A} + \int_V S_\phi dV \quad (1)$$

where: ρ is the fluid density; Γ is the diffusion coefficient; $\vec{u}$ is flow velocity vector; $\vec{u}_g$ is the mesh velocity of the moving mesh; represents the boundary of control volume ∂V; $\vec{A}$ is the face area vector; S_ϕ is the source term of ϕ.

The general scalar transport equation (Eq. 1) is converted into an algebraic equation using control volume method. This method consists of integrating the transport equation about each volume forming a discrete equation that expresses the conservation law on a control volume basis. The update in the change in volume is solved using 1st-order backward difference as:

$$V^{n+1} = V^n + \frac{dV}{dt}\Delta t \tag{2}$$

where: n and $n + 1$ denote the respective quantity at the current and next time level; dV/dt is the volume-time derivative of the control volume.

For sliding mesh approach, the mesh in sliding region remains in its original shape and volume and does not change in time. Thus Eq. 2 can be written as:

$$V^{n+1} = V^n \tag{3}$$

Using Eq. 3, the 1st term in Eq. 1 can be written as:

$$\frac{\partial}{\partial t}\int_V \rho\phi dV = \frac{[(\rho\phi)^{n+1} - (\rho\phi)^n]V}{\Delta t} \tag{4}$$

In order to satisfy the mesh conservation law, the volume-time derivative of the control volume is computed from the dot product of mesh velocity and face area vector $\vec{u}_g \cdot \vec{A}$:

$$\frac{dV}{dt} = \int_{\partial V} \vec{u_g} \cdot d\vec{A} = \sum_j^{n_f} \vec{u_{g,j}} \cdot \vec{A_j} = 0 \tag{5}$$

The results presented in this paper refer to the aerofoil subjected to a harmonic oscillation which is defined as a function of the angle of attack variation, $\alpha(t) = \alpha_m + \alpha_o \sin(\omega t)$, where ω is the oscillation frequency. In the case of oscillating aerofoils, the unsteady motion is characterized by the non-dimensional similarity parameter known as reduced frequency of oscillation, defined by $k = \omega c/2U_\infty$, where U_∞ is the freestream velocity. The aerofoil axis of oscillation is located at ¼ chord point.

The test cases presented are related to aerofoil at post-stall condition with instantaneous angle of attack, $\alpha(t) = 10° + 15° \sin(\omega t)$, and reduced frequency, $k = 0.1$, was chosen. The simulations were run for one and a half cycle with time step $dt = 0.0006$ s. In later discussion, lift and drag forces on aerofoil are described in terms of the non-dimensionalized unit defined as $C_L = F_y/(0.5\rho U_\infty^2 c)$ and $C_D = F_x/(0.5\rho U_\infty^2 c)$, respectively, where F_y and F_x are the lift and drag forces.

RESULTS AND DISCUSSION

In this section, aerodynamic loads in terms of lift and drag coefficients of a NACA 0012 aerofoil undergoing harmonic oscillation at a reduced frequency, $k = 0.1$, are presented with the intention to evaluate the combination of sliding mesh approach and the used k-ε realizable turbulence model.

Figures 2a and 2b show the comparison of the instantaneous lift and drag coefficients hysteresis calculated for a 1 harmonic oscillation cycle. From these figures, the harmonic motion of the aerofoil in the upstroke phase extends the aerofoil stall angle beyond the static stall and up to the angle of attack $\alpha = 24.4°$, as predicted in CFD and the experiment. The predicted maximum lift coefficient is reported to be $C_L = 2.4$ at $\alpha = 24.43°$ and 2.5 times higher than the static stall. The linear lift region also extended up to the angle of attack $\alpha = 20°$ with the gradient of the $dC_L/d\alpha$ maintained similar to the static case. In upstroke phase, CFD predicts that there is slightly loss in lift within $20° \leq \alpha(t) \leq 25°$ (↑). In the dynamic stall region, the wing

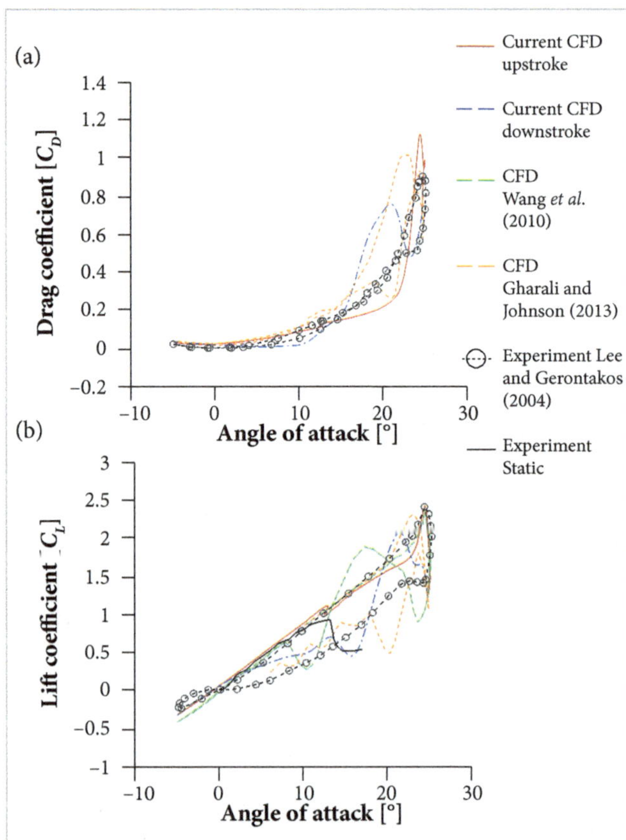

Figure 2. Instantaneous aerodynamic loads hysteresis for test case: $\alpha(t) = 10° + 15° \sin(\omega t)$, $k = 0.1$. (a) Drag coefficient; (b) Lift coefficient.

stall has a similar pattern and magnitude of loss in lift force to that reported in the experiment and in Wang *et al.* (2010). However, the dynamic stall event and the loss in the lift force as predicted in Wang *et al.* (2010) showed that there is continuing loss in lift beyond the dynamic stall angles as reported in the experiment. Gharali and Johnson (2013), on the other hand, show that the used CFD methods inaccurately predict the dynamic stall angle, which was found to occur earlier than reported in the experiment. Similarly to Wang *et al.* (2010), the present CFD analysis predicts a gain in lift as the aerofoil enters in the downstroke phase. Moreover, in Wang *et al.* (2010), the gain in lift appeared to occur at slightly lower incidence angles unlike the present study. As can be seen in Fig. 2b, the present analysis shows that the used CFD method slightly underpredicts the drag coefficient within $15° \leq \alpha(t) \leq 22°$ (↑). A steep increase in the predicted drag can also be seen in the downstroke phase as a result of large increase in the predicted lift (Fig. 2b) but occurs at lower incidence aerofoil angles as compared to the investigation of Gharali and Johnson (2013).

Variation in the predicted lift and drag was found to be strongly associated with the change in the local surface pressure with respect to time. Figures 3 and 4 show the instantaneous field pressure near the aerofoil in upstroke and downstroke phases, respectively. For the purpose of comparison, the field pressures shown in these figures are extracted for the same pressure range. In upstroke phase and at a low angle of attack up to 6°, the field pressure (Fig. 3) and the surface pressure (Fig. 5) suggest that the flow is mostly attached to the surface. As the aerofoil continues moving upward, with the incidence angle reaching approximately 21° (linear region of lift in dynamic condition), the sudden decrease in the suction pressure is apparent on the upper surface, near the leading edge (Fig. 3d). This sudden decrease in the field pressure near the suction area can also be visualized by a large negative pressure gradient near the leading edge of the aerofoil. This phenomenon can also be associated with the generation and propagation of LEV that dominates the upper surface of an aerofoil. As the incidence angle continues to increase, the LEV starts to detach from the surface as well as to propagate and travel downstream towards the trailing edge (Fig. 3e). This LEV carries low pressure field and its close proximity to the surface of an aerofoil greatly alters the value of the aerofoil local surface pressure. As the LEV and trailing edge vortex (TEV) interact with each other, it causes a large drop in field pressure near the trailing edge region (Figs. 3f and 5f).

In the transition phase (from upstroke to downstroke), the used method predicts a steep lost in the lift known as dynamic stall. In this transition phase, the primary LEV and

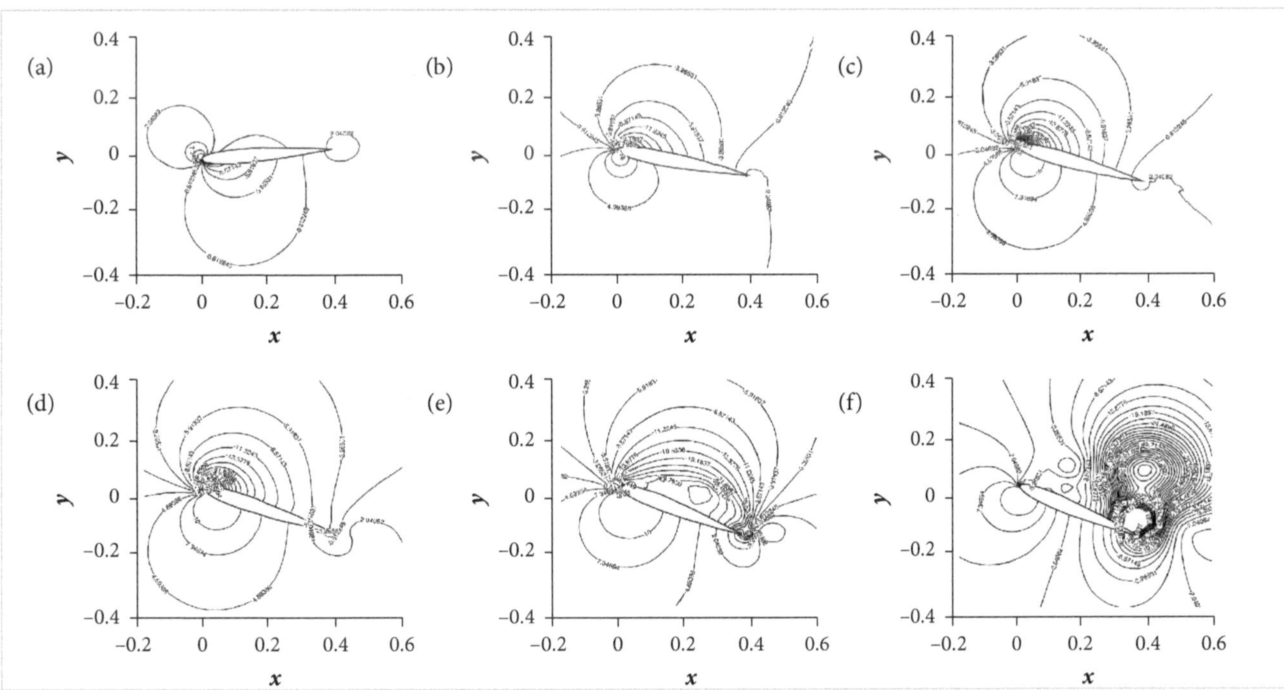

Figure 3. Instantaneous field pressure coefficient in upstroke phase. (a) −4.85°, t = 3,000 s, ↑; (b) 12.42°, t = 100 s, ↑; (c) 16.99°, t = 300 s, ↑; (d) 20.85°, t = 500 s, ↑; (e) 23.58°, t = 700 s, ↑; (f) 24.98°, t = 1,000 s, ↑.

TEV begin to detach from the aerofoil surface, forming vortex shedding in the wake downstream of the aerofoil (Fig. 4a). At this condition, the CFD predicts that a large low pressure area exists on the upper surface of an aerofoil. As the primary

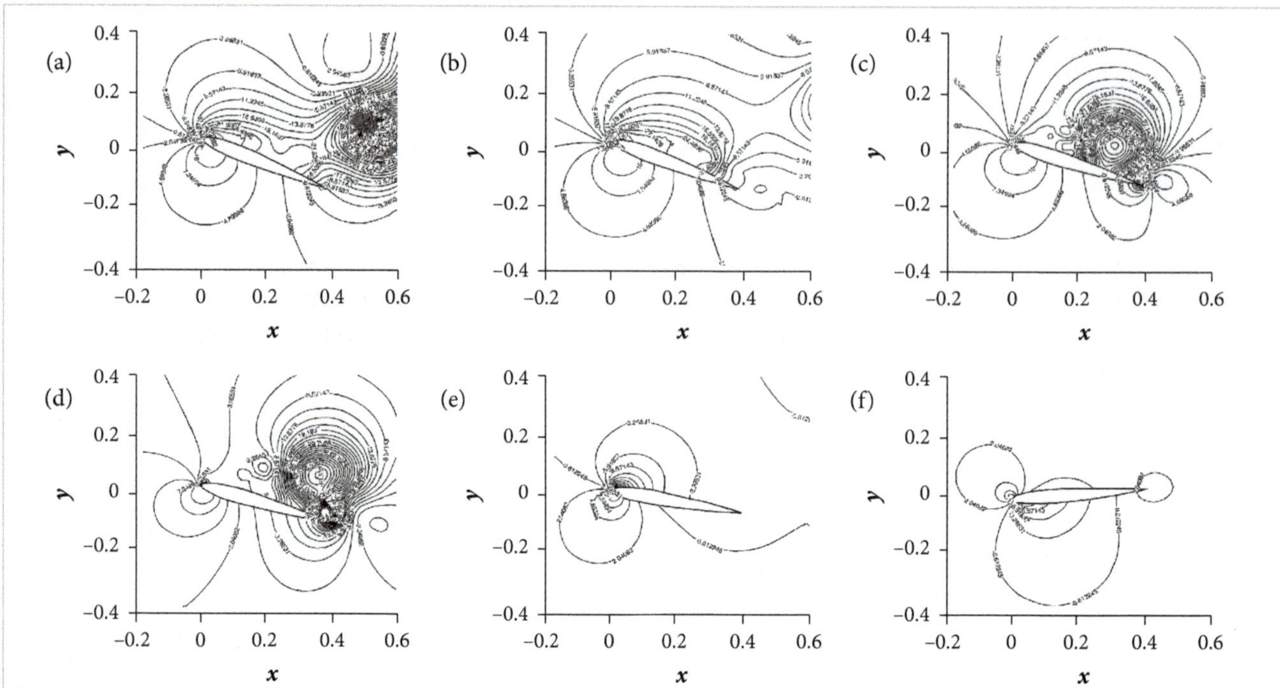

Figure 4. Instantaneous field pressure coefficient in downstroke phase. (a) 23.98°, t = 1,200 s, ↓; (b) 22.93°, t = 1,300 s, ↓; (c) 19.84°, t = 1,500 s, ↓; (d) 17.88°, t = 1,600 s, ↓; (e) 11.03°, t = 1,900 s, ↓; (f) −4.99°, t = 2,900 s, ↓.

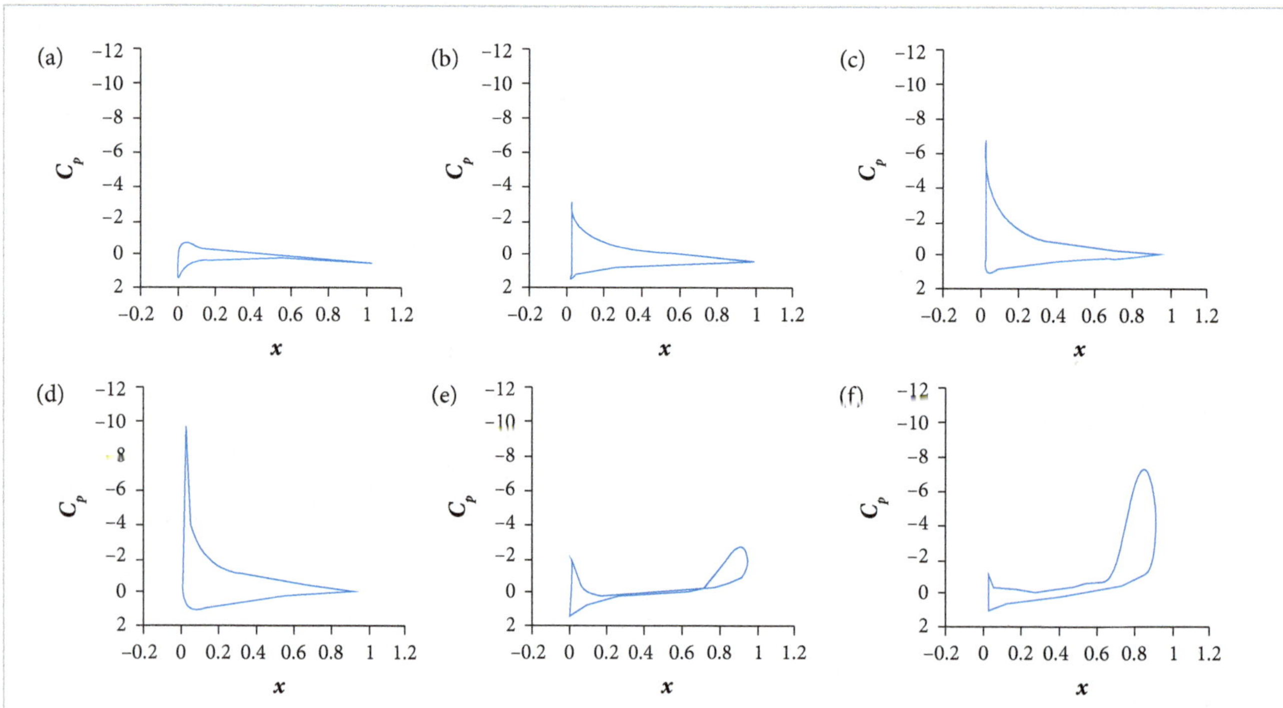

Figure 5. Instantaneous surface pressure coefficient (C_p) distribution in upstroke phase. (a) −4.85°, t = 3,000 s, ↑; (b) 12.42°, t = 100 s, ↑; (c) 16.99°, t = 300 s, ↑; (d) 20.85°, t = 500 s, ↑; (e) 23.58°, t = 700 s, ↑; (f) 24.98°, t = 1,000 s, ↑.

LEV propagates and travels downward, a secondary LEV is developed due to the downward movement of the aerofoil. The progression of LEV from leading to trailing edge of the aerofoil again develops a low pressure area on the upper surface resulting in an increase in the normal and axial forces. This is clearly visible at an aerofoil incidence angle of approximately 20° (↓). This is benefited from a large pressure loss because LEV (clockwise rotation) and TEV (anti-clockwise rotation) activities occur on the upper surface of the aerofoil. The surface static pressure coefficients for an aerofoil in downstroke phase are depicted in Figs. 6a to 6f. These figures also show that, even for a geometrically symmetrical aerofoil, the aerodynamics and flow characteristics in the upstroke phase are heterogeneous to the downstroke phase.

CONCLUSION

The aerodynamic characteristics of a NACA 0012 aerofoil in low Reynolds number flow (Re = 135,000) undergoing harmonic oscillation in a fully unsteady flow condition were numerically simulated using the sliding mesh method and k-ε realizable turbulent model. The purposes of this research were to investigate the potential of sliding mesh and k-ε realizable turbulent model in the simulation of aerofoil in unsteady oscillation, in order to support wind tunnel experiment, to fully understand the aerodynamic behavior of aerofoils in harmonic oscillation and for the design of control system for flexible wing. Overall, the cyclical forces and flow structure, including LEV and TEV interaction, propagation and shedding, agreed very well qualitatively and quantitatively with those reported in literature.

It can also be concluded that, for low Reynolds number problems, the sliding mesh strategy accompanied by the used 2-equation k-ε realizable turbulence model provides better aerodynamic loads prediction compared to the moving mesh approach using the k-ω SST. The k-ε realizable turbulence model is also found to be adequate in capturing the flow field at high incidence angles and during dynamic stall conditions comparable to the experiment. In the upstroke cycle and at high aerofoil incidence angles, the propagation and the close proximity of the LEV to the aerofoil surface cause a large drop in the static

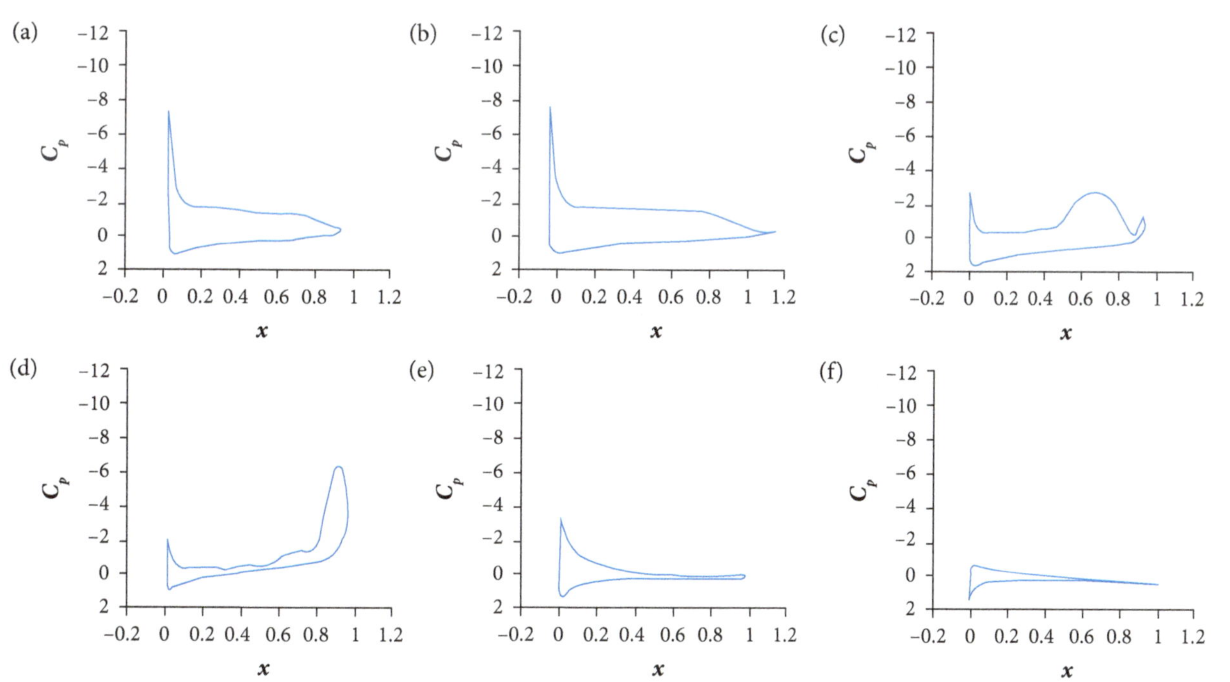

Figure 6. Instantaneous surface pressure coefficient (C_p) distribution in downstroke phase. (a) 23.98°, t = 1,200 s, ↓; (b) 22.93°, t = 1,300 s, ↓; (c) 19.84°, t = 1,500 s, ↓; (d) 17.88°, t = 1,600 s, ↓; (e) 11.03°, t = 1,900 s, ↓; (f) −4.99°, t = 2,900 s, ↓.

pressure on the upper surface of the aerofoil, and this is predicted to occur approximately at ½ of aerofoil chord. The large deficit in static pressure may cause some influence in the aerofoil pitching moment. For future studies, the detail of LEV-TEV interaction, boundary layer profile, separation and attachment will be investigated to fully understand the flow characteristics in the viscous-sublayer region leading to turbulent flow.

ACKNOWLEDGEMENTS

The authors would like to thank the Universiti Teknologi Malaysia, for the financial support to the current study through Flagship Research Grant No: Q.J130000.2424.01G92, as well as Mr. Hasrizam, for preparing mesh and computer setup.

AUTHOR'S CONTRIBUTION

Mohd NARN administered, designed, and made the setup of the numerical simulation as well as wrote the manuscript. Lazim TM and Mansor S provided technical support and conceptual advices. Rahman AHA performed the numerical setup, the data analysis and wrote the manuscript. All authors contributed equally to this paper.

REFERENCES

ANSYS® (2013) ANSYS FLUENT theory guide. Canonsburg: ANSYS, Inc.

Barakos G, Drikakis D (1999) An implicit unfactored method for unsteady turbulent compressible flows with moving boundaries. Comput Fluids 28(8):899-922. doi: 10.1016/S0045-7930(98)00057-7

Coton FN, Galbraith RA (1999) An experimental study of dynamic stall on a finite wing. Aeronaut J 103(1023):229-236. doi: 10.1017/S0001924000027895

Dumlupinar E, Murthy VR (2011) Investigation of dynamic stall of airfoils and wing. Proceedings of the 29th AIAA Applied Aerodynamics Conference; Honolulu, USA.

Ericsson LE, Reding JP (1970) Unsteady airfoil stall review and extension. Proceedings of the 8th AIAA Aerospace Sciences Meeting; New York, USA.

Gaitonde A, Fiddes S (1993) A moving mesh system for the calculation of unsteady flows. Proceedings of the 31st Aerospace Sciences Meeting; Reno, USA.

Gerontakos P (2004) An experimental investigation of flow over an oscillating airfoil (Master's thesis). Montreal: McGill University.

Gharali K, Johnson DA (2013) Dynamic stall simulation of a pitching airfoil under unsteady freestream velocity. J Fluid Struct 42:228-244. doi: 10.1016/j.jfluidstructs.2013.05.005

Hutchinson SR, Brandner PA, Binns JR, Henderson AD, Walker GJ (2010) Development of a CFD model for an oscillating hydrofoil. Proceedings of the 17th Australasian Fluid Mechanics Conference; Auckland, New Zealand.

Kim DH, Chang JW (2009) Reynolds number effects on unsteady boundary layer of an oscillating airfoil. Proceedings of the 27th AIAA Applied Aerodynamics Conference; San Antonio, USA.

Kramer M (1932) Increase in the maximum lift of an airplane wing due to sudden increase in its effective angle of attack resulting from a gust. Washington: National Advisory Committee for Aeronautics .Memorandum No.: TM 678.

Lee T, Gerontakos P (2004) Investigation of flow over an oscillating airfoil. J Fluid Mech 512:313-341. doi: 10.1017/S0022112004009851

Lee T, Su YY (2011) Unsteady airfoil with a harmonically deflected trailing-edge flap. Journal of Fluids and Structures 27(8):1411-1424. doi: 10.1016/j.jfluidstructs.2011.06.008

McCroskey WJ (1981) The phenomenon of dynamic stall. Moffett Field: NASA Ames Research Center. Memorandum No.: TM 81-264.

McCroskey WJ, Carr LW, McAllister KW (1976) Dynamic stall experiments on oscillating airfoils. AIAA J 14(1):57-63. doi: 10.2514/3.61332

McCroskey WJ, Pucci SL (1982) Viscous-inviscid interaction on oscillating airfoils in subsonic flow. AIAA J 20(2):167-174.

Meador WE, Smart MK (2005) Reference enthalpy method developed from solutions of the boundary-layer equations. AIAA J 43(1):135-139.

Panda J, Zaman KBMQ (1994) Experimental investigation of the flow field of an oscillating airfoil and estimation of lift from wake surveys. J Fluid Mech 265:65-95. doi: 10.1017/S0022112094000765

Shih TH, Liou W, Shabbir A, Yang Z, Zhu J (1995) A new k - ε eddy viscosity model for high Reynold number turbulent flows: model development and validation. Comput Fluid 24(3):227-238.

Wang S, Toa Z, Ma L, Ingham D, Pourkashanian M (2010) Numerical investigation on dynamic stall associated with low Reynolds number flows over airfoils. Proceedings of the IEEE 2nd International Conference on Computer Engineering and Technology; Reno, USA.

Wernert P, Geissler W, Raffel M, Kompenhans J (1996) Experimental and numerical investigations of dynamic stall on a pitching airfoil. AIAA J 34(5):982-989. doi: 10.2514/3.13177

Aerofoil Design for Unmanned High-Altitude Aft-Swept Flying Wings

Ahmad Abdulkarim Alsahlan[1], Thurai Rahulan[1]

ABSTRACT: In this paper, 12 new aerofoils with varying thicknesses for an aft-swept flying wing unmanned air vehicle have been designed using a MATLAB tool which has been developed in-house. The tool consists of 2 parts in addition to the aerodynamic solver XFOIL. The first part generates the aerofoil section geometry using a combination of PARSEC and Bezier-curve parameterisation functions. PARSEC parametrisation has been used to represent the camber line while the Bezier-curve has been used to select the thickness distribution. This combination is quite efficient in using an optimisation search process because of the capability to define a range of design variables that can quickly generate a suitable aerofoil. The second part contains the optimisation code using a genetic algorithm. The primary target here was to design a number of aerofoils with low pitching moment, suitable for an aft-swept flying wing configuration operating at low Reynolds number in the range of about 0.5×10^6. Three optimisation targets were set to achieve maximum aerodynamic performance characteristics. Each individual target was run separately to design several aerofoils of different thicknesses that meet the target criteria. According to the set of result obtained so far, the initial observation of the aerodynamic performance of the newly designed aerofoils is that the lift/drag ratio in general is higher than that of the existing ones used in many current-generation high-altitude long-endurance aircraft. Another observation is that increasing the maximum thickness of the aerofoil leads to a decrease in the maximum lift/drag ratio. In addition, as expected, this ratio sharply drops after the maximum value of some of these aerofoils.

KEYWORDS: Aerofoil design, Low Re, Swept flying wing.

INTRODUCTION

The flight profiles of high-altitude long-endurance unmanned air vehicles (HALE UAVs), at high altitudes and moderate flight speeds, represent low-Reynolds number operating conditions. The required lift coefficient is high when compared to more conventional lower-altitude aircraft, due to reduced air density. Therefore, the aerofoil section needs to be optimised to generate high lift coefficients with a profile drag as low as possible. The pitching moment (C_m) should also be considered especially for the flying wing configuration with no tail. An additional constraint in the selection of suitable aerofoil sections is the thickness required to hold fuel tanks or batteries and the required structural elements for adequate stiffness. In previous literature, deferent aerofoils were selected on the basis of a high lift/drag ratio. The swept flying wing configuration is not adopted widely in the literature for high altitude operation. Alsahlani *et al.* (2015) introduced a new aft-swept flying wing solar powered UAV operating at 17 km altitude. An HS520 aerofoil was selected due to its low pitching moment and drag coefficient (C_d). Recently, in July 2015, the Facebook Company revealed its first full-scale solar powered aft-swept flying wing aircraft called Aquila, which is to be used to facilitate internet communications in remote parts of the world (The Guardian 2015). However, there is as yet no detailed information available about this aircraft. Selecting or designing an aerofoil for high altitude aircraft can be affected by several considerations that will be explained briefly:

- Aircraft configuration: stability requirement can affect the aerofoil selection process depending on the layout configuration.

  Unswept flying wing aircraft has no stabiliser. It depends on the pressure distribution along the

1. University of Salford – School of Computing, Science and Engineering – Department of Aeronautical Engineering – Salford – United Kingdom.

Author for correspondence: Ahmad Abdulkarim Alsahlani | University of Salford – School of Computing, Science and Engineering – Department of Aeronautical Engineering | Newton Building | Postal code: M5 4WT – Salford – United Kingdom | Email: ahmad.alsahlani@gmail.com chord-wise direction to achieve 0 pitching moment about its aerodynamic centre. This is achieved by the incorporation of a reflexed the trailing edge in order to generate a negative lift force at the trailing edge, but at the cost of reducing the maximum lift coefficient and increasing the drag (Buckstrom 1979; Qin *et al.* 2004). NASA prototypes such as Helios, Centurion, and Pathfinder are examples of this approach.

Aft-swept flying wing aircraft uses an aerofoil with a less reflexed trailing edge to reduce the reduction in the maximum lift coefficient (C_{lmax}) since a combination of sweep-back and wash-out is used to supplement the generation of the balancing pitching moment. If the pitching moment is large, a huge amount of sweep or wash-out is needed. Therefore, in general, aerofoils with a small pitching moment is preferable for tailless aircraft (Nickel and Wohlfahrt 1994). This approach has been used in manned flying wing aircraft such as the Northrop Grumman B-2 Spirit and the Horten series of aircraft (Mader and Martins 2012; Li *et al.* 2012). This approach will be the main focus of this paper.

Tailed aircraft can use an aerofoil with high lift/drag ratio and the tail is used for stability but at the cost of additional drag from the empennage. QinetiQ Zephyr, Global Observer, X-HALE, and Boeing Phantom Eye are examples of this approach (Rapinett 2009).

- Low Reynolds number: at low Reynolds number the viscous effects are more dominant which generate high drag and at reduced maximum lift while at higher Reynolds number, the lift/drag ratio improves (Greer *et al.* 2000; Lissaman 1983). At low Reynolds number, the performance of an aerofoil is limited by quicker separation of the boundary layer. Therefore, the flow becomes quite unsteady causing the aerodynamic moment and forces to fluctuate with time (Lei *et al.* 2013). A separation bubble may form within boundary layer causing the transition from laminar to turbulent, according to Fig. 1 (Lei *et al.* 2013; Lissaman 1983). The position of this separation bubble and the intensity depends on the shape of the aerofoil, angle of attack (AoA), and Reynolds number (Greer *et al.* 2000). This has the effect of reducing the gradient of the lift curve whilst increasing drag at the same time (Ma and Liu 2009). The majority of the air resistance comes from the pressure drag over the region of the laminar separation bubble. After the separation bubble, the flow reattaches to the aerofoil and becomes turbulent flow or may stay separated. If it remains separated, the lift coefficient will sharply drop with a marked increase in the drag coefficient. This usually happens at a Reynolds number of about 70,000 (Hazra and Jameson 2007). Previous studies have shown that there is a critical Reynolds number at about 7×10^4, where the performance of smooth aerofoils presents a noticeable change in the lift/drag ratio. The smooth aerofoils have a higher lift/drag ratio than rough aerofoils at approximately more than $Re > 10^5$. Thus, below this value, roughness is beneficial because this discontinuity in the surface can help to delay flow separation (McArthur 2007).
- Stiffness of structure and internal volume needed: increasing aerofoil thickness can obviously increase the wing stiffness and reduce wing bending moments sensitivity, which will result in reduced structural weight. Furthermore, it will also provide more space for fuel and payload (Cerra and Katz 2008). In general, thicker wing sections reduce both the maximum lift/drag ratio and the stall angle (Ma and Liu 2009).

Table 1 shows the aerodynamic performance of some aerofoils which are designed for aft-swept and straight flying wing aircraft operating in low Reynolds environment. These results were generated using XFOIL at *Re*. Among these aerofoils, EPPLER 339 and LA2573A have the highest lift/drag ratio and moderate maximum thickness-to-chord ratio (t/c %). Their stall angles seem to occur near the region where the lift/drag ratio reaches a maximum. LA2573A was used in high altitude prototypes such as Helios and Pathfinder aircraft, which sport straight wings (Kroo 1993). This aerofoil has a positive pitching moment to achieve the stability requirement without the need for either wing sweep or twist.

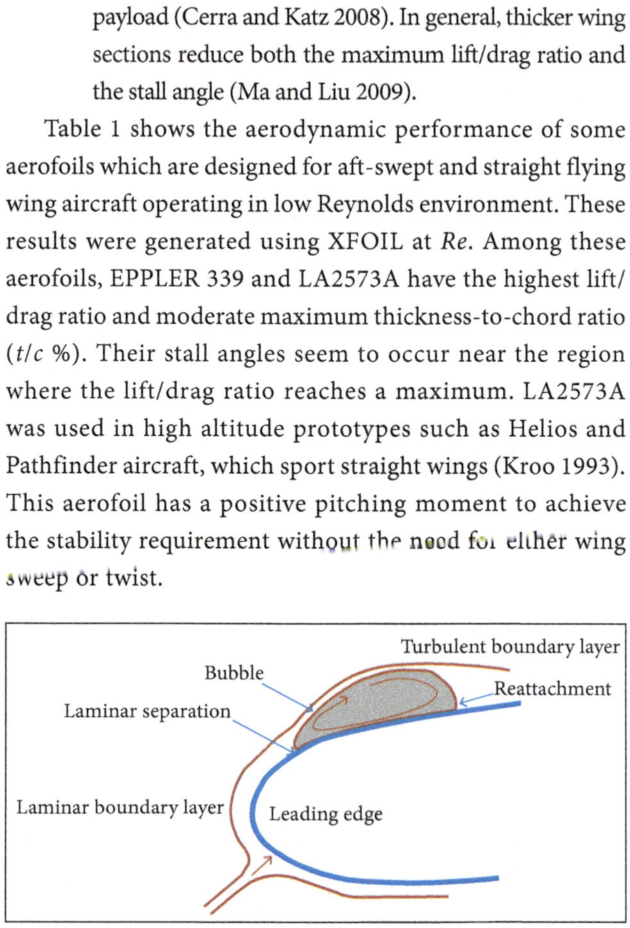

Figure 1. Structure of laminar separation bubble.

Table 1. Number of flying wing aerofoils, at *Re*, evaluated by XFLRE5.

Aerofoil name	At max C_l/C_d					C_{lmax}	C_{lmin}	Stall angle	Thickness t/c %
	C_l/C_d	C_l	C_d	C_m	At AoA ≈				
HS 520	84.37	0.78	0.0093	−0.0022	6	1.283	−0.584	13	9.8
MH 61	84.5	0.65	0.0075	−0.006	5	1.03	−0.62	11	10.28
EH 2.0/12	89.2	1.01	0.0112	−0.011	8	1.2	−0.8	12	12
EH 3.0/12	93.9	1.15	0.0123	−0.005	9	1.25	−0.6	12	12
EPPLER 231	95	0.845	0.009	−0.052	5	1.2	−0.35	13	12.33
MH 81	85.2	1.4	0.0164	−0.001	11	1.65	−0.5	15	13
MH 83	95	1.35	0.0142	−0.064	8	1.9	0.4	15	13.29
EPPLER 339	100	1.35	0.0135	−0.045	9	1.5	−0.3	13	13.5
LA2573A	102	1.247	0.0122	0.0117	10	1.33	−0.52	12	13.7
FAUVEL 14	89.3	1.15	0.0128	0.012	9	1.3	−0.74	11	14
EPPLER 342	89.3	1.39	0.0156	0.013	11	1.484	−0.35	13	14.3
EPPLER 344	94	1.41	0.015	−0.032	10	1.55	−0.29	14	14.7
MH91	78.9	1.07	0.0136	0.012	9	1.38	−0.87	16	14.98
MH95	63.8	0.826	0.013	−0.007	7	1.34	−0.7	17	15.86

In case of using aft-swept configuration, much attention is paid to the pitching moment of the candidate aerofoil because of the absence of a horizontal stabilizer for stability purposes. Adopting this configuration needs the use of an aerofoil with a smaller pitching moment, but not excessively low in order to obtain a high C_l. Since this configuration has not been adopted in application of high altitude aircraft by the other designers as far as is known, this paper will investigate and design new aerofoils for the 3 aerodynamic targets. The expected operational Reynolds number that is appropriate to similar aircraft is about .

In the literature, 2 main approaches are introduced to design aerofoils, which are based on direct numerical optimisation and inverse design. The common aspect between these 2 approaches is that the aerofoil shape is modified until specific goals are satisfied. For the first approach, the goals are usually the aerodynamic performances characteristic such as the lift, drag, and pitching moment coefficients, whereas the inverse design approach searches for an aerofoil shape until the requirements for particular flow characteristic such as pressure distributions and skin friction are fulfilled (Della Vecchia *et al.* 2014).

The aerofoil shape can be represented by several available parametrisation techniques such as the widely used B-splines, Bezier, and PARSEC parametrisations (Della Vecchia *et al.* 2014; Derksen and Rogalsky 2010; Salunke *et al.* 2014). The shape parametrisations can affect the optimisation search and should represent a wide range of aerofoil shapes (Derksen and Rogalsky 2010). In the PARSEC parametrisation, the aerofoil surface can be represented by two curves, one for the upper surface and the other for the lower surface. Each curve can be represent using parameters linked directly to the commonly-defined aerofoil section geometry such as the maximum thickness and its location, radius of the leading edge, thickness and direction angle of the trailing edge. These parameters should be selected carefully to achieve a reasonable aerofoil section, during the optimisation search, capable to be solved by the aerodynamic solver. The range of their values must be selected carefully in order to prevent unnecessary reflex in the upper and lower surfaces generating unreasonable aerofoil shape thus delaying the convergence of optimisation. A methodology for this parametrisation is presented by Sobieczky (1997). Another popular parametrisation is called Bezier curve techniques, representing the aerofoil surfaces by two curves, one for upper surface and the other for lower surface, or one for camber line and the other for thickness distribution. Bezier curve can be designed with using its controlled point which defining the curves shape (Derksen and Rogalsky 2010; Park *et al.* 2008; Salunke *et al.* 2014). Several approaches aiming to enhance the trailing and leading edge shapes representation were introduced using multi curve with number of variables. The number of these curves and their point's coordinates can extremely affect the resolution and optimisation of time consumption (Della Vecchia *et al.* 2014; Salunke *et al.* 2014). Derksen and Rogalsky (2010) introduced a new aerofoil parametrisation (Bezier-PARSEC): the Bezier parametrisation which was used to represent the aerofoil edges, and the PARSEC approach to represent rest of the aerofoil surface. The aim was to enhance the matching of representing a wide number of existing aerofoils in addition to accelerate the speed converge of aerodynamic solver (Derksen and Rogalsky 2010). In this paper, the PARSEC parametrisation is chosen to represent the camber line of the aerofoil and the Bezier curve for the thickness distribution. This combination looks efficient as observed during the optimisation process and the generated aerofoils were reasonable.

AEROFOIL SHAPE OPTIMISATION TOOL

Numerous aerofoils of different thickness have been designed using a MATLAB tool, which has been coded as a part of this paper. This tool consists of 2 parts in addition to the aerodynamic solver (XFOIL), as shown in Fig. 2. The first part generates the aerofoil section geometry using a combination of PARSEC and Bezier-curve parameterisation developed in this paper, involving 9 variables shown in Fig. 3. The second part contains the optimisation code using the canonical genetic algorithm, while several optimisation algorithms can be used within the MATLAB environment.

AEROFOIL SHAPE PARAMETERISATION

A combination of PARSEC and Bezier-Curve has been used to represent the aerofoil surface. In this paper, the PARSEC parametrisation is used to represent the camber line of the aerofoil, while the Bezier-Curve is used to represent the thickness distribution. The reason of using this combination is to achieve reasonable aerofoil geometries during the optimisation process and to prevent unconvergence problem of the aerodynamic solver, thus improving the time consumption and program stability. Moreover, by Bezier-Curve, the thickness distribution can easily be constrained to achieve a positive distribution, in addition to achieving a specific maximum aerofoil thickness. As experience was obtained during this study, it has been found that this combination is efficient to search for the optimised setting because of its capability to define a set of proper design variables leading to a rapid convergence.

Thickness Distribution

Bezier-Curve parametrisation has been used to represent the thickness distribution of the section, and the curve can be represented by associated control points. The generated curve passes through the first and last control points, but does not need to pass through the other control points.

Assume a number of control points (n), with their coordinates $P(xc_i, b_i)$, where b_i will be denoted as Bezier parameters while xc_i are set as in Fig. 3a. The curve can be discretised into a number of segments; each segment length will be denoted as (t),

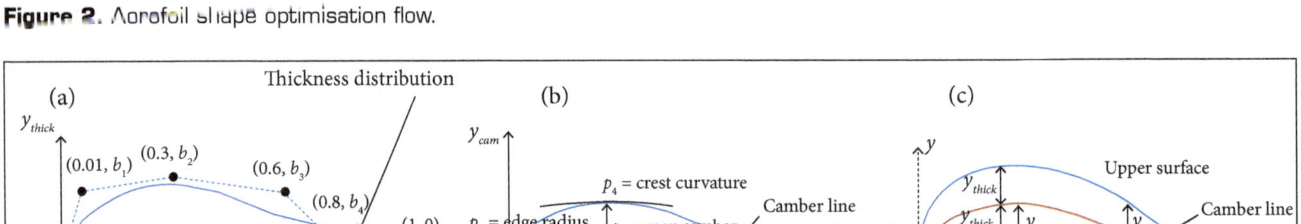

Figure 2. Aerofoil shape optimisation flow.

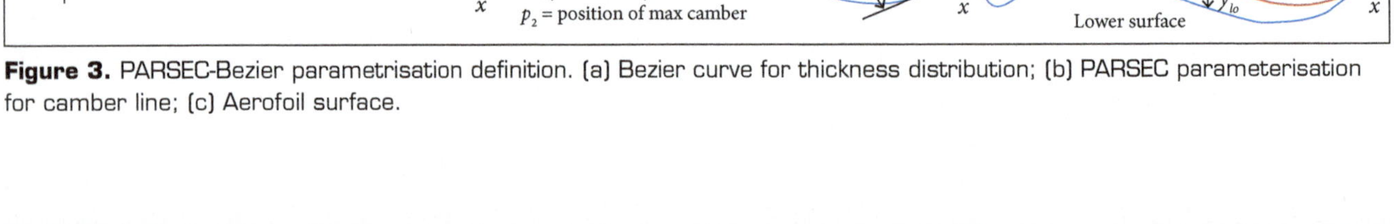

Figure 3. PARSEC-Bezier parametrisation definition. (a) Bezier curve for thickness distribution; (b) PARSEC parameterisation for camber line; (c) Aerofoil surface.

where $0 \le t \le 1$. The blending function of each segment can be evaluated by:

$$B_i^n = \frac{n!}{i!(n-i)!}(1-t)^{n-i} t^i \quad (1)$$

The coordinates of each segment of the curve () associated with the given control points are found by:

$$X_{thick}(t) = \sum_{i=1}^{n} xc_i B_i^n(t)$$

and (2)

$$Y_{thick}(t) = \sum_{i=1}^{n} b_i B_i^n(t)$$

where: Y_{thick} represents the distribution of the half thickness of the aerofoil along the x-coordinate; X_{thick} represents the x-coordinate of the upper and the lower surface.

Camber Line

PARSEC parameterisation has been used to represent the camber line of the aerofoil. In this paper, only 1 curve of the PARSEC parametrisation is taken. This curve can be represented by associated geometry parameters of the camber line (p_1, p_2, p_3, p_4, p_5), as defined in Fig. 3b. According to the PARSEC methodology, the vertical coordinates of the curve at a given (x) coordinate location is given by:

$$Y_{cam}(x) = \sum_{i=1}^{n=5} a_{pa}^{i} . x^{i-\frac{1}{2}} \quad (3)$$

The coefficients a_{pa} are calculated from the 5 parameters ($p_1, p_2, \ldots, p_5$) of the aerofoil camber as follows:

$$b_{pa} = a_{pa} \times C_{pa} \quad (4)$$

where $b_{pa} \wedge C_{pa}$ are evaluated from the following matrices:

$$C_{pa} = \begin{vmatrix} 1 & 1 & 1 & 1 & 1 & 1 \\ p_2^{0.5} & p_2^{1.5} & p_2^{2.5} & p_2^{3.5} & p_2^{4.5} & p_2^{5.5} \\ 0.5 & 1.5 & 2.5 & 3.5 & 4.5 & 5.5 \\ 0.5p_2^{-0.5} & 1.5p_2^{0.5} & 2.5p_2^{1.5} & 3.5p_2^{2.5} & 4.5p_2^{3.5} & 5.5p_2^{4.5} \\ -0.25p_2^{-1.5} & \frac{3}{4}p_2^{-0.5} & \frac{15}{4}p_2^{0.5} & \frac{15}{4}p_2^{0.5} & \frac{63}{4}p_2^{2.5} & \frac{99}{4}p_2^{3.5} \\ 1 & 0 & 0 & 0 & 0 & 0 \end{vmatrix}, \quad (5)$$

$$b_{pa} = \begin{vmatrix} 0 \\ p_3 \\ \tan(p_5/2) \\ 0 \\ p_4 \\ \sqrt{2p_1} \end{vmatrix} \quad (5) \qquad (5)$$

Then, the y-coordinates of the upper and lower surfaces can be calculated as shown in Fig. 3c by:

$$y_{up}(x) = Y_{cam}(x) + Y_{thick}(x) \quad (6)$$

$$y_{lo}(x) = Y_{cam}(x) - Y_{thick}(x) \quad (7)$$

Therefore, the aerofoil surface will be defined regarding the Bezier and PARSEC parameters, with 9 variables: .

The Aerodynamic Solver

XFOIL has been used as the aerodynamic solver of the optimisation tool. It is a 2-D aerodynamic solver widely used in the design and optimisation of aerofoils due to its ability to produce good and rapid solutions that can be obtained to match experimental data. It is a combination of inviscid-viscous flow analyses where some approximations are performed. The input parameters of this solver are the aerofoil coordinates and the flow conditions, such as the AoA and Reynolds number.

Problem Formulation

The primary target was to design a number of aerofoils, with low pitching moment, suitable for an aft-swept flying wing configuration, operating at low Reynolds number of about 0.5×10^6. Three optimisation targets were set to achieve maximum aerodynamic performance characteristics. Each individual target is run separately to design several aerofoils of different thicknesses that meet the target criteria. The set targets are:

- To achieve maximum lift/drag ratio with no less than a pitching moment coefficient of −0.05:

  maximize (C_l/C_d)
  subject: $C_m > -0.05$

- To achieve maximum lift/drag ratio with no less than a pitching moment coefficient of −0.02:

  maximize (C_l/C_d)
  subject: $C_m > -0.02$

- To achieve maximise $(C_l^{1.5}/C_d)$ ratio with no less than a pitching moment coefficient of −0.05:

$$maximize\ (C_l^{1.5}/C_d)$$
$$subject\text{: } C_m > -0.05$$

The design variables of the optimisation problems are the PARSEC and Bezier parameters: $p_1, p_2, p_3, p_4, p_5, b_1, b_2, b_3, b_4$.

In each design target, the thickness of the optimal aerofoil is constrained during the optimisation to offer different aerofoils with different thicknesses. The constraining is done by controlling the upper and the lower bounds of the Bezier parameter values (b_1, b_2, b_3, b_4). The flow condition was set at an AoA of 6° and Reynolds number of 0.5×10^6.

NEWLY DESIGNED AEROFOILS

Table 2 shows the aerodynamic performance of the newly designed aerofoils at the maximum lift/drag ratio in addition to the maximum and minimum lift coefficients and stall angles. Their names and numbers are temporarily used at this stage of study and they have no other meanings. The shapes and aerodynamic characteristics of some of the newly designed aerofoils are shown in Fig. 4 at Reynolds number of 5×10^5.

According to the set of results obtained so far, as tabulated in Table 2 and Fig. 4, the overall observations of the optimised aerofoils can be summarised as follows:

- The lift/drag ratios are higher than corresponding existing aerofoils presented in Table 1, and this is expected because the pitching moment of the new aerofoils is higher than of those in Table 1.
- Increasing the maximum thickness of the aerofoil leads to a decrease in the maximum lift/drag ratio.
- For thicker aerofoils such as ZMR-16, ZMR-18, and ZMR-19, the lift/drag ratio per AoA slops sharply drops after the maximum value.
- The variation of the pitching moment with the AoA is not smooth in some of the optimised aerofoils, but this could be suspicious because the pitching moment of XFOIL results could be inaccurate. Other aerodynamic solver or experimental study will be needed to verify this behaviour.
- The stall angles are not close to the angles in which the maximum lift/drag ratio is achieved.
- Reducing the pitching moments result in reduction of the maximum lift/drag ratio and increased drag (see ZMR-26 and ZMR-27 in Table 2).

It is worth mentioning that the maximum C_l is not employed in the optimisation target and the focus was at an AoA of 6° for maximum lift/drag ratio away from the stall angle. Therefore, most of the newly designed aerofoils would have stalled at about 8° from the angle corresponding to the

Table 2. Aerodynamic performance of the new aerofoils at *Re* = 0.5×10^6, obtained by XFOIL.

Aerofoil name	At max C_l/C_d					C_{lmax}	C_{lmin}	Stall angle	Thickness t/c [%]
	C_l/C_d	C_l	C_d	C_m	At AoA ≈				
				max C_l/C_d $C_m > -0.05$					
ZMR-9	117.9	1.081	0.0091	−0.0497	6	1.34	−0.37	15	9.9
ZMR-12	115.4	0.980	0.0085	−0.0497	6	1.25	−0.418	14	12.71
ZMR-13	109.4	0.96	0.0087	−0.0482	6	1.24	−0.52	15	14.44
ZMR-14	111.1	1.051	0.0094	−0.0493	6	1.3	−0.31	15	16.6
ZMR-15	95.04	0.99	0.0104	−0.0445	6	1.202	−0.41	18	20.91
				max $C_l^{1.5}/C_d$ $C_m > -0.05$					
ZMR-20	120.8	1.135	0.0094	−0.0496	6	1.448	−0.293	17	9.11
ZMR-17	117.6	1.191	0.0101	−0.049	7	1.446	−0.48	15	12.57
ZMR-19	114.7	1.056	0.0092	−0.0498	6	1.388	−0.34	16	15.02
ZMR-18	114	1.149	0.0093	−0.049	7	1.435	−0.49	16	15.83
ZMR-16	95.51	1.017	0.0106	−0.0443	6	1.296	−0.388	17	20.68
				max C_l/C_d $C_m > -0.02$					
ZMR-26	93.91	1.268	0.0139	−0.0181	7	1.495	−1	15	14.91
ZMR-27	80.52	1.234	0.01532	0.0341	10	1.46	−1.152	14	15.57

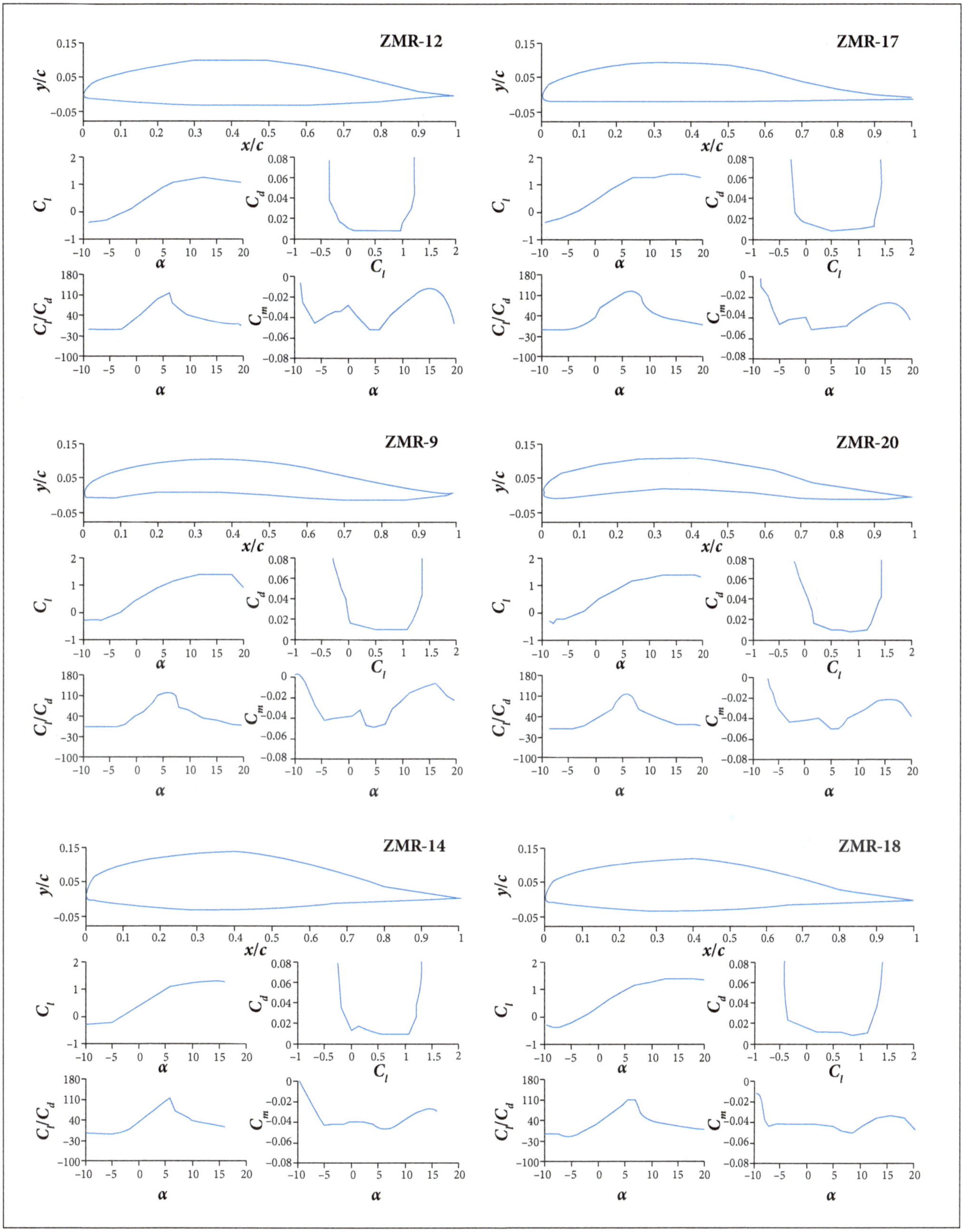

Figure 4. Aerodynamic performance of the newly designed aerofoils at $Re = 0.5 \times 10^6$.

maximum lift/drag ratio. ZMR-26 and ZMR-19 are the most interesting aerofoils for the authors as they will be candidate for an aft-swept aircraft due to their moderate thicknesses and the high maximum lift/drag ratios. They will be validated by a high order analysis solver in the next section.

COMPUTER FLUID DYNAMICS MODEL

High-order Computer Fluid Dynamics (CFD) analysis has been conducted to validate some of the newly designed aerofoils using FLUENT ANSYS. Aerofoils ZMR-19 and ZMR-26 are selected to this study due to their moderate thickness intended to be adopted for a high altitude aft-swept flying wing configuration in future work. A C-type mesh is adopted with a proper mesh density to ensure that Y+ of the first row of cells adjacent to the aerofoil surface is less than one. Figure 5 shows the schematic of a partial domain of ZMR-19. The boundary conditions are the upper and the lower flow domain boundary as a function of the velocity-inlet boundary whilst the downstream is considered as a pressure-outlet boundary. In addition, the aerofoil surface is considered as a wall. The SST k-ω (4eq) model, which can predict laminar-to-turbulent transition, is used due to its accuracy for a wide class of flows (Ansys® 2013). The calculations are carried out at chord-base Reynold number of 0.5×10^6.

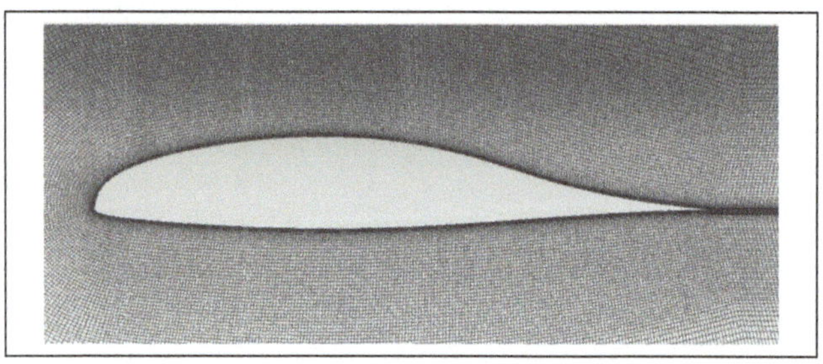

Figure 5. Schematics of the partial domain of ZMR-19.

RESULTS AND DISCUSSION

High-order analyses have been conducted to validate the aerodynamic performance of ZMR-19 and ZMR-26 aerofoils. The XFOIL predictions show that lift, drag, and pitching moment coefficients are in good agreement with FLUENT at low AoAs as shown in Figs. 6 and 7. However, at higher AoAs near the stall region, XFOIL predictions have not shown the same behaviour, especially the pitching moment coefficient.

Figures 8 and 9 show laminar separations (LS) of the ZMR-19 which occur at the upper and lower surfaces of the aerofoil at AoA = 0°. At an AoA equal to 6°, where the maximum lift/drag ratio is achieved, the LS of the lower surface has disappeared, while the separation bubble at the upper surface has slightly migrated forwards. Since the magnitudes of the velocity vectors within the bubble are small, the scales of the vectors shown in the figures were set to 4.

The flow around the ZMR-26 aerofoil also shows similar behaviour as seen in the lower surface at AoA = 0° (Fig. 10). On the other hand, in Fig. 11, at an AoA of 6°, the LS has vanished, and the flow remains attached all the way to the trailing edge. Nevertheless, the LS was also observed in existing aerofoils such as the LA2573A section, which is used in the NASA prototypes Helios and Pathfinder. The velocity contour and vector for the flow around the LA2573A section are shown in Figs. 12 and 13. Figure 12 illustrates the LS at the lower surface, which occurs at AoA = 0°. At 10°, where the maximum lift/drag ratio is achieved, the LS on the upper surface becomes noticeable as shown in Fig. 13.

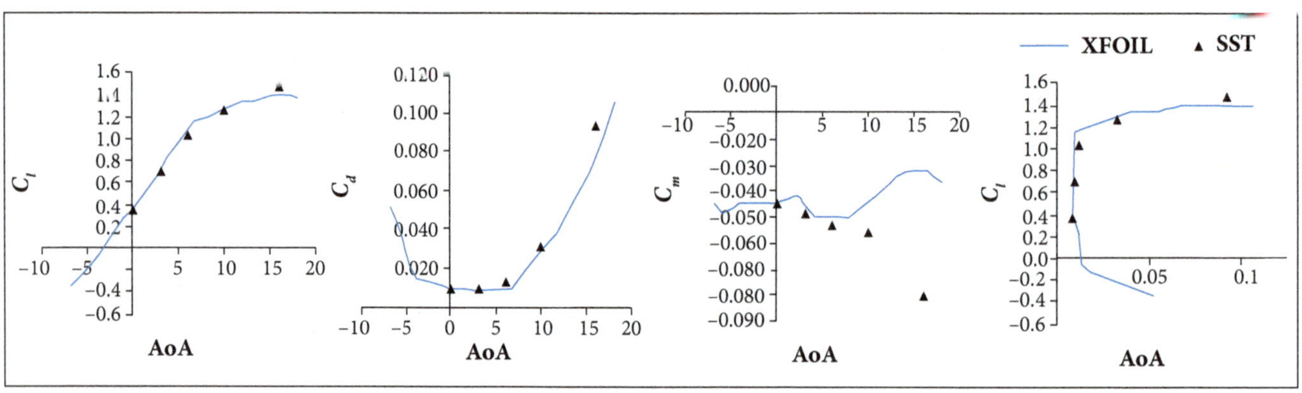

Figure 6. Aerodynamic performance of ZMR-19 aerofoil using SST k-ω (4eq) model and XFOIL at $Re = 0.5 \times 10^6$.

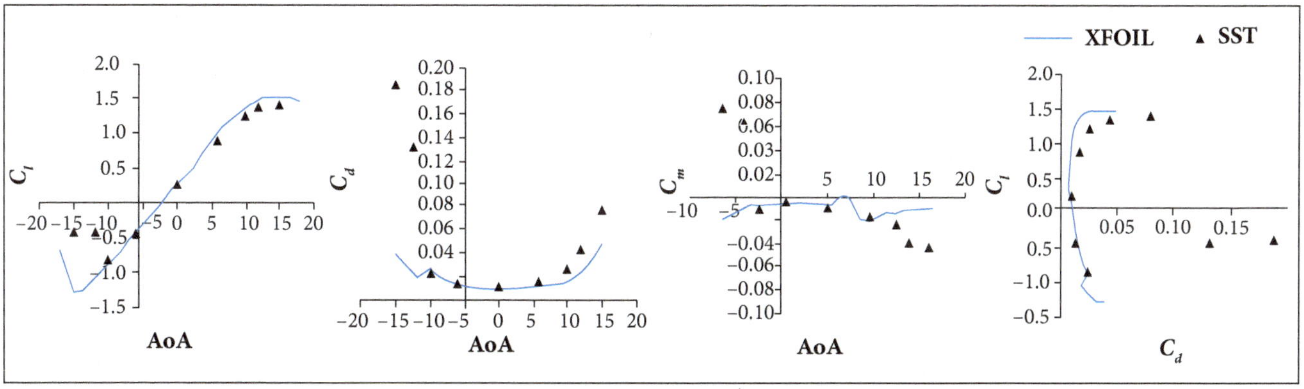

Figure 7. Aerodynamic performance of ZMR-26 aerofoil using SST k-ω (4eq) model and XFOIL at $Re = 0.5 \times 10^6$.

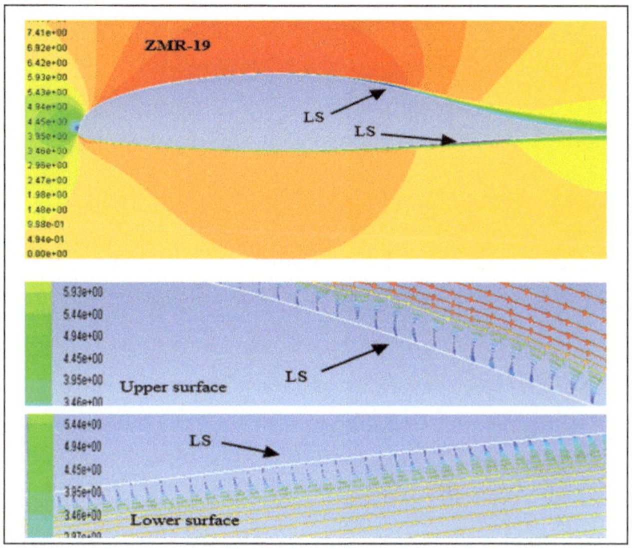

Figure 8. Velocity contour and vector at AoA = 0° of ZMR-19 aerofoil.

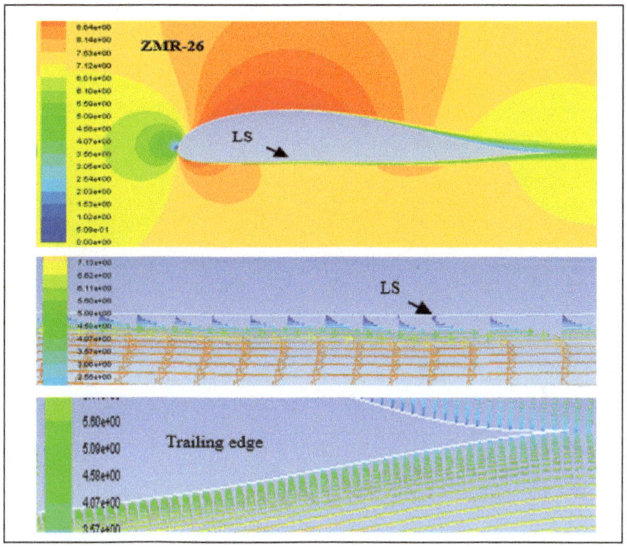

Figure 10. Velocity contour and vector at AoA = 0° of ZMR-26 aerofoil.

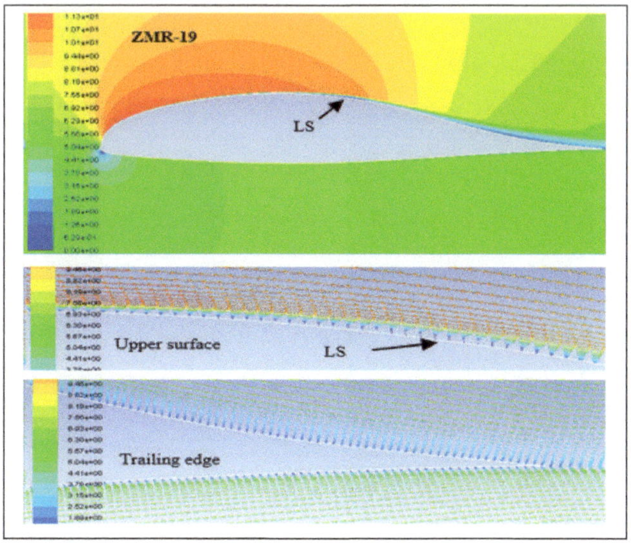

Figure 9. Velocity contour and vector at AoA = 6° of ZMR-19 aerofoil.

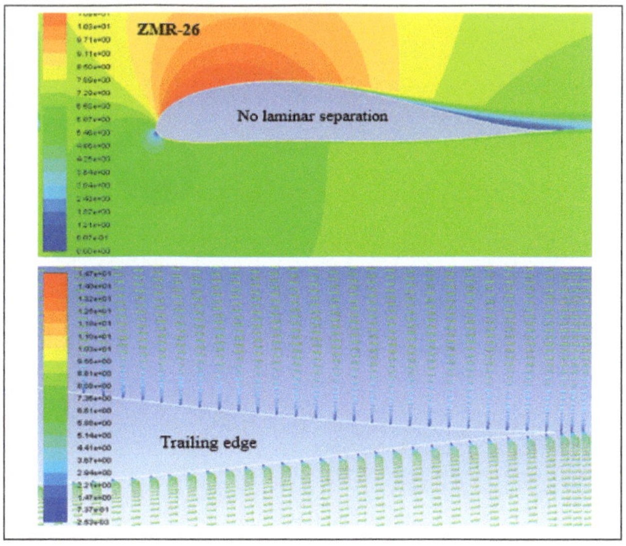

Figure 11. Velocity contour and vector at AoA = 6° of ZMR-26 aerofoil.

It is concluded that, at low Reynolds number, the LS could occur on both surfaces at small AoAs. As the AoA increases, LS bubbles either begin to disappear or move forward. The intensity of the predicted bubbles was small and hence they are not expected to have pronounced effects on performance.

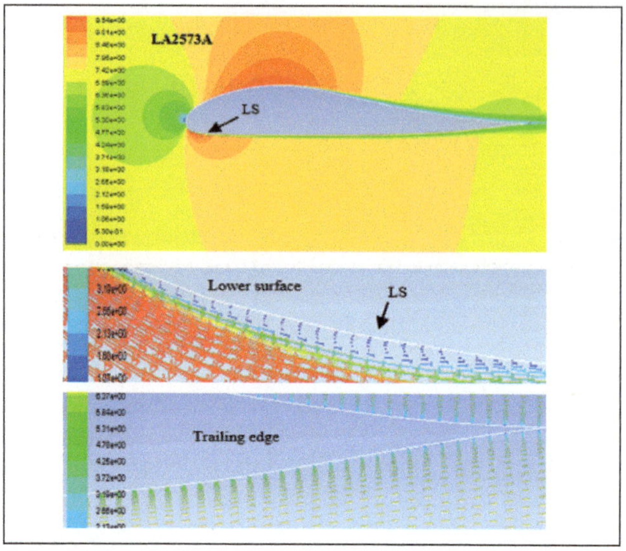

Figure 12. Velocity contour and vector at AoA = 0° of LA2573A aerofoil.

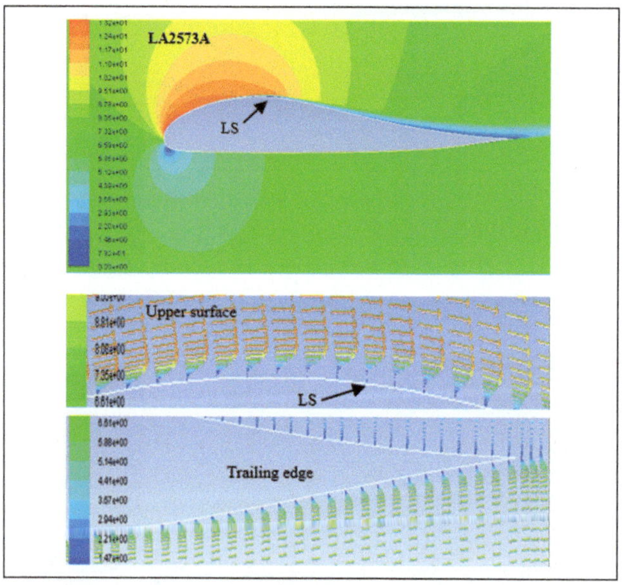

Figure 13. Velocity contour and vector at AoA = 10° of LA2573A aerofoil.

CONCLUSION

A number of new aerofoils with varying thicknesses for an aft-swept flying wing UAV, operating at low Reynolds number of about , have been designed using a developed aerofoil parametrisation employed in the optimisation tool. The aerodynamic performance of the newly designed aerofoil shows that the maximum lift/drag ratio is affected by the thickness-to-chord ratio. Increasing the maximum thickness of the aerofoil leads to a decrease in the maximum lift/drag ratio. In addition, as expected, this ratio sharply drops in value after the maximum is reached. The stall angles of these aerofoils are not close to the corresponding angle of the maximum lift/drag ratio. LS, especially at small AoAs, was observed in the ZMR-19 and ZMR-26 aerofoils. However, this was also observed in existing aerofoils such as the LA2573A. Further optimisation processes and validation by experimental tests will be carried out within the discussed range of Reynolds numbers to instil confidence.

ACKNOWLEDGMENTS

The authors appreciate the Higher Committee of Education of Iraq (HCED-IRAQ) for supporting this study at the University of Salford.

AUTHOR'S CONTRIBUTION

Alsahlan AA conducted the study under the supervision of Rahulan T; both authors discussed the results and commented on the manuscript.

REFERENCES

Alsahlani A, Johnston LJ, Atcliffe PA (2015) Design of a high altitude long endurance flying-wing solar-powered unmanned air vehicle. Proceedings of the 6th European Conference for Aeronautics and Space Sciences; Krakow, Poland.

Ansys® (2013) ANSYS FLUENT theory guide; [accessed 2017 Apr. 25]. http://www.ansys.com

Buckstrom A (1979) The elements of tailless airplane design. Sport Aviation 5:39-44.

Cerra DF, Katz J (2008) Design of a high-lift, thick airfoil for unmanned aerial vehicle applications. J Aircraft 45(5):1789-1793. doi: 10.2514/1.36924

Della Vecchia P, Daniele E, D'Amato E (2014) An airfoil shape optimization technique coupling PARSEC parameterisation and evolutionary algorithm. Aero Sci Tech 32(1):103-110. doi: 10.1016/j.ast.2013.11.006

Derksen R, Rogalsky T (2010) Bezier-PARSEC: An optimized aerofoil parameterisation for design. Adv Eng Software 41(7-8):923-930. doi: 10.1016/j.advengsoft.2010.05.002

Greer D, Hamory P, Edwards C, Krake K, Drela M (2000) Design and predictions for high-altitude (low Reynolds number) aerodynamic flight experiment. J Aircraft 37(4):684-689. doi: 10.2514/2.2652

Hazra SB, Jameson A (2007) Aerodynamic shape optimization of airfoils in ultra-low Reynolds number flow using simultaneous pseudo-time stepping. Aerospace Computing Lab (ACL) Report, 2007-4; [accessed 2017 Apr. 25]. http://aero-comlab.stanford.edu/Papers/lowren_paper.pdf

Kroo I (1993) Tailless aircraft design — recent experiences. Proceedings of the Symposium on Aerodynamics and Acoustics; New Jersey, USA.

Lei J, Guo F, Huang C (2013) Numerical study of separation on the trailing edge of a symmetrical airfoil at a low Reynolds number. Chin J Aeronaut 26(4):918-925. doi: 10.1016/j.cja.2013.06.005

Li P, Zhang B, Chen Y, Yuan C, Lin Y (2012) Aerodynamic design methodology for blended wing body transport. Chin J Aeronaut 25(4):508-516. doi: 10.1016/s1000-9361(11)60414-7

Lissaman P (1983) Low-Reynolds-number airfoils. Annual Review of Fluid Mechanics 15(1):223-239. doi: 10.1146/annurev.fl.15.010183.001255

Ma R, Liu P (2009) Numerical simulation of low-Reynolds-number and high-lift airfoil S1223. Proceedings of the World Congress on Engineering; London, UK.

Mader C, Martins J (2012) Optimal flying wings: a numerical optimization study. Proceedings of the 53rd AIAA/ASME/ASCE/AHS/ASC Structures, Structural Dynamics, and Materials Conference; Honolulu, USA.

McArthur J (2007) Aerodynamics of wings at low reynolds numbers (PhD thesis).Los Angeles: University of Southern California.

Nickel K, Wohlfahrt M (1994) Tailless aircraft in theory and practice. Washington: American Institute of Aeronautics and Astronautics.

Park K, Han JW, Lim HJ, Kim BS, Lee J (2008) Optimal design of airfoil with high aspect ratio in Unmanned Aerial Vehicles. Proceedings of the World Academy of Science: Engineering and Technology 42:182.

Qin N, Vavalle A, Le Moigne A, Laban M, Hackett K, Weinerfelt P (2004) Aerodynamic considerations of blended wing body aircraft. Progr Aero Sci 40(6):321-343. doi: 10.1016/j.paerosci.2004.08.001

Rapinett A (2009) Zephyr: a high altitude long endurance Unmanned Air Vehicle (Master's thesis). Surrey: University of Surrey.

Salunke NP, Juned Ahamad RA, Channiwala SA (2014) Airfoil parameterization techniques: a review. Am J Mech Eng 2(4):99-102. doi: 10.12691/ajme-2-4-1

Sobieczky H (1997) Geometry generator for CFD and applied aerodynamics. Courses and Lectures - International Centre for Mechanical Sciences 366:137-158.

The Guardian (2015) Facebook launches Aquila solar-powered drone for internet access; [accessed 2016 May 3]. https://www.theguardian.com/technology/2015/jul/31/facebook-finishes-aquila-solar-powered-internet-drone-with-span-of-a-boeing-737

A Software-in-the-Loop Simulation Scheme for Position Formation Flight of Multicopters

Davi Ferreira de Castro[1], Davi Antônio dos Santos[1]

ABSTRACT: The cooperative control of small unmanned aerial vehicles such as the multicopters has been extensively investigated worldwide for functionality augmentation and cost reduction with respect to a single larger vehicle. The present paper proposes a software-in-the-loop simulation scheme for performance evaluation and demonstration of formation flight control systems of multicopters. The simulation scheme consists of a computer network where each computer simulates one of the vehicles using the MATLAB/Simulink for implementing the local control system and the X-Plane for simulating the flight dynamics and environment. For cooperation, the local control systems exchange position data by means of the network. In order to illustrate the proposed scheme, a group of 3 octocopters is taken into consideration and a leader-follower strategy is chosen for triangular position formation, with the leader moving in a straight line with constant speed.

KEYWORDS: Software-in-the-loop simulation, Formation control, Multicopter, Micro aerial vehicle.

INTRODUCTION

Technological advances of the last 2 decades have boosted the use of micro aerial vehicles (MAV) for tasks that can be dangerous, inefficient or unsuitable for human beings. Possible applications of these vehicles include search and rescue (Tomic *et al.* 2012, environmental monitoring (Koo *et al.* 2012), buildings inspection (Eschmann *et al.* 2012), border surveillance (Girard *et al.* 2004), fire monitoring (Casbeer *et al.* 2005), powerline inspection (Li *et al.* 2010), traffic monitoring (Puri *et al.* 2007) and surveillance in urban areas (Semsch *et al.* 2009).

Recently, we have seen an increasing interest from industry and academia in MAV formation flight (Hao and Bin 2011; Han *et al.* 2013; Wang and Xin 2013; Zhu *et al.* 2013.) This is motivated by the belief that a cooperative group of small robots can perform better and with a lower cost than a single large vehicle (Wen 2012). The position formation control of a group of vehicles with communication capability can be understood as the position control of each vehicle in such a way that the relative positions of the vehicles converge to a desired disposal.

For evaluating cooperative systems involving MAV, some research groups have built outdoor experimental setups for conducting flight tests (Shim *et al.* 2005; How *et al.* 2004; King *et al.* 2004; Hoffmann *et al.* 2009). In these cases, the success of the tests are dependent on good weather conditions, availability of a safe flight area (Valenti *et al.* 2006) as well as of a team of trained people (How *et al.* 2008). Some other research groups alternatively adopt indoor flight tests relying on pose feedback provided by a stationary camera system (How *et al.* 2008; Lupashin *et al.* 2010; Michael *et al.* 2010; Stirling *et al.* 2012).

1.Departamento de Ciência e Tecnologia Aeroespacial – Instituto Tecnológico de Aeronáutica – Divisão de Engenharia Mecânica – São José dos Campos/SP – Brazil.

Author for correspondence: Davi Ferreira de Castro | Departamento de Ciência e Tecnologia Aeroespacial – Instituto Tecnológico de Aeronáutica – Divisão de Engenharia Mecânica | Praça Marechal Eduardo Gomes, 50 – Vila das Acácias | CEP: 12.228-900 – São José dos Campos/SP – Brazil | Email: davifc@ita.br

In order to maximize the probability of success and save time in flight experiments, preliminary tests based on a realistic simulation environment are often desirable. In particular, the so-called software-in-the-loop (SIL) simulation is an interesting approach. It attempts to gain realism by combining in a simulation scheme 2 or more specilized softwares, each one for modeling a specific part of the system. It can also include the flight computer, which can be evaluated before flight experiments (Pizetta *et al.* 2012). Garcia and Barnes (2010) present a SIL simulation for formation flight of electric helicopters using one computer to implement the control laws in a centralized form and others to simulate the dynamics of the fleet members. The disadvantage of this work, as reported by the authors, is that the processing and communication capabilities of the (centralized) control computer are a bottleneck.

The present paper proposes a distributed SIL simulation scheme for testing position formation control systems of multicopters. The scheme consists of a network where each computer simulates a vehicle of the fleet using the MATLAB/Simulink, for implementing the local control laws, and the flight simulator X-Plane, for simulating the flight dynamics and environment.

The paper is organized in the following manner. The next section presents the multicopter dynamic models. Thus, it is presented a position formation control method. In sequence, the proposed SIL simulation scheme is described. In "Simulation Tests" section, the proposed scheme is illustrated by testing a position formation flight of 3 octocopters. Finally, the last section concludes the paper.

MULTICOPTER MODELING

Consider a multicopter and 2 Cartesian coordinate systems (CCSs), as illustrated in Fig. 1. The body CCS $S_B \triangleq \{X_B, Y_B, Z_B\}$ is fixed on the vehicle's structure with its origin at the vehicle's center of mass (CM). The reference CCS $S_R \triangleq \{X_R, Y_R, Z_R\}$ is fixed on the ground at point O; its Z-axis is aligned with the local vertical axis.

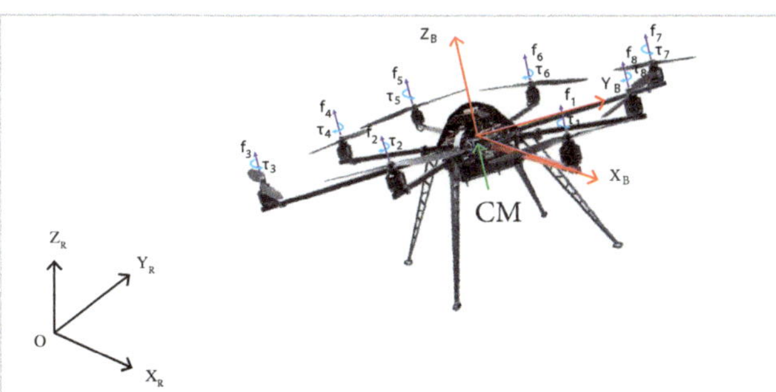

Figure 1. A multicopter and the Cartesian coordinate systems.

ANALYTICAL MODELING

Here we present analytical models for a multicopter. These models will be later compared with the corresponding X-Plane model. The attitude kinematics equation in quaternions $\mathbf{q}$ is given by the following differential one (Wertz 1978):

$$\dot{\mathbf{q}} = \mathbf{\Omega}\mathbf{q}, \tag{1}$$

where:

$$\mathbf{\Omega} = \frac{1}{2}\begin{bmatrix} 0 & -\boldsymbol{\omega}^{\mathrm{T}} \\ \boldsymbol{\omega} & -[\boldsymbol{\omega}\times] \end{bmatrix}, \tag{2}$$

and $\boldsymbol{\omega} \triangleq [\omega_1\ \ \omega_2\ \ \omega_3]^{\mathrm{T}} \in \mathbb{R}^3$ is the S_B representation of the angular velocity of S_B w.r.t. S_R and $[\boldsymbol{\omega}\times]$ is the cross-product matrix of $\boldsymbol{\omega}$, *i.e.*:

$$[\boldsymbol{\omega}\times] \triangleq \begin{bmatrix} 0 & -\omega_3 & \omega_2 \\ \omega_3 & 0 & -\omega_1 \\ -\omega_2 & \omega_1 & 0 \end{bmatrix}. \tag{3}$$

The attitude dynamics is described by the Euler equation:

$$\boldsymbol{\tau} = \dot{\mathbf{h}} + \boldsymbol{\omega} \times \mathbf{h}, \tag{4}$$

where: $\boldsymbol{\tau} \triangleq [\tau_1\ \ \tau_2\ \ \tau_3]^{\mathrm{T}} \in \mathbb{R}^3$ is the S_B representation of the total control (propeller) torque; $\mathbf{h} \triangleq \mathbf{I}\,\boldsymbol{\omega} \in \mathbb{R}^3$ is the S_B representation of the vehicle's angular momentum; $\mathbf{I} \in \mathbb{R}^{3\times3}$ is the vehicle's inertia matrix w.r.t. S_B. It is assumed that the vehicle has a symmetrical structure so that $\mathbf{I}$ can be written as:

$$\mathbf{I} = \begin{bmatrix} I_1 & 0 & 0 \\ 0 & I_2 & 0 \\ 0 & 0 & I_3 \end{bmatrix}. \tag{5}$$

Neglecting perturbations, the translational dynamics model is obtained from the Newton's Second Law as:

$$\ddot{\mathbf{r}} = \frac{1}{m} f\mathbf{v} + \begin{bmatrix} 0 \\ 0 \\ -g \end{bmatrix}, \tag{6}$$

where: $\mathbf{r} \triangleq [r_1\ r_2\ r_3]^T \in \mathbb{R}^3$ is the position of the CM w.r.t. S_R; $\mathbf{v} \triangleq [v_1\ v_2\ v_3]^T \in \mathbb{R}^3$ is the unit vector perpendicular to the plane of rotors; f is the magnitude of the total thrust; g is the gravity acceleration; m is the vehicle's mass.

Now denote the thrust and the reaction torque produced by the j-th rotor by, respectively, f_j and τ_j, $j = 1, \ldots, n$, being n the number of rotors. The usual aerodynamic models for the rotors are:

$$f_j = k_f \varpi_j^2, \tag{7}$$

$$\tau_j = k_\tau \varpi_j^2, \tag{8}$$

where: k_f is the thrust coefficient; k_τ is the reaction torque coefficient; $\bar{\omega}_j$ is the rotation speed of the jth rotor.

Define the vector of individual thrusts by $\mathbf{f} \triangleq [f_1\ f_2\ ...f_n]^T$. We can relate the magnitude f of the total thrust and the resultant torque $\boldsymbol{\tau} \triangleq [\tau_1\ \tau_2\ \tau_3]^T \in \mathbb{R}^3$ with $\mathbf{f}$ by:

$$\begin{bmatrix} f \\ \boldsymbol{\tau} \end{bmatrix} = \mathbf{\Gamma f}, \tag{9}$$

where, for the octocopter configuration taken into account in the present paper (see Fig. 1),

$$\mathbf{\Gamma} \triangleq \begin{bmatrix} 1 & 1 & 1 & 1 & 1 & 1 & 1 & 1 \\ 0 & -\frac{\sqrt{2}}{2}l & -l & -\frac{\sqrt{2}}{2}l & 0 & \frac{\sqrt{2}}{2}l & l & \frac{\sqrt{2}}{2} \\ -l & -\frac{\sqrt{2}}{2} & 0 & \frac{\sqrt{2}}{2} & l & \frac{\sqrt{2}}{2}l & 0 & -\frac{\sqrt{2}}{2}l \\ \kappa & -\kappa & \kappa & -\kappa & \kappa & -\kappa & \kappa & -\kappa \end{bmatrix}, \tag{10}$$

and $\kappa \triangleq k_\tau / k_f$ and l the arm's length.

X-PLANE MODELING

In this study, the multicopter dynamics, its flight environment (gravity, wind, etc.), as well as a graphic visualization of the mission, are modeled in Plane-Maker, the design tool associated with the flight simulator X-Plane. For illustration, we chose the Gyrofly Gyro-200ED-X8 octocopter (Fig. 2a). The corresponding X-Plane model is depicted in Fig. 2b.

The X-Plane was chosen due to the following characteristics: it is certified by the Federal Aviation Administration (FAA); it is able to simulate wind and turbulence; it permits to design customized models by means of the Plane-Marker module; it allows simultaneous visualization of multiple vehicles on the same monitor; and, finally, it can import and export data along the simulation.

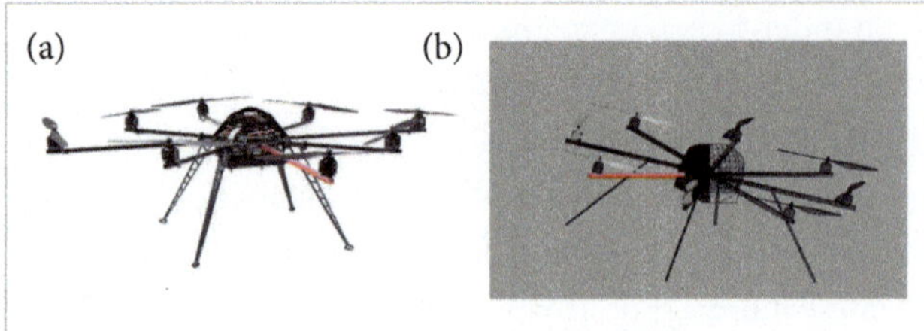

Figure 2. (a) Gyrofly Gyro-200ED-X8 octocopter; (b) X-Plane model.

Based on Blade Elements Theory, the X-Plane reads the geometric shape of the aircraft modeled in Plane-Maker and divides its surfaces into several small elements. Then, it simulates the forces acting on each element. These forces are converted into accelerations that, when integrated over time, result in velocities and translation.

The X-Plane dynamic model is generated numerically as described above. In other words, it does not give the equations of motion neither of some physical parameters such as the inertia matrix and the rotor coefficients. Therefore, for designing of control laws, we need first to identify some parameters of the X-Plane model.

PARAMETER ESTIMATION

The inputs of the analytical modeling (Eqs. 1 to 6) are the magnitude f of the total thrust and the total torque τ produced by the propellers. On the other hand, the inputs of the X-Plane model are rotor throttles $T_j \in [0, 1], j = 1, \ldots, 8$. Therefore, before estimating the rotor coefficients k_f and k_τ, and the moments of inertia I_1, I_2, and I_3, we need to convert commands of f and $\boldsymbol{\tau}$ into T_j. Denote the commands of f, $\boldsymbol{\tau}$, and $\mathbf{f}$ by $\bar{f}$, $\bar{\boldsymbol{\tau}}$, and $\bar{\mathbf{f}}$, respectively. From Eq. 9, one can write:

$$\begin{bmatrix} \bar{f} \\ \bar{\boldsymbol{\tau}} \end{bmatrix} = \mathbf{\Gamma} \bar{\mathbf{f}}, \tag{11}$$

By inverting Eq. 11, we can obtain the individual thrust commands from $\bar{f}$ and $\bar{\boldsymbol{\tau}}$, by:

$$\bar{\mathbf{f}} = \mathbf{\Gamma}^\dagger \begin{bmatrix} \bar{f} \\ \bar{\boldsymbol{\tau}} \end{bmatrix}, \tag{12}$$

where: $\mathbf{\Gamma}^\dagger \triangleq (\mathbf{\Gamma}^T \mathbf{\Gamma})^{-1} \mathbf{\Gamma}^T$ is the pseudo-inverse of $\mathbf{\Gamma}$.

The next step is to convert from $\bar{f}_j$ and $\bar{\tau}_j$ to T_j. For this end, a preliminary test was made, which showed a linear relation between these quantities, *i.e.*:

$$\bar{f}_j = k_f T_j, \tag{13}$$

$$\bar{\tau}_j = k_\tau T_j. \tag{14}$$

In order to estimate the parameters k_f and k_τ, we adopt the SIL simulation scheme that will be presented in "Software-in-the-Loop Simulation" section. The rotors are assumed to be identical and the identification is made only for the first rotor. The vehicle is commanded to hover and an additive random signal is used to excite the X-Plane model through the throttle input $T_1(k)$. Let $\bar{f}_1$ and $\bar{\tau}_1$ denote the measurements of f_1 and τ_1, respectively. We adopt the following measurement model:

$$y(k) = x(k)\Theta + \epsilon(k), \tag{15}$$

where: $y(k) \triangleq [\bar{f}_1(k)\ \bar{\tau}_1(k)]^T \in \mathbb{R}^2$ is the output measurement vector; $x(k) \triangleq [T_1(k)] \in \mathbb{R}$ is the input measurement; $\Theta \triangleq [k_f k_\tau]^T \in \mathbb{R}^2$ is the parameter vector; $\epsilon(k) \triangleq [\epsilon_1(k)\ \epsilon_2(k)]^T \in \mathbb{R}^2$ is the modeling error vector.

Applying the data set took at instants $k = 1, 2, \ldots, N$ in Eq. 15, one obtains a set of linear equations with unknown Θ. These equations can be compactly represented by the following matrix equation:

$$\mathbf{Y} = \mathbf{X}\Theta + \mathbf{E}, \tag{16}$$

where: $\mathbf{Y} \triangleq [y(1)^T y(2)^T \ldots y(N)^T]^T$, $\mathbf{X} \triangleq [x(1)^T x(2)^T \ldots x(N)]^T$, and $\mathbf{E} \triangleq [\epsilon(1)^T \epsilon(2)^T \ldots \epsilon(N)^T]^T$

Using the measurement model (Eq. 16) and the data set represented in **X** and **Y**, one can immediately obtain a least-squares estimate $\hat{\boldsymbol{\Theta}}$ of $\boldsymbol{\Theta}$ by (Ljung 1987):

$$\widehat{\boldsymbol{\Theta}} = (\mathbf{X}^T\mathbf{X})^{-1}\mathbf{X}^T\mathbf{Y}. \tag{17}$$

A similar procedure is adopted for estimating the moments of inertia I_1, I_2, and I_3 from tests with the SIL simulation scheme. In this case, each rotation DOF is consider separately, resulting in the estimation of the corresponding moment of inertia. The vehicle is initially commanded to hover at a point and a random signal is added to the attitude command about the considered axis. The attitude command is sampled to obtain the input measurements **X**, while the true attitude computed by the X-Plane gives the output measurements **Y**. Finally, on the basis of a measurement model similar to Eq. 16, a least-squares estimate of the moment of inertia is obtained using Eq. 17. Table 1 presents the obtained estimates.

Table 1. Estimation of the parameters k_f, k_τ, I_1, I_2, and I_3.

k_f	k_τ	I_1	I_2	I_3
28.0269 N	0.3869 N m	0.0628 kg m^2	0.0762 kg m^2	0.1014 kg m^2

A POSITION FORMATION CONTROL METHOD

PRELIMINARY DEFINITIONS

This section presents a position formation flight control method based on the leader-follower strategy (Santos 2013). To begin with, consider a team of N multicopters and the CCSs of Fig. 3. The body's CCSs are denoted by $S_B^i = \{X_B^i, Y_B^i, Z_B^i\}$, $i \in I \triangleq \{1, 2, \ldots. N\}$ The ith body CCS S_B^i is assumed to be attached to the structure of the i-th multicopter and centered at its center of mass CM^i. The reference CCS is denoted by $S_R = \{X_R, Y_R, Z_R\}$. It is assumed to be fixed on the ground at point O.

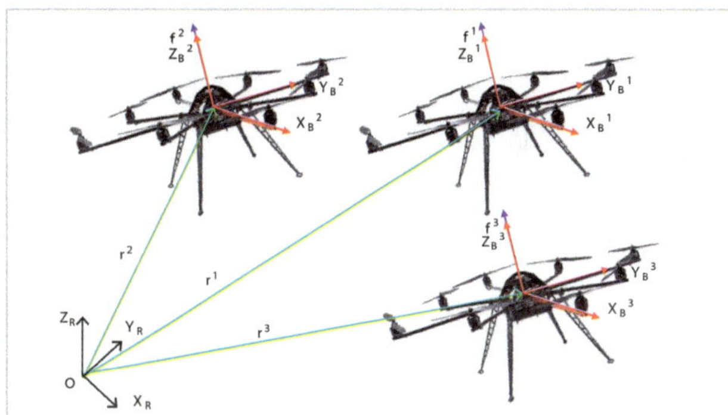

Figure 3. A team of N multirotor vehicles and the Cartesian coordinate systems.

Assume that S_R is an inertial frame and neglect any disturbance force. The direct application of the Second Newton's Law gives the following translational dynamic models represented in S_R.

$$\ddot{\mathbf{r}}^i = \frac{1}{m^i} f^i \mathbf{v}^i + \begin{bmatrix} 0 \\ 0 \\ -g \end{bmatrix}, \forall i \in \mathcal{I}, \tag{18}$$

where: $\mathbf{r}^i \triangleq [r_1^i\ r_2^i\ r_3^i]^T \in \mathbb{R}^3$ denotes the three-dimensional position of CM^i; $\mathbf{v}^i \in \mathbb{R}^3$ is the unit vector normal to the rotor plane of the ith multicopter; m^i is the mass of the ith multicopter; f^i is the magnitude of the total thrust produced by the rotors of the ith multicopter.

Define the total thrust vector $\mathbf{f}^i \in \mathbb{R}^3$ of the ith vehicle by:

$$\mathbf{f}^i \triangleq f^i \mathbf{v}^i, \forall i \in \mathcal{I}. \tag{19}$$

and the inclination angle ϕ^i of $\mathbf{v}^i$ as:

$$\phi^i \triangleq \cos^{-1} v_3^i, \forall i \in \mathcal{I}. \tag{20}$$

Note that ϕ^i consists of the angle between $\mathbf{v}^i$ and Z_R.

Let $\mathbf{r}_d^i \triangleq [r_{d,1}^i \; r_{d,2}^i \; r_{d,3}^i]^T \in \mathbb{R}^3$ denote the position command for the i-th vehicle and define the corresponding tracking error $\tilde{\mathbf{r}}^i \in \mathbb{R}^3$ by:

$$\tilde{\mathbf{r}}^i \triangleq \mathbf{r}^i - \mathbf{r}_{\mathrm{d}}^i, \forall i \in \mathcal{I}. \tag{21}$$

Let the position formation be described by the set:

$$\mathcal{F} \triangleq \{\mathbf{r}_{\mathrm{r}}, \boldsymbol{\rho}^1, \boldsymbol{\rho}^2, ..., \boldsymbol{\rho}^N\}, \tag{22}$$

where: $\mathbf{r}_r \in \mathbb{R}^3$ is defined as the reference position for the formation; $\boldsymbol{\rho}_i \in \mathbb{R}^3$, $\forall_i \in I$, are the actual relative positions of the vehicles w.r.t. $\mathbf{r}_r$. Using the same notation, one can represent the corresponding desired formation by:

$$\mathcal{F}_{\mathrm{d}} \triangleq \{\mathbf{r}_{\mathrm{rd}}, \boldsymbol{\rho}_{\mathrm{d}}^1, \boldsymbol{\rho}_{\mathrm{d}}^2, ..., \boldsymbol{\rho}_{\mathrm{d}}^N\}, \tag{23}$$

where: $\mathbf{r}_{rd} \in \mathbb{R}^3$ is the desired relative reference position of the formation; $\boldsymbol{\rho}_d^i \in \mathbb{R}$, $\forall_i \in I$, are the desired positions of the vehicles w.r.t. $\mathbf{r}_{rd}$.

For all $i \in I$, let $\phi_{max}^i \in \mathbb{R}$ be the maximum allowable value of the inclination angle ϕ^i and $f_{min}^i \in \mathbb{R}$ and $f_{max}^i \in \mathbb{R}$ be, respectively, the minimum and the maximum allowable values of the total thrust magnitudes f^i. The position formation control problem is to find N feedback control laws for $\mathbf{f}^i$, $\forall_i \in I$, which make $F \to F_d$ while respecting the constraints $f_{min}^i \leq f^i \leq f_{max}^i$ and $\phi^i \leq \phi_{max}^i$, $\forall_i \in I$.

LOCAL POSITION CONTROL

In order to solve the afore-defined problem, we first establish the local position control law to be adopted by the i-th multicopter. This control law is partitioned into an altitude controller and a horizontal position controller.

Altitude Control

Let $r_3^i \in \mathbb{R}$, $v_3^i \in \mathbb{R}$, and $r_{d,3}^i \in \mathbb{R}$ denote the vertical projections of $\mathbf{r}^i$, $\mathbf{v}^i$, and $\mathbf{r}_d^i$, respectively. Consider the minimum $f_{min}^i \geq 0$ and maximum $f_{max}^i > f_{min}^i$ allowable values for the thrust magnitude f^i. Assume that m^i, g, and $v_3^i \neq 0$ are exactly known. The altitude feedback control law is given by (Santos *et al.* 2013):

$$f^i = \begin{cases} f_{\mathrm{min}}^i, & \alpha_3^i < f_{\mathrm{min}}^i \\ \alpha_3^i, & \alpha_3^i \in [f_{\mathrm{min}}^i, f_{\mathrm{max}}^i] \\ f_{\mathrm{max}}^i, & \alpha_3^i > f_{\mathrm{max}}^i \end{cases}, \tag{24}$$

with

$$\alpha_3^i \triangleq \frac{m^i}{v_3^i}\left(g - k_1\left(r_3^i - r_{\mathrm{d},3}^i\right) - k_2 \dot{r}_3^i\right) \tag{25}$$

The above control law respects the total thrust magnitude constraint $f_{min}^i \leq f^i \leq f_{max}^i$ and is such that, if $f_{min}^i \leq \alpha_3^i \leq f_{max}^i$, then the vehicle altitude r_3^i responds to the altitude command $r_{d,3}^i$ as if it were governed by a second-order linear time-invariant model. The gains k_1 and k_2 are design parameters and have effect of proportional and derivative actions, respectively.

Horizontal Position Control

Let $\mathbf{r}_{12}^i \triangleq [r_1^i \; r_2^i]^T \in \mathbb{R}^2$, $\mathbf{v}_{12}^i \triangleq [v_1^i \; v_2^i]^T \in \mathbb{R}^2$, and $\mathbf{r}_{d,12}^i \in \mathbb{R}^2$ denote the horizontal projections of $\mathbf{r}^i$, $\mathbf{v}^i$, and $\mathbf{r}_d^i$, respectively. Consider the maximum allowable value $\phi_{max}^i > 0$ for the inclination angle ϕ^i. Assume that m^i and $f^i \neq 0$ are exactly known. The horizontal-position feedback control law is given by (Santos *et al.* 2013):

$$\mathbf{v}_{12}^i = \begin{cases} \boldsymbol{\alpha}_{12}^i, & \|\boldsymbol{\alpha}_{12}^i\| \leq \sin\phi_{\mathrm{max}}^i \\ \sin\phi_{\mathrm{max}}^i \frac{\boldsymbol{\alpha}_{12}^i}{\|\boldsymbol{\alpha}_{12}^i\|}, & \|\boldsymbol{\alpha}_{12}^i\| > \sin\phi_{\mathrm{max}}^i \end{cases}, \tag{26}$$

with

$$\boldsymbol{\alpha}_{12}^i \triangleq \frac{m^i}{f^i}\left[k_3\left(\mathbf{r}_{12}^i - \mathbf{r}_{\mathrm{d},12}^i\right) + k_4 \dot{\mathbf{r}}_{12}^i\right]. \tag{27}$$

The above control law respects the inclination angle constraint $\phi^i \leq \phi_{max}^i$ and is such that, if $\|\alpha_{12}^i\| \leq \sin\phi_{max}^i$, then r_{12}^i responds to the horizontal position command $\mathbf{r}_{d,12}^i \triangleq [r_{d,1}^i \; r_{d,2}^i]^T$ as if it were governed by a pair of second-order linear time-invariant models. The gains k_3 and k_4 are design parameters and have effect of proportional and derivative actions, respectively.

The control law (Eq. 26) provides the horizontal projection $\mathbf{v}_{12}^i$ of the normal vector $\mathbf{v}^i$. Since $\mathbf{v}^i$ is a unit vector, one can thus write:

$$\mathbf{v}^i = \begin{bmatrix} \mathbf{v}_{12}^i \\ \sqrt{1 - \|\mathbf{v}_{12}^i\|^2} \end{bmatrix}. \tag{28}$$

Therefore, Eqs. 24 and 28 give together a solution $\mathbf{f}^i = f^i\mathbf{v}^i$ to the local control problem of the ith multicopter.

LEADER-FOLLOWER STRATEGY

After establishing the stable local position controllers, it remains to define the position formation strategy. Here we adopt a leader-follower strategy with a single leader.

Consider $\mathbf{r}_r \equiv \mathbf{r}^{(1)}$, which corresponds to attribute the leadership to vehicle $i = 1$. Assume that the local control laws $\mathbf{f}^i = f^i\mathbf{v}^i, \forall_i \in I$, are asymptotically stable, *i.e.* $\bar{\mathbf{r}}^i \to 0, \forall_i \in I$, for an arbitrary position command $\mathbf{r}_d^i, \forall_i \in I$. Therefore, if the position commands of the individual multicopters are chosen to be:

$$\mathbf{r}_d^1 = \mathbf{r}_{rd}, \tag{29}$$

$$\mathbf{r}_d^q = \mathbf{r}^{(1)} + \rho_d^q, \; q = 2, 3, ..., N, \tag{30}$$

then the control laws $\mathbf{f}^i, \forall_i \in I$, make $F \to F_d$ (Santos 2013).

In order to specify the translation of the fleet, the designer needs to choose the desired leader position $\mathbf{r}_{rd}$ as an appropriate time-function. On the other hand, for specifying the form of the formation, the designer needs to establish the desired relative positions of the followers ρ_d^q w.r.t. the leader.

SOFTWARE-IN-THE-LOOP SIMULATION

The present section proposes a software-in-the-loop simulation scheme for evaluation and demonstration of position formation control of multicopters. Particularly, we take into consideration a leader-follower formation with one leader and 2 followers. Each one of the vehicles is simulated by a separate desktop computer, where the control laws are implemented in MATLAB/Simulink and the flight dynamics and environment are simulated in the flight simulator X-Plane.

The 3 computers are interconnected by an Ethernet network (Fig. 4). Computer 1 simulates the leader multicopter, while computers 2 and 3 simulate the followers 1 and 2, respectively. The data exchaged in the network throughtout the simulation is the leader's position $\mathbf{r}^1$, which is sent from the leader to the 2 followers.

The use of a separate computer for simulating each multicopter is justified by the X-Plane's inability to simulate more than 1 vehicle in a single computer. Nevertheless, despite the use of a distributed simulation architecture, the X-Plane still allows simultaneous visualization of up to 20 vehicles on the same monitor by means of its multiplayer function. The proposed scheme uses this function to show the formation flight on a larger screen (Fig. 4b).

For communicating with external processes and machines, the X-Plane adopts the User Datagram Protocol (UDP). Unlike the Transmission Control Protocol (TCP), the UDP does not assure that data packages will arrive completetly and orderly (Garcia and Barnes 2010). However, the communication via UDP is extremely fast and makes efficient use of the bandwidth (Liu *et al.* 2002), which are advantageous characteristics for the simulation under consideration.

In the proposed scheme, the UDP interface between MATLAB/Simulink and X-Plane is used in 2 directions. On one hand, the X-Plane sends to Simulink the vehicle's position, linear velocity, Euler angles, and angular velocity. On the other hand, the Simulink sends rotor throttle commands to the X-Plane. The present study considers a sampling frequency of 90 Hz.

To enable the X-Plane exchanging data with Simulink, both softwares running in the same computer, a loopback network is adopted, *i.e.* the computer sends data via the UDP network

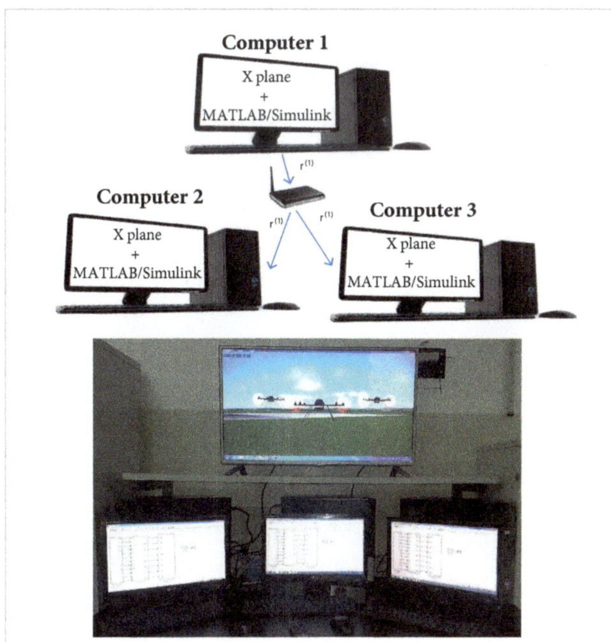

Figure 4. (a) SIL simulation scheme; (b) Real simulation environment.

to its own address. Ribeiro and Oliveira (2012) and Garcia and Barnes (2010) present more details about UDP communication between MATLAB/Simulink and X-Plane.

To perform the communication between computers it was set an Ethernet network with star topology, by assigning a fixed IP address for each computer. This network was assembled through a standard router that has an internal switch. The switch allows data to be sent directly to the desired port, avoiding excessive data traffic on the network. The transmission is performed through twisted pair cables.

SIMULATION TESTS

In order to validate the SIL scheme, a pure MATLAB simulation using the analytical models given in "Analytical Modeling" section is also implemented for comparison. In the pure MATLAB simulation, numerical integrations are performed using the 4th-order Runge-Kutta method with step of 0.002 s. The multicopter's initial positions are chosen to be: $\mathbf{r}^1 = [0\ 0\ 2]^T$ m, $\mathbf{r}^2 = [-2\ 0\ 2]^T$ m, and $\mathbf{r}^3 = [-4\ 0\ 2]^T$ m. The vehicles' mass is $m = 2.132$ kg and the gravity acceleration is $g = 9.796$ m/s^2. The parameters k_1, k_2, k_3, and k_4 of the local position control laws are tuned for a time response with overshoot of $M_P \approx 0.02$ m and peak time of $t_P \approx 2$ s. Moreover, the constraint parameters of the local controllers are set at $\varphi^i_{max} = 30°$, $f^i_{max} = 40$ N, and $f^i_{min} = 2$ N.

For illustrating the proposed SIL simulation scheme, the desired formation F_d (see the notation defined in Eq. 23) is chosen with:

$$\begin{aligned} \mathbf{r}_{rd}(t) &= [\lambda t\ 0\ h]^T \\ \boldsymbol{\rho}^1_d(t) &= [0\ 0\ 0]^T \\ \boldsymbol{\rho}^2_d(t) &= [-\eta\ \ -\gamma\ \ 0]^T \\ \boldsymbol{\rho}^3_d(t) &= [-\eta\ \ \gamma\ \ 0]^T \end{aligned} \quad (31)$$

where: $t \geq 0$ denotes the continuous time; $\lambda = 0.5$ m/s is the longitudinal velocity; $h = 2$ m is the formation altitude; $\eta = 2.6$ m is the longitudinal distance from the CM of the leader vehicle to the CM of the follower vehicles; $\gamma = 1.5$ m is the transversal distance from the CM of the leader vehicle to the CM of the follower vehicles.

Note that the afore-defined F_d is a triangular formation (Fig. 5). Moreover, it is worth mentioning that, by Eq. 31, the leader vehicle is commanded to follow a straight line on the horizontal plane, with constant speed λ and with an altitude h.

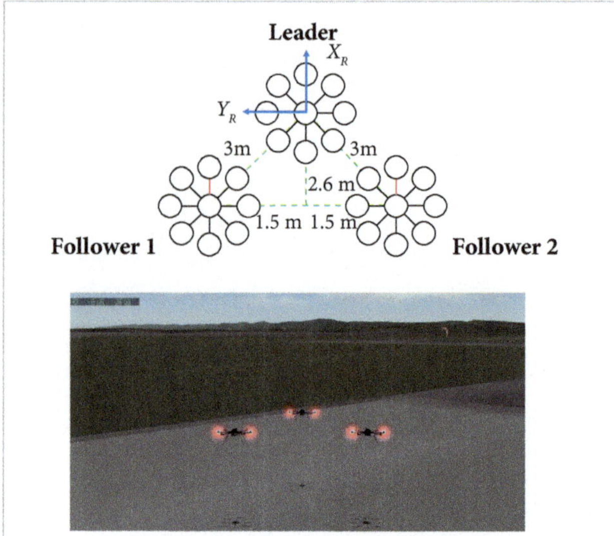

Figure 5. A triangular desired formation flight. (a) Geometry of the desired formation; (b) X-Plane visualization of the desired formation.

In order to quantify the formation error and compare the SIL simulation with the corresponding pure MATLAB simulation, the following accuracy figure of merit is adopted:

$$\mathrm{I}_a(t) \triangleq \sum_{i=1}^{3} \|\mathbf{r}_{rd}(t) + \boldsymbol{\rho}^i_d(t) - \mathbf{r}^i(t)\|. \quad (32)$$

Note that $\mathbf{I}_a(t) \geq 0, \forall_t$, is such that, if $F \rightarrow F_d$, then $\mathbf{I}_a(t) \rightarrow 0$. First of all, we conducted a test with the SIL simulation scheme in order to evaluate the performance of the network's data exchange. In this test, data are exchanged between computers of the network and compared to check if they are equal in the origin and destination. As a result, it was verified that the communication stays synchronized, as it shows no delays and no data loss. The observed characteristics are essential for a consistent operation of the proposed SIL simulation scheme.

Figure 6 shows the time response of the vehicles' translation considering the desired formation given in Eq. 31. This figure compares the SIL simulation with the pure MATLAB simulation. In both cases, the true formation converges to the desired one with a transient of less than 5 s and an steady-state error of less than 0.02 m. These results also show that the performance observed

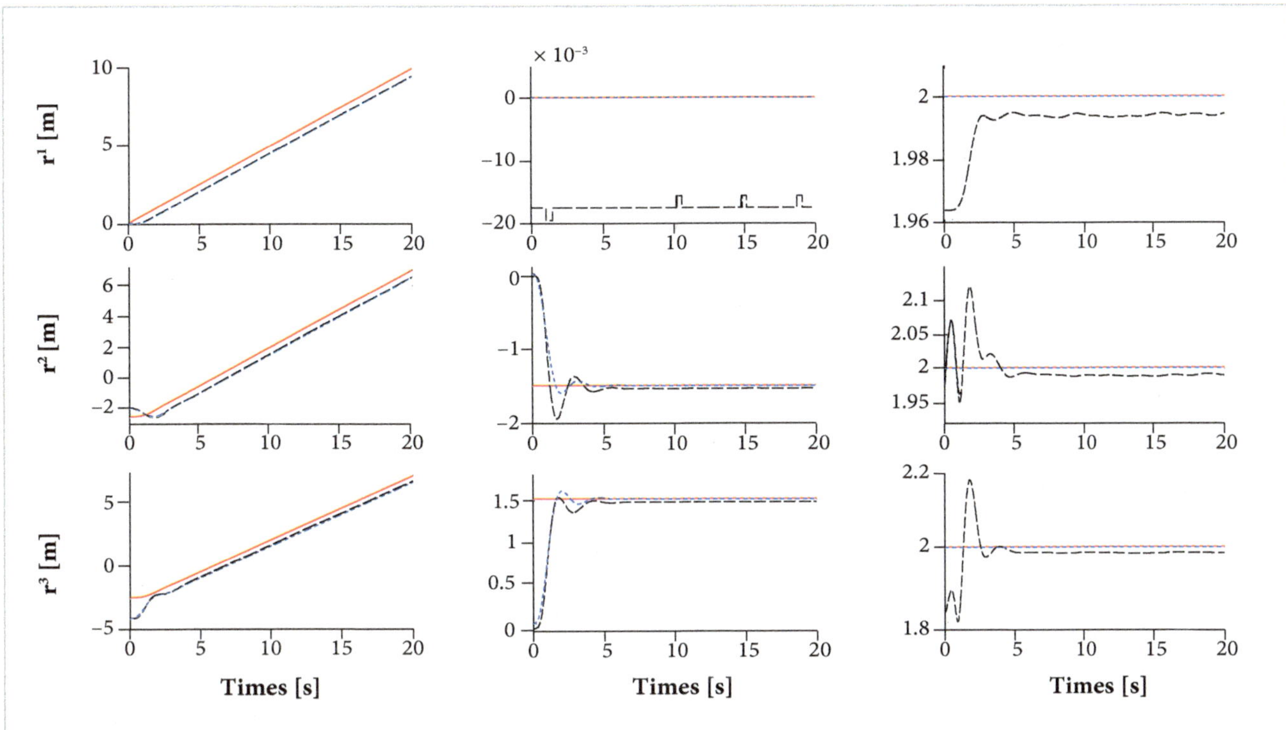

Figure 6. Time response to the formation command (Eq. 31). The solid red lines represent the components of $\mathbf{r}_d^{[1]}$, $\mathbf{r}_d^{[2]}$, and $\mathbf{r}_d^{[3]}$. The dashed black lines represent the components of $\mathbf{r}^{[1]}$, $\mathbf{r}^{[2]}$, and $\mathbf{r}^{[3]}$ obtained in the SIL simulation. The dashed blue lines represent the components of $\mathbf{r}^{[1]}$, $\mathbf{r}^{[2]}$, and $\mathbf{r}^{[3]}$ obtained in the pure MATLAB simulation.

in the SIL simulation is quite similar to that obtained using the pure MATLAB simulation, which, in some sense, validates the model of "Analytical Modeling" section with parameters identified, as described in "Parameter Estimation" section.

For a better visualization of the position formation, Fig. 7 illustrates the trajectory of the multicopters on the X-Y plane.

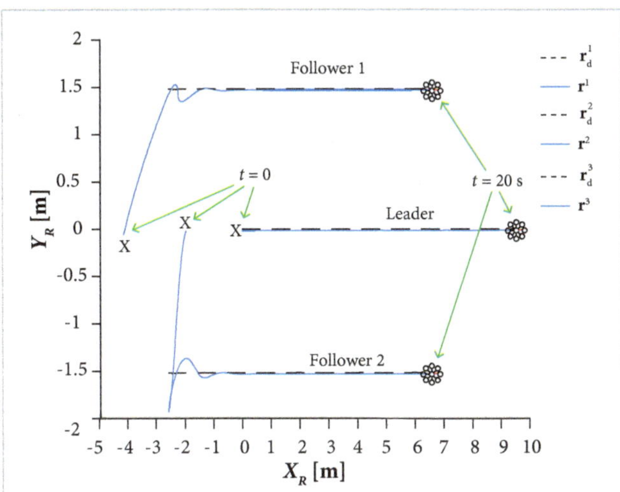

Figure 7. The position formation on the x-y plane, obtained in the SIL simulation.

Each curve shows the trajectory of 1 vehicle, starting on the left and finishing at $t = 20$ s on the right extremity. In this figure, we can see the convergence of F towards F_d.

Finally, we illustrate the position formation in the SIL simulation scheme under constant wind disturbances. Figure 8 compares the performance of the position formation using the figure of merit $I_a(t)$ for 3 different values of the wind speed v. Here, it is possible to see the impact of the wind on the perfomance of the formation flight.

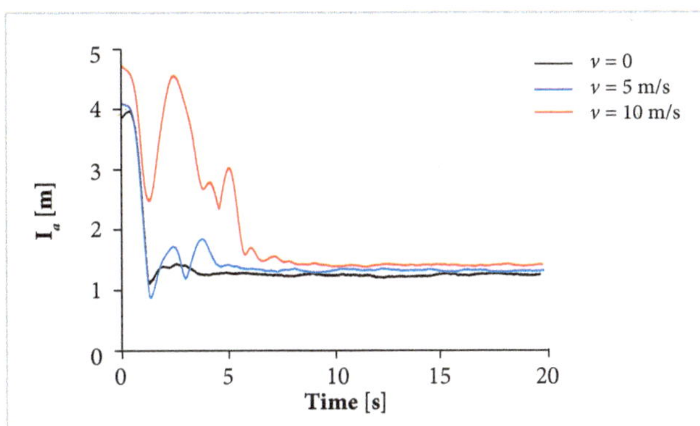

Figure 8. The position formation error for different wind velocities.

CONCLUSIONS

The paper proposed a SIL simulation scheme for demonstration and evaluation of position formation control of multicopters. The scheme is implemented in a star-topology computer network, where each computer simulates 1 multicopter of the fleet. The MATLAB/Simulink is used for implementing the control laws, while the flight simulator X-Plane models the environment, the vehicles' dynamics and provides a graphical visualization as well.

For illustrating the SIL simulation, a leader-follower strategy together with distributed linear saturated position control laws were considered. The simulation tests showed that the SIL simulation scheme proved itself to be effective and reliable to simulate a fleet of multicopters in formation flight, allowing a good visualization of the scene and providing realistic data.

The proposed SIL simulation scheme can be used effectively for preliminary evaluation of position formation control systems of multicopters, before conducting real flight tests. We argue that, by adopting such a simulation, it is possible to reduce costs, avoid risks, and save time in the development of such a fleet of micro aerial vehicles.

ACKNOWLEDGEMENTS

The first author would like to thank the Fundação de Amparo à Pesquisa do Estado do Amazonas (FAPEAM), for providing his master's scholarship during the project. Furthermore, we are grateful to Conselho Nacional de Desenvolvimento Científico e Tecnológico (CNPq), for supporting the project under the Grant 475251/2013.

AUTHOR'S CONTRIBUTION

Castro DF and Santos DA conceived the idea and co-wrote the text. Castro DF worked on the simulations results.

REFERENCES

Casbeer DW, Li SM, Beard RW, Mehra RK, McLain TW (2005) Forest fire monitoring with multiple small UAVs. Proceedings of the American Control Conference; Portland, USA.

Eschmann C, Kuo CM, Kuo CH, Boller C (2012) Unmanned aircraft systems for remote building inspection and monitoring. Proceedings of the European Workshop on Structural Health Monitoring; Dresden, Germany.

Garcia R, Barnes LE (2010) Multi-UAV simulator utilizing X-Plane. J Intell Robot Syst 57:393-406. doi: 10.1007/s10846-009-9372-4

Girard AR, Howell AS, Hedrick JK (2004) Border patrol and surveillance missions using multiple unmanned air vehicles. Proceedings of the IEEE Conference on Decision and Control; Paradise Island, Bahamas.

Han J, Hong Wang C, Xing Yi G (2013) Cooperative control of UAV based on multi-agent system. Proceedings of the IEEE 8th Conference on Industrial Electronics and Applications; Melbourne, Australia.

Hao Y, Bin J (2011) Fault tolerant cooperative control with application to UAV team longitudinal flight. Proceedings of the IEEE 30th Chinese Control Conference; Yantai, China.

Hoffmann GM, Waslander SL, Vitus MP, Huang H, Gillula JH, Pradeep V, Tomlin CJ (2009) Stanford testbed of autonomous rotorcraft for multi-agent control. Proceedings of the IEEE/RSJ International Conference on Intelligent Robots and Systems; St Louis, USA.

How J, King E, Kuwata Y (2004) Flight demonstrations of cooperative control for UAV teams. Proceedings of the AIAA 3rd "Unmanned Unlimited"Technical Conference, Workshop and Exhibit; Chicago, USA.

How JP, Behihke B, Frank A, Dale D, Vian J (2008) Real-time indoor autonomous vehicle test environment. IEEE Contr Syst Mag 28(2):51-64. doi: 10.1109/MCS.2007.914691

King E, Kuwata Y, Alighanbari M, Bertuccelli L, How J (2004) Coordination and control experiments on a multi-vehicle testbed. Proceedings of the IEEE American Control Conference; Boston, USA.

Koo VC, Chan YK, Gobi V, Chua MY, Lim CH, Lim CS, Thum CC, Lim TS, Ahmad Z, Mahmood KA, Shahid MH, Ang CY, Tan WQ, Tan PN, Yee1 KS, Cheaw WG, Boey HS, Choo AL, Sew BC (2012). A new unmanned aerial vehicle synthetic aperture radar for environmental monitoring. PIER 122:245-268. doi: 10.2528/PIER11092604

Li Z, Li Y, Walker R, Hayward R, Zhang J (2010) Towards automatic power line detection for a UAV surveillance system using pulse coupled neural filter and an improved Hough transform. Mach Vis Appl 21:677-686. doi: 10.1007/s00138-009-0206-y

Liu PX, Meng M, Ye X, Cui J (2002) An UDP-based protocol for internet robots. Proceedings of the 4th World Congress on Intelligent Control and Automation; Shanghai, China.

Ljung L (1987) System identification: theory for the user. 2nd edition. Englewood Cliffs: Prentice Hall.

Lupashin S, Schollig A, Sherback M, D'Andrea R (2010) A simple learning strategy for high-speed quadrocopter multi-flips. Proceedings of the IEEE International Conference on Robotics and Automation; Anchorage, USA.

Michael N, Mellinger D, Lindsey Q, Kumar V (2010) The GRASP multiple micro-UAV testbed. IEEE Robot Autom Mag 17(3):56-65. doi: 10.1109/MRA.2010.937855

Pizetta IHB, Brandão AS, Sarcinelli Filho M, Carelli R (2012) Uma plataforma hardware-in-loop para vants de asas rotativas. Proceedings of the XIX Congresso Brasileiro de Automática; Campina Grande, Brazil.

Puri A, Valavanis KP, Kontitsis M (2007) Statistical profile generation for traffic monitoring using real-time UAV based video data. Proceedings of the IEEE Mediterranean Conference on Control and Automation; Athens, Greece.

Ribeiro LR, Oliveira NMF (2012) UAV autopilot controllers test platform using Matlab/Simulink and X-Plane. Proceedings of the ASEE/IEEE Frontiers in Education Conference; Seattle, USA.

Santos DA (2013) Automatic position control of multirotor helicopters. Paris: ESIEE. Technical report.

Santos DA, Saotome O, Cela A (2013) Trajectory control of multirotor helicopters with thrust vector constraints. Proceedings of the IEEE 21st Mediterranean Conference on Control and Automation; Chania, Greece.

Semsch E, Jakob M, Pavlicek D, Pechoucek M (2009) Autonomous UAV surveillance in complex urban environments. Proceedings of the International Joint Conference on Web Intelligence and Intelligent Agent Technology; Stevenage, United Kingdom.

Shim DH, Chung H, Kim HJ, Sastry S (2005) Autonomous exploration in unknown urban environments for unmanned aerial vehicles. Proceedings of the AIAA Guidance, Navigation, and Control Conference and Exhibit; San Francisco, USA.

Stirling T, Roberts J, Zufferey JC, Floreano D (2012) Indoor navigation with a swarm of flying robots. Proceedings of the IEEE International Conference on Robotics and Automation; St Paul, USA.

Tomic T, Schmid K, Lutz P, Domel A, Kassecker M, Mair E, Grixa IL, Ruess F, Suppa M, Burschka D (2012) Toward a fully autonomous UAV: Research platform for indoor and outdoor urban search and rescue. IEEE Robot Autom Mag 19(3):46-56. doi: 10.1109/MRA.2012.2206473

Valenti M, Bethke B, Fiore G, How JP, Feron E (2006) Indoor multi-vehicle flight testbed for fault detection, isolation and recovery. Proceedings of the AIAA Guidance, Navigation and Control Conference and Exhibit; Keystone, USA.

Wang J, Xin M (2013) Integrated optimal formation control of multiple unmanned aerial vehicles. IEEE Trans Contr Syst Tech 21(5):1731-1744. doi: 10.1109/TCST.2012.2218815

Wen G (2012) Distributed cooperative control for multi-agent systems (PhD thesis). Villeneuve-d'Ascq: École Centrale de Lille.

Wertz JR (1978) Spacecraft attitude determination and control. Dordrecht, Boston: Reidel.

Zhu S, Wang D, Low CB (2013) Cooperative control of multiple UAVs for moving source seeking. Proceedings of the IEEE International Conference on Unmanned Aircraft Systems; Grand Hyatt, USA.

The Theoretical Research for the Rotor/Fuselage Unsteady Aerodynamic Interaction Problem

Liu Dawei[1], Xin Ji[2], Huang Jun[1]

ABSTRACT: Based on coupled unsteady panel/free-wake method, a universal analysis model was established, which provides a good prediction for the rotor/fuselage unsteady aerodynamic interaction. Considering the deficiencies of the traditional time-marching rotor free-wake algorithms, notably on stability and efficiency, the CB3D algorithm with 3rd-order accuracy is proposed. For solving the problem that part of the wake vortices may penetrate the fuselage, a "material line" rectification method with 3rd-order accuracy is proposed. An analysis for the model accuracy was then conducted to validate the accuracy of the new model, and a comparison against the available experimental data is performed. The simulated results show a good agreement with these experimental data. With the new model, several simulations are conducted for the typical rotor/fuselage aerodynamic interaction, and the results are analyzed.

KEYWORDS: Aerodynamic interaction, Panel method, Free-wake, Rotor/fuselage interaction.

INTRODUCTION

The rotor unsteady aerodynamic interaction is one of the research hotspots of rotorcraft. It not only has a great influence on aerodynamic general layout designing of helicopter, but also plays a more and more important role in helicopter flight dynamic modeling and simulation. Many scholars from different countries have carried out researches on this issue. This paper focuses on the prediction model about the typical aerodynamic interaction of rotor-fuselage interaction.

Helicopter rotor flow-field simulation methods can be divided into 2 categories: Computational Fluid Dynamics (CFD) and free-wake. Serious non-physical dissipation happens when the CFD method is used to simulate the blade vortex interaction. In addition, a huge amount of computation is needed for CFD method to accurately predict the rotor flow-field. By contrast, the problems mentioned will not happen in free-wake model, so it is more suitable for the calculation of unsteady aerodynamics interaction in engineering application.

In the free-wake prediction approaches, comparing to relaxation iteration algorithm (Bagai 1995), the time-stepping free-wake method can not only simulate the steady flight, but also predict the flow-field of a transient flying rotor. However, at present, the widely used PC2B algorithm (Bhagwat 2003) is a kind of predict-correction method. It has a relatively lower efficiency when considering the influence of other boundary such as the fuselage. Recently a one-step CB2D algorithm was proposed (Li and Chen 2012) and it has a higher efficiency. But divergence of free-wake structure may happen when it is used to simulate the rotor interaction problems. So, considering the deficiencies

1.Beijing University of Aeronautics and Astronautics – School of Aeronautic Science and Engineering – Beijing – China. 2.China Helicopter Research and Developing Institute – Jingdezhen – Jiangxi – China.

Author for Correspondence: Liu Dawei | Beijing University of Aeronautics and Astronautics – School of Aeronautic Science and Engineering | Xueyuan Road 37 – Haidian | Beijing – China | Email: buaapebble@sina.com of traditional CB2D algorithm on stability, an explicit CB3D algorithm with 3rd-order accuracy is proposed in this paper. CB3D eliminates the anti-dissipation effect caused by 2nd-order error terms in CB2D algorithm, which could improve the numerical stability, analysis accuracy and computational efficiency in the simulation of rotor aerodynamic interaction problems.

In the research of rotor-fuselage interaction, the literature (Bi and Leishman 1990; Leishman and Bi 1990; Smith and Betzina 1986) introduced several experiments on this issue, and the pressure distribution on fuselage surface was measured. In the theoretical analysis field, some papers (Rand 1988; Lorber and Egolf 1990; Crouse *et al.* 1992) used a prescribed wake method to predict the interaction flow-field. In view of the particularity of rotor-fuselage aerodynamic interaction, many empirical and ideal components were added into the analytical model, which limited its universal property and accuracy. The literature (Mavris *et al.* 1989; Xu *et al.* 2002; Quackenbush *et al.* 1999) used the more advanced free-wake method to simulate the rotor flow-field. In the free-wake model, only a few used tip vortices (Katz 2001) and others used a full-span free-wake model. For solving the problem that some of the wake vortices discrete points may move not physically into the fuselage surface, Lorber and Egolf (1990) placed the vortices near the fuselage as half-circle vortex segments around it, and some scholars placed the wake points which move into the fuselage in a mirror-image location. These rectification approaches lacked theoretical basis and then had a low rectification accuracy, which may lead to a less accurate aerodynamic interaction analysis model.

A coupled unsteady panel/time-stepping free-wake method is established to predict the helicopter rotor interaction flow-field. The panel method is used to simulate the influence of the helicopter fuselage. A "material line" rectification method is proposed to solve the problem that part of wake vortices may penetrate the fuselage surface. To validate the accuracy of this model in rotor interaction, a comparison of free-wake structure, rotor aerodynamic performance and the pressure on fuselage against the available experimental data is carried out, and the simulated results agree well with the experimental data.

PREDICTION MODEL

ROTOR AND FREE-WAKE AERODYNAMIC MODEL

The circulation of the blade bound vortex and the aerodynamic performance are calculated using the lifting surface theory. A series of vortex lattices are divided into the spanwise and chordwise direction, respectively. The sketch of blade lifting surface is illustrated in Fig. 1.

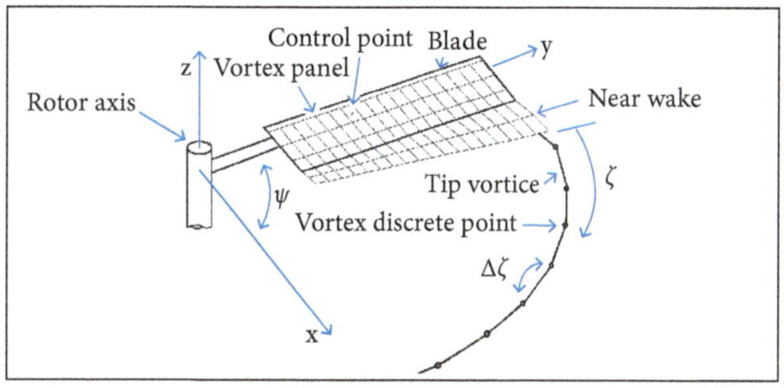

Figure 1. Figure of blade lifting and tip vortices.

From the experimental observations, the trailing vortices will roll up to a single tip vortex filament within a few chords behind the rotor blades (about 30 - 60 degrees behind blade trailing edge). The tip vortex has been documented to be the most dominant structure in the rotor flow-field. The inboard trailers have a negligible influence on the rotor performance.

But in rotor-fuselage interaction flow-field, since most part of the slim fuselage is just below the rotor blade root, the free-wake model only keeping tip vortex will be less accurate in the prediction of the pressure on fuselage surface. Thus more sophisticated model is required to determine the rotor/fuselage interaction. Provision has been made to use a full-span free-wake model including 5 - 7 trailing vortices, as shown in Fig. 2. Circulation of the vortices is determined by the maximum circulations on both sides of the vortex release point.

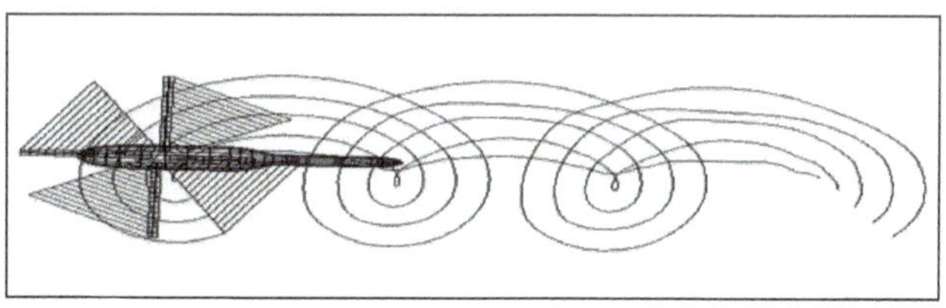

Figure 2. Full-span free-wake geometry trailing from one blade.

In the free-wake method, the vortices are force-free and convect with the local air velocity. The motion is governed by the following equation:

$$\frac{\partial \boldsymbol{r}}{\partial \psi}+\frac{\partial \boldsymbol{r}}{\partial \zeta}=\frac{1}{\Omega}\cdot[\boldsymbol{u}_{\infty}+\boldsymbol{u}_{freewake-induce}(\psi,\zeta)+ +\boldsymbol{u}_{fuselage-induce}(\psi,\zeta)]=\frac{1}{\Omega}\cdot\boldsymbol{u} \tag{1}$$

where: r is the position vector of a wake collocation point;

ψ is the azimuth angle; ζ is the vortex age; Ω is the rotor rotational speed; $\boldsymbol{u}$ is the wake velocity; $\boldsymbol{u}_\infty$ is the free stream velocity; $\boldsymbol{u}_{freewake\text{-}induce}(\psi, \zeta)$ and $\boldsymbol{u}_{fuselage\text{-}induce}(\psi, \zeta)$ are the velocities induced by free-wake and fuselage, respectively, on vortices points.

CB3D NUMERICAL INTEGRATION SCHEME

The original CB2D algorithm is shown in Eq. 2, where $\boldsymbol{r}_{l,k}$ is the position vector of the kth vortex discrete point on a vortex line shed by blade located at the l-th azimuth angle; $\boldsymbol{u}(\boldsymbol{r}_{l,k})$ is the velocity of the vortex point, and the term with a γ coefficient is the artificial dissipative one proposed by Li and Chen (2012):

$$\frac{\boldsymbol{r}_{l+1,k+1}-\boldsymbol{r}_{l,k}}{\Delta\psi}=\frac{1}{\Omega}\left[\frac{1}{2}(3\boldsymbol{u}(\boldsymbol{r}_{l,k})-\boldsymbol{u}(\boldsymbol{r}_{l-1,k-1}))\right]+ \\ +\frac{\gamma}{2}(\boldsymbol{r}_{l+1,k+1}-2\boldsymbol{r}_{l+1,k}+\boldsymbol{r}_{l+1,k-1}+\boldsymbol{r}_{l,k+1}- \\ -2\boldsymbol{r}_{l,k}+\boldsymbol{r}_{l,k-1}) \tag{2}$$

Equation 2 is obtained based on the assumption that $\Delta\psi = \Delta\zeta$, and its modified equation (expanded at the point $(l + 1/2, k + 1/2)$) is as follows ($\boldsymbol{u}_{\psi,\zeta} = \partial_2 \boldsymbol{u}/\partial\psi\partial\zeta$):

$$\frac{\partial \boldsymbol{r}}{\partial\psi}+\frac{\partial \boldsymbol{r}}{\partial\zeta}=\frac{1}{\Omega}\cdot\boldsymbol{u}-\frac{5}{12}\cdot\frac{1}{\Omega}\cdot\boldsymbol{u}_{\psi\psi}\cdot\Delta\psi^2-\frac{5}{6}\cdot\frac{1}{\Omega}\cdot\boldsymbol{u}_{\psi\zeta}\cdot \\ \cdot\Delta\psi^2-\frac{1}{3}\cdot\frac{1}{\Omega}\cdot\boldsymbol{u}_{\zeta\zeta}\cdot\Delta\psi^2+\gamma\Delta\psi^2\boldsymbol{r}_{\zeta\zeta}+ \\ +O(\Delta\psi^3) \tag{3}$$

Since the principal error terms on the right side of the equation are the 2nd-order ones, the CB2D is a 2nd-order accuracy algorithm.

According to Eq. 3 we can transform the error terms $((-5/12)g(1/\Omega)g\boldsymbol{u}_{\psi\psi}g\Delta\psi^2 - (1/3)g(1/\Omega)g\boldsymbol{u}_{\zeta\zeta}g\Delta\psi^2)$ into $1/12\ \Omega\cdot(\boldsymbol{u}_r\boldsymbol{r}_\zeta + \boldsymbol{u}_{rr}\boldsymbol{r}_\zeta^2 - 2(\boldsymbol{u}\boldsymbol{u}_r)_r\boldsymbol{r}_\zeta - 4\boldsymbol{u}(\boldsymbol{u}\boldsymbol{u}_r)_r)\Delta\psi^2$. As analyzed in the literature (Li and Chen 2012; Warming and Hyett 1974), the terms containing the 2nd-order partial derivatives $\boldsymbol{r}_{\zeta\zeta}$ and $\boldsymbol{r}_\zeta^2$, which mainly play an anti-dissipation role in the free-wake iteration, because of their coefficient variables $\boldsymbol{u}_r$, $\boldsymbol{u}_{rr}$, often become minus. So in the literature the artificial dissipative term $\gamma\Delta\psi^2\boldsymbol{r}_{\zeta\zeta}(\gamma \in (0, \Delta\psi))$ with a constant positive coefficient is added to strengthen the numerical dissipative effect, improving the stability of the numerical scheme. In the present study, an approach is proposed to further improve stability of the scheme as follows.

Note the following existing relations:

$$\boldsymbol{u}_{l,k}+\boldsymbol{u}_{l-1,k-1}-\boldsymbol{u}_{l-1,k}-\boldsymbol{u}_{l,k-1}=\Delta\psi^2\cdot(\boldsymbol{u}_{\psi\zeta})_{l+\frac{1}{2},k+\frac{1}{2}}+ \\ +O(\Delta\psi^3)$$

$$\boldsymbol{u}_{l,k+1}-2\boldsymbol{u}_{l,k}+\boldsymbol{u}_{l,k-1}=\Delta\psi^2\cdot(\boldsymbol{u}_{\zeta\zeta})_{l+\frac{1}{2},k+\frac{1}{2}}+O(\Delta\psi^3) \tag{4}$$

$$\boldsymbol{u}_{l,k+1}-2\boldsymbol{u}_{l-1,k+1}+\boldsymbol{u}_{l-2,k+1}=\Delta\psi^2\cdot(\boldsymbol{u}_{\psi\psi})_{l+\frac{1}{2},k+\frac{1}{2}}+ \\ +O(\Delta\psi^3)$$

The 2nd-order terms on the right side of the equations are just the error terms in Eq. 3. If the polynomials on the left side of Eq. 4 are added to the right side of Eq. 2, according to the coefficients of the corresponding terms in Eq. 3, then a new algorithm is obtained:

$$\frac{\boldsymbol{r}_{l+1,k+1}-\boldsymbol{r}_{l,k}}{\Delta\psi}=\frac{1}{\Omega}\left[\frac{1}{2}(3\boldsymbol{u}(\boldsymbol{r}_{l,k})-\boldsymbol{u}(\boldsymbol{r}_{l-1,k-1}))\right]+\frac{5}{12}\cdot\frac{1}{\Omega}\cdot \\ \cdot\frac{1}{\Omega}\cdot(\boldsymbol{u}_{l,k+1}-2\boldsymbol{u}_{l-1,k+1}+\boldsymbol{u}_{l-2,k+1})+\frac{5}{6}\cdot \\ \cdot\frac{1}{\Omega}\cdot(\boldsymbol{u}_{l,k}+\boldsymbol{u}_{l-1,k-1}-\boldsymbol{u}_{l-1,k}-\boldsymbol{u}_{l,k-1})+ \\ +\frac{1}{3}\cdot\frac{1}{\Omega}\cdot(\boldsymbol{u}_{l,k+1}-2\boldsymbol{u}_{l,k}+\boldsymbol{u}_{l,k-1})+ \\ +\frac{\gamma}{2}(\boldsymbol{r}_{l+1,k+1}-2\boldsymbol{r}_{l+1,k}+\boldsymbol{r}_{l+1,k-1}+\boldsymbol{r}_{l,k+1}- \\ -2\boldsymbol{r}_{l,k}+\boldsymbol{r}_{l,k-1}) \tag{5}$$

The modified equation (expanded at the point $(l + 1/2, k + 1/2)$) of Eq. 5 is as follows:

$$\frac{\partial \boldsymbol{r}}{\partial\psi}+\frac{\partial \boldsymbol{r}}{\partial\zeta}=\frac{1}{\Omega}\cdot\boldsymbol{u}+\gamma\Delta\psi^2\boldsymbol{r}_{\zeta\zeta}+O(\Delta\psi^3) \tag{6}$$

Given that $\gamma \in (0, \Delta\psi)$ (Li and Chen 2012), Eq. 5 is an algorithm of 3rd-order accuracy. This algorithm is referred to as CB3D scheme in this study. Since the 2nd-order anti-dissipative terms in Eq. 3 have been eliminated, the stability of the new scheme should be strengthened.

Next the CB2D and CB3D algorithms will be used to determine the flow-field of an isolated rotor (Li and Chen 2012), and the comparisons of iteration convergence speed and wake geometry are shown in Figs. 3a and 3b.

From Fig. 3 we can see that the convergence speed of CB3D is faster than CB2D, and the wake geometry predicted by the 2 algorithms have got a little difference, which could be attributed to the different error terms in the modified equations of the 2 algorithms.

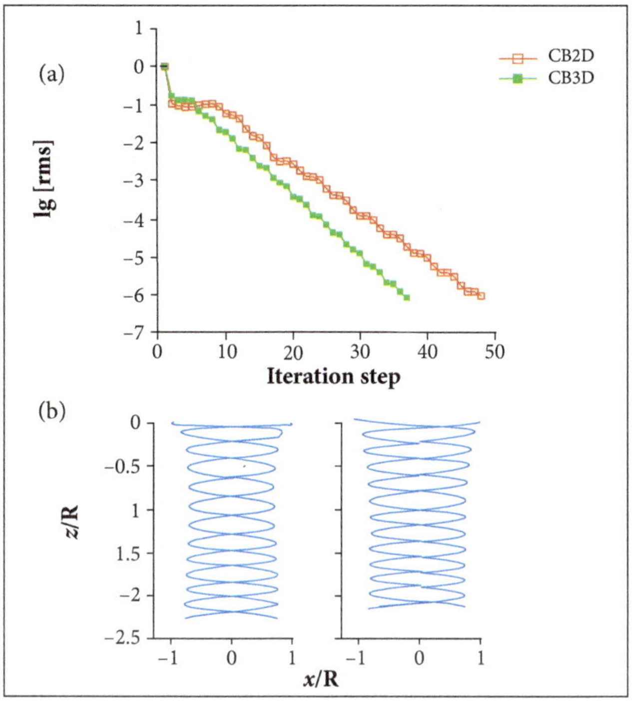

Figure 3. Comparisons of iterated convergence speed and wake geometry predicted by CB2D and CB3D. (a) Convergence speed of CB2D and CB3D; (b) Wake geometry predicted by CB2D and CB3D.

FUSELAGE MODEL

The interaction effects of fuselage on rotor flow-field are all added by the panel method. According to the low-speed aerodynamics (Katz 2001), the velocity potential ϕ at a point in space induced by the panels is determined by the distribution of doublets μ and sources σ, as follows:

$$\phi = \frac{1}{4\pi}\int_{Sf} \mu \cdot \vec{n} \cdot \nabla(\frac{1}{r})dS - \frac{1}{4\pi}\int_{Sf} \sigma \cdot (\frac{1}{r})dS \tag{7}$$

where: Sf represents the whole boundary surface; r is the distance from the doublet or source to the measured point; $\vec{n}$ is the local outward normal vector of the boundary surface.

The sources and doublets are uniformly distributed on every surface panel element. The discrete fuselage model is shown in Fig. 4.

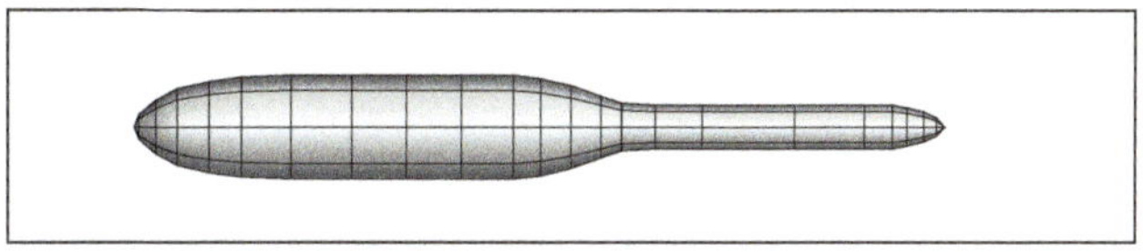

Figure 4. Discrete fuselage with panel method.

According to the research made by Katz (2001), the control point is set at the centroid of every element. The flow-field velocity on every control point needs to satisfy the no-penetration boundary conditions. For solving the doublet and source strength of every panel, the source panel strength can be written as Eq. 8:

$$\sigma = \vec{n} \cdot (\vec{V}_\infty + \vec{V}_{freewake-induce} - \vec{V}_{fuselage}) \tag{8}$$

where: $\vec{V}_\infty$ is the free stream velocity; $\vec{V}_{freewake-induce}$ is the induced velocity at control point induced by the far-wake vortex; $\vec{V}_{fuselage}$ is the motion velocity of fuselage at control point.

Based on the relationship between velocity potential and velocity, combined with the no-penetration boundary condition on solid surface, the velocity at the ith control point should satisfy the equation:

$$\begin{aligned} \vec{n} \cdot \nabla \phi_i &= \vec{n} \cdot (\sum_{k=1}^{N} C_{i,k}\mu_k + \sum_{k=1}^{N} B_{i,k}\sigma_k) + \\ &+ \vec{n} \cdot (\vec{V}_{\infty,i} + \vec{V}_{freewake-induce,i} - \vec{V}_{fuselage,i}) = 0 \end{aligned} \tag{9}$$

where: $B_{i,k}$ and $C_{i,k}$ are the induced velocity at the ith panel element induced by the unit strength source or doublet distributed at the kth panel element.

We can get N equations of N panel elements like Eq. 9. The source strengths σ_k can be achieved from Eq. 8, so the doublet strengths can be obtained by solving the N-dimensional linear equations. Then the induced velocity and the induced velocity potential at any point in the flow-field can be calculated.

Considering that the flow on the fuselage is unsteady under the influence of rotor interaction, the pressure and load on the fuselage is obtained through solving the unsteady Bernoulli Equation in:

$$\begin{aligned} P &= \frac{\rho}{2} \cdot (|\vec{V}_\infty|^2 - |\vec{V}|^2) + P_\infty - \rho \cdot \frac{\partial \phi}{\partial t} \\ C_p &= 100 \cdot (\frac{P - P_\infty}{0.5 \cdot \rho(\Omega R)^2}) \end{aligned} \tag{10}$$

where: P, $\vec{V}$ and ϕ are, respectively, the pressure, velocity and potential of the calculating point; ρ is the air density, in kg·m^{-3}; R is the blade radius, in m; P_∞ is far-field pressure; C_p is defined as the non-dimensional pressure coefficient in this paper.

The studies of Crouse *et al.* (1992) and Quackenbush *et al.* (1999) demonstrated that, in fuselage surface pressure determination, the unsteady term cannot be ignored. So, the unsteady term $\partial\phi/\partial t$ is the *aporia* in unsteady flow-field pressure prediction. If r is the distance from the potential panel to the predicted point, that is $r = \sqrt{(x-x')^2+(y-y')^2+(z-z')^2}$, considering, $\partial\phi/\partial x = (\partial\phi/\partial r)$ g $(\partial r/\partial x)$, we have, $\partial\phi/\partial x' = (\partial\phi/\partial r)$ g $(\partial r/\partial x') = 1$g $(\partial\phi/\partial r)$ g $(\partial r/\partial x)$, and the calculation formula of $\partial\phi/\partial t$ can be written as:

$$\frac{\partial\phi}{\partial t} = \frac{\partial\phi}{\partial x}\frac{\partial x}{\partial t} + \frac{\partial\phi}{\partial y}\frac{\partial y}{\partial t} + \frac{\partial\phi}{\partial z}\frac{\partial z}{\partial t} + \frac{\partial\phi}{\partial x'}\frac{\partial x'}{\partial t} + \frac{\partial\phi}{\partial y'}\frac{\partial y'}{\partial t} + \frac{\partial\phi}{\partial z'}\frac{\partial z'}{\partial t} = \vec{V}_{induced}(\vec{x})\cdot\vec{V}(\vec{x}) + (-1)\cdot\vec{V}_{induced}(\vec{x})\cdot\vec{V}_{source}(\vec{x}') \quad (11)$$

In Eq. 11, $\vec{V}_{induce}(\vec{x})$ is the induced velocity of the potential panel on the predicted point; $\vec{V}(\vec{x})$ is the motion velocity of the predicted point; $\vec{V}_{source}(\vec{x}\,')$ is the motion velocity of the potential panel. So, in the case of a stationary fuselage, we can get $\partial\phi/\partial t = -1$g $(\vec{V}_{induce}(\vec{x}))$ g $(\vec{V}_{source}(\vec{x}\,'))$.

"MATERIAL LINE" RECTIFICATION METHOD

The description of flow-field by the free-wake mathematical model is based on the Lagrange method. When updating the free-wake geometry on discrete temporal and spatial step, part of the wake vortices may not physically penetrate the fuselage surface as shown in Fig. 5.

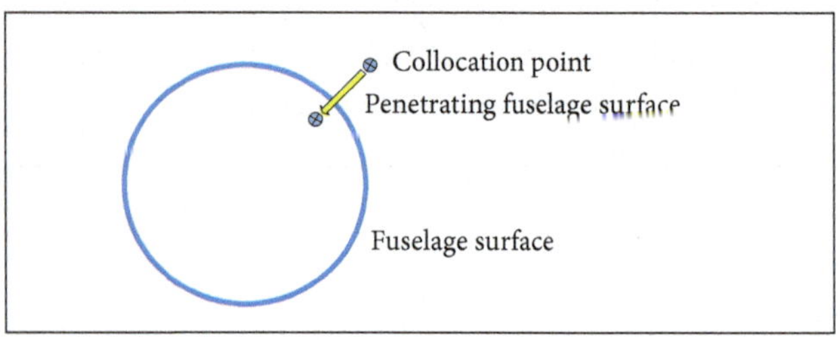

Figure 5. The rotor wake collocation point "penetrates" fuselage surface.

In view of this problem, a "material line" rectification method is proposed in this section. The method is based on the fact that the fluid micelles at the panel control points will not move inward solid surface to rectify the positions of the not physically located free-wake collocation points by interpolation.

As shown in Fig. 6, in the rectification, first, at the last discrete time step when the wake collocation point is still outside the solid surface, a panel control point near the collocation point should be selected as an auxiliary point. The auxiliary point and the collocation point are connected with a straight line, then extend the line segment to another space auxiliary point, which is demanded to not penetrate the solid surface after one time step and get as close to the collocation point as possible.

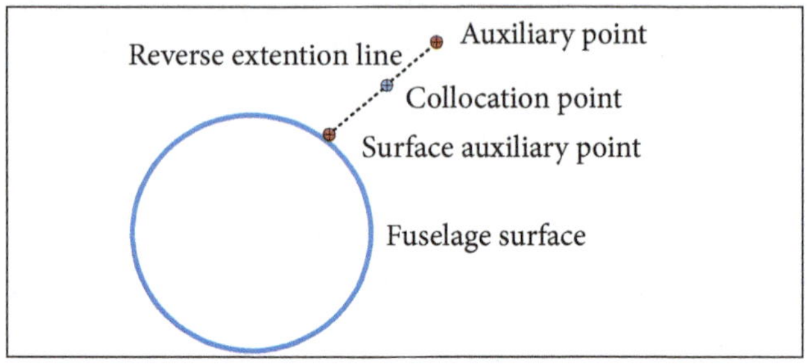

Figure 6. Normal located wake collocation point and chosen auxiliary point at the last time step.

Considering the extended line segment as a fluid material line, since one discrete time step is very short, the material line segment can be thought as a straight line segment after one time step (Fig. 7). So, if the 2 auxiliary points do not penetrate the solid surface, all the points on the material line segment do not penetrate the solid surface, and the rectified location of collocation point could be predicted by interpolation between 2 auxiliary points on the line segment.

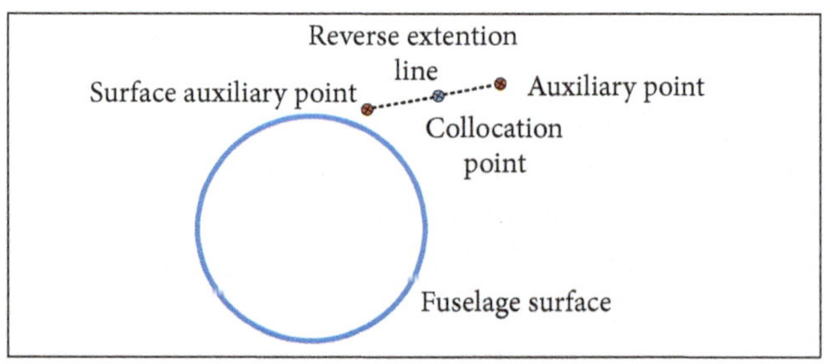

Figure 7. Location of the rotor wake collocation point after rectification.

If we write the collocation point coordinates as (x_{p0}, y_{p0}, z_{p0}), the fuselage surface auxiliary point coordinates as (x_{10}, y_{10}, z_{10}), and the auxiliary point coordinates on the extended line as (x_{20}, y_{20}, z_{20}), after one time step, the coordinates of each of the abovementioned 3 points can be written as (x_{pt}, y_{pt}, z_{pt}), (x_{1t}, y_{1t}, z_{1t}), and (x_{2t}, y_{2t}, z_{2t}), and the collocation point coordinates (x_{pt}, y_{pt}, z_{pt}) after rectification can be predicted by the following functions:

$$\begin{cases} x_{pt} = x_{1t} + \dfrac{x_{p0} - x_{10}}{x_{20} - x_{10}} \cdot (x_{2t} - x_{1t}) \\ y_{pt} = y_{1t} + \dfrac{y_{p0} - y_{10}}{y_{20} - y_{10}} \cdot (y_{2t} - y_{1t}) \\ z_{pt} = z_{1t} + \dfrac{z_{p0} - z_{10}}{z_{20} - z_{10}} \cdot (z_{2t} - z_{1t}) \end{cases} \tag{12}$$

Next, an accuracy analysis for the "material line" rectification method will be given. Let us choose the rectification formula as a representative sample for the analysis. If we consider $k = (z_{p0} - z_{10})/(z_{20} - z_{10})$, then we have:

$$\begin{aligned} z_{p0} &= k \cdot z_{20} + (1-k) \cdot z_{10} \\ z_{pt} &= k \cdot z_{2t} + (1-k) \cdot z_{1t} \end{aligned} \tag{13}$$

At the last time step, the velocity on z direction can be written as V_z, then we have:

$$\begin{aligned} k \cdot V_{z2} + (1-k) \cdot V_{z1} &= k \cdot \left(\frac{z_{2t} - z_{20}}{\Delta t}\right) + (1-k) \cdot \left(\frac{z_{1t} - z_{10}}{\Delta t}\right) \\ &= \frac{[k \cdot z_{2t} + (1-k) \cdot z_{1t}] - [k \cdot z_{20} + (1-k) \cdot z_{10}]}{\Delta t} \end{aligned} \tag{14}$$

So the interpolation for collocation point location is an interpolation for its velocity in essence. Next, this section will give an accuracy analysis for point velocity interpolation.

If we set the surface auxiliary point as original point (subscript 1) and set l axis along the "material line", the coordinates of space auxiliary point (subscript 2) in this coordinate system can be written as Δl. According to the proportional relation in Eq. 13, expanding V_{z2} and V_{z1} at location of the collocation point with a Taylor series, we can get:

$$\begin{aligned} k \cdot V_{z2} = k \cdot V_{zp} &+ k \cdot \left.\frac{dV_z}{dl}\right|_p \cdot [(1-k) \cdot \Delta l] + \\ &+ k \cdot \frac{1}{2!} \cdot \left.\frac{d^2 V_z}{dl^2}\right|_p \cdot [(1-k) \cdot \Delta l]^2 + \cdots \end{aligned} \tag{15}$$

$$\begin{aligned} (1-k) \cdot V_{z1} = (1-k) \cdot V_{zp} &+ (1-k) \cdot \left.\frac{dV_z}{dl}\right|_p \cdot [-k \cdot \Delta l] + \\ &+ (1-k) \cdot \frac{1}{2!} \cdot \left.\frac{d^2 V_z}{dl^2}\right|_p \cdot [-k \cdot \Delta l]^2 + \cdots \end{aligned} \tag{16}$$

Adding Eq. 15 and Eq. 16, we obtain:

$$V_{zp} = k \cdot V_{z2} + (1-k) \cdot V_{z1} + 0 + O(\Delta l^2) \tag{17}$$

According to the former auxiliary points method chosen, the magnitude orders of the material line length Δl and $(V_{20} g\Delta t)$, that is $(V_{20} g(\Delta\psi/\Omega))$, are close. From Eq. 17, we can see that the interpolation accuracy of "material line" rectification method for velocity is of 2nd-order. So the interpolation accuracy of the method for location is $O(\Delta l^2) g\Delta t$, whose 3rd-order is in temporal discrete scale.

RESULTS AND DISCUSSION

SIMULATION AND ANALYSIS FOR ROTOR-FUSELAGE AERODYNAMIC INTERACTION

Leishman team conducted a series of experiments (Bi and Leishman 1990; Leishman and Bi 1990) about rotor-fuselage interaction phenomenon; the outlook of fuselage model they used is similar to the actual helicopter fuselage as shown in the left part of Fig. 4. In this section, the rotor/fuselage aerodynamic interaction of Maryland model in forward flight is simulated and compared with measured data. Finally, the influences of advanced ratio on unsteady aerodynamics interaction are analyzed. The simulated typical free-wake structure in interaction is shown in Fig. 8.

As can be seen in Fig. 8, in forward flight situation, the free-wake vortex approaches the rear of fuselage. This indicates that the vortex-panel interactions become more intensive as the free-wake moves towards the backward of fuselage.

The distribution of pressure coefficient (C_p) along the fuselage left centerline at various advanced ratios is shown in Fig. 9. The abscissa in Fig. 9 means the dimensionless lengthwise position on the left centerline — X_f is the lengthwise position which is 0 at the airframe nose and L is the total length of the fuselage. As can be seen, the simulated results compare well with the experimental data, which further validates the accuracy of aerodynamic interaction model developed in this paper.

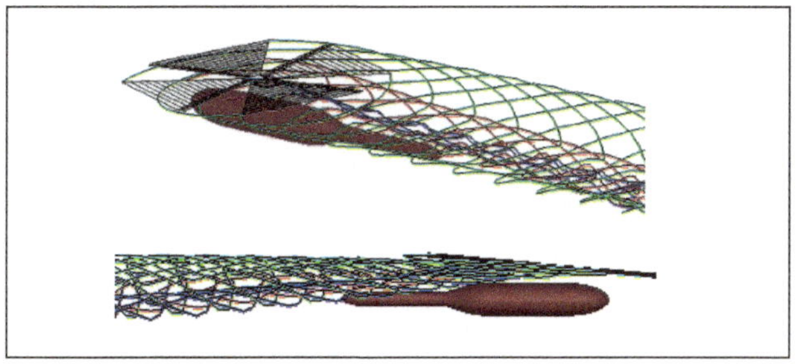

Figure 8. Space diagram and side elevation of rotor wake geometry at an advance ratio of 0.2.

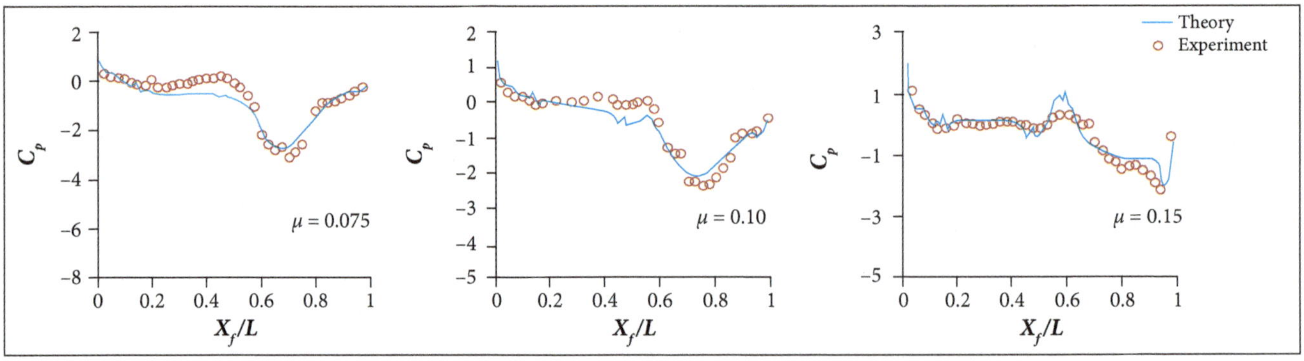

Figure 9. Comparison of predicted and experimental time average pressure on the left centerline of the fuselage surface at advance ratios of 0.075, 0.10 and 0.15 (C_T/σ = 0.085, α_s = −6°, where C_T is the rotor thrust coefficient and α_s is the rotor shaft tilt angle).

From Fig. 9 we can also point out that, as the free stream velocity increases, the pressure coefficient on the fuselage nose increases. The negative peak of pressure coefficient emerges at the rear of the fuselage, which reflects the location of vortex/panel interaction and moves backward on the fuselage as the advanced ratio increases.

CONCLUSIONS

In this paper, an unsteady panel/time-stepping free-wake method is established to predict the helicopter rotor flow-field in aerodynamic interaction conditions. The panel method is applied to simulate the influence of helicopter fuselage on flow-field. A "material line" rectification method is proposed to solve the problem that part of the wake vortices may penetrate the fuselage in numerical simulation. Considering the problem that the stability of traditional time-stepping free-wake algorithms is deficient for the simulation of unsteady aerodynamics in rotor interaction, an explicit CB3D algorithm with 3rd-order accuracy is proposed. According to this study, a few general conclusions can be drawn:

- The coupled panel/time-stepping free-wake method is proved to be accurate and efficient for predicting the rotor flow-field of rotor-fuselage interaction. The simulated results are in good agreement with available experimental data.
- The "material line" rectification method is flexible and has a 3rd-order precision. It is applicable to solve the non-physical motion problem encountered by the interaction between rotor wake geometry and solid surface.
- Full-span free-wake model improves the simulation accuracy of rotor/fuselage interaction phenomenon. As the free stream velocity increases, the pressure coefficient on the head of fuselage increases, and the negative peak of pressure coefficient at the rear of fuselage moves backward.

REFERENCES

Bagai A (1995) Contributions to the mathematical modeling of rotor flow-fields using a pseudo-implicit free-wake analysis (PhD thesis). College Park: University of Maryland.

Bhagwat M (2003) Mathematical modeling of the transient dynamics of helicopter rotor wakes using a time-accurate free-vortex methods (PhD thesis). College Park: University of Maryland.

Bi N, Leishman J (1990) Experimental study of rotor/body aerodynamic interactions. J Aircraft 27(9):779-788. doi: 10.2514/3.45938

Crouse L, Leishman J, Bi N (1992) Theoretical and experimental study of unsteady rotor/body aerodynamic interactions. J Am Helicopter Soc 37(1):55-65. doi: 10.4050/JAHS.37.55

Katz J (2001) Low-speed aerodynamics. 2nd edition. Cambridge: Cambridge University Press.

Leishman J, Bi N (1990) Aerodynamic interactions between a rotor and a fuselage in forward flight. J Am Helicopter Soc 35(3):22-31. doi: 10.4050/JAHS.35.22

Li P, Chen R (2012) Rotor unsteady aerodynamics model using an efficient free-vortex method. Aircr Eng Aerosp Tec 84(5):311-320. doi: 10.1108/00022661211255494

Lorber P, Egolf T (1990) An unsteady helicopter rotor-fuselage aerodynamic interaction analysis. J Am Helicopter Soc 35(3):32-42. doi: 10.4050/JAHS.35.32

Mavris D, Komerath N, McManhon H (1989) Prediction of aerodynamic rotor-airframe interactions in forward flight. J Am Helicopter Soc 34(4):37-46. doi: 10.4050/JAHS.34.37

Quackenbush T, Keller J, Wachspress D (1999) Reduced order free wake modeling for near real time simulation of rotorcraft flight mechanics. Proceedings of the 55th Annual Forum of the American Helicopter Society; Montreal, Canada.

Rand O (1988) Influence of interactional aerodynamics on helicopter rotor/fuselage coupled response in hover and forward flight. J Am Helicopter Soc 34(4):28-36. doi: 10.4050/JAHS.34.28

Smith C, Betzina M (1986) Aerodynamic loads induced by a rotor on a body of revolution. J Am Helicopter Soc 31(1):29-36. doi: 10.4050/JAHS.31.29

Warming R, Hyett B (1974) The modified equation approach to the stability and accuracy analysis of finite-difference methods. Journal of Computational Physics 14(2):159-179. doi: 10.1016/0021-9991(74)90011-4

Xu G, Zhao Q, Gao Z, Zhao J (2002) Prediction of aerodynamic interactions of helicopter rotor on its fuselage. Chin J Aeronaut 15(1):12-17. doi: 10.1016/S1000-9361(11)60123-4

LCL-T Resonant Converter Based on Dual Active Bridge Topology in Solar Energy Applications

Alexander Vladimirovich Osipov[1], Yury Alexandrovich Shinyakov[1], Vadim Nikolaevich Shkolniy[2], Michael Sergeevich Sakharov[1]

ABSTRACT: Resonant LCL-T converter can operate as stable voltage source, being fed from current, for instance, the photovoltaic battery. It is shown that LCL-T resonant tank has intrinsic ability to convert stable AC current into stable AC voltage thus parametrically regulating output voltage at a fixed value. This mode of operation is made possible by the use of active (synchronous) rectifier to recoup energy from the output back to the LCL-T resonant tank. Basic characteristics of resonant LCL-T converter regulated by phase shift between inverter and rectifier regardless of a solar battery current drift have been defined. It is shown that phase control guarantees 0 voltage and 0 current on switching; however, turn-off current could be substantial. Calculations and assumptions made in this study have been confirmed by simulation and hardware prototype.

KEYWORDS: Spacecraft power supply system, Resonant converter, Bidirectional dual active bridge, LCL resonant converter, ZVS, ZCS.

INTRODUCTION

Advancements in spacecraft (SC) power supply unit (PSU) are mostly related to enhancements in specific energy parameters, notably size, weight, and efficacy. Contemporary power converters in PSU are presented by non-isolated boost types controlled by pulse width modulation (PWM). The most prohibitive feature of such converters is hard commutation of switching devices, being a source of high losses as well as electromagnetic interference (EMI). The mitigation of hard switching shortcomings by introduction of various snubbing circuits to provide resonant switching transitions is not helpful due to added complexity and need for additional active elements to be commutated in the very same way (Pavlovic et al. 2012; Shiva Kumar et al. 2015; Bodur et al. 2003; Akın 2014; Goryashin and Khoroshko 2011). The commutation itself becomes too long and quite often creates unwanted oscillations thus increasing safety margin required and limiting conversion frequency as well as regulation range. All of this have led to the use of dual active bridge (DAB) resonant converters (Hillers et al. 2012; Selvaperumal et al. 2009; Sowjanya and Raghavendran 2013; Krismer and Kolar 2009; Osipov et al. 2015; Shivaraja 2015), including series LC tank, which is a resonant circuit. Sinusoidal waveform of inverter current automatically provides soft switching without axillary components. However, the energy source of SC PSU is usually a photovoltaic (PV) battery. It operates as current source while voltage is usually limited by isolation breakdown level in vacuum. At the same time load can vary a lot. Typical resonant LC converter cannot be used as a voltage regulator

1.Tomsk State University of Control Systems and Radioelectronics – Institute of Space Technologies – Research and Development Institute of Space Technologies – Tomsk/Tomsk – Russia. 2.Reshetnev Company – Information Satellite Systems – Department 600 – Zheleznogorsk/Krasnoyarsk – Russia.

Author for correspondence: Alexander Vladimirovich Osipov | Tomsk State University of Control Systems and Radioelectronics – Institute of Space Technologies | 40 Lenina Prospect | 634050 – Tomsk/Tomsk – Russia | Email: ossan@mail.ru just because no-load condition will force the PV battery to operate in voltage source mode, creating high voltage across its terminals and increasing the possibility of breakdown.

The solution for this problem is the use of LCL-T converters (usually referred as "inductive-capacitive converter" in the Russian literature), typically applied as alternating current (AC) source and fed by stable AC voltage. The performance of such converters was analyzed in Russia (Milach *et al.* 1964; Dozorov 2013) and elsewhere (Borage *et al.* 2005; Borage and Tiwari 2012; Zouggar *et al.* 2000). However LCL converters can operate just in opposite way converting current source into voltage source, which is exactly the task the converter should perform in PV battery fed by PSU. In that case input current defines resonant tank 1, which, in turn, makes the value of AC voltage at resonant capacitor near constant regardless the load. Interestingly enough no investigations of such mode of operation could be referredto. To date published research papers describe the operation of LCL-T converter with diode-based (passive or uncontrollable) rectifier in output current stabilization mode, for instance, Borage *et al.* (2005). Therefore, the purpose of this study was to analyze resonant LCL-T converter in DAB configuration as voltage source mode being fed by PV battery operating in current source mode.

METHODOLOGY

LCL-T RESONANT CONVERTER OPERATION IN CURRENT-TO-VOLTAGE CONVERSION MODE

Transistors in resonant converters are commutated at the frequency near LC tank resonant 1, which provides sinusoidal current with near 0-value at switching instances, minimizing losses. Resonant tank can be loaded in different ways depending on the input source impedance. If the PV battery operates in voltage mode then the load is connected in series with the resonant tank (typical resonant DCDC converter) (Fig. 1a). If PV battery operates as current source the load is connected in parallel to the capacitor in resonant tank creating the so-called Boucherot circuit, in which circuit inverter's output voltage is square by waveform. If the frequency equals to resonance 1 the current has sinusoidal waveform allowing switching transistors with minimal losses.

Regarding Boucherot circuit being fed by current source, capacitor's current I_{Cn} amplitude is stable and actually does not depend on the load; correspondingly output voltage is stable, picked up at terminals of the capacitor C_n, and the value is defined by the equation

$$U_{out} = I_{PV} \cdot \rho \tag{1}$$

where: $\rho = \sqrt{L_n/C_n}$ is the tank's characteristic impedance; L_n is the inductance value of resonant inductor L_n; C_n is the capacitance value of resonant capacitor C_n.

Accordingly, varying parameters of the resonant tank 1 may match the current level of PV battery to the output voltage required without the use of transformer allowing to have output voltage lower than the input one, *i.e.* $R_L < \rho$, where R_L means load resistance.

In Busherot circuit (Fig. 1a) load current is the difference between capacitor and inverter ones, while capacitor's current is stable by amplitude and is in phase with inverter's voltage; inverter current is lagging voltage U_{1M} by angle α, depending on the load (Fig. 1b):

$$\alpha = \text{arctg}\frac{\sqrt{Ln/Cn}}{R_L} = \text{arctg}\frac{\rho}{R_L} \tag{2}$$

Voltage across the load U_{Cn} is shifted in respect to inverters one by $\pi/2$.

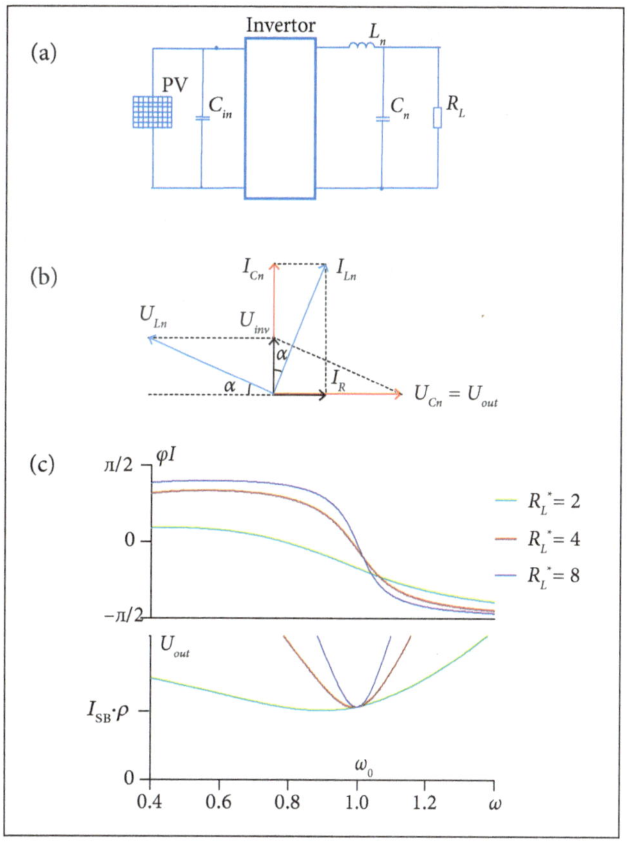

Figure 1. Series LC resonant converter. (a) Circuit; (b) Vector diagram; (c) Bode plot for phase and gain.

Frequency-related behavior of the resonant converter, plotted for different normalized load values $R_L^* = R_L/\rho$, is shown in Fig. 1c. It is clear that the shunting influence of the load shifts resonant frequency of the tank in accordance with

$$\omega_{res} = \sqrt{1/LnCn - 1/R_L^2Cn^2} \tag{3}$$

If the frequency is fixed as $\omega_0 = \sqrt{1/L_nC_n}$ the load increase yields the phase shift defined in Eq. 2; in that case output voltage of an angular frequency ω_0 becomes stable due to compensation by the inductor current. Bode plot shows that load increase leads to the loss of converter resonance properties. To reduce phase shift due to the load change LCL-T topology may be used (Fig. 2).

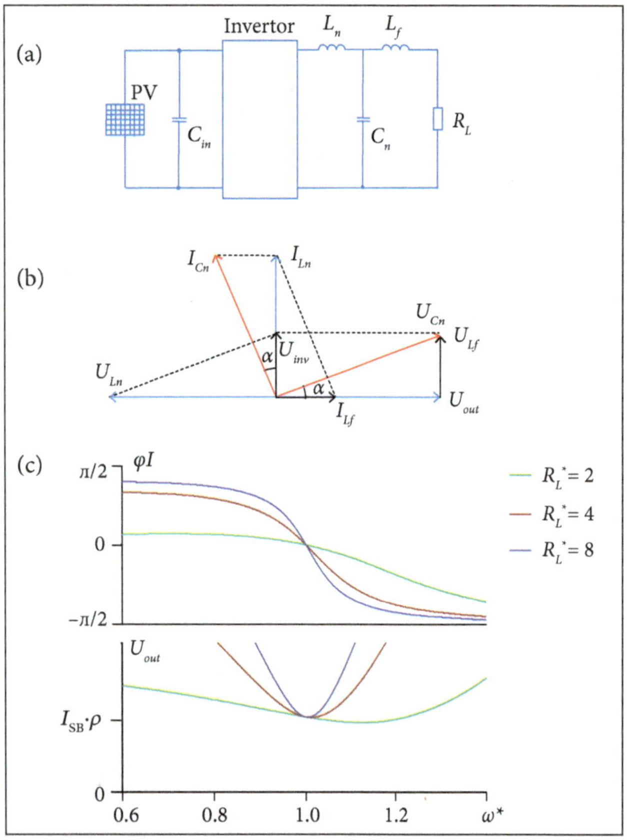

Figure 2. Resonant LCL-T converter. (a) Schematics; (b) Vector diagram; (c) Phase and amplitude responses.

The vector diagram of LCL-T converter (Fig. 2b) shows that in case of $L_n = L_f$ an additional voltage drop across L_f allows forming an angle equal to $\pi/2$ between the vectors of load current and inverter one regardless of the load. This condition can be met by the equality of right triangles formed by the vectors U_{inv}, U_{Ln}, U_{Cn}, and U_R, U_{Lf}, U_{Cn}, which leads to $U_{Ln} = U_{out}$; the stability of output voltage because the change in U_{inv} is compensated by U_{Lf}.

In that case the phase of capacitor current is shifted in regard to the inverter's one by the angle α defined by the load

$$tg\alpha = \frac{U_{PV}}{U_R} = \frac{I_L}{I_{inv}} = \frac{\rho}{R_L} \tag{4}$$

where: U_{PV} represents photovoltaic voltage.

Frequency-related parameters of LCL-T converter are shown in Fig. 2c. It can be observed that LCL-T tank allows stabilizing output voltage parametrically operating at the resonant frequency for the full load range.

LCL-T CONVERTER OPERATION ANALYSIS IN CURRENT-TO-VOLTAGE CONVERSION MODE WITH OUTPUT RECTIFIER

In order to supply constant output voltage one may need to use rectifier, bridge type, for instance (Fig. 3a), which substantially changes the converter behavior. Before anything else, the first harmonic experiences least resistance

$$R_{1M} = \frac{8}{\pi^2} R_L \tag{5}$$

as a result, the output voltage can be defined as

$$U_{out} = \frac{\pi^2}{8} I_{PV}\rho \tag{6}$$

So the resonant tank values could be defined as

$$C_n = \frac{I_{PV}}{U_{out}} \frac{\pi^2}{8} \frac{1}{\omega_0} \quad \text{and} \quad L_n = \frac{U_{out}}{I_{PV}} \frac{8}{\pi^2} \frac{1}{\omega_0} \tag{7}$$

The simulation for fixed output voltage $U_{out} = 100$ V, input current $I_{PV} = 8$ A, $f = 50$ kHz, and resonant tank parameters taken in accordance with Eq. 7 shows that output current is discontinuous, distorting rectifier input voltage U_{rect}, having $I_{Lf} = 0$ and $U_{rect} = U_{Cn}$ (Fig. 3b). The converter operates in non-resonant mode, and output voltage value does not comply with calculated value $U_{out} = 100$ V, increasing along with the load resistance. Idle operation is not possible. These problems can be partially solved by operating frequency adjustment (Borage *et al.* 2005; Zouggar *et al.* 2000).

If one uses fully-controlled switches for rectifier, operation in continuous conduction mode could be forced and the aforementioned problems fixed by recouping energy from

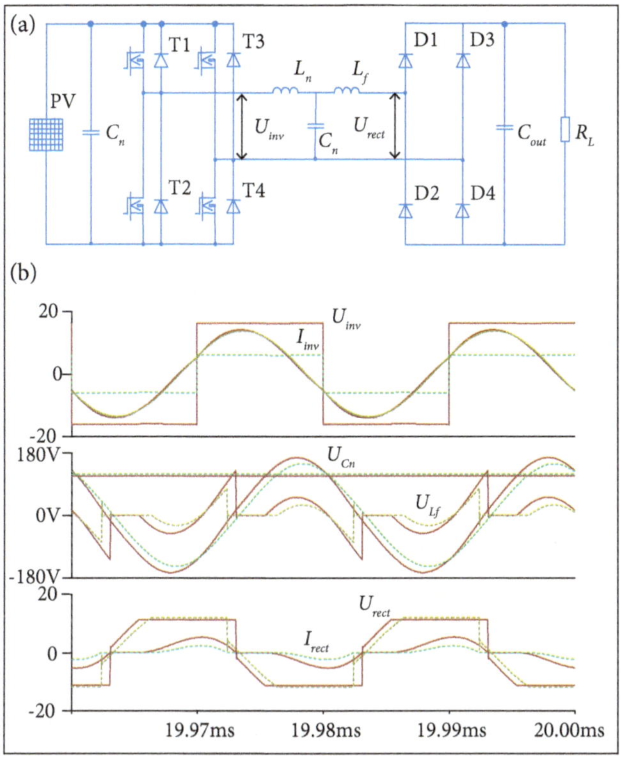

Figure 3. (a) Resonant LCL-T converter with non-active rectifier; (b) Operational waveforms at U_{out} = 100 V, I_{PV} = 8 A, and f = 50 kHz.

the output capacitor back to the resonant tank (Fig. 4a) thus making topology commonly referred as DAB. For proper operation output bridge control should be shifted by π/2 in respect to inverter. This is exactly the phase between inverter and load currents (Fig. 2b).

From the simulation it is clearly verified that, in order to stabilize output voltage at calculated value of U_{out} = 100 V, regardless of the load, one needs to operate at resonant point in continuous conduction mode for active rectifier I_{rect} (Fig. 4b). Rectifier current I_{rect} is not sinusoidal, being result of the application of 2 voltages to output inductor, rectangular U_{rect}, and sinusoidal U_{Cn}, which is specific to that topology. As a result rectifier current contains sinusoidal as well as saw tooth components:

$$I_{Lf}(t) = I_{Lf\_saw}(t) + I_{Lf\_sin}(t) = \frac{U_{out}}{4L_f f}\cdot\frac{2t}{T} + \frac{4}{\pi}\frac{U_{out}}{\omega L_f}\frac{1}{\cos\alpha}\cdot\cos(\omega t+\alpha) \quad (8)$$

Rectifier current diagrams for a different load values are shown in Fig. 5. For instance, at no load or idle capacitor's, voltage is in phase with rectifier's voltage, α = 0; saw tooth

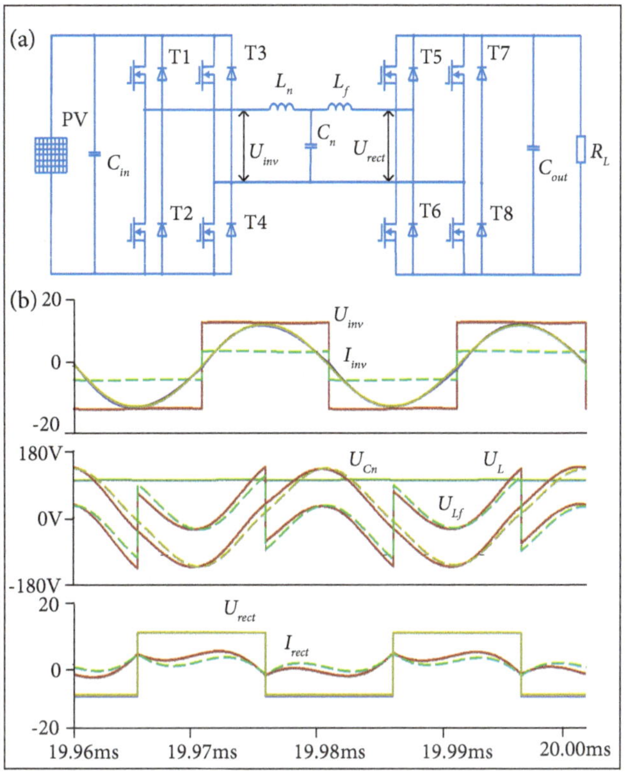

Figure 4. (a) Resonant LCL converter in DAB configuration schematics; (b) Operational waveforms at U_{out} = 100 V, I_{PV} = 8 A, and f = 50 kHz.

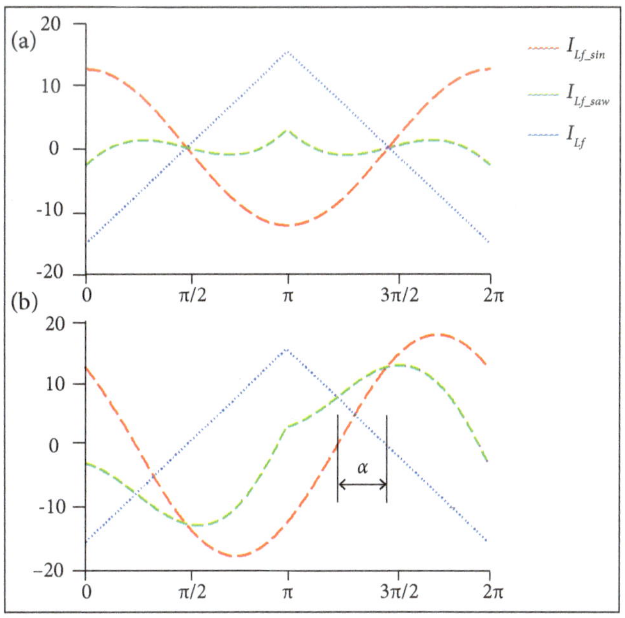

Figure 5. Rectifier input current at the idle. (a) At normalized load (no-load operation); (b) Nominal load operation: $R_L = \pi 2/8\rho$.

as well as sinusoidal components in current do have equal fundamental harmonic and, correspondingly, there is no fundamental harmonic in the output current. Besides, averaged rectified current is equal to 0 (Fig. 5a).

As we increase the load phase shift between capacitor and rectifier voltages, α starts to appear, and capacitor's voltage starts (Fig. 5b). In the case when $R_H = \pi^2/8\rho$, *i.e.*, at the value$R_H^* = 1$, phase shift $\alpha = \pi/4$ and rectifier's current contains fundamental harmonic

$$I_{\text{rect\_1M}} = \frac{\pi}{2} I_{\text{PV}} \tag{9}$$

This way output current depends on load phase angle α.

PHASE SHIFT CONTROL OF LCL-T RESONANT CONVERTER

One of the most important PV battery features is instability of the output current level, which depends on quite a few external conditions like illumination change during shade-light transition, efficiency degradation due to radiation, etc. Such effects can be compensated by the shifting control of rectifier U_{rect} in respect to inverter voltage U_{inv} by phase β. For classic DAB implementation, such method of control is described in Pavlovic *et al.* (2012). As a result of current recoup into a resonant tank, output voltage would rise. The process of control is shown in details in the vector diagram of Fig. 6a.

If we shift rectifier control by the angle β it shifts rectifier current phase to the very same value in respect to rectifier input voltage, forcing output voltage U_{out} to rise in accordance with

$$U_{\text{out}} = \frac{\pi^2}{8} I_{\text{PV}}\rho \cdot \frac{1}{\cos\beta} \tag{10}$$

It is very important to hold phase shift of π/2 between inverter voltage U_{inv} and rectifier current I_{rect} to provide conditions for parametric stabilization of the output voltage, because it will preserve the equality of triangles formed by the vectors U_{inv}, U_{Ln}, U_{Cn} and U_{rect}, U_{Lf}, U_{Cn}, respectively.

Angle α is determined by the load value similarly to the previous case and could be calculated as follows:

$$\alpha = \beta - \text{arctg}\left(\text{tg}\,\beta - \frac{1}{R_L^* \cos^2\beta}\right) \tag{11}$$

where: $R_L^* = (8/\pi^2)(RL/\rho)$ is the normalized load value.

Angle α, in turn, defines voltage across C_n:

$$U_{\text{Cn}} = \frac{\pi}{2} \cdot \frac{1}{\cos(\beta-\alpha)} I_{\text{PV}}\rho \quad \text{and}$$

$$U_{\text{Cn}}^* = \frac{\pi}{2} \cdot \frac{1}{\cos(\beta-\alpha)} \tag{12}$$

Regulation curves are shown in Fig. 7. It is clear that the angle of control α depends on the control angle β, while minimum capacitor voltage occurs if $\alpha = \beta$, and it corresponds to $U_{Cn}^* = (\pi/2)(1/\cos\beta)$; in case of $\beta = \pi/4$ and $R_L^* \to \infty$, the voltage across C_n would be $U_{Cn}^* = (\pi/\sqrt{2})$.

For a given case there are diagrams shown in Fig. 6b. If PV battery current is decreased to $I_{PV} = 5.6$ A, $\sigma I_{PV} = \sqrt{2}$ and $R_L^* = 2$, in accordance with Eq. 10; to produce previous output voltage level, control angle $\beta = \pi/4$ is necessary, meaning $\alpha = \beta$. Thus using Eq. 12 we can derive $U_{Cn} = (2\sqrt{2}/\pi)U_{out}$, which is shown in the diagram of Fig. 6b.

Commutation mode of switches in LCL-T converter using phase regulation: the condition necessary for ZVS is the lag of rectifier voltage in respect to inverter's voltage by $\pi/2 + \beta$.

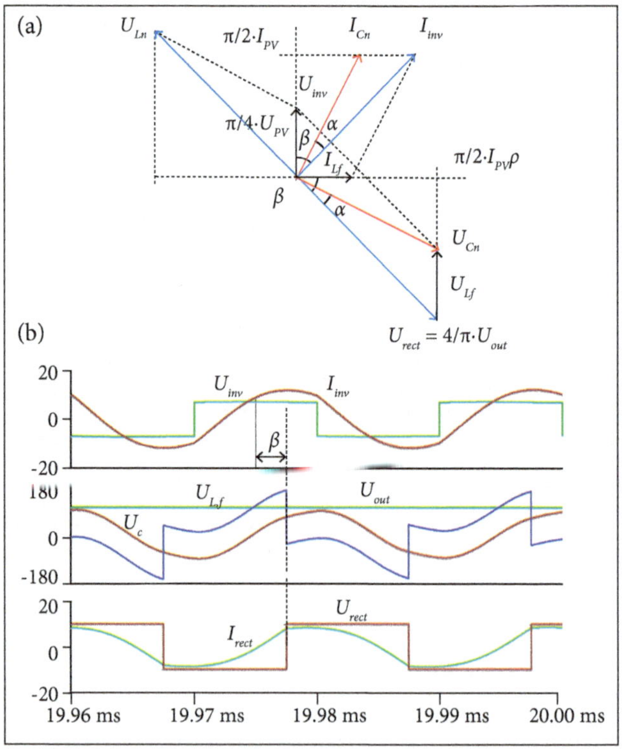

Figure 6. Phase regulation in resonant LCL-T converter with active rectifier. (a) LCL-T converter vector diagram of phase regulation process; (b) Operation diagram at $I_{PV} = 5.6$ A, $\beta = \pi/4$, $R_L = 25\,\omega$, and $R_L^* = 2$.

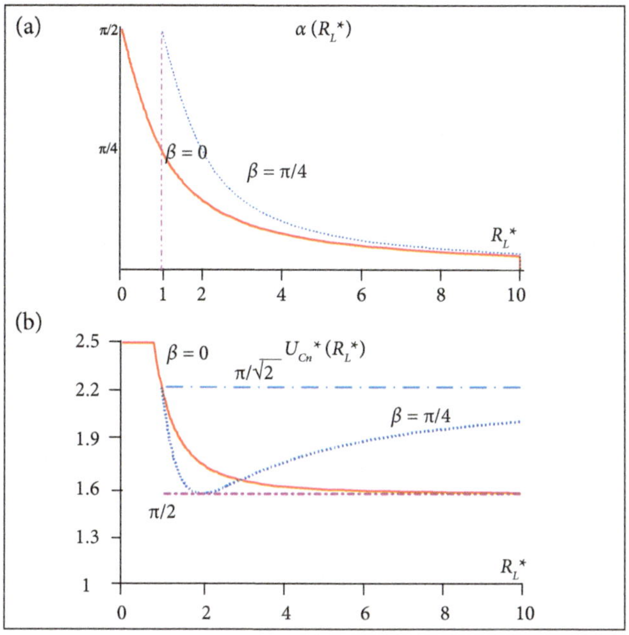

Figure 7. (a) Control angle α; (b) Resonant capacitor voltage, U_{Cn}, versus normalized load in the LCL converter.

Such control mode provides conditions for turning transistors ON. Commutation transitions depend on a bridge place. In the inverter, one has to make commutation before the current changes the sign (Fig. 8b), which provides turning in the complementary switch at reverse bias or 0 voltage across. In rectifier, one has to make commutation after the change of the current sign (Fig. 8). This way, in order to provide ZVS, one has to supply a leading current phase of the inverter and a lagging one for the rectifier.

Turn off transition occurs at rather high current; however, proper snubbing capacitor across the switch or even transistor's output capacitance together with dead time adjustment does provide proper commutation.

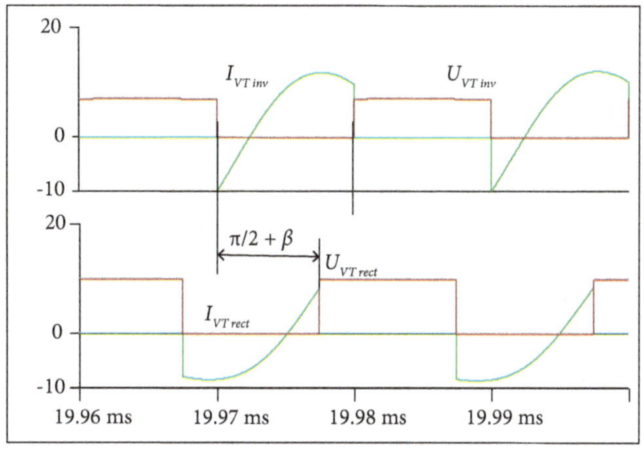

Figure 8. Current and voltage waveforms of rectifier and inverter in resonant LCL-T converter using phase control.

RESULTS

The components used in hardware prototype are shown in Table 1. To verify simulation data hardware prototype has been built (Fig. 4a) Actual operation diagrams are presented in Fig. 9. The parametric stabilization obtained 2%, and soft commutation could be observed in all cases.

It is shown that, if input current amplitude is held constant at 0.86 A (using hardware PV battery simulator IPV-200/7-4),

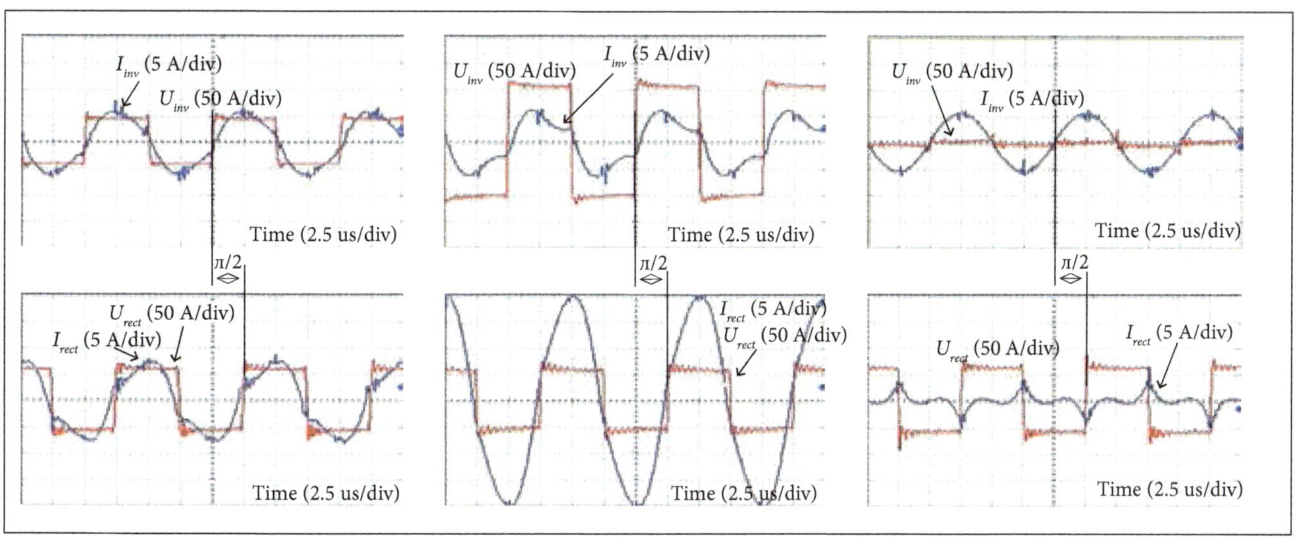

Figure 9. Operating diagrams for different load values of the LCL-T converter. (a) R_L = 18 ω; (b) R_L = 6 ω; (c) Idle (R_L = ∞).

then output voltage stays at 60 V. Variation of the load resistance does change input current consumed.

Table 1. Components used in the hardware prototype.

Part	Value
T1… T8	IRFP4668
$L_n = L_f$	15.6 uH*
C_n	0.165 uF**
Conversion frequency	102 kHz
Primary power source type	IPV-200/7-4
Input current I_{PV}	3.2 A
Load resistance	6… ∞ Ω

*ETD34/17/11 N = 28 turns; **К78-2-315В-0.033 uF 5 pcs.

CONCLUSION

Resonant LCL-T converter has the ability to convert the current source into the voltage one, which best serves for SC PSU where the primary source is PV battery. Through the use of an active rectifier, LCL-T converter provides high accuracy of parametric stabilization of the output voltage from the idle up to the full load. Moreover, for operation at the idle unlike the state-of-the-art prototypes (Pavlovic *et al.* 2012; Hillers *et al.* 2012; Selvaperumal *et al.* 2009), there is no need for frequency adjustment, which in turn eases control effort.

ACKNOWLEDGMENTS

The study was performed during the execution of a complex project (number 02.G25.31.0182) with the financial support of the Russian Government (Ministry of Education of Russia).

AUTHOR'S CONTRIBUTION

Osipov AV, Shinyakov YA, and Shkolniy VN conceived the idea and co-wrote the main text; Osipov AV and Sakharov MS performed the experiments; Sakharov MS prepared the figures. All authors discussed the results and commented on the manuscript.

REFERENCES

Akın B (2014) An improved ZVT–ZCT PWM DC–DC boost converter with increased efficiency. IEEE Trans Power Electron 29(4):1919-1926. doi: 10.1109/TPEL.2013.2269172

Bodur H, Bakan AF, Baysal M (2003) A detailed analytical analysis of a passive resonant snubber cell perfectly constructed for a pulse width modulated d.c.–d.c. buck converter. Electr Eng 85(1):45-52. doi: 10.1007/s00202-002-0141-7

Borage M, Tiwari S (2012) AC analysis of resonant converters using PSpice. A quicker approach. Asian Power Electronics Journal 6(2):1-6.

Borage M, Tiwari S, Kotaiah S (2005) Analysis and design of an LCL-T resonant converter as a constant-current power supply. IEEE Trans Ind Electron 52(6):1547-1554. doi: 10.1109/TIE.2005.858729

Dozorov SA (2013) Research and development of inductive-capacitive power source (PhD thesis). Saint Petersburg: Saint Petersburg Electrotechnical University "LETI". In Russian.

Goryashin NN, Khoroshko AY (2011) On increasing the energy efficiency of a pulsed voltage converter with a resonant switching. Bulletin SibSAU 4(37):20-24.

Hillers A, Christen D, Biela J (2012) Design of a highly efficient bidirectional isolated LLC resonant converter. Proceedings of the 15th International Power Electronics and Motion Control Conference.

Krismer F, Kolar JW (2009) Accurate small-signal model for the digital control of an automotive bidirectional dual active bridge. IEEE Trans Power Electron 24(12):2756-2768. doi: 10.1109/TPEL.2009.2027904

Milach AN, Kubyshkin BE, Volkov IV (1964) Inductive-capacitive voltage converters in current sources. Kiev: Naukova Dumka.

Osipov AV, Shinyakov Yu A, Chernaya MM, Tkachenko AA (2015) Resonant converters Solar energy. Proceedings of the 19th International Scientific Conference "Reshetnev's Readings". Krasnoyarsk; Russia.

Pavlovic Z, Oliver J, Alou P, Garcia O, Cobos J (2012) Bidirectional dual active bridge series resonant converter with pulse modulation [accessed 2016 Jun 3]. https://www.researchgate.net/publication/254019572_Bidirectional_Dual_Active_Bridge_Series_Resonant_Converter_with_pulse_modulation

Selvaperumal S, Christober C, Rajan A (2009) Embedded control of LCL resonant converter analysis, design, simulation and experimental results. Engineering 1(1):7-15. doi: 10.4236/eng.2009.11002

Shiva Kumar S, Panda AK, Ramesh T (2015) A ZVT–ZCT PWM synchronous buck converter with a simple passive auxiliary circuit for reduction of losses and efficiency enhancement. Ain Shams Engineering Journal 6(2):491-500. doi: 10.1016/j.asej.2014.10.018

Shivaraja L (2015) Modeling and simulation of LLC resonant converter for photovoltaic systems. International Journal of Emerging Technology and Innovative Engineering 1(4):175-179.

Sowjanya K, Raghavendran P (2013) Isolated bidirectional full-bridge DC-DC converter with a flyback snubber. International Journal of Soft Computing and Engineering 3(2):16-25.

Zouggar SH, Charif N, Azizi M (2000) Neural control and transient analysis of the LCL-type resonant converter. Eur Phys J Appl Phys 11(1):21-27. doi: 10.1051/epjap:2000142

21

Multidisciplinary Design Optimization of UAV Under Uncertainty

Majid Hosseini[1], Mehran Nosratollahi[1], Hossein Sadati[1]

ABSTRACT: Uncertainty-based multidisciplinary design optimization considers probabilistic variables and parameters and provides an approach to account for sources of uncertainty in design optimization. The aim of this study was to apply a decoupling uncertainty-based multidisciplinary design optimization method without any dependence on probability mathematics. Existing approaches of uncertainty-based multidisciplinary design optimization are based on probability mathematics (transformation to standard space), calculating an approximation of the constraint functions in standard space and finding the most probable point, which is the best possible one. The current approach used in this paper was inspired on interval modeling, so it is good when there is insufficient data to develop a good estimate of the probability density function shape or parameters. This approach has been implemented for an existing Unmanned Aerial Vehicle (UAV, Global Hawk) designed for purposes of comparison and validation. The advantages of the provided approach are independence of probability mathematics, appropriate when there is insufficient data to approximate the uncertainties variables, appropriate speed to calculate the best reliable response, and proper success rate in the presence of uncertainties.

KEYWORDS: Uncertainty-based multidisciplinary design optimization, MDO, Systemic design, Unmanned Aerial Vehicles design.

INTRODUCTION

Uncertainty existing at the early phases of the design process influences the system reliability. It is important to manage error early in the design process to decrease the redesign likelihood. Designing complicated and large systems such as aerospace vehicles requires appropriate compromise for gaining balance between multiple coupled targets. The targets include high performance and low costs. The sooner these compromises are understood in design process, the more technology, programming, and cost-related risks can be minimized. There are complicated relationships existing between assignment requirements, constraints, design sub-systems and contradicted targets, which could be coordinated using a suitable strategy of optimization. Multidisciplinary design optimization (MDO) or coordination between multidisciplinary analyses makes understandable more effective solutions during design and optimization of complicated systems. This allows system engineers to look for a vast scope of compromise in a systematic and thoughtful way and to consider more structures in conceptual design phase and before concentrating on final design.

Preliminary application of optimization in aerospace industries is accompanied by optimization of sub-systems or components such as aerodynamic shape, orbital path, as well as optimization of sub-systems altogether. Anyway, an optimized systemic compound will not be always created through optimization of sub-systems. In aerospace engineering, MDO was first applied in design of airplanes and to date the method has been used in academic papers and manufactured airplanes (Geethaikrishnan 2003).

In Olds (1993), González *et al.* (2005), Morris and Kroo (1990), Raymer (2002), Çavuş (2009), Neufeld (2010), Hendrich

1.Malek Ashtar University of Technology – Faculty of Aerospace – Department of Aerospace – Tehran/Tehran – Iran.

Author for correspondence: Majid Hosseini | Malek Ashtar University of Technology – Faculty of Aerospace – Department of Aerospace | Babaii Highway | Zip code: 15875-1774 – Tehran/Tehran – Iran | Email: Mjdhosseini@mut.ac.ir

(2011), Giunta (1997), Buonanno (2005), Iqbal (2009), Mattos and Secco (2013), Eisler (2003), Rowell *et al.* (1999), Goraj (2005), Lee *et al.* (2007), Tianyuan and Xiongqing (2009), Jaeger *et al.* (2013), Ahn *et al.* (2002), Perez *et al.* (2004), Sóbester and Keane (2006), Lee *et al.* (2009), Choi *et al.* (2010), Zill *et al.* (2011), and Tekinlap and Cavus (2012), the airplane conceptual designing is done by taking advantage of MDO. In these references, all fighter, passenger, and conventional planes, as well as Unmanned Aerial Vehicles (blended wing body and conventional) are considered. In a systemic attitude towards previous studies, the applied design algorithms are: MDOs in different frameworks, such as All at Once (AAO), Multiple Discipline Feasible (MDF), Collaborative (CO), and Bi-level Integrated System Synthesis (BLISS), which are made compatible to different optimization methods like evolutionary algorithms and Steepest Descent. Mostly, design criteria are minimum-cost, minimum-weight, and minimum-drag under constraints of scenario performance and functional capability. New researches started to study and develop novel design methods, applying optimized and efficient frameworks in various fields. Aerospace science is included in a way that the application of MDO methods with various single- or multi-level frameworks in aerospace vehicles — such as airplane, launch vehicles, and satellites — is seen referring to reliable papers, being considered as a current theme up to date. Aircraft design under uncertainty has been the subject of some recent studies too. In Ahn and Kwon (2006), it was introduced a BLISS based on Reliability-Based Design Optimization (RBDO) framework to design a simplified supersonic transport problem (Ahn and Kwon 2006; Sobieszczanski-Sobieski *et al.* 2000). The study assumed normal distributions with coefficients of variation equal to 0.3 (the ratio of the mean to standard deviation) on each of the 10 design variables considered such as wing area, span, and others describing aircraft geometry. In Smith and Mahadevan (2003), it was solved a spacecraft conceptual optimization problem using RBDO to consider uncertain design variables, reflecting the possibility of minor design changes later in the design process. Probabilistic error terms were added to the responses of the aerodynamics and structural analysis output with assumed values of 10%. The optimization problems were solved with several MDO architectures and First-Order Reliability Method (FORM) based reliability analysis methods. The aforementioned studies consider uncertainties in the design variables or parameters such as atmospheric conditions or material properties.

This study aimed to introduce the decoupling of Uncertainty-based Multidisciplinary Design Optimization (UMDO) method, applying it to design UAVs as a case study. The following section introduces UMDO methods. Then, it is developed UAV decoupling UMDO algorithm, based on MDO method in single-level frameworks (MDF), using genetic algorithm optimization and sequential quadratic programming (SQP). In the section "Implementation of Uncertainty-Based Multidisciplinary Design Optimization Methods and Their Comparison", the redesign of Global Hawk UAV was made using prepared algorithm, and the comparison was carried out between MDO and UMDO while validating the results.

UMDO METHODS

MDO is a branch of engineering science that applies optimization methods for solving design problems with multiple contexts and themes with coupled parameters. The method is called multidisciplinary optimization or Multidisciplinary System Design Optimization (MSDO). The method allows designers to consider related themes simultaneously (Olds 1993). Any deviation in the designer's assumptions (*i.e.*, the material strength or the manufacturing precision of a structural member) or approximate analysis methods may result in the failure of the optimized design because the results take place in feasibility bound.

Any design variable, parameter, or any output from analysis codes in a given optimization problem can be considered as uncertain quantities provided that the uncertainty can be mathematically represented. Uncertainty at the early phases of the development exists due to the limited knowledge concerning the system characteristics and due to the low-fidelity simulations and analyses performed. UMDO methods are recent, still under development, and partially applied in conceptual phases. Many methods currently exist for quantifying the behavior of aleatory and epistemic uncertainty. The methods most often applied in design optimization are interval analysis, fuzzy numbers, and probability theory. The choice of the method is driven by the quantity of information available to the designer about the source of uncertainty. In general, sources of uncertainty in which there is insufficient data to accurately estimate a probability density function (PDF), interval analysis or fuzzy numbers are preferred (Hu and Qiu 2010; Schueller and Jensen 2008; Hajela 2002; Vittal and Hajela 2003). UMDO process relies on 2 steps (Fig. 1): uncertain system modeling and UMDO procedure.

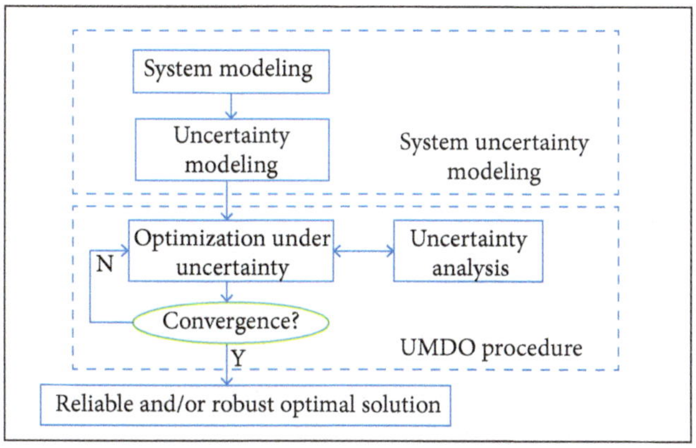

Figure 1. UMDO process (Yao *et al.* 2011).

The first phase of the process consists on desired system and mathematical uncertainty modeling; the second phase includes optimization in the presence of uncertainty, resistance, and uncertainty analysis, as well as the reliability of the answer. UMDO should match the mathematical uncertainty modeling (probability theory, possibility, etc.) and type of design algorithm (AAO, MDF, CO, etc.). In an optimum problem designing, the failure of the plan will be determined by the problem constraints. If these constraints are identified by uncertainty variable, the output of the constraints will have some uncertainty. There are 2 ways to resolve the issue through reliability strategies: nested and decoupling approaches.

In the first method (nested approach), reliability analysis is carried out through optimization cycles for feasibility in the presence of uncertainty. For this purpose, at each iteration of optimizer, reliability analysis is carried out through uncertainty modeling in the parameters in mind. One of the theories available can be used for uncertainty modeling (probability theory, possibility, etc.). Depending on the use of existing theories for uncertainty modeling, there are 2 cases (Rao and Cao 2002):

- RBDO: in this method, the probability theory is used for uncertainty modeling. This technique allows the propagation of uncertainty (given probability density function of variables) in the design process to determine their influence on the final answer (Rao 1992). For uncertainty analysis in RBDO strategy, there are various methods such as first- and second-order analysis (first-order reliability method — FORM and second-order reliability method — SORM). In this method, all design variables are mapped to normal distribution space, then the minimum distance between the design point and feasibility boundary is calculated. The point in the border of feasibility nearest to the design point is known as MPP. The distance between the design point and MPP is called reliability (β) of the desirable value which is determined by the designer and must have a minimum value. Several methods are used to calculate the MPP and β such as performance measure approach (PMA) and reliability index approach (RIA). These methods are based on FORM but differ in methods of finding MPP (Yu *et al.* 1997; Tu *et al.* 1999).
- Possibility-Based Design Optimization (PBDO): when there is not enough information about the uncertainty of design variables, PBDO is appropriate. In this method, the interval or fuzzy theory is used for uncertainty modeling, and interval or FORM-based optimization methods are applied. In general, the results of this method compared to RBDO are the more stable (Du *et al.* 2006).

Finally, to check the reliability of the obtained optimal solution, uncertainty analysis can be used.

In the second method (decoupling approach), optimization cycles and reliability analysis are separated. For this purpose, 4 steps are carried out:

1. Optimization without uncertainty.
2. Uncertainty analysis on the found answer from step 1 by applying the uncertainty on the desired parameters of the issue.
3. Convergence: if the uncertainties violate the possibility of the answer, a reliable optimal solution is obtained.
4. Lack of convergence: if the uncertainties violate the possibility of the answer, a shifting vector of the answers must be found for feasibility. Then step 3 is carried out again.

The important thing in this method is finding the shifting vector of answers in a way that the least number of repetitions are needed. One of the ways to find shifting vector is using PMA method to find MPP for each constraint. In this case, the difference between optimal solution without uncertainty and the answer located at the MPP is considered as a shifting vector, and this cycle continues to achieve reliable optimal solution (Du and Chen 2004). In finding shifting vector, the designer's experience can be very effective.

UAV DECOUPLING UMDO ALGORITHM

Before presenting the design algorithm, it is necessary that various parts of the UAV design be identified, modeled, and transformed into a software code. Other subjects involved in design are: aerodynamic, structure, propulsion, and path simulation.

INTEGRATED DESIGN ALGORITHM WITH UMDO METHOD

UAV design includes, respectively, 11, 15, and 11 common, uncommon, and coupled parameters (a total of 37) in addition to 33 constraints. The above design algorithm, mission definition, and flight scenario exist within which the following parameters are determined: payload geometrical specification and mass; cruise phase speed, altitude, and range; loiter phase speed, altitude, and duration; in addition to stall speed. Optimization methods applied in system level is a combination of genetic algorithm and SQP, taking the number of design parameters and related constraints into consideration. Combining these 2 methods makes the optimized point resulted from genetic algorithm using SQP method more accurate.

Figure 2 presents UAV's multidisciplinary design optimization algorithm in MDF structure, in which an optimizer is located at system level within which UAV design parameters (a total of 37) are achieved in a way that, observing the problem's constraints (a total of 33), optimization criterion (overall mass of the UAV) becomes minimum.

In order to expedite design algorithm and maintain a better convergence, constraints are prioritized, and a certain value is added to the criterion function against any of these priorities not being observed. This way, optimization algorithm will satisfy constraints while minimizing criterion function. Constraints are prioritized according to their importance: 1) geometrical constraints; 2) constraints related to minimum required thrust; 3) constraints related to minimum require lift; 4) static stability constraints; 5) load coefficient constraints; 6) propulsion mass constraints; and 7) constraints related to performing an mission.

In this framework, disciplines would be feasible through multidisciplinary analyses and internal cycles between coupled sub-systems. In other words, internal cycles between coupled sub-systems are continued till all disciplines are feasible in each optimizer evaluation.

The uncertainty of some of these parameters, such as the aerodynamic ones, according to the theory used, is available in the references; however, for the rest of the parameters, there is no authoritative reference. Thus, according to the authors' experience, sources of uncertainty and their values (3σ) are considered in this work as follows:

- Environmental uncertainties include gravity (5% of nominal value) and density (5% of nominal value).
- Uncertainties of the appliance model include mass (5% of nominal value), fuel consumption (5% of nominal value), aerodynamic coefficients (10% of nominal value [5, 17]), and thrust (5% of nominal value).

To provide design algorithm in the presence of uncertainties, it is necessary to identify the sources of the uncertainty. According to the given description, all design methods are based on reliability on

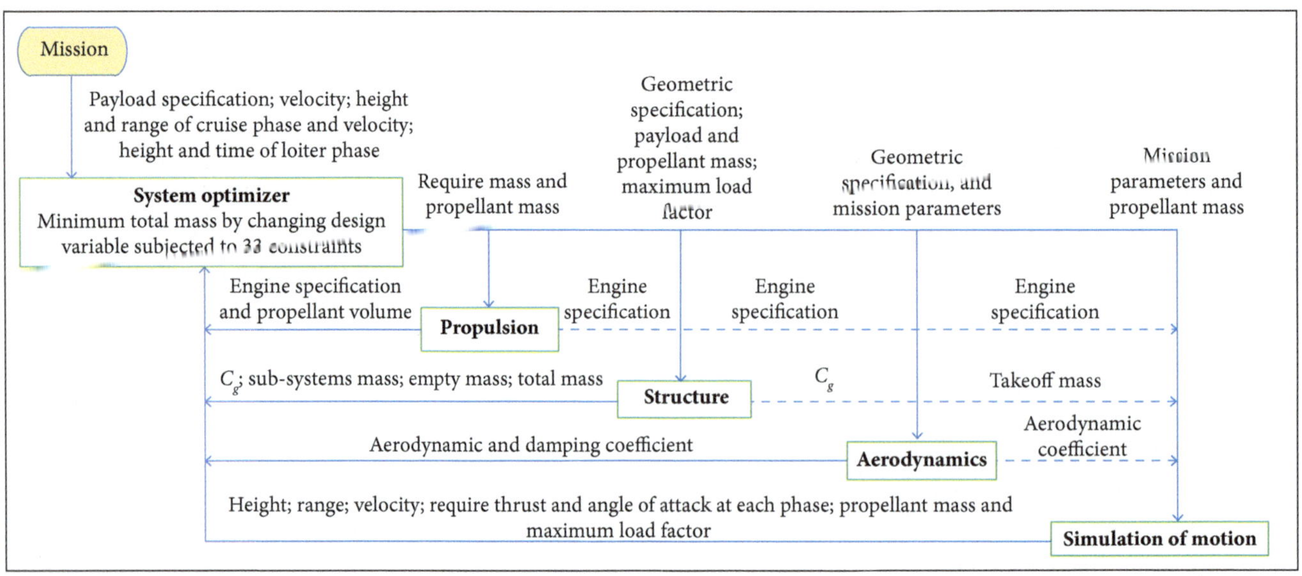

Figure 2. Multidisciplinary design optimization of UAV.

the basis of the mathematics of uncertainty (mapping to a normal distribution space), calculating the border of feasibility and the nearest point from it with a point (MPP). Since the constraints defined in the current problem of UAV are in a way that there is no clear feasibility border, using such methods is very difficult. Here it is presented the method for UAV multidisciplinary optimal design in the presence of uncertainties, which is based on the evolutionary algorithm. The advantage of this method compared to introduction ones is its independence from mathematics of uncertainty and no need for calculating the feasibility border. To perform the design in the presence of uncertainties, the following steps should be taken separately:

1. Optimization without uncertainty (Table 1-3).
2. Uncertainty modeling by the found answer from step 1 as follows: the worst case of uncertainty in a definite range for any constraint is calculated by an optimization algorithm as below. The parameters of optimization algorithm are uncertainty values, and optimization criteria maximize the impossibility of any constraint (for example, minimizing the duration of the flight or an increase in instability). The optimization algorithm in this method, like the original optimization algorithm, is a combination of genetic algorithm and SQP. Therefore, its output will be the set of the worst uncertainties for each constraint):
   Maximize constraint(i), (i = No. of constraints)
   By changing (7 parameters): $U_T, U_{SFC}, U_{We}, U_{CD}, U_{CL}, U_g, U_p$
3. Convergence: uncertainty analysis by applying the set of the worst uncertainties on the parameters of the problem. If the uncertainties do not violate the possibility of the answer, a reliable optimal solution is obtained.

Table 1. Optimization without uncertainty (step 1). Fitness function.

Fitness function	Function name
Minimize takeoff weight in order to below mission parameters	
V_{Cruise}	Cruise velocity
H_{Cruise}	Cruise height
R	Range
V_{Loiter}	Loiter velocity
H_{Loiter}	Loiter height
E	Endurance
Payload specifications	
V_S	Stall speed
R_N	Nose radius

Calculate & check constraint(i), (i = No. of constraints)
in order to mission parameters ($V_{Cruise}, H_{Cruise}, R, V_{Loiter}, H_{Loiter}, E$, Payload Specifications, V_S, R_N)
& Uncertainty value ($U_i = [U_T, U_{SFC}, U_{We}, U_{CD}, U_{CL}, U_g, U_p]$)
for $i = 1, 2, \ldots, n_c$
& Optimum design variable: $X_W, C_{rW}, C_{tW}, b_W, R_B, L_B, \Lambda_W, C_{rH}, C_{tH}, b_H, \Lambda_H, C_{rV}, C_{tV}, b_V, \Lambda_V, i_P, Z_W, i_W, i_H, L_N, L_A, \Theta_W, \Gamma_W, \Theta_H, \Gamma_H, Z_{CG}, W_F, T, n$

Table 2. Optimization without uncertainty (step 1). Design variable.

Design variable	Variable name
C_{rW}	Wing root chord
C_{tW}	Wing tip chord
b_W	Wing span
R_B	Body radius
L_B	Body length
Λ_W	Wing sweep
C_{rH}	Horizontal tail root chord
C_{tH}	Horizontal tail tip chord
b_H	Horizontal tail span
Λ_H	Horizontal tail sweep
C_{rV}	Vertical tail root chord
C_{tV}	Vertical tail tip chord
b_V	Vertical tail span
Λ_V	Vertical tail sweep
i_P	Propulsion system incidence
Z_W	Wing vertical position
i_W	Wing incidence
i_H	Horizontal tail incidence
L_N	Nose body length
L_A	Aft body length
Θ_W	Wing twist
Γ_W	Wing dihedral
Θ_H	Horizontal tail twist
Γ_H	Horizontal tail dihedral
Z_{CG}	Vertical position of gravity center
W_F	Fuel weight
T	Thrust
N	Load factor

4. Lack of convergence: if the uncertainties violate the possibility of the answer, a shifting vector of the answers, appropriate for the violated constraint, must be found for feasibility — for example, if the constraints related to flight time requirements is not met. The difference between the required fuel and the available one will be added to the latter (required fuel is the output of movement simulation), then step 3 is carried out again.

After designing in the presence of uncertainty, its analysis using the Monte Carlo method is performed to check the final answer. The UAV multidisciplinary optimal design algorithm in the presence of uncertainties is presented in Fig. 3.

UAV design algorithm in the presence of uncertainties was transformed into a software code; through it, the results of re-designing UAV — Global Hawk (RQ-4B) — in the presence of the uncertainties are presented. The results of UAV design have been compared without the presence of uncertainties.

Table 3. Optimization without uncertainty (step 1). Constraints.

Constraints
$V_{PL} + V_F < V_W + V_b$
$b_W > 2R_B$
$b_H > 2R_B$
$b_V > 2R_B$
$C_{tW} < C_{rW}$
$C_{tH} < C_{rH}$
$C_{tV} < C_{rV}$
$Z_W < R_B$
$X_W + C_{rW} < 0.8L_B$
$L_N < 0.5L_B$
$L_A < 0.5L_B$
$C_{m_\alpha} < 0$ without fuel and external payload
$C_{n_\beta} > 0$ without fuel and external payload
$C_{m_\alpha} < 0$ without fuel and with external payload
$C_{n_\beta} > 0$ without fuel and with external payload
$C_{m_\alpha} < 0$ with fuel and without external payload
$C_{n_\beta} > 0$ with fuel and without external payload
$C_{m_\alpha} < 0$ with fuel and external payload
$C_{n_\beta} > 0$ with fuel and external payload
$W_{F\,Opt} > W_{F\,Required}$
$n_{Opt} > n_{Calculated}$
Lift > W_{TO} in stall speed
Lift > W_{TO} in cruise speed
Lift > $W_{TO\text{ in loiter speed}}$
$T > T_{\text{Calculated in stall speed}}$
$T > T_{\text{Calculated in cruise speed}}$
$T > T_{\text{Calculated in loiter speed}}$
$V_{Cruise} = V_{Cruise\text{ calculated in simulation}}$
$H_{Cruise} = H_{Cruise\text{ calculated in simulation}}$
$R = R_{\text{Calculated in Simulation}}$
$V_{Loiter} = V_{Loiter\text{ calculated in simulation}}$
$H_{Loiter} = H_{Loiter\text{ calculated in simulation}}$
$E = E_{\text{Calculated in simulation}}$

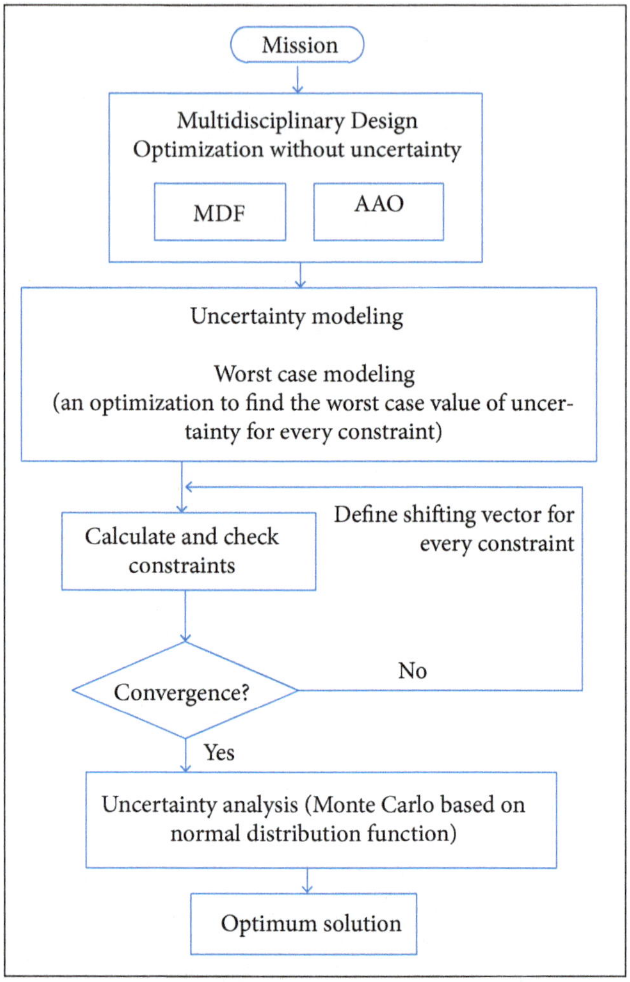

Figure 3. UAV MDO algorithm in the presence of uncertainties.

IMPLEMENTATION OF UMDO METHODS AND THEIR COMPARISON

The results of the MDO algorithm in MDF structure and decoupled method are provided for Global Hawk UAV in the rest of the paper, and parameters such as total mass, runtime code, and the percentage of success have been compared.

To do so, mission information of this UAV was considered as design code input and the output extracted. Then, outputs were compared to real UAV information. Code inputs for Global Hawk redesign are: 575 km/h of velocity cruise, 28 h of flight endurance, maximum altitude of 18 km, and payload mass of 1,360 kg.

In Table 4, by the percentage of success, we wonder what is the percentage of likelihood of fulfillment of all constraints (as explained in previous seasons) in the presence of uncertainty for the final design. As seen in this table, result differences in MDF method with existing information from Global Hawk are between 13 – 23% in mass and less than 9% in geometrical specifications. The following points could be referred to as the reasons for these differences:

- Lack of sufficient information about Global Hawk flight scenario.
- Lack of information about airfoil and lifting instruments at the time of takeoff and landing.
- Lack of information about accurate stall speed and some of functional parameters.
- Lack of information about payload geometric specifications, sub-systems layout, and fuel.
- Lack of consideration of uncertainties.

Convergence procedure in MDF frameworks is shown in Fig. 4. As can be seen, the success rate of optimal response from MDF algorithm (without uncertainty) is 51 and 100% for decoupled method. The run time of this code in decoupled method is 1,600 s more than the MDF algorithm

Table 4. Outputs of design code for Global Hawk redesign in presence of uncertainty.

Section	Value		
	Real value	MDF	Decoupling approach
Wing span (m)	39.9	39	39
Wing area (m^2)	63.02	68.2	68.25
Body diameter (m)	1.42	1.28	1.3
Body length (m)	14.5	14.5	14.5
Empty mass (kg)	5,868	5,229	5,538
Takeoff mass (kg)	14,628	12,159	14,133
Propellant mass (kg)	7,400	5,570	7,235
Time of optimization(s)	-	21,507	23,107
Percentage of success	-	51	100

(without uncertainty). The total mass obtained from the decoupled method is 1,974 kg more than that of the MDF algorithm.

The total mass obtained from the optimal design algorithm in the presence of uncertainties in decoupled method is 14.1 t. This means that, to compensate the failure probability in multidisciplinary optimal algorithm without uncertainties, the mass is increased in order to increase the success chance from 51 to 100%, considering that the uncertainties have also resulted in less difference between the obtained responses and the real values. Regarding the obtained result, although designing without uncertainties in a more optimal way, it is not reliable.

In Figs. 5 and 6, the results of the Monte Carlo analysis are presented for thousands of performances. In Fig. 7, it is observed the 3-D view; in Fig. 8, the presented charts resulted from redesigned UAV simulation of motion.

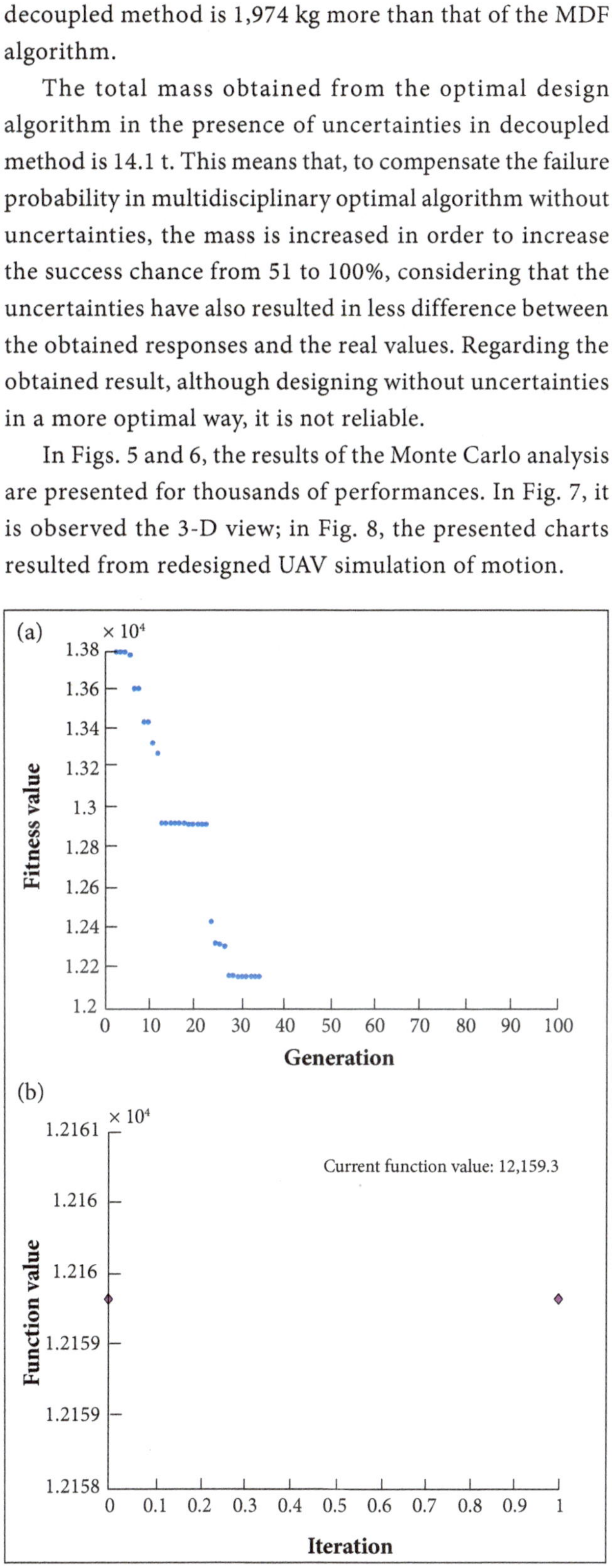

Figure 4. (a) Convergence trend in GA algorithm ; (b) SQP.

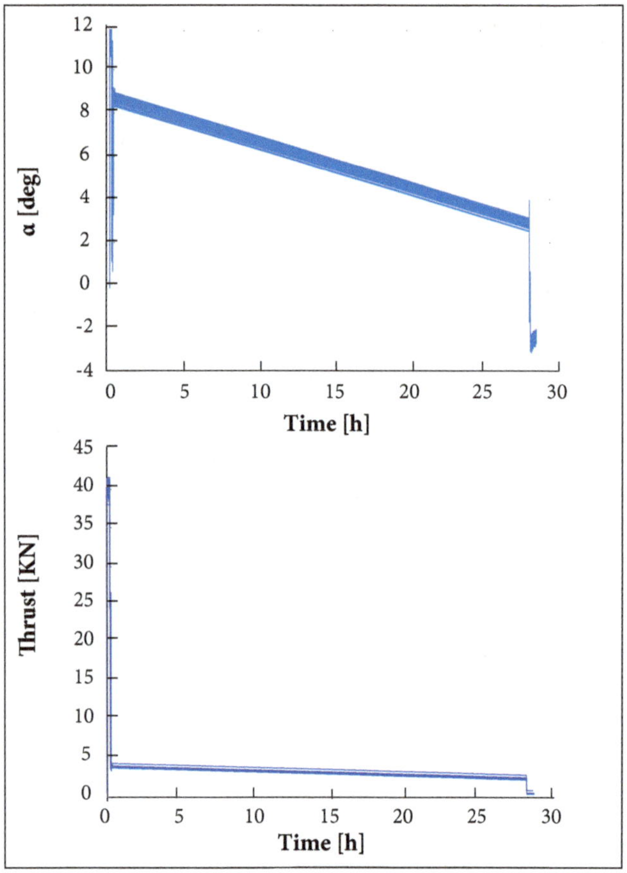

Figure 5. Monte Carlo analysis: angle of attack and thrust *versus* time.

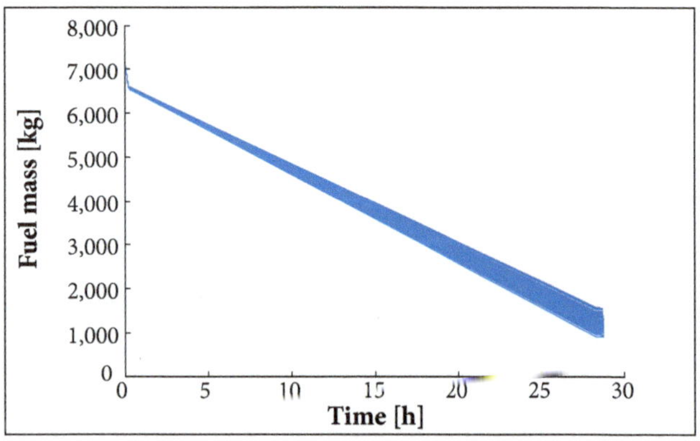

Figure 6. Monte Carlo analysis: fuel consumption *versus* time.

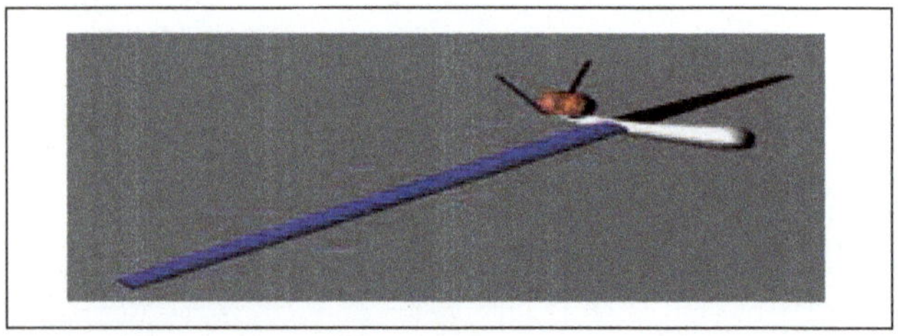

Figure 7. 3-D view of the redesigned UAV.

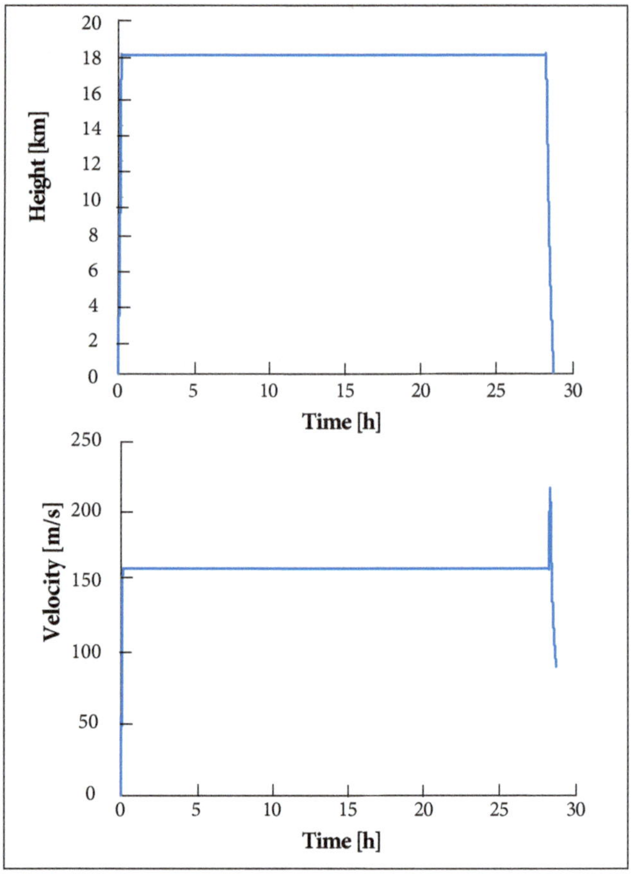

Figure 8. Height and velocity *versus* time.

CONCLUSION

The present study introduced MDO and UMDO frameworks in UAVs, implemented and compared as a case study. To this aim, sub-systems and disciplines involved in design were modeled, and a proposed algorithm for conceptual design of UAV was developed based on multidisciplinary optimization method in MDF framework, using genetic algorithm optimization method and SQP. This algorithm includes 11 common parameters, 15 uncommon parameters, 11 coupled parameters (a total of 37) and 33 constraints.

Regarding the results achieved from MDO in the presence of uncertainties for operational UAV, the results are explained as:

- The comparison of the code outputs, results of the movement simulation, and Monte Carlo analysis for multidisciplinary optimal design code output in the presence of uncertainties indicates the correctness and reliability of results of the proposed algorithm.

- To compensate the failure probability in multidisciplinary optimal algorithm without uncertainties, the mass is increased in order to increase the success chance to 100%.

The advantages of the provided approach are: independence from probability mathematics, appropriate when there is insufficient data to approximate the uncertainties variables or develop a good estimate of the probability density function shape, appropriate speed to calculate the best reliable response, and proper success rate in the presence of uncertainties.

Totally, regarding the presence of the uncertainties in the real world, it is better to consider the mass increase penalty in order to increase the reliability of the design. With the help of this algorithm, we can, according to the defined operation for the UAV and the level of uncertainties of the proposed models in the design, achieve a reliable and optimal answer in the conceptual design phase with the least time and highest accuracy.

AUTHOR'S CONTRIBUTION

Hosseini M and Nosratollahi M, conceived the idea and co-wrote the main text; Hosseini M, prepared the tables and figures. All authors discussed the results and commented on the manuscript.

REFERENCES

Ahn J, Kwon JH (2006) An efficient strategy for reliability-based multidisciplinary design optimization using BLISS. Structural and Multidisciplinary Optimization 31(5):363-372. doi: 10.1007/s00158-005-0565-6

Ahn J, Lee S, Kim JS. (2002). A robust approach to pre-concept design of UCAV considering survivability. Proceedings of the 9th AIAA/ISSMO Symposium on Multidisciplinary Analysis and Optimization; Atlanta, USA.

Buonanno MA (2005) A method for aircraft concept exploration using multicriteria interactive genetic algorithms (PhD thesis). Georgia Institute of Technology.

Çavuş N (2009) Multidisciplinary and multiobjective design optimization of an Unmanned Combat Aerial Vehicle (UCAV) (Master's thesis). Ankara: Middle East Technical University.

Choi SM, Nguyen NV, Kim WS, Kim S, Lee JW, Byun YH (2010) Multidisciplinary Unmanned Combat Air Vehicle (UCAV) system design using multi-fidelity models. Proceedings of the 48th AIAA Aerospace Sciences Meeting Including the New Horizons Forum and Aerospace Exposition; Orlando, USA.

Du L, Choi KK, Youn BD (2006) Inverse possibility analysis method for possibility-based design optimization. AIAA J 44(11):2682-2690. doi: 10.2514/1.16546

Du X, Chen W (2004) Sequential optimization and reliability assessment method for efficient probabilistic design. J Mech Des 126(2):225-233. doi: 10.1115/1.1649968

Eisler CA (2003) Multidisciplinary optimization of conceptual aircraft design (Master's thesis). Ottawa: Carleton University.

Geethaikrishnan C (2003) Multidisciplinary design optimization strategy in multi-stage launch vehicle conceptual design (PhD thesis). Bombay: Indian Institute of Technology.

Giunta A (1997) Aircraft multidisciplinary design optimization using design of experiments theory and response surface modeling methods (PhD thesis). Blacksburg: Virginia Polytechnic Institute and State University.

González LF, Srinivas K, Périaux J, Whitney EJ (2005) Multidisciplinary design optimization of Unmanned Aerial Vehicles (UAV) using multi-criteria evolutionary algorithms. Proceedings of the 6th World Congress of Structural and Multidisciplinary Optimization; Rio de Janeiro, Brazil.

Goraj Z (2005) Design challenges associated with development of a new generation UAV. Aircr Eng Aerosp Tec 77(5):361-368. doi: 10.1108/00022660510617086

Hajela P (2002) Soft computing in multidisciplinary aerospace design - new directions for research. Progr Aero Sci 38(1):1-21. doi: 10.1016/S0376-0421(01)00015-X

Hendrich TJM (2011) Multidisciplinary design optimization in the conceptual design phase creating a conceptual design of the blended wing-body with the BLISS optimization strategy (Master's thesis). Delft: Delft University of Technology.

Hu J, Qiu Z (2010) Non-probabilistic convex models and interval analysis method for dynamic response of a beam with bounded uncertaint. Appl Math Model 34(3):725-734. doi: 10.1016/j.apm.2009.06.013

Iqbal LU (2009) Multidisciplinary design and optimization (MDO) methodology for the aircraft conceptual design (PhD thesis). West Lafayette: Purdue University.

Jaeger L, Gogu C, Segonds S, Bes C (2013) Aircraft multidisciplinary design optimization under both model and design variables uncertainty. J Aircraft 50(2):528-538. doi: 10.2514/1.C031914

Lee DS, Gonzalez LF, Srinivas K, Auld DJ, Wong KC (2007) Aerodynamic shape optimisation of Unmanned Aerial Vehicles using hierarchical asynchronous parallel evolutionary algorithms. Int J Comput Intell Res 3(3):229-250.

Lee JW, Nguyen NV, Choi SM, Kim WS, Jeon KS, Byun YH (2009) Multidisciplinary Unmanned Combat Air Vehicle-UCAV design optimization using variable complexity modeling. Proceedings of the 9th AIAA Aviation Technology, Integration, and Operations Conference (ATIO); Hilton Head, USA.

Mattos BS, Secco NR (2013) An airplane calculator featuring a high-fidelity methodology for tailplane sizing. J Aerosp Technol Manag 5(4):371-386. doi: 10.5028/jatm.v5i4.254

Morris SJ, Kroo I (1990) Aircraft design optimization with dynamic performance constraints. J Aircraft 27(12):1060-1067. doi: 10.2514/3.45982

Neufeld D (2010) Multidisciplinary aircraft conceptual design optimization considering fidelity uncertainties (PhD thesis). Toronto: Ryerson University.

Olds JR (1993) Multidisciplinary design techniques applied to conceptual aerospace vehicle design (PhD thesis). Raleigh: North Carolina State University.

Perez RE, Liu HHT, Behdinan K (2004) Evaluation of multidisciplinary optimization approaches for aircraft conceptual design. Proceedings of the 10th AIAA/ISSMO Multidisciplinary Analysis and Optimization Conference; Albany, USA.

Rao SS (1992) Reliability-based design. New York: McGraw-Hill.

Rao SS, Cao L (2002) Optimum design of mechanical systems involving interval parameters. J Mech Des 124(3):465-472. doi: 10.1115/1.1479691

Raymer DP (2002) Enhancing aircraft conceptual design using multidisciplinary optimization (PhD thesis). Stockholm: Royal Institute of Technology.

Rowell LF, Braun RD, Olds JR, Unal R (1999) Multidisciplinary conceptual design optimization of space transportation systems. J Aircraft 36(1):218-226. doi: 10.2514/2.2428

Schueller G, Jensen H (2008) Computational methods in optimization considering uncertainties - an overview. Comput Meth Appl Mech Eng 198(1):2-13. doi: 10.1016/j.cma.2008.05.004

Smith, Mahadevan S (2003) Probabilistic methods for aerospace system conceptual design. J Spacecraft Rockets 40(3):411-418. doi: 10.2514/2.3961

Sóbester A, Keane AJ (2006) Multidisciplinary design optimization of UAV airframes. Proceedings of the 47th AIAA/ASME/ASCE/AHS/ASC Structures, Structural Dynamics, and Materials Conference; Newport, USA.

Sobieszczanski-Sobieski J, Agte JS, Sandusky RR (2000) Bilevel integrated system synthesis. AIAA J 38(1):164-172. doi: 10.2514/2.937

Tekinlap O, Cavus N (2012) Multiobjective conceptual design of an Unmanned Combat Air Vehicle. Proceedings of the 12th AIAA Aviation Technology, Integration, and Operations (ATIO) Conference and 14th AIAA/ISSMO Multidisciplinary Analysis and Optimization Conference; Indianapolis, USA.

Tianyuan H, Xiongqing Y (2009) Aerodynamic/stealthy/structural multidisciplinary design optimization of Unmanned Combat Air Vehicle. Chin J Aeronaut 22(4):380-386. doi: 10.1016/S1000-9361(08)60114-4

Tu J, Choi KK, Park YH (1999) A new study on reliability-based design optimization. J Mech Des 121(4):557-564. doi: 10.1115/1.2829499

Vittal S, Hajela P (2003) Probabilistic design using empirical distributions. Proceedings of the 44th AIAA/ASME/ASCE/AHS/ASC Structures, Structural Dynamics, and Materials Conference; Norfolk, USA.

Yao W, Chen X, Luo W, van Tooren M, Guo J (2011) Review of uncertainty-based multidisciplinary design optimization methods for aerospace vehicles. Progr Aero Sci 47(6):450-479. doi: 10.1016/j.paerosci.2011.05.001

Yu X, Choi KK, Chang KH (1997) A mixed design approach for probabilistic structural durability. Struct Optim 14(2):81-90. doi: 10.1007/BF01812509

Zill T , Böhnke D, Nagel B (2011) Preliminary aircraft design in a collaborative multidisciplinary design environment. Proceedings of the 11th AIAA Aviation Technology, Integration, and Operations (ATIO) Conference; Virginia Beach, USA.

22

Production and Characterization of Activated Carbon Fiber from Textile PAN Fiber

Miguel Angelo do Amaral Junior[1], Jorge Tadao Matsushima[1], Mirabel Cerqueira Rezende[2], Emerson Sarmento Gonçalves[3], Jossano Saldanha Marcuzzo[1], Maurício Ribeiro Baldan[1]

ABSTRACT: This paper presents the preparation and characterization of carbon fiber felt and activated carbon fiber felt from textile polyacrylonitrile fiber. Carbon fibers are usually related to aircraft manufacturing or high mechanical purposes. Activated carbon fibers are known as excellent adsorbent materials. Despite all advantages, carbon fiber and activated carbon fiber are expensive materials because of their raw material cost. On the other hand, in this study, carbon fiber felt and activated carbon fiber felt were produced from textile polyacrylonitrile fiber, which is cheaper than their precursor, polyacrylonitrile fiber, and can be converted into carbon fiber felt and activated material with high micropore content and surface area. This research describes the transformation of textile polyacrylonitrile fiber in its oxidized form. After that, the oxidized material was transformed in felt and, in the sequence, converted into carbon fiber felt and activated carbon felt. The carbon fiber felt and activated carbon fiber felt were characterized by X-ray photoelectron spectroscopy, Raman spectroscopy and scanning electron microscope. N_2 isotherms were performed to qualify the material obtained for further electrochemical applications. The main result was the conversion dynamics of textile polyacrilonitrile fiber into carbon fiber in felt form and activated carbon fiber in felt with high surface area and high micropores content.

KEYWORDS: Carbon fiber, Activated carbon fiber, Raman spectroscopy.

INTRODUCTION

Activated Carbon Fiber Felt (ACFF) has special characteristics when compared with common activated carbons (granular or powder). It can be transformed into fabric, woven or yarn forms which gives them self-sustainable characteristics. In addition, ACFF shows well-defined pore structures on its surface providing a high and fast adsorption capacity for specific components, such as for specific components with small molecular dimensions as ions, metals or organic. One of the most important characteristics, which makes ACFF a very special adsorbent material, is its pore size distribution (Suzuki 1994). In the ACFF case, a large amount of micropores can be found directly connected on their surface leading to a faster and less energetic adsorption mechanism, especially for gases (Cuña *et al.* 2014). ACFF is widely used in many applications such as air purification and water treatment, chemical (adsorption and desorption for organic compounds and solvents), military area such as protection garment and masks (Marsh and Reinoso 2006). The density of ACFF is considerably lower than the regular carbon fiber, making it ideal for applications requiring low weight. Despite all the advantages of Carbon Fiber Felt (CFF) and ACFF application, their use has been limited due to their relatively high cost.

The use of the textile type polyacrilonitrile (PAN) fiber to produce CFF and ACFF is a way to produce a material having an attractive surface area and some studies show this possibility (Carrott *et al.* 2001; Nabais *et al.* 2005; Marcuzzo *et al.* 2012). The use of standard textile fiber as raw material to make CFF

1.Instituto Nacional de Pesquisas Espaciais – Coordenação de Laboratórios Associados – Laboratório Associado de Sensores e Materiais – São José dos Campos/SP – Brazil. 2.Universidade Federal de São Paulo – Instituto de Ciência e Tecnologia – Curso de Engenharia de Materiais – São José dos Campos/SP – Brazil. 3.Departamento de Ciência e Tecnologia Aeroespacial – Instituto de Aeronáutica e Espaço – Divisão de Materiais – São José dos Campos/SP – Brazil.

Author for correspondence: Miguel Angelo do Amaral Junior | Instituto Nacional de Pesquisas Espaciais – Coordenação de Laboratórios Associados – Laboratório Associado de Sensores e Materiais | Avenida dos Astronautas, 1.758 – Jardim da Granja | CEP: 12.227-010 – São José dos Campos/SP – Brazil | Email: miguel.junior.mat@hotmail.com

and ACFF is a way to make this material economically attractive (Yoon *et al.* 2004). Containing more than 95% of carbon, CFF and ACFF made from textile PAN have low electric resistance and they show potential application for catalytic process, especially when combined with metallic particles or in the form of films (Chung 2004; Pierozynski 2012). The CFF and ACFF can also be used as electrodes. The aim of this study is to present the production and characterization of CFF and ACFF made from a standard textile PAN.

MATERIALS AND METHODS

MATERIAL

The commercial 200 ktex tow of 5.0 dtex textile PAN fibers was thermal oxidized in a laboratory scale oven set by, aiming the production of flame resistant fibers. The oxidation process was performed in 2 steps, the first at 200 °C and the second at 300 °C. The total time process was 50 min for each step. After that, the oxidized PAN produced was used as a raw material to produce felt with 200 g/m^2 and about 3 mm thickness. The oxidized PAN fiber felt was carbonized, resulting into CFF. After that, the CFF was submitted to activation process to obtain the ACFF (Marcuzzo *et al.* 2012).

During the carbonization process, the oxidized PAN loses about 50% in mass and shrinks linearly about 10%. The shrinkage is an important parameter and must be controlled because an inadequate shrinkage result in poor mechanical characteristics and the fiber can not be handled. For this purpose, the oxidized PAN fiber felt sample was cut into pieces of about 0.7 × 0.25 m and placed in a special sample holder that can control the sample shrinkage in 2 dimensions.

The set was then introduced in an electrical furnace. Both ends of the furnace tube were closed by flanges, which allowed the insertion and the purge of processing gas to provide an atmosphere condition necessary for the carbonization and activation. The carbonization was performed in argon atmosphere at a final temperature of 1,000 °C at a heating rate of 30 °C/min. The process time at maximum temperature was set in 20 min to complete the carbonization process. After finishing the carbonization process, the furnace was turned off and maintained in Ar atmosphere. This condition of inert atmosphere was maintained until the room temperature inside the furnace reactor was reached.

The CFFs were carbonized at 900 °C, prior to activation, which was immediately accomplished at 1,000 °C, replacing the Ar gas by CO_2 gas, during 50 min. The activation time was previously defined by tests where the mechanical evaluation was done (Marcuzzo *et al.* 2012). The activation time of 50 min was fixed to guarantee the minimal mechanical condition to complete the carbonization process.

METHODS

The characterization was performed by gas adsorption aiming the measurements of surface area and pore size distribution function. The isotherm was performed by Beckman Coulter SA 3100 equipment. The Brunauer, Emmett, Teller (BET) method was applied to determine the total surface area and the pore size distribution was estimated by applying the Non-Local Density Function Theory (NLDFT) method over the adsorption isotherm, and the micropore volume was also estimated by the same method (Marsh and Reinoso 2006). The burn-off was estimated weighing the sample before and after the activation process.

The structural characterization of the CFF and ACFF were carried out by micro-Raman scattering spectroscopy (Renishaw Microscopy - 2000). Using the 514.5 nm line of an Ar ion laser taking the spectra at the range from ~ 800 to ~ 1,800 cm^{-1}. The optical microscope was used to excitation laser beam onto the sample and to collect the backscattered light. The diameter of the laser spot on the sample surface was 0.45 μm for the fully focused laser beam at 50X objective magnification. The instrument was calibrated using a pure diamond crystal. From the Raman spectroscopy, it was possible, quantitatively, to evaluate the change in the graphite organization through fitting by a combination of 4 Lorentzian bands (L, D_1, D_2, D_4) and a Gaussian band (D3) (Sadezky *et al.* 2005).

The morphological aspects of the CFF and ACFF were evaluated by scanning electron microscopy (SEM) using a JEOL JSM microscope. All the X-ray photoelectron spectroscopy (XPS) measurements were carried out with a Kratos Axis Ultra XPS spectrometer using a monochromatic Al-K_{alpha} (1,486.5 eV) X-ray radiation with power of 15 kV at 150 W. The emitted photoelectrons were detected using a hemispherical analyzer at 15 μm spatial resolution. The vacuum system was maintained at approximately 10^{-9} Torr during all the experiments. Survey scans were collected from 0 to 1,100 eV with a passing energy equal to 160 eV with step size of 1 eV to identify the elements present on the surface. Using the X-ray diffraction, it was possible to obtain information related to the crystalline structure. To carry out these measures, we used a PANalytical diffractometer serie X'PertPRO.

RESULTS AND DISCUSSION

SURFACE TEXTURE ANALYSIS

The aim of this topic is to show the characterization of CFF and ACFF. The carbonized fiber felt material, in theory, has about no significant surface area. The nitrogen isotherm technique was not able to show information about CFF surface area; consequently, it was considered smooth and without pores. On the other hand, for ACFF, the N_2 adsorption at 77 K isotherm showed significant information about surface texture. The isotherm is shown in Fig. 1 and is typically type I isotherm with no hysteresis and the saturation occurring at 0.2 relative working pressure and atmospheric pressure (P/P_0).

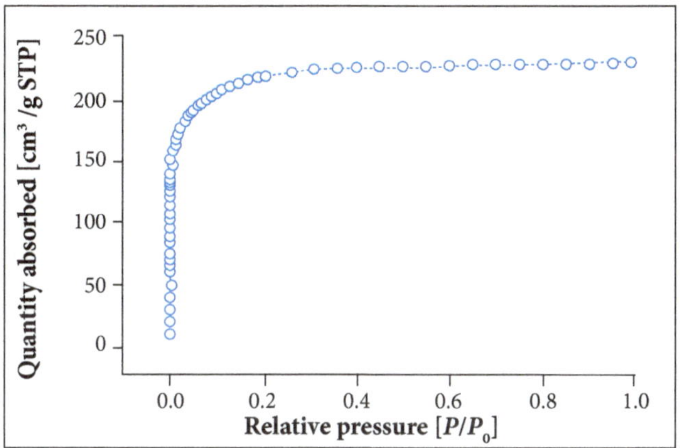

Figure 1. Nitrogen isotherm of ACFF at 77 K.

These results indicate that this activated material is predominantly populated with micropores. The pore size distribution curve is shown in Fig. 2, and it clearly shows that the maximum pore width presented is around 2 nm and the predominant pores are sized at around 1.2 nm. This technique does not give information about pores sized less than 1.0 nm, because of N_2 penetration limit. However, it can be clearly observed in this curve that the distribution in the region for a width less than 1 nm is ascendant in the direction of origin. This fact infers that pore volume and surface area of micropores may be bigger than those calculated by using these isotherms. The main ACFF characteristics are presented in Table 1.

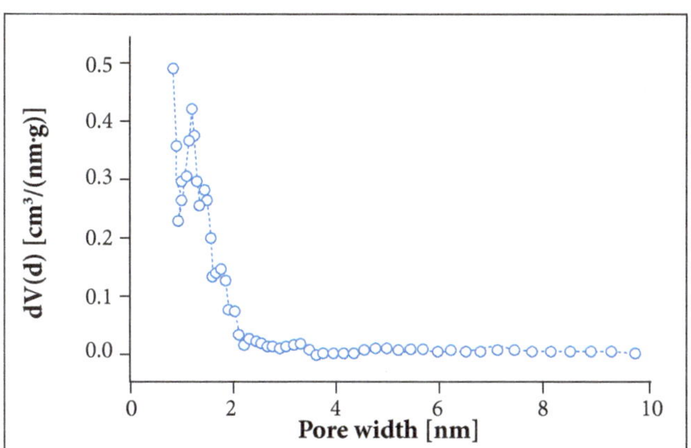

Figure 2. Pore size distribution curve of ACFF by NLDFT.

Table 1. Surface characteristics of ACFF.

Sample	Burn off (w%)	BET (m²/g)	$V_{micropore}$	V_{total}
ACFF	35	1,260	0.53	0.59

X-RAY PHOTOELECTRON SPECTROSCOPY DATA

XPS has proved a powerful method for the investigation of carbon surfaces (Kim and Park 2011; Ma *et al.* 2013). XPS surface characterization method combines surface sensitivity with the ability to quantitatively obtain both elemental and chemical state information. The surface chemistry of carbons is determined by the distribution and the nature of the surface functional groups. Many different functional groups can be identified on carbon surface. The information of all elements can be obtained from the survey scan spectrum of XPS. Vision software was used to calculate atomic concentrations. From survey spectra, atomic concentrations were calculated using peak areas of elemental lines.

X-Ray Photoelectron Spectroscopy Survey Spectra

Figure 3 shows the XPS survey spectra from which the elemental composition of the most external surface of the CFF

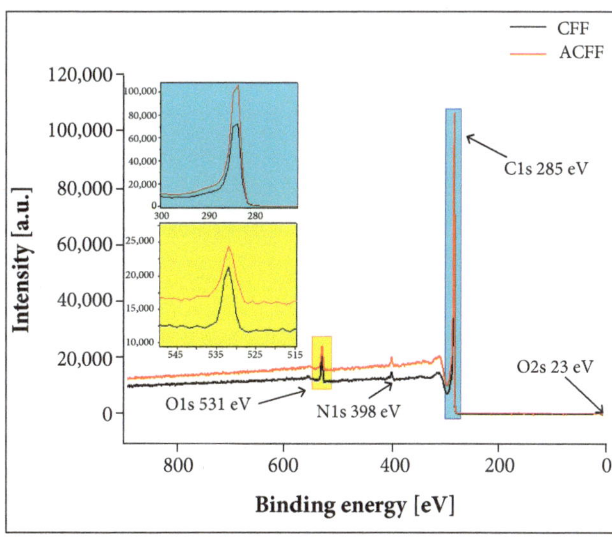

Figure 3. XPS survey spectra (a) CFF and (b) ACFF.

and ACFF samples was revealed. The spectra indicate that the major peaks in the spectra were due to the C 1s and O 1s, and a smaller N 1s peak is also discernible.

The contribution of the XPS technique in this work is the precise identification of the elements present on the surface of CFF and ACFF samples. The elemental composition (in %) on the CFF and ACFF samples were summarized in Table 2. XPS survey spectra showed the presence of less than 4.0% of oxygen on the surface of CFF and ACFF. The ACFF displayed O/C ratio of 2.42% while CFF showed higher O/C ratio of 4.05%. The same behavior was observed in the relative percentages of N/C atoms decreasing from 2.79 to 1.75% after the activation process.

Table 2. Elemental composition and N/C and O/C ratio for the CFF and ACFF samples.

Sample	C (%)	O (%)	N (%)	N/C	O/C
CFF	93.61	3.79	2.61	2.79	4.05
ACFF	96.00	2.32	1.68	1.75	2.42

The elemental composition and N/C as well as O/C ratio show a chemical surface change after the activation process. Besides, a slight increase in the amount of carbon was observed. The contribution of the XPS technique in this work is the precise identification of the elements present on the surface of CFF and ACFF samples. As expected, carbon, oxygen and nitrogen have been found on the surface of both fibers. In Fig. 3, it was observed that the carbon peak intensity is higher for ACFF. On the other hand, it may be observed a decrease in the oxygen amount after the activation process. In other words, the activation process contributed to the removal of the oxygen layer on the CFF, increasing the amount of carbon on the surface of ACFF.

RAMAN ANALYSIS

Raman spectroscopy is a technique widely used for analyzing carbon-based materials due to its sensitivity to different carbon structures, which produce distinctive Raman peaks for various forms of carbon. For example, the exact frequencies of the Raman bands of diamond, graphite, and amorphous forms of carbon depend on the crystallite size and stress present in the different carbon domains. The graphitic materials consist of a large number of peaks. However, the most intense broad bands are the *G* and *D*. In this paper, the *D* band is assigned as D_1. Raman spectra have been analyzed following Sadezky *et al.* (2005). Figure 4 shows the Raman spectra of the CFF and ACFF samples. The spectra consist of 2 major peaks attributed to D_1 and *G*. The *G* peak corresponds to the 1st-order Raman. The band frequency *G* (~ 1,584 cm^{-1}) is the E_{2g} 1st-order mode at the Brillouim zone center (gamma point). The *G* peak is due to the C-C bond stretching of all pairs of sp^2 atoms in both rings and chains. In particular, the G peak is the main Raman signature for sp^2 carbon. The spectra also exhibit additional first-order band characterized by a disorder that represents a zone-edge A_{1g} mode. Since the D_1 band is activated by defects, its intensity can be used to quantify disorders. Besides, it is used to characterize the microcrystallite size (L_a), which can be estimated from the ratio between D_1 and *G* bands by using the method of Tuinstra and Koenig (1970). Thomsen *et al.* (2004) were the first to explain the Raman spectra of the *D* mode using the concept of double resonances for a given laser energy and phonon branch. The origin and dispersion of the *D* band in carbon materials were also investigated by Matthews *et al.* (1999). To explain the physical basis, these authors discussed the 2-D electron and phonon dispersion curves for graphite and concluded that the electronic transition only occurs in the

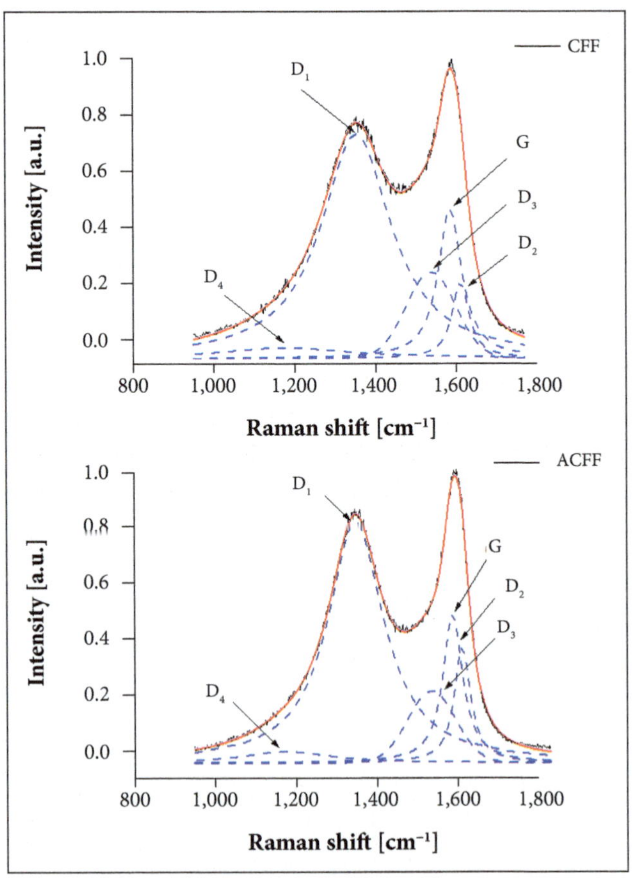

Figure 4. Raman spectra of the CFF and ACFF.

vicinity of the K point in the Brillouin zone. In the literature, it has been accepted that the relationship between band intensities is proportional to the degree of organization of carbonaceous materials. However, this relationship has been used in different forms as $(I_{D1}/I_G)_{A,H}$ or $(I_{D1}/(I_G+I_{D1}))_{A,H}$. The subscripts *A* and *H* indicate that the ratio is based on integrated intensities or peak heights, respectively. Beyssac *et al.* (2003) proposed to characterize the organization using $R_2 = (I_{D1}/(I_G+I_{D1}+I_{D2}))_{A,H}$. In this study, we tested the 3 different forms used in the literature. Besides, Lobo *et al.* (2005) considered the intensity values obtained from the integration of the *D* and *G* bands instead of using the ratio of the peak heights. In other words, the effect of line broadening is included in their calculation.

Special care should be taken to define a baseline to compare one spectrum to another. It is very important to point out that the decomposition is very sensitive to the choice of baseline. A linear baseline was chosen for both spectra to be the most appropriate. Raman spectra of the CFF and ACFF samples were submitted to deconvolution to separate *G* and D_1 peak in the first order by fitting as a sum of the Lorentzian-shaped *G*, D_1, D_2 and D_4 bands, as well as the Gaussian-shaped D_3 band. The reason for the D_3 band being Gaussian-shaped is because it presents better fit with the experimental data (Sadezky *et al.* 2005) and this study is also in agreement with this result. The fitting accounts the totality of the signal. It is noteworthy that both spectra *G* and *D* bands overlap each other to some extent. The overlapping is very important when we consider the integrated intensities. The spectra were recorded at different positions on the surface and looked very similar. The proportions in which the band participates are reported in Fig. 4 and the parameters used to calculus in Table 3. There is an appreciable difference in the absolute value of the 3 different forms to characterize carbonaceous materials. However, when we analyze each one individually, the ratio based on integrated intensities or peak heights did not present significant difference. By comparing the 3 different forms presented (Fig. 5), it is possible to verify that $I_{D1}/(I_G+I_{D1})_{A,H}$ and $I_{D1}/(I_G+I_{D1}+I_{D2})$ are more dispersive than I_{D1}/I_G. Several authors (Tuinstra and Koenig 1970; Thomsen *et al.* 2004; Matthews *et al.* 1999; Beyssac *et al.* 2003) discuss the *D* and *G* band intensity ratios (I_{D1}/I_G) and full width at half maximum for *G*-band (FWHM-*G*) and for *D*-band (FWHM-*D*) decrease with the increasing degree of crystalline. By considering the integrated intensities of *D* and *G*, our results reveal that (I_{D1}/I_G) for both CFF and ACFF are very close. However, according to Table 3, the FWHM-D_1 changes from 210.87 for CFF to 167.01 cm^{-1} for ACFF present a decrease of approximately 21%. Besides, the FWHM-*G* decreased after the activation processing by 17%. The decrease can be interpreted as some kind of transition related with the temperature during the activation process. Probably, it is related with the decrease in oxygen during the activation process as can be observed in XPS measurements. This indicates that defects were not introduced

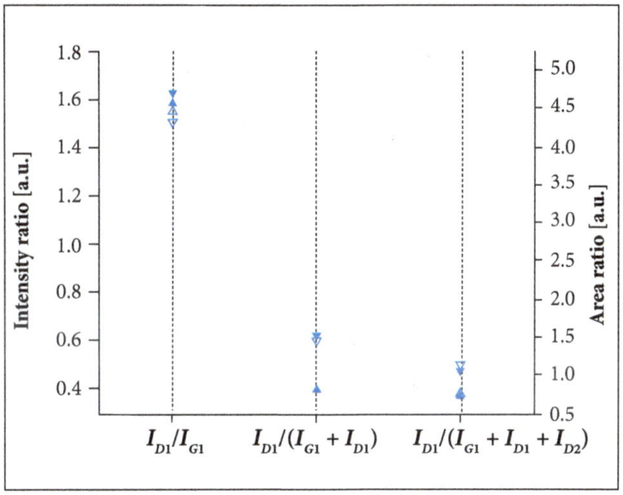

Figure 5. Calculation of the different ratio involving I_G, I_{D1} and I_{D2} bands.

Table 3. Fitting parameters of the *G*, D_1, D_2, D_3 and D_4 bands for the CFF and ACFF samples.

Band	Line shape	CFF - Sample			ACFF - Sample		
		Area	Center	FWHM	Area	Center	FWHM
D_4	Lorentz	25	1,173.71	401.34	16.29	1,173.14	257.66
D_1	Lorentz	266.70	1,351.36	210.87	225.04	1,344.24	167.01
D_3	Gaussian	48.27	1,536.53	146.32	37.88	1,530.21	139.14
G	Lorentz	59.44	1,584.32	71.09	48.91	1,582.81	59.01
D_2	Lorentz	22	1,610.46	51.38	31.43	1,606.48	49.01

during the process of activation. In fact, this indicates a small organization of the ACFF. Recently, Mallet-Ladeira *et al.* (2014) and Hu *et al.* (2015) reported different fitting procedures; the authors claim that their procedures are the most appropriate. There is a considerable controversy regarding fitting procedures. In this study, we consider that the most appropriated procedure for our results is to deconvolute the spectra utilizing multi-bands as reported in Sadezky *et al.* (2005). After analyzing the 3 different forms, we conclude that the variations of FWHM are no reflected in the (I_{D1}/I_G) form. On the other hand, the relation $I_{D1}/(I_G+I_{D1})_{A,H}$ and $I_{D1}/(I_G+I_{D1}+I_{D2})$ indicates that the structure of the CFF was changed after the activation process, which may be related to the production of surface defects after the removal of surface radicals from the sample.

X-RAY PHOTOELECTRON SPECTROSCOPY ANALYSIS

Figure 6 illustrates the XRD spectra for the CFF and ACFF samples. The diffratogram pattern of these samples exhibits 2 distinctive broad reflections in the θ angles located approximately at 25 and 44°. The peak broadening is an indicative of an amorphous structural aspect. The broad reflection in the 2θ at 25° and 44° are associated with (002) and (001) diffraction line, respectively.

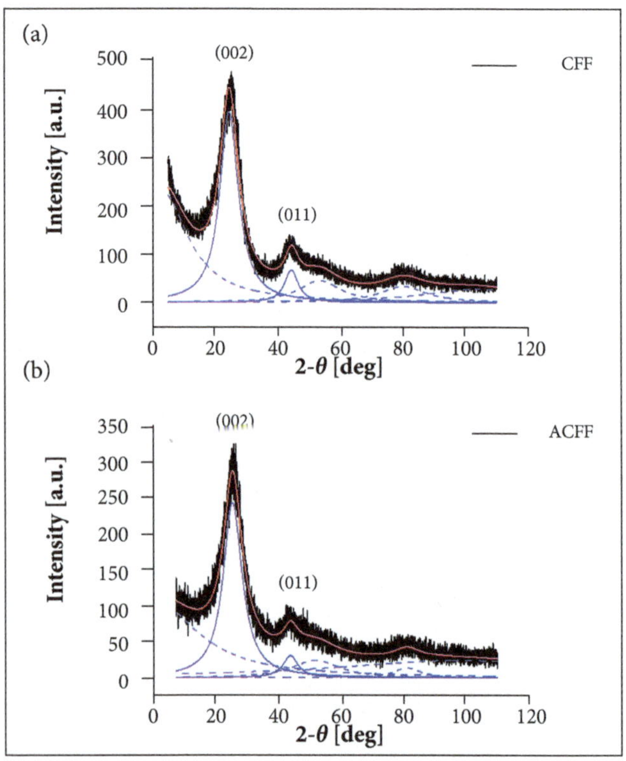

Figure 6. XRD spectra for (a) CFF and (b) ACFF samples.

From the X-ray diffraction (XRD) patterns of the CFF and ACFF, the lateral size (L_a) and stacking height (L_c) parameters were calculated using the Scherrer equation (Warren 1969) — Eqs.1 and 2, respectively. For calculation of L_a and L_c, we applied the Scherrer constant k_c equal to 0.9 for (002) diffraction line and k_a equal to 1.77 for (011) diffraction line, where λ (0.15 nm) is the wavelength of the X-ray used; β_{011} and β_{002} are full width at half maximum of the diffraction line (002) and (011). The results are summarized in Table 4.

$$Lc = \frac{k_c \lambda}{\beta_{002}\, cos\theta} \quad (1)$$

$$La = \frac{k_a \lambda}{\beta_{011}\, cos\theta} \quad (2)$$

Table 4. Crystallite size of the CFF and ACFF.

Sample	L_c (nm)	L_a (nm)
CFF	4.52	12.88
ACFF	4.44	11.71

It can be noticed a small change in L_a and L_c after the activation process. This can be explained by the formation of pores on the CFF surface after the activation process. Although the changes in L_a and L_c are not considerable, there is a significant increase in the crystallite size (Table 4).

SCANNING ELECTRON MICROSCOPY ANALYSES

The SEM images for the CFF and ACFF samples are presented in Fig. 7. From the SEM images obtained from the magnification of 1,000X, it is possible to observe the production of samples consisting of fibers of about 10 – 15μm in diameter. The SEM images also revealed the presence of smooth fibers, without

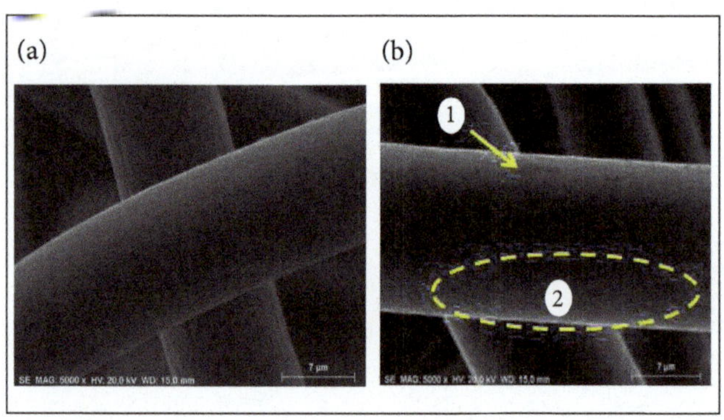

Figure 7. (a) CFF and (b) ACFF SEM images (1: Dark zones; 2: Groove region).

damages, grooves and holes. The visualization of the fiber surface with more details was possible through SEM images obtained from magnification of 5,000X. Based on these images, a smooth, clean and damage-free surface can be visualized for the CFF sample. On the other hand, the ACFF sample presents a surface containing dark zones and grooves. Pores with nano-dimension are not identified by SEM in Fig. 7. However, the presence of pores is confirmed according with Fig. 2. This morphological difference related to surface damages may be associated with the activation process. In fact, these types of damages were expected.

CONCLUSION

Activated carbon fibers produced from PAN fibers textile were successfully achieved. These studies had an important contribution regarding the understanding of the activation process on the CFF samples. The activation process plays an important role for the production of fibers with high surface and pore distribution, which are determinant to increase the exposed area. These characteristics were revealed through nitrogen adsorption isotherm analysis. XPS analysis reported the decrease in the heteroatoms quantity present in the sample, causing defects on the surface. The defects were observed by Raman spectroscopy analysis. The presence of surface imperfections was revealed through SEM images. The X-ray diffractograms showed the CFF sample had higher crystallites when compared to the ACFF one. This decrease in the crystallite size may be associated with the surface tension resulting from the activation process. Besides, the release of heteroatoms is responsible by the creation of defect, causing damage on the surface.

ACKNOWLEDGEMENTS

The authors would like to thank Fundação de Amparo à Pesquisa do Estado de São Paulo (FAPESP), Coordenação de Aperfeiçoamento de Pessoal de Nível Superior (CAPES) and Conselho Nacional de Desenvolvimento Científico e Tecnológico (CNPq) for financial support.

AUTHOR'S CONTRIBUTION

Conceptualization, Amaral Junior MA, Marcuzzo JS, and Baldan MR; Characterization, Amaral Junior MA,Marcuzzo JS, Baldan MR, and Gonçalves ES; Materials Production, Marcuzzo JS and Baldan MR; Methodology, Amaral Junior MA, Matsushima JT, Gonçalves ES, and Rezende MC; Writing – Original Draft, Amaral Junior MA and Baldan MR; Writing – Review & Editing, Amaral Junior MA, Baldan MR, and Rezende MC; Funding Acquisition, Baldan MR; Resources, Baldan MR and Marcuzzo JS; Supervision, Amaral Junior MA, Rezende MC, and Baldan MR.

REFERENCES

Beyssac O, Goffé B, Petitet JP, Froigneux E, Moreau M, Rouzaud JN (2003) On the characterization of disordered and heterogeneous carbonaceous materials by Raman spectroscopy. Spectrochimica Acta Part A: Molecular and Biomolecular Spectroscopy 59(10):2267-2276. doi: 10.1016/S1386-1425(03)00070-2

Carrott PJM, Nabais JMV, Carrott MR, Pajares JA (2001). Preparation of activated carbon fibres from acrylic textile fibres. Carbon 39(10):1543-1555. doi: 10.1016/S0008-6223(00)00271-2

Chung DDL (2004) Review electrical applications of carbon materials. J Mater Sci 39:2645-2661. doi: 10.1023/B:JMSC.0000021439.18202.ea

Cuña A, Tancredi N, Bussi J, Deiana AC, Sardella MF, Barranco V, Rojo JME (2014) Grandis as a biocarbons precursos for supercapacitor electrode application. Wast and Biomass Valor 5(1):305-313.

Hu C, Sedghi S, Silvestre-Albero A, Andersson GG, Sharma A, Pendleton P, Biggs MJ (2015) Raman spectroscopy study of the transformation of the carbonaceous skeleton of a polymer-based nanoporous carbon along the thermal annealing pathway. Carbon 85:147-158. doi: 10.1016/j.carbon.2014.12.098

Kim YK, Park H (2011) Light-harvesting multi-walled carbon nanotubes and CdS hybrids: application to photocatalytic hydrogen production from water. Energy & Environmental Science 4(3):685-694. doi: 10.1039/C0EE00330A

Lobo AO, Martin AA, Antunes EF, Trava-Airoldi VJ, Corat EJ (2005) Caracterização de materiais carbonosos por espectroscopia Raman. Revista Brasileira de Aplicações de Vácuo 24(2):98-103. doi: 10.17563/rbav.v24i2.99

Ma X, Yuan C, Liu X. (2013) Mechanical, microstructure and surface characterizations of carbon fibers prepared from cellulose after liquefying and curing. Materials 7(1):75-84. doi: 10.3390/ma7010075

Mallet-Ladeira P, Puech P, Toulouse C, Cazayous M, Ratel-Ramond N, Weisbecker P, Monthioux M (2014). A Raman study to obtain crystallite size of carbon materials: a better alternative to the Tuinstra-Koenig Law. Carbon 80:629-639.

Marcuzzo JS, Otani C, Polidoro HA, Otani S (2012) Influence of thermal treatment on porosity formation on carbon fiber from textile PAN. Mater Res 16(1):136-144.

Marsh H, Reinoso FR (2006) Activated carbon. Amsterdam: Elsevier. Chapter 4, Characterization of activated carbon; p. 143-242.

Matthews MJ, Pimenta MA, Dresselhaus G, Dresselhaus MS, Endo M. (1999) Origin of dispersive effects of the Raman D band in carbon materials. Phys Rev B 59(10):R6585. doi: 10.1103/PhysRevB.59.R6585

Nabais JV, Carrott PJM, Carrott MR (2005) From commercial textile fibres to activated carbon fibres: chemical transformations. Mater Chem Phys 93(1):100-108. doi: 10.1016/j.matchemphys.2005.02.031

Pierozynski B (2012) Electrodeposition of nickel onto 12K carbon fibre tow in a continuous manner. Croat Chem Acta 85(1):1-8. doi: 10.5562/cca1743

Sadezky A, Muckenhuber H, Grothe H, Niessner R, Pöschl U (2005) Raman microspectroscopy of soot and related carbonaceous materials: spectral analysis and structural information. Carbon 43(8):1731-1742. doi: 10.1016/j.carbon.2005.02.018

Suzuki M (1994) Activated carbon fiber: fundamentals and applications. Carbon 32(4):577-586. doi: 10.1016/0008-6223(94)90075-2

Thomsen C, Reich, S, Maultzsch J (2004) Resonant Raman spectroscopy of nanotubes. Philosophical Transactions of the Royal Society of London A: Mathematical, Physical and Engineering Sciences 362(1824):2337-2359. doi: 10.1098/rsta.2004.1444

Tuinstra F, Koenig, JL (1970). Raman spectrum of graphite. J Chem Phys 53(3):1126-1130. doi: 10.1063/1.1674108

Warren BE (1969) X-ray diffraction. North Chelmsford: Courier Corporation.

Yoon SH, Lim S, Song Y, Ota Y, Qiao W, Tanaka A, Mochida I (2004) KOH activation of carbon nanofibers. Carbon 42(8):1723-1729. doi: 10.1016/j.carbon.2004.03.006

Permissions

All chapters in this book were first published in JATM, by Departamento de Ciência e Tecnologia Aeroespacial; hereby published with permission under the Creative Commons Attribution License or equivalent. Every chapter published in this book has been scrutinized by our experts. Their significance has been extensively debated. The topics covered herein carry significant findings which will fuel the growth of the discipline. They may even be implemented as practical applications or may be referred to as a beginning point for another development.

The contributors of this book come from diverse backgrounds, making this book a truly international effort. This book will bring forth new frontiers with its revolutionizing research information and detailed analysis of the nascent developments around the world.

We would like to thank all the contributing authors for lending their expertise to make the book truly unique. They have played a crucial role in the development of this book. Without their invaluable contributions this book wouldn't have been possible. They have made vital efforts to compile up to date information on the varied aspects of this subject to make this book a valuable addition to the collection of many professionals and students.

This book was conceptualized with the vision of imparting up-to-date information and advanced data in this field. To ensure the same, a matchless editorial board was set up. Every individual on the board went through rigorous rounds of assessment to prove their worth. After which they invested a large part of their time researching and compiling the most relevant data for our readers.

The editorial board has been involved in producing this book since its inception. They have spent rigorous hours researching and exploring the diverse topics which have resulted in the successful publishing of this book. They have passed on their knowledge of decades through this book. To expedite this challenging task, the publisher supported the team at every step. A small team of assistant editors was also appointed to further simplify the editing procedure and attain best results for the readers.

Apart from the editorial board, the designing team has also invested a significant amount of their time in understanding the subject and creating the most relevant covers. They scrutinized every image to scout for the most suitable representation of the subject and create an appropriate cover for the book.

The publishing team has been an ardent support to the editorial, designing and production team. Their endless efforts to recruit the best for this project, has resulted in the accomplishment of this book. They are a veteran in the field of academics and their pool of knowledge is as vast as their experience in printing. Their expertise and guidance has proved useful at every step. Their uncompromising quality standards have made this book an exceptional effort. Their encouragement from time to time has been an inspiration for everyone.

The publisher and the editorial board hope that this book will prove to be a valuable piece of knowledge for researchers, students, practitioners and scholars across the globe.

List of Contributors

Igor Afonso Acampora Prado ,Davi Ferreira de Castro and Davi Antônio dos Santos
Departamento de Ciência e Tecnologia Aeroespacial - Instituto Tecnológico de Aeronáutica - Divisão de Engenharia Mecânica - São José dos Campos/ SP - Brazil

Vahid Nasr Esfahani and Mahmoud Mani
Amirkabir University of Technology - Aerospace Engineering Department - Tehran/Iran

George Barakos
University of Glasgow - School of Engineering - CFD Laboratory - Glasgow - Scotland

João Otávio Falcão Arantes Filho
Departamento de Ciência e Tecnologia Aeroespacial - Instituto de Pesquisas e Ensaios em Voo - Divisão de Ensaios em Voo - São José dos Campos/SP - Brazil

Donizeti de Andrade
Departamento de Ciência e Tecnologia Aeroespacial - Instituto Tecnológico de Aeronáutica - Departamento de Projetos - São José dos Campos/ SP - Brazil

Mahzad Khoshlessan and Seyed Mohammad Hossein Karimian
Amirkabir University of Technology - Aerospace Engineering Department - Center of Excellence in Computational Aerospace Engineering - Tehran/ Iran

Lester de Abreu Faria, Caio Augusto de Melo Silvestre and Marcelino Aparecido Feitosa Correia
Departamento de Ciência e Tecnologia Aeroespacial - Instituto Tecnológico de Aeronáutica - Divisão de Engenharia Eletrônica - São José dos Campos/ SP - Brazil

Maria Alice Carvalho Mazzeu
Departamento de Ciência e Tecnologia Aeroespacial - Instituto Tecnológico de Aeronáutica - Programa de Pós-Gradução em Ciência e Tecnologias Espaciais - São José dos Campos/SP - Brazil
Departamento de Ciência e Tecnologia Aeroespacial - Instituto de Fomento e Coordenação Industrial - Divisão de Certificação de Sistema de Gestão - São José dos Campos/SP - Brazil

Lohana Komorek Faria
Departamento de Ciência e Tecnologia Aeroespacial - Instituto de Aeronáutica e Espaço - Divisão de Materiais - São José dos Campos/SP - Brazil
Universidade do Vale do Paraíba - Faculdade de Engenharia, Arquitetura e Urbanismo - São José dos Campos/SP - Brazil

Andreza de Moura Cardoso and Adriana Medeiros Gama
Departamento de Ciência e Tecnologia Aeroespacial - Instituto de Aeronáutica e Espaço - Divisão de Materiais - São José dos Campos/SP - Brazil

Emerson Sarmento Gonçalves
Departamento de Ciência e Tecnologia Aeroespacial - Instituto Tecnológico de Aeronáutica - Programa de Pós-Gradução em Ciência e Tecnologias Espaciais - São José dos Campos/SP - Brazil
Departamento de Ciência e Tecnologia Aeroespacial - Instituto de Aeronáutica e Espaço - Divisão de Materiais - São José dos Campos/SP - Brazil

Daniel Consoli Silveira and Edson Cocchieri Botelho
Universidade Estadual Paulista Júlio de Mesquita Filho - Faculdade de Engenharia de Guaratinguetá - Departamento de Materiais e Tecnologia - Guaratinguetá/SP - Brazil

Newton Gomes
Departamento de Ciência e Tecnologia Aeroespacial - Instituto Tecnológico de Aeronáutica - Laboratório de Guerra Eletrônica - São José dos Campos/SP - Brazil

Mirabel Cerqueira Rezende
Departamento de Ciência e Tecnologia Aeroespacial - Instituto Tecnológico de Aeronáutica - Laboratório de Guerra Eletrônica - São José dos Campos/SP - Brazil

Universidade Federal de São Paulo - Instituto de Ciência e Tecnologia - Curso de Engenharia de Materiais - São José dos Campos/SP - Brazil

Karthik Reddy Buchireddy Sri, Poondla Aneesh, Kiran Bhanu and M Natarajan
VIT University - School of Mechanical Engineering Mechanical Department - Vellore/Tamil Nadu - India

Nelson Javier Pedraza Betancourth, Julio Enoc Parra Villamarin and John Jairo Vaca Rios
Fundación Universitaria Los Libertadores - Facultad de Ingeniería Aeronáutica - Bogotá - Colombia

Pedro David Bravo-Mosquera and Hernán Darío Cerón-Muñoz
Universidade de São Paulo - Escola de Engenharia de São Carlos - Departamento de Engenharia Aeronáutica - São Carlos/SP - Brazil

Igor Dabkiewicz, Ítalo Vinícius Soares e Silva, Jossano Saldanha Marcuzzo and Rita de Cassia Mendonça Sales Contini
Centro Estadual Tecnológico Paula Souza - Faculdade de Tecnologia de São José dos Campos - Prof. Jessen Vidal - São José dos Campos/SP - Brazil

Graham Wild, Kellie Gavin, Jose Silva and Glenn Baxter
RMIT University - School of Engineering - Aerospace Engineering and Aviation - Melbourne/ Victoria - Australia.

John Murray
RMIT University - School of Engineering - Aerospace Engineering and Aviation - Melbourne/ Victoria - Australia
Edith Cowan University - School of Engineering - Aviation - Perth/Western Australia - Australia

Yalin Pan, Jun Huang and Chuxiong Yan
Beihang University - School of Aeronautic Science and Engineering - Beijing - China

Feng Li
China Academy of Aerospace Aerodynamics - Beijing - China

Paulo Pedro Kenedi
Centro Federal de Educação Tecnológica Celso Suckow da Fonseca - Departamento de Engenharia Mecânica - Laboratório de Compósitos e Adesivos - Rio deJaneiro/RJ - Brazil
Centro Federal de Educação Tecnológica Celso Suckow da Fonseca - Programa de Pós-Graduação em Engenharia Mecânica e Tecnologia de Materiais - Laboratório de Compósitos e Adesivos - Rio de Janeiro/RJ - Brazil

Lucas Lisboa Vignoli
Universidade Federal do Rio de Janeiro - Centro de Mecânica Não-Linear - Departamento de Engenharia Mecânica - Rio de Janeiro/RJ, Brazil

Brenno Tavares Duarte
Centro Federal de Educação Tecnológica Celso Suckow da Fonseca - Programa de Pós-Graduação em Engenharia Mecânica e Tecnologia de Materiais - Laboratório de Compósitos e Adesivos - Rio de Janeiro/RJ - Brazil

Fernando Cesar de Abreu Matos and Humberto Oberosler Terço Dias
Centro Federal de Educação Tecnológica Celso Suckow da Fonseca - Departamento de Engenharia Mecânica - Laboratório de Compósitos e Adesivos - Rio deJaneiro/RJ - Brazil

Mauricio Ferrapontoff Lemos
Instituto de Pesquisas da Marinha - Grupo de Tecnologia de Materiais - Rio de Janeiro/RJ - Brazil

Günter Mussbach
Bayern-Chemie GmbH - Aschau am Inn - Germany

Manfred August Bohn
Fraunhofer Institute for Chemical Technology - Pfinztal - Germany

Aiman Hakim Abdul Rahman, Nik Ahmad Ridhwan Nik Mohd, Tholudin Mat Lazim and Shuhaimi Mansor
Universiti Teknologi Malaysia - Faculty of Mechanical Engineering - Aeronautical Laboratory - Skudai/Johor - Malaysia

Ahmad Abdulkarim Alsahlan and Thurai Rahulan
University of Salford - School of Computing, Science and Engineering - Department of Aeronautical Engineering - Salford - United Kingdom

Liu Dawei and Huang Jun
Beijing University of Aeronautics and Astronautics - School of Aeronautic Science and Engineering - Beijing - China

Xin Ji
China Helicopter Research and Developing Institute - Jingdezhen - Jiangxi - China

Alexander Vladimirovich Osipov, Yury Alexandrovich Shinyakov and Michael Sergeevich Sakharov
Tomsk State University of Control Systems and Radioelectronics - Institute of Space Technologies - Research and Development Institute of Space Technologies - Tomsk/Tomsk - Russia

Vadim Nikolaevich Shkolniy
Reshetnev Company - Information Satellite Systems - Department 600 - Zheleznogorsk/Krasnoyarsk - Russia

Majid Hosseini, Mehran Nosratollahi and Hossein Sadati
Malek Ashtar University of Technology - Faculty of Aerospace - Department of Aerospace - Tehran/Tehran - Iran

Miguel Angelo do Amaral Junior, Jorge Tadao Matsushima, Jossano Saldanha Marcuzzo and Maurício Ribeiro Baldan
Instituto Nacional de Pesquisas Espaciais - Coordenação de Laboratórios Associados - Laboratório Associado de Sensores e Materiais - São José dos Campos/SP -Brazil

Index